Student Solutions Manual

Laurel Technical Services

Intermediate Algebra

SECOND EDITION

A Graphing Approach

K. Elayn Martin-Gay

Margaret Greene

Prentice Hall

Upper Saddle River, NJ 07458

Executive Editor: Karin E. Wagner
Project Manager: Mary Beckwith
Assistant Managing Editor, Math Media Production: John Matthews
Production Editor: Barbara A. Till
Supplement Cover Manager: Paul Gourhan
Supplement Cover Designer: PM Workshop Inc.
Manufacturing Buyer: Alan Fischer

© 2001, 1997 by Prentice-Hall, Inc.
Upper Saddle River, NJ 07458

Printed in the United States of America

10   9   8   7   6   5   4   3

ISBN   0-13-017333-9

Prentice-Hall International (UK) Limited, London
Prentice-Hall of Australia Pty. Limited, Sydney
Prentice-Hall Canada, Inc., Toronto
Prentice-Hall Hispanoamericana, S.A., Mexico City
Prentice-Hall of India Private Limited, New Delhi
Pearson Education Asia Pte. Ltd., Singapore
Prentice-Hall of Japan, Inc., Tokyo
Editora Prentice-Hall do Brazil, Ltda., Rio de Janeiro

Table of Contents

Chapter 1

Exercise Set 1.1

1. $5x = 5 \cdot 7 = 35$

3. $9.8z = 9.8 \cdot 3.1 = 30.38$

5. $ab = \dfrac{1}{2} \cdot \dfrac{3}{4} = \dfrac{3}{8}$

7. $3x + y = 3 \cdot 6 + 4 = 18 + 4 = 22$

9. $400t = 400 \cdot 5 = 2000$
The aircraft traveled 2000 miles in 5 hours.

11. $lw = 5.1 \cdot 4 = 20.4$
The display needs 20.4 square feet of floor space.

13. $7098t = 7098 \cdot 5.2 = 36{,}909.60$
The total cost to operate the aircraft for 5.2 hours is $36,909.60.

15. $\{1, 2, 3, 4, 5\}$

17. $\{11, 12, 13, 14, 15, 16\}$

19. $\{0\}$

21. $\{0, 2, 4, 6, 8\}$

23.

25.

27.

29. Answers may vary.

31. $\sqrt{36} = 6$
$\{3, 0, \sqrt{36}\}$

33. $\sqrt{36} = 6$
$\{3, \sqrt{36}\}$

35. $\{\sqrt{7}\}$

37. $-11 \in \{x \mid x \text{ is an integer}\}$

39. $0 \notin \{x \mid x \text{ is a positive integer}\}$

41. $12 \notin \{1, 3, 5, \ldots\}$

43. $0 \notin \{1, 2, 3, \ldots\}$

45. True; every integer is a real number.

47. True; -1 is an integer.

49. False; 0 is not a natural number.

51. False; $\sqrt{5}$ is an irrational number.

53. True; every natural number is an integer.

55. False; the number $\sqrt{7}$, for example, is a real number, but it is not a rational number.

57. Answers may vary.

59. $-|2| = -2$ since $|2| = 2$ and 2 is located 2 units from 0 on the number line.

61. $|-4| = 4$ since -4 is located 4 units from 0 on the number line.

63. $|0| = 0$ since 0 is located 0 units from 0 on the number line.

65. $-|-3| = -3$ since $|-3| = 3$ and -3 is located 3 units from 0 on the number line.

67. Answers may vary.

69. The opposite of -6.2 is $-(-6.2) = 6.2$.

71. The opposite of $\dfrac{4}{7}$ is $-\dfrac{4}{7}$.

73. The opposite of $-\dfrac{2}{3}$ is $-\left(-\dfrac{2}{3}\right) = \dfrac{2}{3}$.

75. The opposite of 0 is 0.

77. $2x$

79. $2x + 5$

81. $x - 10$

83. $x + 2$

85. $\dfrac{x}{11}$

87. $3x + 12$

89. $x - 17$

91. $2(x + 3)$

93. $\dfrac{5}{4 - x}$

95. China = 137
USA = 102
France = 93
Spain = 71
Hong Kong = 59

97. Answers may vary.

Exercise Set 1.2

1. $-3 + 8 = 5$

3. $-14 + (-10) = -24$

5. $-4.3 - 6.7 = -11$

7. $13 - 17 = -4$

9. $\dfrac{11}{15} - \left(-\dfrac{3}{5}\right) = \dfrac{11}{15} + \dfrac{9}{15} = \dfrac{20}{15} = \dfrac{4}{3}$

11. $19 - 10 - 11 = 9 - 11 = -2$

13. Since the signs of the two numbers are different, the product is negative. Thus $(-5)(12) = -60$.

15. Since the signs of the two numbers are the same, the product is positive. Thus $(-8)(-10) = 80$.

17. Since the signs of the two numbers are the same, the quotient is positive. Thus $\dfrac{-12}{-4} = 3$.

19. $\dfrac{0}{-2} = 0$

21. $(-4)(-2)(-1) = 8(-1) = -8$

23. $\dfrac{-6}{7} \div 2 = \dfrac{-6}{7} \cdot \dfrac{1}{2} = \dfrac{-3}{7} = -\dfrac{3}{7}$

25. $\left(-\dfrac{2}{7}\right)\left(-\dfrac{1}{6}\right) = \dfrac{2}{42} = \dfrac{1}{21}$

27. $-7^2 = -(7 \cdot 7) = -49$

29. $(-6)^2 = (-6)(-6) = 36$

31. $(-2)^3 = (-2)(-2)(-2) = 4(-2) = -8$

33. Answers may vary.

35. $\sqrt{49} = 7$ since 7 is positive and $7^2 = 49$.

37. $\sqrt{\dfrac{1}{9}} = \dfrac{1}{3}$ since $\dfrac{1}{3}$ is positive and $\left(\dfrac{1}{3}\right)^2 = \dfrac{1}{9}$.

39. $\sqrt[3]{64} = 4$ since $4^3 = 64$.

41. $\sqrt[4]{81} = 3$ since 3 is postive and $3^4 = 81$.

43. $3(5 - 7)^4 = 3(-2)^4 = 3(16) = 48$

45. $-3^2 + 2^3 = -9 + 8 = -1$

47. $\dfrac{3-(-12)}{-5} = \dfrac{3+12}{-5} = \dfrac{15}{-5} = -3$

49. $|3.6-7.2| + |3.6+7.2| = |-3.6| + |10.8|$
$$= 3.6 + 10.8$$
$$= 14.4$$

51. $\dfrac{(3-\sqrt{9})-(-5-1.3)}{-3} = \dfrac{(3-3)-(-6.3)}{-3}$
$$= \dfrac{0+6.3}{-3}$$
$$= \dfrac{6.3}{-3}$$
$$= -2.1$$

53. $\dfrac{|3-9|-|-5|}{-3} = \dfrac{6-5}{-3} = \dfrac{1}{-3} = -\dfrac{1}{3}$

55. $(-3)^2 + 2^3 = 9 + 8 = 17$

57. $\dfrac{3(-2+1)}{5} - \dfrac{-7(2-4)}{1-(-2)} = \dfrac{3(-1)}{5} - \dfrac{-7(-2)}{1+2}$
$$= \dfrac{-3}{5} - \dfrac{14}{3}$$
$$= \dfrac{-9}{15} - \dfrac{70}{15}$$
$$= -\dfrac{79}{15}$$

59. $\dfrac{\frac{-3}{10}}{\frac{42}{50}} = -\dfrac{3}{10} \cdot \dfrac{50}{42} = -\dfrac{5}{14}$

61. $\dfrac{-1.682 - 17.895}{(-7.102)(-4.691)} = \dfrac{-19.577}{33.315482} \approx -0.5876$

63. $x^2 + z^2 = (-2)^2 + 3^2 = 4 + 9 = 13$

65. $-5(-x+3y) = -5[-(-2)+3(-5)]$
$$= -5(2-15)$$
$$= -5(-13)$$
$$= 65$$

67. $\dfrac{3z-y}{2x-z} = \dfrac{3(3)-(-5)}{2(-2)-3}$
$$= \dfrac{9+5}{-4-3}$$
$$= \dfrac{14}{-7}$$
$$= -2$$

69. $3x - 2y = 3(1.4) - 2(-6.2)$
$$= 4.2 + 12.4$$
$$= 16.6$$

71. $\dfrac{|x-y|}{2y} = \dfrac{|1.4-(-6.2)|}{2(-6.2)}$
$$= \dfrac{|1.4+6.2|}{-12.4}$$
$$= \dfrac{|7.6|}{-12.4}$$
$$= \dfrac{7.6}{-12.4}$$
$$\approx -0.61$$

73. $-3\left(x^2+y^2\right) = -3\left[(1.4)^2 + (-6.2)^2\right]$
$$= -3(1.96 + 38.44)$$
$$= -3(40.4)$$
$$= -121.2$$

75. $x = -2.7;\ y = 0.9;\ x^2 + 2y = 9.09$

77. $A = 4;\ B = -2.5;\ A^2 + B^2 = 22.25$

79. a. $(1/2)x = 1;\ 1/2x = 1$

   b. The results are the same. Answers may vary.

81. a.

y	5	7	10	100
$8 + 2y$	18	22	28	208

   b. The perimeter increases as the width remains the same and the length increases. Answers may vary.

83. a.

x	10	100	1000
$\dfrac{100x + 5000}{x}$	600	150	105

   b. The cost decreases as the number of bookshelves manufactured increases. Answers will vary.

85. $1 - \left(\dfrac{1}{5} + \dfrac{3}{7} \right) = 1 - \left(\dfrac{7}{35} + \dfrac{15}{35} \right)$

$\qquad\qquad = \dfrac{35}{35} - \dfrac{22}{35}$

$\qquad\qquad = \dfrac{13}{35}$

87. $10{,}203 - 5998 = 4205$
The volcano is 4205 meters above sea level.

89. b; The spinner has equal blue, red, and yellow areas.

91. As player 1, you should choose the spinner with the smallest yellow region, which is spinner d.

93. Yes. Two players have 6 points each (third player has 0 points), or two players have 5 points each (third has 2 points).

95. $\sqrt{273} \approx 16.5227$

97. $\sqrt{19.6} \approx 4.4272$

99. 13.2%

101. $12.9\% - 2.1\% = 10.8\%$

103. Answers may vary.

Exercise Set 1.3

1. $4c = 7$

3. $3(x + 1) = 7$

5. $\dfrac{n}{5} = 4n$

7. $z - 2 = 2z$

9. $0 > -2$ since 0 lies to the right of -2 on the number line.

11. $\dfrac{12}{3} = \dfrac{8}{2}$ since $\dfrac{12}{3} = 4$ and $\dfrac{8}{2} = 4$.

13. $-7.9 < -7.09$ since -7.9 lies to the left of -7.09 on the number line.

15. $7x \le -21$

17. $-2 + x \ne 10$

19. $2(x - 6) > \dfrac{1}{11}$

21. $y - 7 = 6$

23. $2(x - 6) = -27$

25. The opposite of -8 is $-(-8) = 8$.
The reciprocal of -8 is $-\dfrac{1}{8}$.

27. The opposite of $-\dfrac{1}{4}$ is $-\left(-\dfrac{1}{4} \right) = \dfrac{1}{4}$.
The reciprocal of $-\dfrac{1}{4}$ is -4.

29. The opposite of 0 is 0.
The reciprocal of 0 is undefined.

31. The opposite of $\dfrac{7}{8}$ is $-\dfrac{7}{8}$.
The reciprocal of $\dfrac{7}{8}$ is $\dfrac{8}{7}$ because
$\dfrac{7}{8} \cdot \dfrac{8}{7} = 1.$

33. Negative; positive

35. Zero. Answers may vary.

37. $7x + y = y + 7x$

39. $z \cdot w = w \cdot z$

41. $\dfrac{1}{3} \cdot \dfrac{x}{5} = \dfrac{x}{5} \cdot \dfrac{1}{3}$

43. No, subtraction is not commutative. Answers may vary.

45. $5 \cdot (7x) = (5 \cdot 7)x$

47. $(x + 1.2) + y = x + (1.2 + y)$

49. $(14z) \cdot y = 14(z \cdot y)$

51. $12 - (5 - 3) = 10;\ (12 - 5) - 3 = 4$
Because $10 \neq 4$, subtraction is not associative.

53. $3(x + 5) = 3 \cdot x + 3 \cdot 5 = 3x + 15$

55. $-(2a + b) = -1(2a + b) = -2a - b$

57. $2(6x + 5y + 2z) = 12x + 10y + 4z$

59. $2y - 6 = \dfrac{1}{8}$

61. $\dfrac{n + 5}{2} > 2n$

63. $6 + 3x$

65. 0

67. 7

69. $(10 \cdot 2)y$

71. Associative property of addition

73. Commutative property of multiplication

75. In words: | One hundred twelve | minus | x |
Translate: $112 - x$

77. In words: | Ninety degrees | minus | $5x$ |
Translate: $90° - 5x$

79. In words: | Cost of a book | times | y |
Translate: $35.61y$

81. The next even integer would be 2 more than the given even integer.
In words: | The first even integer | plus | two |
Translate: $2x + 2$

83. $5y - 14 + 7y - 20y = 5y + 7y - 20y - 14$
$\qquad = (5 + 7 - 20)y - 14$
$\qquad = -8y - 14$

85. $-11c - (4 - 2c) = -11c - 4 + 2c$
$\qquad = -11c + 2c - 4$
$\qquad = (-11 + 2)c - 4$
$\qquad = -9c - 4$

87. $(8 - 5y) - (4 + 3y) = 8 - 5y - 4 - 3y$
$\qquad = -5y - 3y + 8 - 4$
$\qquad = (-5 - 3)y + (8 - 4)$
$\qquad = -8y + 4$
$\qquad = 4 - 8y$

89. $-4(y + 3) - 7y + 1 = -4y - 12 - 7y + 1$
$\qquad = -4y - 7y - 12 + 1$
$\qquad = (-4 - 7)y + (-12 + 1)$
$\qquad = -11y - 11$

91. $-(8 - t) + (2t - 6) = -8 + t + 2t - 6$
$\qquad = t + 2t - 8 - 6$
$\qquad = (1 + 2)t + (-8 - 6)$
$\qquad = 3t - 14$

93. $5(2z - 6) + 10(3 - z) = 10z - 30 + 30 - 10z$
$\qquad = 10z - 10z - 30 + 30$
$\qquad = (10 - 10)z + 0$
$\qquad = 0$

95. $7n + 3(2n - 6) - 2 = 7n + 6n - 18 - 2$
$\qquad = (7 + 6)n - 20$
$\qquad = 13n - 20$

97. $5.8(-9.6 - 31.2y) - 18.65$
$\qquad = -55.68 - 180.96y - 18.65$
$\qquad = -180.96y - 55.68 - 18.65$
$\qquad = -180.96y - 74.33$

99. $6.5y - 4.4(1.8x - 3.3) + 10.95$
$= 6.5y - 7.92x + 14.52 + 10.95$
$= 6.5y - 7.92x + 25.47$

101. No. Answers may vary.

103. 70 million

105. 35 million

107. $(4.1\%) \cdot 3 = 12.3\%$

Section 1.4

Mental Math

1. $3x + 5x + 6 + 15 = 8x + 21$

2. $8y + 3y + 7 + 11 = 11y + 18$

3. $5n + n + 3 - 10 = 6n - 7$

4. $m + 2m + 4 - 8 = 3m - 4$

5. $8x - 12x + 5 - 6 = -4x - 1$

6. $4x - 10x + 13 - 16 = -6x - 3$

Exercise Set 1.4

1. $-3x = 36$
$\dfrac{-3x}{-3} = \dfrac{36}{-3}$
$x = -12$
The solution is -12.

3. $x + 2.8 = 1.9$
$x + 2.8 - 2.8 = 1.9 - 2.8$
$x = -0.9$
The solution is -0.9.

5. $5x - 4 = 26$
$5x - 4 + 4 = 26 + 4$
$5x = 30$
$\dfrac{5x}{5} = \dfrac{30}{5}$
$x = 6$
The solution is 6.

7. $-4 = 3x + 11$
$-4 - 11 = 3x + 11 - 11$
$-15 = 3x$
$\dfrac{-15}{3} = \dfrac{3x}{3}$
$-5 = x$
The solution is -5.

9. $-4.1 - 7z = 3.6$
$-4.1 - 7z + 4.1 = 3.6 + 4.1$
$-7z = 7.7$
$\dfrac{-7z}{-7} = \dfrac{7.7}{-7}$
$z = -1.1$
The solution is -1.1.

11. $5y + 12 = 2y - 3$
$5y + 12 - 12 = 2y - 3 - 12$
$5y = 2y - 15$
$5y - 2y = 2y - 15 - 2y$
$3y = -15$
$\dfrac{3y}{3} = \dfrac{-15}{3}$
$y = -5$
The solution is -5.

13. $8x - 5x + 3 = x - 7 + 10$
$3x + 3 = x + 3$
$2x = 0$
$x = 0$
The solution is 0.

15. $5x + 12 = 2(2x + 7)$
$5x + 12 = 4x + 14$
$x = 2$
The solution is 2.

17. $3(x - 6) = 5x$
$3x - 18 = 5x$
$-18 = 2x$
$-9 = x$
The solution is -9.

19. $-2(5y-1)-y=-4(y-3)$
$-10y+2-y=-4y+12$
$-11y+2=-4y+12$
$-7y=10$
$y=-\dfrac{10}{7}$

The solution is $-\dfrac{10}{7}$.

21. a. $4(x+1)+1=4x+4+1=4x+5$

b. $4(x+1)+1=-7$
$4x+5=-7$
$4x=-12$
$x=-3$
The solution is -3.

c. Answers may vary.

23. $\dfrac{x}{2}+\dfrac{2}{3}=\dfrac{3}{4}$

$12\left(\dfrac{x}{2}+\dfrac{2}{3}\right)=12\left(\dfrac{3}{4}\right)$

$6x+8=9$

$6x=1$

$x=\dfrac{1}{6}$

The solution is $\dfrac{1}{6}$.

25. $\dfrac{3t}{4}-\dfrac{t}{2}=1$

$4\left(\dfrac{3t}{4}-\dfrac{t}{2}\right)=4(1)$

$3t-2t=4$

$t=4$
The solution is 4.

27. $\dfrac{n-3}{4}+\dfrac{n+5}{7}=\dfrac{5}{14}$

$28\left(\dfrac{n-3}{4}+\dfrac{n+5}{7}\right)=28\left(\dfrac{5}{14}\right)$

$7(n-3)+4(n+5)=2(5)$
$7n-21+4n+20=10$
$11n-1=10$
$11n=11$
$n=1$

The solution is 1.

29. $0.6x-10=1.4x-14$
$10(0.6x-10)=10(1.4x-14)$
$6x-100=14x-140$
$40=8x$
$5=x$

The solution is 5.

31. $4(n+3)=2(6+2n)$
$4n+12=12+4n$
$4n+12=4n+12$
$0=0$

This is true for all n. Therefore, all real numbers are solutions.

33. $3(x-1)+5=3x+7$
$3x-3+5=3x+7$
$3x+2=3x+7$
$3x+2-3x=3x+7-3x$
$2=7$

This is false for any x. Therefore, the solution set is $\varnothing$.

35. Answers may vary.

37. $6x-5=4x-21$
$x=-8$

39. $3(N-3)+2=-1+N$
$N=3$

41. $6x+9=51$
$6x=42$
$x=7$
The solution is 7.

43. $-5x + 1.5 = -19.5$
$-5x = -21$
$x = 4.2$
The solution is 4.2.

45. $x - 10 = -6x + 4$
$7x = 14$
$x = 2$
The solution is 2.

47. $3x - 4 - 5x = x + 4 + x$
$-2x - 4 = 2x + 4$
$-4x = 8$
$x = -2$
The solution is -2.

49. $5(y + 4) = 4(y + 5)$
$5y + 20 = 4y + 20$
$y = 0$
The solution is 0.

51. $0.7x + 9 = 2.3x - 11$
$0.7x + 20 = 2.3x$
$20 = 1.6x$
$\dfrac{20}{1.6} = x$
$12.5 = x$
The solution is 12.5.

53. $6x - 2(x - 3) = 4(x + 1) + 4$
$6x - 2x + 6 = 4x + 4 + 4$
$4x + 6 = 4x + 8$
$6 = 8$
This is false for any x. Therefore, the solution set is $\varnothing$.

55. $\dfrac{3}{8} + \dfrac{b}{3} = \dfrac{5}{12}$
$24\left(\dfrac{3}{8} + \dfrac{b}{3}\right) = 24\left(\dfrac{5}{12}\right)$
$9 + 8b = 10$
$8b = 1$
$b = \dfrac{1}{8}$
The solution is $\dfrac{1}{8}$.

57. $z + 3(2 + 4z) = 6(z + 1) + 5z$
$z + 6 + 12z = 6z + 6 + 5z$
$13z + 6 = 11z + 6$
$2z = 0$
$z = 0$
The solution is 0.

59. $\dfrac{3t + 1}{8} = \dfrac{5 + 2t}{7} + 2$
$56\left(\dfrac{3t + 1}{8}\right) = 56\left(\dfrac{5 + 2t}{7} + 2\right)$
$7(3t + 1) = 8(5 + 2t) + 112$
$21t + 7 = 40 + 16t + 112$
$21t + 7 = 16t + 152$
$5t = 145$
$t = 29$
The solution is 29.

61. $\dfrac{m - 4}{3} - \dfrac{3m - 1}{5} = 1$
$15\left(\dfrac{m - 4}{3} - \dfrac{3m - 1}{5}\right) = 15(1)$
$5(m - 4) - 3(3m - 1) = 15$
$5m - 20 - 9m + 3 = 15$
$-4m - 17 = 15$
$-4m = 32$
$m = -8$
The solution is -8.

63. $5(x - 2) + 2x = 7(x + 4) - 38$
$5x - 10 + 2x = 7x + 28 - 38$
$7x - 10 = 7x - 10$
$0 = 0$
This is true for all x. Therefore, all real numbers are solutions.

65. $y + 0.2 = 0.6(y + 3)$
$y + 0.2 = 0.6y + 1.8$
$0.4y = 1.6$
$y = 4$
The solution is 4.

67.
$$2y + 5(y - 4) = 4y - 2(y - 10)$$
$$2y + 5y - 20 = 4y - 2y + 20$$
$$7y - 20 = 2y + 20$$
$$5y = 40$$
$$y = 8$$
The solution is 8.

69.
$$2(x - 8) + x = 3(x - 6) + 2$$
$$2x - 16 + x = 3x - 18 + 2$$
$$3x - 16 = 3x - 16$$
This is true for all x. Therefore, all real numbers are solutions.

71.
$$\frac{3x - 1}{9} + x = \frac{3x + 1}{3} + 4$$
$$9\left(\frac{3x - 1}{9} + x\right) = 9\left(\frac{3x + 1}{3} + 4\right)$$
$$(3x - 1) + 9x = 3(3x + 1) + 36$$
$$3x - 1 + 9x = 9x + 3 + 36$$
$$12x - 1 = 9x + 39$$
$$3x = 40$$
$$x = \frac{40}{3}$$
The solution is $\frac{40}{3}$.

73.
$$1.5(4 - x) = 1.3(2 - x)$$
$$10[1.5(4 - x)] = 10[1.3(2 - x)]$$
$$15(4 - x) = 13(2 - x)$$
$$60 - 15x = 26 - 13x$$
$$-2x = -34$$
$$x = 17$$
The solution is 17.

75.
$$-2(b - 4) - (3b - 1) = 5b + 3$$
$$-2b + 8 - 3b + 1 = 5b + 3$$
$$-5b + 9 = 5b + 3$$
$$-10b = -6$$
$$b = \frac{6}{10} = \frac{3}{5}$$
The solution is $\frac{3}{5}$.

77.
$$\frac{1}{4}(a + 2) = \frac{1}{6}(5 - a)$$
$$12 \cdot \frac{1}{4}(a + 2) = 12 \cdot \frac{1}{6}(5 - a)$$
$$3(a + 2) = 2(5 - a)$$
$$3a + 6 = 10 - 2a$$
$$5a = 4$$
$$a = \frac{4}{5}$$
The solution is $\frac{4}{5}$.

79.
$$3.2x + 4 = 5.4x - 7$$
$$3.2x + 4 - 4 = 5.4x - 7 - 4$$
$$3.2x = 5.4x - 11$$
From this we see that $K = -11$.

81.
$$\frac{x}{6} + 4 = \frac{x}{3}$$
$$6\left(\frac{x}{6} + 4\right) = 6\left(\frac{x}{3}\right)$$
$$x + 24 = 2x$$
From this we see that $K = 24$.

83.
$$2.569x = -12.48534$$
$$\frac{2.569x}{2.569} = \frac{-12.48534}{2.569}$$
$$x = -4.86$$
The solution is -4.86.

85.
$$2.86z - 8.1258 = -3.75$$
$$2.86z = 4.3758$$
$$\frac{2.86z}{2.86} = \frac{4.3758}{2.86}$$
$$z = 1.53$$
The solution is 1.53.

87. Not a fair game

89. $7x^2 + 2x - 3 = 6x(x+4) + x^2$

$7x^2 + 2x - 3 = 6x^2 + 24x + x^2$

$7x^2 + 2x - 3 = 7x^2 + 24x$

$2x - 3 = 24x$

$-3 = 22x$

$-\dfrac{3}{22} = x$

The solution is $-\dfrac{3}{22}$.

91. $x(x+1) + 16 = x(x+5)$

$x^2 + x + 16 = x^2 + 5x$

$x + 16 = 5x$

$16 = 4x$

$4 = x$

The solution is 4.

Exercise Set 1.5

1. $y + y + y + y = 4y$

3. $z + (z+1) + (z+2) = 3z + 3$

5. $5x + 10(x+3) = 5x + 10x + 30 = 15x + 30$

The total amount of money is $(15x + 30)$ cents.

7. $4x + 3(2x+1) = 4x + 6x + 3 = 10x + 3$

9. $4(x-2) = 2 + 6x$

$4x - 8 = 2 + 6x$

$-10 = 2x$

$-5 = x$

The number is -5.

11. Let x = one number, then $5x$ = the other number.

$x + 5x = 270$

$6x = 270$

$x = 45$

$5x = 5(45) = 225$

The numbers are 45 and 225.

13. $30\% \cdot 260 = 0.30 \cdot 260 = 78$

15. $12\% \cdot 16 = 0.12 \cdot 16 = 1.92$

17. $29\% \cdot 2271 = 0.29 \cdot 2271 = 658.59$

Approximately 658.59 million acres are federally owned.

19. $47\% \cdot 110{,}000 = 0.47 \cdot 110{,}000 = 51{,}700$

You would expect 51,700 homes to have computers.

21. $100 - 12 - 39 - 8 - 21 = 20$

20% of credit union loans are for credit cards and other unsecured loans.

23. $39\% \cdot 300 = 0.39 \cdot 300 = 117$

You would expect 117 of the loans to be automobile loans.

25. $3x + 17.5 = 199$

$3x = 181.5$

$x = 60.5$

$x + 7.7 = 60.5 + 7.7 = 68.2$

$x + 9.8 = 60.5 + 9.8 = 70.3$

The Dallas/Ft. Worth airport has 60.5 million annual arrivals and departures. The Atlanta airport has 68.2 million annual arrivals and departures. The Chicago airport has 70.3 million annual arrivals and departures.

27. Let x = the number of seats in the B737-200, then $x + 104$ = the number of seats in the B767-300ER.

$x + x + 104 = 328$

$2x + 104 = 328$

$2x = 224$

$x = 112$

$x + 104 = 112 + 104 = 216$

The B737-200 aircraft has 112 seats. The B767-300ER aircraft has 216 seats.

29. Let x = the price before taxes.

$x + 0.08x = 464.40$

$1.08x = 464.40$

$x = 430$

The price was \$430 before taxes.

31. a. Let x = the number of telephone company operators in 1996.

$$x - 0.47x = 26,000$$
$$0.53x = 26,000$$
$$x \approx 49,057$$

There were approximately 49,057 operators in 1996.

b. Answers may vary.

33. $20 - 75\% \cdot 20 = 20 - 0.75 \cdot 20$
$$= 20 - 15$$
$$= 5$$
The lifespan is about 5 years.

35. Let x = the number (in millions) of returns electronically filed in 1998, then
$x + 0.24x$ = number in 1999.
$$x + 0.24x = 21.1$$
$$1.24x = 21.1$$
$$x \approx 17$$
There were approximately 17 million electronically filed returns in 1998.

37. Let x = the length of a side of the square, then
$x + 6$ = the length of a side of the triangle.
$$4x = 3(x + 6)$$
$$4x = 3x + 18$$
$$x = 18$$
$$x + 6 = 24$$
The square's sides are 18 centimeters.
The triangle's sides are 24 centimeters.

39. Let x = the width of the room, then
$2x + 2$ = the length of the room.
$$x + 2x + 2 + x + 2x + 2 = 40$$
$$6x + 4 = 40$$
$$6x = 36$$
$$x = 6$$
$2x + 2 = 2(6) + 2 = 14$
The width is 6 centimeters and the length is 14 centimeters.

41. Let x = the width, then $5(x + 1)$ = the height.
$$x + 5(x + 1) = 55.4$$
$$x + 5x + 5 = 55.4$$
$$6x + 5 = 55.4$$
$$6x = 50.4$$
$$x = 8.4$$
$5(x + 1) = 47$
The width is 8.4 meters and the height is 47 meters.

43. Let x = the measure of the second angle, then $2x$ = the measure of first angle, and $3x - 12$ = the measure of the third angle.
$$x + 2x + 3x - 12 = 180$$
$$6x - 12 = 180$$
$$6x = 192$$
$$x = 32$$
$2x = 2 \cdot 32 = 64$
$3x - 12 = 3 \cdot 32 - 12 = 84$
The angles measure 64°, 32°, and 84°.

45. $x + 20 + x = 180$
$$2x + 20 = 180$$
$$2x = 160$$
$$x = 80$$
$x + 20 = 80 + 20 = 100$
The angles measure 80° and 100°.

47. $x + 5x = 90$
$$6x = 90$$
$$x = 15$$
$5x = 5 \cdot 15 = 75$
The angles measure 15° and 75°.

49. Let x = the measure of the angle, then
$180 - x$ = the measure of its supplement.
$$x = 3(180 - x) + 20$$
$$x = 540 - 3x + 20$$
$$4x = 560$$
$$x = 140$$
$180 - x = 180 - 140 = 40$
The angles measure 140° and 40°.

51. Let x = the first integer, then
$(x + 1)$ = the next consecutive integer, and
$(x + 2)$ = the third consecutive integer.
$$x + (x + 1) + (x + 2) = 228$$
$$3x + 3 = 228$$
$$3x = 225$$
$$x = 75$$
$x + 1 = 75 + 1 = 76$
$x + 2 = 75 + 2 = 77$
The integers are 75, 76 and 77.

53. Let x = the first integer, then $x + 2$ = the
second integer, and $x + 4$ = the third
integer.
$$2x + x + 4 = 268,222$$
$$3x + 4 = 268,222$$
$$3x = 268,218$$
$$x = 89,406$$
$x + 2 = 89,406 + 2 = 89,408$
$x + 4 = 89,406 + 4 = 89,410$
Fallon's zip code is 89406, Fernley's
is 89408, and Gardnerville Ranchos'
is 89410.

55. Let x = the first integer, then $x + 1$ = the
second integer, and $x + 2$ = the third
integer.
$$x + x + 1 + x + 2 = 3(x + 1)$$
$$3x + 3 = 3x + 3$$
Since this is an identity, any three
consecutive integers will work.

57. $24x = 100 + 20x$
$\quad\ 4x = 100$
$\quad\ \ x = 25$
The manufacturer needs to produce and
sell 25 skateboards to break even.

59. $7.50x = 4.50x + 2400$
$\quad\ \ 3x = 2400$
$\quad\ \ \ x = 800$
800 books should be produced and sold to
break even.

61. Answers may vary.

63. Let x = the number (in millions) of trees'
worth of newsprint recycled, then
$x + 30$ = the number (in millions) of trees'
worth of newsprint either recycled or
discarded.
$$x = 0.27(x + 30)$$
$$x = 0.27x + 8.1$$
$$0.73x = 8.1$$
$$x \approx 11$$
About 11 million trees' worth of
newsprint is recycled.

Section 1.6

Mental Math

1. f

2. a

3. b

4. e

5. d

6. c

Exercise Set 1.6

1.

X	Y₁	
7	343	
8	512	
9	729	
10	1000	
11	1331	

X=

Side length of a cube	x	7	8	9	10	11
Volume	x^3	343	512	729	1000	1331

   a. The volume of a cube whose side measures 9 centimeters is 729 cubic centimeters.

   b. The volume of a cube whose side measures 11 feet is 1331 cubic feet.

   c. If a cube has volume 343 cubic inches, the length of its side is 7 inches.

3.

X	Y₁	
35	270	
37	284	
39	298	
39.5	301.5	
40	305	
40.5	308.5	
44	333	

X=44

Hours worked	x	39	39.5	40	40.5
Gross pay (dollars)	$25 + 7x$	298	301.50	305	308.50

   a. If the gross pay is $333, the number of hours worked is 44.

   b. If the gross pay is $270, the number of hours worked is 35.

   c. If the gross pay is $284, the number of hours worked is 37.

5. **a.** $y_1 = 5.25x$

   b. $y_2 = 7x$

   c.

Hours	1	1.25	1.5	1.75	2	2.25	2.5	2.75	3
Rough draft	5.25	6.56	7.88	9.19	10.50	11.81	13.13	14.44	15.75
Manuscript	7.00	8.75	10.50	12.25	14.00	15.75	17.50	19.25	21.00

   d. $10 \text{ pages} \cdot \dfrac{0.25 \text{ hr}}{2 \text{ pages}} \cdot \dfrac{\$5.25}{1 \text{ hr}} = \$6.56$

   She charges $6.56 for a 10-page rough draft.

   e. $10.75 \text{ hr} \cdot \dfrac{\$7}{1 \text{ hr}} = \$75.25$

   She charges $75.25 for typing the bound manuscript.

7.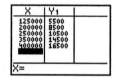

Radius of circle	x	2	5	21	94.2
Circumference	$2\pi x$	12.57	31.42	131.95	591.88
Area	πx^2	12.57	78.54	1385.44	27,877.36

   a. From the calculator table, when the radius is 30 yards, the circumference is 188.50 yards and the area is 2827.43 square yards.

   b. From the calculator table, when the radius is 78.5 millimeters, the circumference is 493.23 millimeters and the area is 19,359.28 square millimeters.

9. a. $y_1 = 500 + 0.04x$

X	Y₁
125000	5500
200000	8500
250000	10500
350000	14500
400000	16500

X=

Sales (in thousands)	125	200	250	350	400
Gross pay	5500	8500	10,500	14,500	16,500

   b. Her gross pay would range between $5500 and $16,500.

11. a. $y_1 = 12.96 + 1.10x$

   b.

Gallons of water used (thousands)	0	1	2	3	4	5	6
Monthly cost (dollars)	12.96	14.06	15.16	16.26	17.36	18.46	19.56

13. $y_1 = 1.5x$

Ingredient	Granulated sugar	Brown sugar	Butter	Flour	Soda
Recipe	$\frac{3}{4}$ cup	$\frac{1}{4}$ cup	$\frac{1}{2}$ cup	3 cups	$\frac{1}{8}$ teaspoon
New Amount	$1\frac{1}{8}$ cups	$\frac{3}{8}$ cup	$\frac{3}{4}$ cup	$4\frac{1}{2}$ cups	$\frac{3}{16}$ teaspoon

15. a. $y_1 = 14,500 + 1000x$, $y_2 = 20,000 + 575x$

Note that Kelsey's salary during year 5 is her salary after 4 raises, so $x = 4$ corresponds to 5 years.

Years	5	10	15	20
First Job	$18,500	$23,500	$28,500	$33,500
Second Job	$22,300	$25,175	$28,050	$30,925

b. The first job pays $14,500 + $15,500 + $\cdots$ + $33,500 = $480,000.
The second job pays $20,000 + $20,575 + $\cdots$ + $30,925 = $509,250.
The second job pays more.

17. a. $y_1 = 15 + 0.08924x$

b.

Kilowatt-hours	0	500	1000	1500	2000
Monthly cost (dollars)	15	59.62	104.24	148.86	193.48

19. a. $y_1 = (1 - 0.35)x$ or $y_1 = 0.65x$

b.

Item	Blouse	Skirt	Shorts	Shoes	Purse	Earrings	Backpack
Price tag	$29.95	$35.95	$19.25	$39.95	$17.95	$9.95	$25.75
Sale price	$19.47	$23.37	$12.51	$25.97	$11.67	$6.47	$16.74

c. The sale prices add up to $116.20.

d. $(116.20)(1.07) = 124.33$
The total after taxes is $124.33.

e. Sum of price tags = $178.75
$178.75 - 116.20 = 62.55$
She saved $62.55.

21. $y_1 = 3500 + 55x$, $y_2 = 75x$

x calculators	100	125	150	175	200
Cost	9000	10,375	11,750	13,125	14,500
Revenue	7500	9375	11,250	13,125	15,000

The break-even point occurs when 175 calculators are made and sold.

23. When 200 calculators are produced and sold, the profit is $15,000 - $14,500 = $500.

25.

Diameter	Area	Circumference	Cost (Area)	Cost (Circumference)
6	28.27	18.85	$1.41	$4.71
12	113.10	37.70	$5.65	$9.42
18	254.47	56.55	$12.72	$14.14
24	452.39	75.40	$22.62	$18.85
30	706.86	94.25	$35.34	$23.56
36	1017.88	113.10	$50.89	$28.27

Preferred prices may vary.

27. a. $y(1) = 1569$, $y(2) = 1606$, $y(3) = 1611$, $y(4) = 1584$, $y(5) = 1525$

b.

X	Y₁
2.4	1611.8
2.5	1612.5
2.6	1612.8
2.7	1612.9
2.8	1612.6
2.9	1611.9
3	1611

Y₁=1612.86

The rocket's maximum height is 1613 feet.

29. a. Anne: $25 + 0.08(300) = 49.00$
Michelle: $35 + 0.05(300) = 50.00$
Anne's rental charge is $49.00 and Michelle's rental charge is $50.00.

b. $25 + 0.08d = 35 + 0.05d$
$$0.03d = 10$$
$$d = 333\frac{1}{3}$$

Anne's rental agreement is cheaper than Michelle's for distances less than $333\frac{1}{3}$ miles.

31. a. $C = 90 + 75x$

b.

Number of hours	2	4	6	8
Total fee (dollars)	240	390	540	690

33.

X	Y₁
4.7	3.4874
4.8	3.4944
4.9	3.4986
5	3.5
5.1	3.4986
5.2	3.4944
5.3	3.4874

$Y_1 = 3.5$

   a. The maximum height is 3.5 feet.

   b. The horizontal distance is 5 feet.

   c. y is zero again when $x = 10$ feet.

35. a. $y(2) \approx 6.73 = 6\dfrac{73}{100}$

     $6 + \dfrac{43.8}{60} \approx 6:44$

     Sunrise will be at 6.73 or about 6:44 A.M.

   b. y is greatest for $x = 4$ (then $y = 7.23$). This corresponds to July 1.

   c. $7.23 = 7\dfrac{23}{100}$

     $7 + \dfrac{13.8}{60} \approx 7:14$

     Sunrise will be at about 7:14 A.M.

   d. $y(6.5) = 6.28 = 6\dfrac{28}{100}$

     $6 + \dfrac{16.8}{60} \approx 6:17$

     Sunrise will be at about 6:17 A.M.

   e. No. Answers may vary.

Section 1.7

Mental Math

1. $2x + y = 5$
     $y = 5 - 2x$

2. $7x - y = 3$
     $7x - 3 = y$ or $y = 7x - 3$

3. $a - 5b = 8$
     $a = 5b + 8$

4. $7r + s = 10$
     $s = 10 - 7r$

5. $5j + k - h = 6$
     $k = h - 5j + 6$

6. $w - 4y + z = 0$
     $z = 4y - w$

Exercise Set 1.7

1. $D = rt$

   $\dfrac{D}{r} = \dfrac{rt}{r}$

   $\dfrac{D}{r} = t$

   $t = \dfrac{D}{r}$

3. $I = PRT$

   $\dfrac{I}{PT} = \dfrac{PRT}{PT}$

   $\dfrac{I}{PT} = R$

   $R = \dfrac{I}{PT}$

5. $9x - 4y = 16$
   $9x - 4y - 9x = 16 - 9x$
         $-4y = 16 - 9x$

   $\dfrac{-4y}{-4} = \dfrac{16 - 9x}{-4}$

      $y = \dfrac{9x - 16}{4}$

7. $P = 2L + 2W$
   $P - 2L = 2W$

   $\dfrac{P - 2L}{2} = \dfrac{2W}{2}$

   $\dfrac{P - 2L}{2} = W$

     $W = \dfrac{P - 2L}{2}$

9.
$$J = AC - 3$$
$$J + 3 = AC$$
$$\frac{J+3}{C} = \frac{AC}{C}$$
$$\frac{J+3}{C} = A$$
$$A = \frac{J+3}{C}$$

11.
$$W = gh - 3gt^2$$
$$W = g(h - 3t^2)$$
$$\frac{W}{h-3t^2} = \frac{g(h-3t^2)}{h-3t^2}$$
$$\frac{W}{h-3t^2} = g$$
$$g = \frac{W}{h-3t^2}$$

13.
$$T = C(2 + AB)$$
$$T = 2C + ABC$$
$$T - 2C = ABC$$
$$\frac{T-2C}{AC} = \frac{ABC}{AC}$$
$$\frac{T-2C}{AC} = B$$
$$B = \frac{T-2C}{AC}$$

15.
$$C = 2\pi r$$
$$\frac{C}{2\pi} = \frac{2\pi r}{2\pi}$$
$$\frac{C}{2\pi} = r$$
$$r = \frac{C}{2\pi}$$

17.
$$E = I(r + R)$$
$$E = Ir + IR$$
$$E - IR = Ir$$
$$\frac{E-IR}{I} = \frac{Ir}{I}$$
$$r = \frac{E-IR}{I}$$

19.
$$s = \frac{n}{2}(a + L)$$
$$2 \cdot s = 2 \cdot \frac{n}{2}(a + L)$$
$$2s = n(a + L)$$
$$2s = an + Ln$$
$$2s - an = Ln$$
$$\frac{2s-an}{n} = \frac{Ln}{n}$$
$$L = \frac{2s-an}{n}$$

21.
$$N = 3st^4 - 5sv$$
$$N - 3st^4 = -5sv$$
$$\frac{N-3st^4}{-5s} = \frac{-5sv}{-5s}$$
$$\frac{3st^4 - N}{5s} = v$$
$$v = \frac{3st^4 - N}{5s}$$

23.
$$S = 2LW + 2LH + 2WH$$
$$S - 2LW = 2LH + 2WH$$
$$S - 2LW = H(2L + 2W)$$
$$\frac{S-2LW}{2L+2W} = \frac{H(2L+2W)}{2L+2W}$$
$$\frac{S-2LW}{2L+2W} = H$$
$$H = \frac{S-2LW}{2L+2W}$$

25. $A = P\left(1 + \dfrac{r}{n}\right)^{nt} = 3500\left(1 + \dfrac{0.03}{n}\right)^{10n}$

n	1	2	4	12	365
A	\$4703.71	\$4713.99	\$4719.22	\$4722.74	\$4724.45

27. Choose the account that is compounded 12 times a year because you earn more interest.

29. $A = P\left(1 + \dfrac{r}{n}\right)^{nt} = 6000\left(1 + \dfrac{0.04}{n}\right)^{5n}$

    a. $n = 2$

$$A = 6000\left(1 + \frac{0.04}{2}\right)^{5 \cdot 2} \approx 7313.97$$

The amount in the account is \$7313.97.

    b. $n = 4$

$$A = 6000\left(1 + \frac{0.04}{4}\right)^{5 \cdot 4} \approx 7321.14$$

The amount in the account is \$7321.14.

    c. $n = 12$

$$A = 6000\left(1 + \frac{0.04}{12}\right)^{5 \cdot 12} \approx 7325.98$$

The amount in the account is \$7325.98.

31. $C = \dfrac{5}{9}(F - 32)$

$C = \dfrac{5}{9}(104 - 32)$

$C = \dfrac{5}{9}(72)$

$C = 40$

The day's high temperature was $40°\text{C}$.

33. $d = rt$ or $t = \dfrac{d}{r}$, where $d = 2(90) = 180$ and $r = 50$, so $t = \dfrac{180}{50} = 3.6$.

Thus, she takes 3.6 hours or 3 hours, 36 minutes to make the round trip.

35. $A = s^2$, where $s = 64$, so $A = 64^2 = 4096$.

Thus, $\dfrac{4096}{24} \approx 171$ packages of tiles should be bought.

37. 1.3 miles is equivalent to
$1.3 \cdot 5280 = 6864$ feet.
$$V = \pi r^2 h$$
$$3800 = \pi r^2 (6864)$$
$$0.42 \approx r$$
The radius of the hole is approximately 0.42 feet.

39. $\dfrac{168 \text{ miles}}{1 \text{ hour}} \cdot \dfrac{5280 \text{ feet}}{1 \text{ mile}} \cdot \dfrac{1 \text{ hour}}{60 \text{ minutes}} \cdot \dfrac{1 \text{ minutes}}{60 \text{ seconds}}$
$= 246.4$ feet per second
$$d = rt$$
$$60.5 = 246.4t$$
$$t \approx 0.25$$
The ball would reach the plate in approximately 0.25 second.

41. $C = \pi d$
$$= \pi(41.125)$$
$$= 41.125\pi \text{ ft}$$
$$\approx 4.125(3.14)$$
$$\approx 129.1325$$
The circumference of Earth is $41.125\pi \approx 129.1325$ feet.

43. Cost per person $= \dfrac{\$1.7 \text{ billion}}{250 \text{ million people}}$
$$= \dfrac{\$1,700,000,000}{250,000,000 \text{ people}}$$
$$= \$6.80 \text{ per person}$$

45. $A = P\left(1 + \dfrac{r}{n}\right)^{nt}$
$$A = 10,000\left(1 + \dfrac{0.085}{4}\right)^{4 \cdot 2}$$
$$\approx \$11,831.96$$
$11,831.96 - \$10,000 = \$1831.96$
The interest is $\$1831.96$.

47. The area of one pair of walls is
$2 \cdot 14 \cdot 8 = 224$ square feet and the area of the other walls is $2 \cdot 16 \cdot 8 = 256$ square feet, for a total of 480 square feet. Multiplying by 2, the number of coats, yields 960 square feet. Dividing this by 500 yields 1.92. Thus, 2 gallons should be purchased.

49. a. $V = \pi r^2 h$
$$= \pi(4.2)^2(21.2)$$
$$\approx 1174.86$$
The volume of the cylinder is 1174.86 cubic meters.

b. $V = \dfrac{4}{3}\pi r^3$
$$= \dfrac{4}{3}\pi(4.2)^3$$
$$\approx 310.34$$
The volume of the sphere is 310.34 cubic meters.

c. $V = 1174.86 + 310.34 = 1485.20$
The volume of the tank is 1485.20 cubic meters.

51. $x = $ length (parallel to house)
$$A = lw = x\left(\dfrac{56 - x}{2}\right)$$

a. Make a table using $y_1 = \dfrac{x(56 - x)}{2}$.
y_1 is largest when $x = 28$.
The dimensions are 14 feet by 28 feet.

b. $14 \times 28 = 392$
The largest area is 392 square feet.

53. $d = rt$ or $t = \dfrac{d}{r}$
$$t = \dfrac{135}{60} = 2.25$$
Thus, it takes him 2.25 hours, or 2 hours and 15 minutes.

55. 80.5 months $\cdot \dfrac{30 \text{ days}}{1 \text{ month}} \cdot \dfrac{24 \text{ hours}}{1 \text{ day}}$
$= 57,960$ hours
$$d = rt \text{ or } r = \dfrac{d}{t}$$
$$r = \dfrac{2,000,000,000}{57,960} \approx 34,507$$
The average speed is 34,507 miles per hour.

57.

Planet	AU from Sun
Mercury	0.388
Venus	0.723
Earth	1.000
Mars	1.523
Jupiter	5.202
Saturn	9.538
Uranus	19.193
Neptune	30.065
Pluto	39.505

59. $\dfrac{1}{4}$

61. $\dfrac{3}{8}$

63. $\dfrac{3}{8}$

65. $\dfrac{3}{4}$

67. 1

69. 1

Exercise Set 1.8

1. $\dfrac{164 + 397 + 520 + 604 + 684}{5} = 473.8$
The mean is \$473.80.

3. $\dfrac{195 + 368 + 464 + 526 + 587}{5} = 428$
The mean is \$428.

5. First, write the numbers in order.
3.2, 9.7, 10.4, 11, 11.1
The median is the middle number, 10.4%.

7. First, write the numbers in order.
11.1, 11.5, 13.1, 14.1, 18
The median is the middle number, 13.1%.

9. The mode is 85 since 85 occurs most often.

11. Growth/Income: (\$15,000)(0.35) = \$5250
Small Co./Aggressive:
(\$15,000)(0.25) = \$3750
International: (\$15,000)(0.20) = \$3000
Bonds: (\$15,000)(0.20) = \$3000

13. Answers may vary.

15. $\dfrac{1454 + 1368 + 1362 + 1250 + 1136}{5} = 1314$
The mean height is 1314 feet.

17. $\dfrac{1127 + 1136}{2} = 1131.5$
The median height for the ten tallest buildings is 1131.5 feet.

19. $\dfrac{7.8 + 6.9 + 7.5 + 4.7 + 6.9 + 7.0}{6} = 6.8$
The mean is 6.8 seconds.

21. The mode is 6.9 since 6.9 seconds occurs most often.

23. First, write the numbers in order.
74, 77, 85, 86, 91, 95
The median is the mean of the two middle numbers, or $\dfrac{85 + 86}{2} = 85.5$.

25. $\dfrac{\text{sum of values}}{\text{number of values}} = \dfrac{1095}{15} = 73$
The mean is 73.

27. The modes are 70 and 71 since each occurs two times.

29. Refer to the ordered numbers in exercise 26. There are 9 rates lower than the mean of 73.

31. 21 has to occur at least twice, and 20 must occur at least once because there are an odd number of values and the median is 20. The missing numbers are 21, 21, and 20.

33. A police officer walks the farthest because this profession has the longest graph.

35. The length of the graph for a nurse is about 940. A nurse walks an average of 940 miles per year.

37. Answers may vary.

39. The Democratic Party shows the greatest percent of the adult population for all the years shown because its graph is the highest for each year.

41. The greatest difference occurs in 1964 because the distance between the two points representing Democrats and Republicans is the greatest.

43. Estimate the height of the dots directly above the year 1992 for each of the parties. The percents of the adult population identified as Democrat, Republican, and Independent are 50%, 38%, and 10%, respectively.

45. Answers may vary.

47. $363(0.58) = 210.54$
The average amount of money spent on purchasing clothes was $210.54.

49. $210.54(0.16) \approx 33.69$
The average amount paid by credit card for clothes was approximately $33.69.

51. $\dfrac{28.7 + 20.8 + 23.4 + 27.4 + 21.2}{5} = 24.3$
The mean of the 5 most populous cities is 24.3 million people.

53. The middle value (eighth in descending order) is 17.6. The median is 17.6 million people.

55.

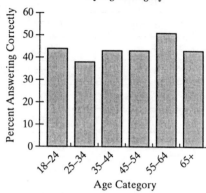

Correct Answer to History Question by Age Category

57. Answers may vary.

59. No, because the mean must lie within the range.

61. The mean is lowered; the median and mode are unchanged.

Chapter 1 Review

1. $7x = 7(3) = 21$

2. $st = (1.6)(5) = 8$

3. One hour is $60(60) = 3600$ seconds.
$90t = 90(3600) = 324,000$
The hummingbird has 324,000 wing beats per hour.

4. $\{-1, 1, 3\}$

5. $\{-2, 0, 2, 4, 6\}$

6. $\varnothing$

7. $\varnothing$

8. $\{6, 7, 8, ...\}$

9. $\{..., -1, 0, 1, 2\}$

10. True

11. False; $9 \in B$.

12. True, since $\sqrt{169} = 13$.

13. True, since zero is not an element of the empty set.

14. False, since π is irrational.

15. True, since π is a real number.

16. False, since $\sqrt{4} = 2$.

17. True, since -9 is a rational number.

18. True

19. True, since C is not a subset of B.

20. False, since all integers are rational numbers.

21. True, since the empty set is a subset of all sets.

22. True, since every set is a subset of itself.

23. True, since every element of D is also an element of C.

24. True, since every integer is a real number.

25. True, since every irrational number is also a real number.

26. False, since B does not contain the set $\{5\}$.

27. True, since $\{5\}$ is a subset of B.

28. $\left\{5, \dfrac{8}{2}, \sqrt{9}\right\}$, since $\sqrt{9} = 3$.

29. $\left\{5, \dfrac{8}{2}, \sqrt{9}\right\}$, since $\sqrt{9} = 3$.

30. $\left\{5, -\dfrac{2}{3}, \dfrac{8}{2}, \sqrt{9}, 0.3, 1\dfrac{5}{8}, -1\right\}$

31. $\left\{\sqrt{7}, \pi\right\}$

32. $\left\{5, -\dfrac{2}{3}, \dfrac{8}{2}, \sqrt{9}, 0.3, \sqrt{7}, 1\dfrac{5}{8}, -1, \pi\right\}$

33. $\left\{5, \dfrac{8}{2}, \sqrt{9}, -1\right\}$

34. The opposite of $-\dfrac{3}{4}$ is $-\left(-\dfrac{3}{4}\right) = \dfrac{3}{4}$.

35. The opposite of 0.6 is -0.6.

36. The opposite of 0 is $-0 = 0$.

37. The opposite of 1 is -1.

38. $\left|-\dfrac{3}{4}\right| = \dfrac{3}{4}$ since $-\dfrac{3}{4}$ is located $\dfrac{3}{4}$ unit from 0 on the number line.

39. $|0.6| = 0.6$ since 0.6 is located 0.6 unit from 0 on the number line.

40. $|0| = 0$ since 0 is located 0 units from 0 on the number line.

41. $|1| = 1$ since 1 is located 1 unit from 0 on the number line.

42. $-7 + 3 = -4$

43. $-10 + (-25) = -35$

44. $5(-0.4) = -2$

45. $(-3.1)(-0.1) = 0.31$

46. $-7 - (-15) = -7 + 15 = 8$

47. $9 - (-4.3) = 9 + 4.3 = 13.3$

48. $(-6)(-4)(0)(-3) = 0$

49. $(-12)(0)(-1)(-5) = 0$

50. $(-24) \div 0$ is undefined.

51. $0 \div (-45) = 0$

52. $(-36) \div (-9) = 4$

53. $(60) \div (-12) = -5$

54. $\left(-\dfrac{4}{5}\right)-\left(-\dfrac{2}{3}\right)=-\dfrac{12}{15}+\dfrac{10}{15}=-\dfrac{2}{15}$

55. $\dfrac{5}{4}-\left(-2\dfrac{3}{4}\right)=\dfrac{5}{4}+\dfrac{11}{4}=\dfrac{16}{4}=4$

56. $1-\dfrac{1}{4}-\dfrac{1}{3}=\dfrac{12}{12}-\dfrac{3}{12}-\dfrac{4}{12}=\dfrac{5}{12}$

57. $\begin{aligned}-5+7-3-(-10)&=2-3+10\\&=-1+10\\&=9\end{aligned}$

58. $\begin{aligned}8-(-3)+(-4)+6&=8+3-4+6\\&=11-4+6\\&=7+6\\&=13\end{aligned}$

59. $3(4-5)^{4}=3(-1)^{4}=3(1)=3$

60. $6(7-10)^{2}=6(-3)^{2}=6(9)=54$

61. $\left(-\dfrac{8}{15}\right)\cdot\left(-\dfrac{2}{3}\right)^{2}=-\dfrac{8}{15}\cdot\dfrac{4}{9}=-\dfrac{32}{135}$

62. $\left(-\dfrac{3}{4}\right)^{2}\cdot\left(-\dfrac{10}{21}\right)=\left(\dfrac{9}{16}\right)\left(-\dfrac{10}{21}\right)=-\dfrac{15}{56}$

63. $\begin{aligned}\dfrac{-\frac{6}{15}}{\frac{8}{15}}&=-\dfrac{6}{15}\div\dfrac{8}{25}\\&=-\dfrac{6}{15}\cdot\dfrac{25}{8}\\&=-\dfrac{150}{120}\\&=-\dfrac{5}{4}\end{aligned}$

64. $\begin{aligned}\dfrac{\frac{4}{9}}{-\frac{8}{45}}&=\dfrac{4}{9}\div\left(\dfrac{8}{45}\right)\\&=\dfrac{4}{9}\cdot\left(-\dfrac{45}{8}\right)\\&=-\dfrac{180}{72}\\&=-\dfrac{5}{2}\end{aligned}$

65. $\begin{aligned}-\dfrac{3}{8}+3(2)\div 6&=-\dfrac{3}{8}+6\div 6\\&=-\dfrac{3}{8}+1\\&=-\dfrac{3}{8}+\dfrac{8}{8}\\&=\dfrac{5}{8}\end{aligned}$

66. $\begin{aligned}5(-2)-(-3)-\dfrac{1}{6}+\dfrac{2}{3}&=-10+3-\dfrac{1}{6}+\dfrac{2}{3}\\&=-7-\dfrac{1}{6}+\dfrac{2}{3}\\&=-\dfrac{42}{6}-\dfrac{1}{6}+\dfrac{4}{6}\\&=-\dfrac{39}{6}\\&=-6\dfrac{1}{2}\end{aligned}$

67. $\begin{aligned}\left|2^{3}-3^{2}\right|-|5-7|&=|8-9|-|-2|\\&=|-1|-2\\&=1-2\\&=-1\end{aligned}$

68. $\begin{aligned}\left|5^{2}-2^{2}\right|+|9\div(-3)|&=|25-4|+|-3|\\&=|21|+3\\&=21+3\\&=24\end{aligned}$

69. $\begin{aligned}(2^{3}-3^{2})-(5-7)&=(8-9)-(-2)\\&=-1+2\\&=1\end{aligned}$

70. $(5^2 - 2^4) + [9 \div (-3)] = (25 - 16) + (-3)$
$$= 9 + (-3)$$
$$= 6$$

71. $\dfrac{(8-10)^3 - (-4)^2}{2 + 8(2) \div 4} = \dfrac{(-2)^3 - 16}{2 + 16 \div 4}$
$$= \dfrac{-8 - 16}{2 + 4}$$
$$= \dfrac{-24}{6}$$
$$= -4$$

72. $\dfrac{(2+4)^2 + (-1)^5}{12 \div 2 \cdot 3 - 3} = \dfrac{6^2 + (-1)}{6 \cdot 3 - 3}$
$$= \dfrac{36 - 1}{18 - 3}$$
$$= \dfrac{35}{15}$$
$$= \dfrac{7}{3}$$

73. $\dfrac{(4-9) + 4 - 9}{10 - 12 \div 4 \cdot 8} = \dfrac{-5 + 4 - 9}{10 - 3 \cdot 8}$
$$= -\dfrac{-1 - 9}{10 - 24}$$
$$= \dfrac{-10}{-14}$$
$$= \dfrac{5}{7}$$

74. $\dfrac{3 - 7 - (7-3)}{15 + 30 \div 6 \cdot 2} = \dfrac{-4 - (4)}{15 + 5 \cdot 2}$
$$= \dfrac{-8}{15 + 10}$$
$$= -\dfrac{8}{25}$$

75. $\dfrac{\sqrt{25}}{4 + 3 \cdot 7} = \dfrac{5}{4 + 21} = \dfrac{5}{25} = \dfrac{1}{5}$

76. $\dfrac{\sqrt{64}}{24 - 8 \cdot 2} = \dfrac{8}{24 - 16} = \dfrac{8}{8} = 1$

77. $x^2 - y^2 + z^2 = 0^2 - 3^2 + (-2)^2$
$$= 0 - 9 + 4$$
$$= -5$$

78. $\dfrac{5x + z}{2y} = \dfrac{5(0) + (-2)}{2(3)} = \dfrac{0 - 2}{6} = \dfrac{-2}{6} = -\dfrac{1}{3}$

79. $\dfrac{-7y - 3z}{-3} = \dfrac{-7(3) - 3(-2)}{-3}$
$$= \dfrac{-21 - (-6)}{-3}$$
$$= \dfrac{-21 + 6}{-3}$$
$$= \dfrac{-15}{-3}$$
$$= 5$$

80. $(x - y + z)^2 = (0 - 3 + (-2))^2$
$$= (-3 - 2)^2$$
$$= (-5)^2$$
$$= 25$$

81. a.

r	1	10	100
$2\pi r$	6.28	62.83	628.32

    b. As the radius increases, the circumference increases.

82. $5xy - 7xy + 3 - 2 + xy$
$$= 5xy - 7xy + xy + 3 - 2$$
$$= (5 - 7 + 1)xy + (3 - 2)$$
$$= (-1)xy + 1$$
$$= -xy + 1$$

83. $4x + 10x - 19x + 10 - 19$
$$= (4 + 10 - 19)x + (10 - 19)$$
$$= -5x + (-9)$$
$$= -5x - 9$$

84. $6x^2 + 2 - 4(x^2 + 1) = 6x^2 + 2 - 4x^2 - 4$
$$= 6x^2 - 4x^2 + 2 - 4$$
$$= (6 - 4)x^2 + (2 - 4)$$
$$= 2x^2 - 2$$

85. $-7(2x^2 - 1) - x^2 - 1$
$$= -14x^2 + 7 - x^2 - 1$$
$$= -14x^2 - x^2 + 7 - 1$$
$$= (-14 - 1)x^2 + (7 - 1)$$
$$= -15x^2 + 6$$

86. $(3.2x - 1.5) - (4.3x - 1.2)$
$= 3.2x - 1.5 - 4.3x + 1.2$
$= 3.2x - 4.3x - 1.5 + 1.2$
$= (3.2 - 4.3)x - 0.3$
$= -1.1x - 0.3$

87. $(7.6x + 4.7) - (1.9x + 3.6)$
$= 7.6x + 4.7 - 1.9x - 3.6$
$= 7.6x - 1.9x + 4.7 - 3.6$
$= (7.6 - 1.9)x + 1.1$
$= 5.7x + 1.1$

88. $12 = -4x$

89. $n + 2n = -15$

90. $4(y + 3) = -1$

91. $6(t - 5) = 4$

92. $z - 7 = 6$

93. $9x - 10 = 5$

94. $x - 5 \geq 12$

95. $-4 < 7y$

96. $\frac{2}{3} \neq 2\left(n + \frac{1}{4}\right)$

97. $t + 6 \leq -12$

98. Associative property of addition

99. Distributive property

100. Additive inverse property

101. Commutative property of addition

102. Associative and commutative properties of multiplication
To see this:
$(XY)Z = X(YZ) = (YZ)X$

103. Multiplicative inverse property

104. Additive identity property

105. Associative property of multiplication

106. Commutative property of addition

107. Distributive property

108. $5(x - 3z) = 5x - 15z$

109. $(7 + y) + (3 + x) = (3 + x) + (7 + y)$ is one possible solution.

110. $0 = 2 + (-2)$, for example

111. $1 = 2 \cdot \frac{1}{2}$, for example

112. $[(3.4)(0.7)]5 = (3.4)[(0.7)5]$

113. $7 = 7 + 0$

114. $-9 > -12$ since -9 lies to the right of -12 on the number line.

115. $0 > -6$ since 0 lies to the right of -6 on the number line.

116. $-3 < -1$ since -3 lies to the left of -1 on the number line.

117. $7 = |-7|$ since $|-7| = 7$.

118. $-5 < -(-5)$ since $-(-5) = 5$.

119. $4(x - 5) = 2x - 14$
$4x - 20 = 2x - 14$
$2x = 6$
$x = 3$
The solution is 3.

120. $x + 7 = -2(x + 8)$
$x + 7 = -2x - 16$
$3x = -23$
$x = -\frac{23}{3}$
The solution is $-\frac{23}{3}$.

121. $3(2y - 1) = -8(6 + y)$
$6y - 3 = -48 - 8y$
$14y = -45$
$y = -\frac{45}{14}$
The solution is $-\frac{45}{14}$.

122. $-(z+12) = 5(2z-1)$
$-z - 12 = 10z - 5$
$-7 = 11z$
$-\dfrac{7}{11} = z$
The solution is $-\dfrac{7}{11}$.

123. $0.3(x-2) = 1.2$
$10[0.3(x-2)] = 10(1.2)$
$3(x-2) = 12$
$3x - 6 = 12$
$3x = 18$
$x = 6$
The solution is 6.

124. $1.5 = 0.2(c - 0.3)$
$1.5 = 0.2c - 0.06$
$100(1.5) = 100(0.2c - 0.06)$
$150 = 20c - 6$
$156 = 20c$
$7.8 = c$
The solution is 7.8.

125. $-4(2 - 3h) = 2(3h - 4) + 6h$
$-8 + 12h = 6h - 8 + 6h$
$-8 + 12h = 12h - 8$
$-8 = -8$
The solution is all real numbers.

126. $6(m - 1) + 3(2 - m) = 0$
$6m - 6 + 6 - 3m = 0$
$3m = 0$
$m = 0$
The solution is 0.

127. $6 - 3(2g + 4) - 4g = 5(1 - 2g)$
$6 - 6g - 12 - 4g = 5 - 10g$
$-6 - 10g = 5 - 10g$
$-6 = 5$
There is no solution.

128. $20 - 5(p + 1) + 3p = -(2p - 15)$
$20 - 5p - 5 + 3p = -2p + 15$
$15 - 2p = -2p + 15$
$15 = 15$
The solution is all real numbers.

129. $\dfrac{x}{3} - 4 = x - 2$
$3\left(\dfrac{x}{3} - 4\right) = 3(x - 2)$
$x - 12 = 3x - 6$
$-6 = 2x$
$-3 = x$
The solution is -3.

130. $\dfrac{9}{4}y = \dfrac{2}{3}y$
$12\left(\dfrac{9}{4}y\right) = 12\left(\dfrac{2}{3}y\right)$
$3(9y) = 4(2y)$
$27y = 8y$
$19y = 0$
$y = 0$
The solution is 0.

131. $\dfrac{3n}{8} - 1 = 3 + \dfrac{n}{6}$
$24\left(\dfrac{3n}{8} - 1\right) = 24\left(3 + \dfrac{n}{6}\right)$
$9n - 24 = 72 + 4n$
$5n = 96$
$n = \dfrac{96}{5}$
The solution is $\dfrac{96}{5}$.

132. $\dfrac{z}{6} + 1 = \dfrac{z}{2} + 2$
$6\left(\dfrac{z}{6} + 1\right) = 6\left(\dfrac{z}{2} + 2\right)$
$z + 6 = 3z + 12$
$-6 = 2z$
$-3 = z$
The solution is -3.

133.
$$\frac{b-2}{3} = \frac{b+2}{5}$$
$$15\left(\frac{b-2}{3}\right) = 15\left(\frac{b+2}{5}\right)$$
$$5b - 10 = 3b + 6$$
$$2b = 16$$
$$b = 8$$
The solution is 8.

134.
$$\frac{2t-1}{3} = \frac{3t+2}{15}$$
$$15\left(\frac{2t-1}{3}\right) = 15\left(\frac{3t+2}{15}\right)$$
$$5(2t-1) = 3t + 2$$
$$10t - 5 = 3t + 2$$
$$7t = 7$$
$$t = 1$$
The solution is 1.

135.
$$\frac{x-2}{5} + \frac{x+2}{2} = \frac{x+4}{3}$$
$$30\left(\frac{x-2}{5} + \frac{x+2}{2}\right) = 30\left(\frac{x+4}{3}\right)$$
$$6(x-2) + 15(x+2) = 10(x+4)$$
$$6x - 12 + 15x + 30 = 10x + 40$$
$$21x + 18 = 10x + 40$$
$$11x = 22$$
$$x = 2$$
The solution is 2.

136.
$$\frac{2z-3}{4} - \frac{4-z}{2} = \frac{z+1}{3}$$
$$12\left(\frac{2z-3}{4} - \frac{4-z}{2}\right) = 12\left(\frac{z+1}{3}\right)$$
$$3(2z-3) - 6(4-z) = 4(z+1)$$
$$6z - 9 - 24 + 6z = 4z + 4$$
$$12z - 33 = 4z + 4$$
$$8z = 37$$
$$z = \frac{37}{8}$$
The solution is $\frac{37}{8}$.

137.
$$2(x-3) = 3x + 1$$
$$2x - 6 = 3x + 1$$
$$-7 = x$$
The number is -7.

138. Let x = the smaller number, then
$x + 5$ = the larger number.
$$x + x + 5 = 285$$
$$2x = 280$$
$$x = 140$$
$$x + 5 = 145$$
The numbers are 140 and 145.

139. $40\% \cdot 130 = 0.40 \cdot 130 = 52$

140. $1.5\% \cdot 8 = 0.015 \cdot 8 = 0.12$

141. Let x = 1998 earnings for a high school graduate.
$$x + 0.3047x = 29,872$$
$$1.3047x = 29,872$$
$$x \approx 22,896$$
Rounded to the nearest dollar, the average annual salary for a high school graduate in 1998 was $22,896.

142. Let n = the first integer, then
$n + 1$ = the second integer,
$n + 2$ = the third integer, and
$n + 4$ = the fourth integer.
$$(n+1) + (n+2) + (n+3) - 2n = 16$$
$$n + 6 = 16$$
$$n = 10$$
Therefore, the integers are 10, 11, 12, and 13.

143. Let x = the smaller odd integer, then
$x + 2$ = the larger odd integer.
$$5x = 3(x+2) + 54$$
$$5x = 3x + 6 + 54$$
$$2x = 60$$
$$x = 30$$
Since this is not odd, no such consecutive odd integers exist.

144. Let w = the width of the playing field, then
$2w - 5$ = the length of the playing field.
$$2w + 2(2w - 5) = 230$$
$$2w + 4w - 10 = 230$$
$$6w = 240$$
$$w = 40$$
Then $2w - 5 = 2(40) - 5$
$$= 80 - 5$$
$$= 75$$
Therefore, the field is 75 meters long and 40 meters wide.

145. Let n = the number of miles driven.
$$2(29.95) + 0.15(n - 200) = 83.6$$
$$59.9 + 0.15n - 30 = 83.6$$
$$29.9 + 0.15n = 83.6$$
$$0.15n = 53.7$$
$$n = 358$$
He drove 358 miles.

146. Solve $R = C$.
$$16.50x = 4.50x + 3000$$
$$12x = 3000$$
$$x = 250$$
Thus, 250 calculators must be produced and sold in order to break even.

147. Solve $R = C$.
$$40x = 20x + 100$$
$$20x = 100$$
$$x = 5$$
$R = 40 \cdot 5 = 200$
She will break even if she sells 5 plants.
The revenue will be $200.

148.

X	Y₁
1	10.472
1.5	23.562
2	41.888
2.5	65.45
3	94.248

X=

Radius of cone	x	1	1.5	2	2.5	3
Volume (if height is 10 inches)	$\frac{10}{3}\pi x^2$	10.47	23.56	41.89	65.45	94.25

‹9.

X	Y₁
2	34
4.68	39.36
9.5	49
12.68	55.36

X=

Radius of cone	x	2	4.68	9.5	12.68
Perimeter (if length is 15 units)	$30 + 2x$	34	39.36	49	55.36

150. $y_1 = 5.75x$

X	Y₁
5	28.75
10	57.5
15	86.25
20	115
25	143.75
30	172.5

X=

Hours	5	10	15	20	25	30
Gross pay (dollars)	28.75	57.50	86.25	115.00	143.75	172.50

151. $y_1 = 8 + 1.1(x - 4)$, $y_2 = 12 + 1.5(x - 5)$

X	Y₁	Y₂
5	9.1	12
6	10.2	13.5
7	11.3	15
8	12.4	16.5

X=

Gallons (in thousands used)	5	6	7	8
Coast charge (dollars)	9.1	10.2	11.3	12.4
Cross Gates charge (dollars)	12.0	13.5	15.0	16.5

152. a.

Seconds	0	1	2	3	4	5
Height in feet	500	524	516	476	404	300

X	Y1	
0	500	
1	524	
2	516	
3	476	
4	404	
5	300	
X=		

b.

X	Y1	
1	524	
1.1	524.64	
1.2	524.96	
1.3	524.96	
1.4	524.64	
1.5	524	
1.6	523.04	
Y1=524.96		

At the maximum, $y \approx 525$, so the maximum height of the rock is 525 feet.

c.

X	Y1	
6.5	84	
6.6	67.04	
6.7	49.76	
6.8	32.16	
6.9	14.24	
7	-4	
7.1	-22.56	
X=7		

When $y = 0$, $x \approx 7.0$, so the rock hits the ground after approximately 7.0 seconds.

153. $V = lwh$

$$\frac{V}{lh} = \frac{lwh}{lh}$$

$$\frac{V}{lh} = w$$

$$w = \frac{V}{lh}$$

154. $C = 2\pi r$

$$\frac{C}{2\pi} = \frac{2\pi r}{2\pi}$$

$$\frac{C}{2\pi} = r$$

$$r = \frac{C}{2\pi}$$

155. $5x - 4y = -12$

$$5x + 12 = 4y$$

$$y = \frac{5x + 12}{4}$$

156. $5x - 4y = -12$

$$5x = 4y - 12$$

$$x = \frac{4y - 12}{5}$$

157. $y - y_1 = m(x - x_1)$

$$\frac{y - y_1}{x - x_1} = \frac{m(x - x_1)}{x - x_1}$$

$$\frac{y - y_1}{x - x_1} = m$$

$$m = \frac{y - y_1}{x - x_1}$$

158. $y - y_1 = m(x - x_1)$

$$y - y_1 = mx - mx_1$$

$$y - y_1 + mx_1 = mx$$

$$\frac{y - y_1 + mx_1}{m} = x$$

$$x = \frac{y - y_1 + mx_1}{m}$$

159.
$$E = I(R + r)$$
$$E = IR + Ir$$
$$E - IR = Ir$$
$$\frac{E - IR}{I} = r$$
$$r = \frac{E - IR}{I}$$

160.
$$S = vt + gt^2$$
$$S - vt = gt^2$$
$$\frac{S - vt}{t^2} = g$$
$$g = \frac{S - vt}{t^2}$$

161.
$$T = gr + gvt$$
$$T = g(r + vt)$$
$$\frac{T}{r + vt} = g$$
$$g = \frac{T}{r + vt}$$

162.
$$I = Prt + P$$
$$I = P(rt + 1)$$
$$\frac{I}{rt + 1} = P$$
$$P = \frac{I}{rt + 1}$$

163.
$$A = \frac{h}{2}(B + b)$$
$$2A = hB + hb$$
$$2A - hb = hB$$
$$\frac{2A - hb}{h} = B$$
$$B = \frac{2A - hb}{h}$$

164.
$$V = \frac{1}{3}\pi r^2 h$$
$$3V = \pi r^2 h$$
$$\frac{3V}{\pi r^2} = h$$
$$h = \frac{3V}{\pi r^2}$$

165.
$$R = \frac{r_1 + r_2}{2}$$
$$2R = r_1 + r_2$$
$$r_1 = 2R - r_2$$

166.
$$\frac{V_1}{T_1} = \frac{V_2}{T_2}$$
$$T_2 V_1 = T_1 V_2$$
$$T_2 = \frac{T_1 V_2}{V_1}$$

167. $A = P\left(1 + \dfrac{r}{n}\right)^{nt} = 3000\left(1 + \dfrac{0.03}{n}\right)^{7n}$

   a. $A = 3000\left(1 + \dfrac{0.03}{2}\right)^{14} \approx 3695.27$

      The amount in the account is
$3695.27.

   b. $A = 3000\left(1 + \dfrac{0.03}{52}\right)^{364} \approx 3700.81$

      The amount in the account is
$3700.81.

168. Let $x = $ the width, then $x + 2 = $ the length.
$$(x + 4)(x + 2 + 4) = x(x + 2) + 88$$
$$(x + 4)(x + 6) = x^2 + 2x + 88$$
$$x^2 + 10x + 24 = x^2 + 2x + 88$$
$$8x = 64$$
$$x = 8$$
$x + 2 = 10$
The width is 8 inches and the length is
10 inches.

169. a.

Celsius	−40	−15	10	60
Fahrenheit	−40	5	50	140

   b. $\dfrac{9(100) + 160}{5} = 212$

      100°C is 212°F.

   c. $\dfrac{9(0) + 160}{5} = 32$

      0°C is 32°F.

170. a.

Hours	0	0.25	0.50	0.75	1	1.25	1.5
Miles	0	13.75	27.5	41.25	55	68.75	82.5

b. $\dfrac{192.5}{55} = 3.5$

It took 3.5 hours to travel 192.5 miles at 55 miles per hour.

171. Area $= 18 \times 21 = 378$ square feet.

Packages $= \dfrac{378}{24} = 15.75$

16 packages are needed to cover the floor.

172. $V_{box} = lwh = 8(5)(3) = 120$

$V_{cyl} = \pi r^2 h = \pi \cdot 3^2 \cdot 6 = 54\pi \approx 170$

Therefore, the cylinder holds more ice cream.

173. $d = rt$

$130 = r(2.25)$

$57.8 \approx r$

Her average speed is 58 miles per hour.

174.

Height h	6	6	6	6	6
Radius x	1	1.5	2	2.25	3
Volume	6.28	14.14	25.13	31.81	56.55

175.

Height h	10	10	10	10	10
Radius x	1.5	2.1	2.75	3	3.5
Volume	23.56	46.18	79.19	94.25	128.28

176. mean: $\dfrac{21 + 28 + 16 + 42 + 38}{5} = 29$

median: 28 is the middle value.

mode: none

177. mean: $\dfrac{42 + 35 + 36 + 40 + 50}{5} = 40.6$

median: 40 is the middle value.

mode: none

178. mean: $\dfrac{7.6+8.2+8.2+9.6+5.7+9.1}{6} \approx 8.07$

median: $\dfrac{8.2+8.2}{2} = 8.2$

mode: 8.2 occurs twice.

179. mean: $\dfrac{4.9+7.1+6.8+6.8+5.3+4.9}{6} \approx 5.97$

median: $\dfrac{5.3+6.8}{2} = 6.05$

mode: 4.9 and 6.8 each occur twice.

180. mean: $\dfrac{0.2+0.3+0.5+0.6+0.6+0.9+0.2+0.7+1.1}{9} \approx 0.57$

median: 0.6 is the middle value.
mode: 0.2 and 0.6 each occur twice.

181. mean: $\dfrac{0.6+0.6+0.8+0.4+0.5+0.3+0.7+0.8+0.1}{9} \approx 0.53$

median: 0.6 is the middle value.
mode: 0.6 and 0.8 each occur twice.

182. Federal loans, because this portion of the circle graph is the largest.

183. Federal loans: $(\$50.3)(0.57) \approx \$28.67$ billion
Institutional: $(\$50.3)(0.20) = \$10.06$ billion
Federal grants/work study: $(\$50.3)(0.17) \approx \$8.55$ billion
State grants: $(50.3)(0.06) \approx \$3.02$ billion

184. The height of the bar for 11–24 employees is 44%. Therefore, 44% of employees receive health benefits for companies with 11–24 employees.

185. The shortest bar, which has a height to 29%, is for companies with less than 10 employees. Therefore, companies with less than 10 employees have the lowest percent of employees receiving health benefits. This percent is about 29%.

186. The tallest bar, which has a height of 75%, is for companies with 1000 and over employees. Therefore, companies with 1000 and over employees have the highest percent of employees receiving health benefits. This percent is about 75%.

187. Answers may vary.

Chapter 1 Test

1. True; -2.3 lies to the right of -2.33 on the number line.

2. False; $-6^2 = -36$, while $(-6)^2 = 36$.

3. False; $-5 - 8 = -13$, while $-(5 - 8) = -(-3) = 3$.

4. False; $(-2)(-3)(0) = 0$, while $\dfrac{(-4)}{0}$ is undefined.

5. True; natural numbers are positive integers.

6. False; for example, $\dfrac{1}{2}$ is a rational number that is not an integer.

7. $5 - 12 \div 3(2) = 5 - 4 \cdot 2 = 5 - 8 = -3$

8. $\begin{aligned}|4 - 6|^3 - (1 - 6^2) &= |-2|^3 - (1 - 36) \\ &= 2^3 - (-35) \\ &= 8 + 35 \\ &= 43\end{aligned}$

9. $\begin{aligned}(4 - 9)^3 - |-4 - 6|^2 &= (-5)^3 - |-10|^2 \\ &= -125 - 10^2 \\ &= -125 - 100 \\ &= -225\end{aligned}$

10. $\begin{aligned}\left[3|4 - 5|^5 - (-9)\right] \div (-6) \\ = \left[3|-1|^5 + 9\right] \div (-6) \\ = (3 \cdot 1^5 + 9) \div (-6) \\ = (3 + 9) \div (-6) \\ = 12 \div (-6) \\ = -2\end{aligned}$

11. $\begin{aligned}\dfrac{6(7 - 9)^3 + (-2)}{(-2)(-5)(-5)} &= \dfrac{6(-2)^3 - 2}{10(-5)} \\ &= \dfrac{6(-8) - 2}{-50} \\ &= \dfrac{-48 - 2}{-50} \\ &= \dfrac{-50}{-50} \\ &= 1\end{aligned}$

12. $q^2 - r^2 = 4^2 - (-2)^2 = 16 - 4 = 12$

13. $\dfrac{5t - 3q}{3r - 1} = \dfrac{5(1) - 3(4)}{3(-2) - 1} = \dfrac{5 - 12}{-6 - 1} = \dfrac{-7}{-7} = 1$

14. **a.** When $x = 1$, $5.75x = 5.75(1) = 5.75$.
    When $x = 3$, $5.75x = 5.75(3) = 17.25$.
    When $x = 10$,
    $5.75x = 5.75(10) = 57.50$.
    When $x = 20$,
    $5.75x = 5.75(20) = 115.00$.

x	1	3	10	20
$5.75x$	5.75	17.25	57.50	115.00

    b. As the number of adults increases, the total cost increases.

15. $2|x + 5| = 30$

16. $\dfrac{(6 - y)^2}{7} < -2$

17. $\dfrac{9z}{|-12|} \neq 10$

18. $3\left(\dfrac{n}{5}\right) = -n$

19. $20 = 2x - 6$

20. $-2 = \dfrac{x}{x+5}$

21. Distributive property

22. Associative property of addition

23. Additive inverse property

24. Multiplication property of zero

25. $0.05n + 0.1d$

26. $4y^2 + 10 - 2(y^2 + 10) = 4y^2 + 10 - 2y^2 - 20 = 4y^2 - 2y^2 + 10 - 20 = (4-2)y^2 - 10 = 2y^2 - 10$

27. $(8.3x - 2.9) - (9.6x - 4.8) = 8.3x - 2.9 - 9.6x + 4.8$
$$= 8.3x - 9.6x - 2.9 + 4.8$$
$$= (8.3 - 9.6)x + 1.9$$
$$= -1.3x + 1.9$$

28. $0.2(-3.1)^3 + 5(-3.1)^2 - 6.2(-3.1) + 3 = 64.3118$

29.

Diameter	d	2	3.8	10	14.9
Radius	r	1	1.9	5	7.45
Circumference	πd	6.28	11.94	31.42	46.81
Area	πr^2	3.14	11.34	78.54	174.37

30. $8x + 14 = 5x + 44$
$$3x = 30$$
$$x = 10$$
The solution is 10.

31. $3(x+2) = 11 - 2(2-x)$
$$3x + 6 = 11 - 4 + 2x$$
$$3x + 6 = 7 + 2x$$
$$x = 1$$
The solution is 1.

32. $3(y-4) + y = 2(6+2y)$
$$3y - 12 + y = 12 + 4y$$
$$4y - 12 = 12 + 4y$$
$$-12 = 12$$
There is no solution.

33. $7n - 6 + n = 2(4n - 3)$
$8n - 6 = 8n - 6$
$-6 = -6$
The solution is all real numbers.

34. $\dfrac{z}{2} + \dfrac{z}{3} = 10$

$6\left(\dfrac{z}{2} + \dfrac{z}{3}\right) = 6(10)$

$3z + 2z = 60$
$5z = 60$
$z = 12$
The solution is 12.

35. $\dfrac{7w}{4} + 5 = \dfrac{3w}{10} + 1$

$20\left(\dfrac{7w}{4} + 5\right) = 20\left(\dfrac{3w}{10} + 1\right)$

$35w + 100 = 6w + 20$
$29w = -80$
$w = -\dfrac{80}{29}$

The solution is $-\dfrac{80}{29}$.

36. $3x - 4y = 8$
$3x - 8 = 4y$
$\dfrac{3x - 8}{4} = y$
$y = \dfrac{3x - 8}{4}$

37. $4(2n - 3m) - 3(5n - 7m) = 0$
$8n - 12m - 15n + 21m = 0$
$9m - 7n = 0$
$9m = 7n$
$\dfrac{9m}{7} = n$
$n = \dfrac{9m}{7}$

38. $12\% \cdot 80 = 0.12 \cdot 80 = 9.6$

39. Let x = the number of employees in 1996.
$x + 1.18x = 461{,}000$
$2.18x = 461{,}000$
$x \approx 211{,}468$
The number of people employed in these occupations in 1996 was 211,468.

40. Recall that $C = 2\pi r$. Here $C = 78.5$.
$78.5 = 2\pi r$
$r = \dfrac{78.5}{2\pi} = \dfrac{39.25}{\pi}$
Also, recall that $A = \pi r^2$.
$A = \pi\left(\dfrac{39.25}{\pi}\right)^2 \approx \dfrac{39.25^2}{3.14} \approx 490.63$
Dividing this by 60 yields approximately 8.18. Therefore, about 8 hunting dogs could safely be kept in the pen.

41. Solve $R = C$.
$7.4x = 3910 + 2.8x$
$4.6x = 3910$
$x = 850$
Therefore, more than 850 sunglasses must be produced and sold in order to break even.

42. $A = P\left(1 + \dfrac{r}{n}\right)^{nt}$

$= 2500\left(1 + \dfrac{0.035}{4}\right)^{4 \cdot 10}$

≈ 3542.27
The amount of money in the account is $3542.27.

43. a. Use $y_1 = 1500 + 0.05x$

Sales (dollars)	8000	9000	10,000	11,000	12,000
Gross monthly pay (dollars)	1900	1950	2000	2050	2100

b. If she has sales of $11,000 per month, her gross monthly pay is $2050, so her gross annual pay is $12 \cdot \$2050 = \$24,600$.

c. If $2200 = 1500 + 0.05x$, then $x = 14,000$. She must sell $14,000 per month to earn a gross monthly pay of $2200.

44. $A = lw = l\left(\dfrac{80-l}{2}\right)$

$y_1 = x((80-x)/2)$

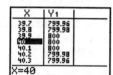

From the calculator table, the maximum value of A is 800 square feet.

45. a.

X	Y₁	
0	20	
1	84	
2	116	
3	116	
4	84	
5	20	
6	-76	

X=2

At $x = 2$ seconds and 3 seconds, the height of the rocket is 116 feet.

b.

X	Y₁	
2.2	118.56	
2.3	119.36	
2.4	119.84	
2.5	120	
2.6	119.84	
2.7	119.36	
2.8	118.56	

Y₁=120

The rocket's maximum height is 120 feet.

c. At $x = 2.5$ seconds, the rocket reaches its maximum height.

d. $2.5 + 0.5 = 3.0$

At $x = 3.0$ seconds, the rocket explodes.

46. $\dfrac{\text{sum of values}}{\text{number of values}} = \dfrac{57}{12} = 4.75$

The mean number of books checked out per week is 4.75.

47. The median is the mean of the two middle numbers, or $\dfrac{5+5}{2} = 5$ books.

48. 5 is the mode because it occurs 3 times.

49. In 1970, about 80 pounds of fresh fruit per person were consumed.

50. In 1995, about 119 pounds of fresh vegetables per person were consumed.

51. It occured in 1990 because the difference between the two graphs is the greatest.

52. Answers may vary.

Chapter 2

Section 2.1

Mental Math

1. Point A is (5, 2).

2. Point B is (2, 5).

3. Point C is (3, –1).

4. Point D is (–1, 3).

5. Point E is (–5, –2).

6. Point F is (–3, 5).

7. Point G is (–1, 0).

8. Point H is (0, –3).

Exercise Set 2.1

1. (3, 2) is in quadrant I.

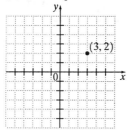

3. (–5, 3) is in quadrant II.

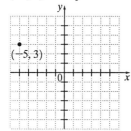

5. $\left(5\frac{1}{2}, -4\right)$ is in quadrant IV.

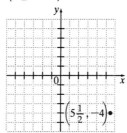

7. (0, 3.5) is on the y-axis.

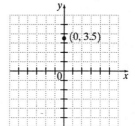

9. (–2, –4) is in quadrant III.

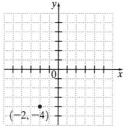

11. $\left(x, -y\right)$ is in quadrant IV.

13. $\left(x, 0\right)$ is on the x-axis.

15. $\left(-x, -y\right)$ is in quadrant III.

17. The point is (2, 8). It lies in quadrant I.

19. The point is (1, –4) It lies in quadrant IV.

21. Possible answer:
[–10, 10, 1] by [0, 20, 1]

39

23. Possible answer:
$[-100, 100, 10]$ by $[-100, 100, 10]$

25. c

27. d

29. Let $x = 0$, $y = 5$.
$y = 3x - 5$
$5 \overset{?}{=} 3 \cdot 0 - 5$
$5 = -5$ False
$(0, 5)$ is not a solution.
Let $x = -1$, $y = -8$.
$y = 3x - 5$
$-8 \overset{?}{=} 3(-1) - 5$
$-8 \overset{?}{=} -3 - 5$
$-8 = -8$ True
$(-1, -8)$ is a solution.

31. Let $x = 1$, $y = 0$.
$-6x + 5y = -6$
$-6(1) + 5(0) \overset{?}{=} -6$
$-6 + 0 \overset{?}{=} -6$
$-6 = -6$ True
$(1, 0)$ is a solution.

Let $x = 2$, $y = \dfrac{6}{5}$.
$-6x + 5y = -6$
$-6(2) + 5\left(\dfrac{6}{5}\right) \overset{?}{=} -6$
$-12 + 6 \overset{?}{=} -6$
$-6 = -6$ True
$\left(2, \dfrac{6}{5}\right)$ is a solution.

33. Let $x = 1$, $y = 2$.
$y = 2x^2$
$2 \overset{?}{=} 2(1)^2$
$2 = 2$ True
$(1, 2)$ is a solution.
Let $x = 3$, $y = 18$
$y = 2x^2$
$18 \overset{?}{=} 2(3)^2$
$18 \overset{?}{=} 2(9)$
$18 = 18$ True
$(3, 18)$ is a solution.

35.

Equation	Linear or non-linear	Shape (Line, Parabola, Cubic, V-shaped)
$y - x = 8$	linear	line
$y = 6x$	linear	line
$y = x^2 + 3$	non-linear	parabola
$y = 6x - 5$	linear	line
$y = -|x| + 2$	non-linear	V-shaped
$y = 3x^2$	non-linear	parabola
$y = -4x + 2$	linear	line
$y = -|x|$	non-linear	V-shaped
$y = x^3$	non-linear	cubic

37. $2x + y = 10$
$2x + y - 2x = -2x + 10$
$y = -2x + 10$

39. $-7x - 3y = 4$
$-7x - 3y + 7x = 7x + 4$
$-3y = 7x + 4$
$\dfrac{-3y}{-3} = \dfrac{7x + 4}{-3}$
$y = \dfrac{7x + 4}{-3}$
$y = -(7x + 4)/3$

41.

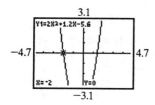

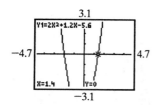

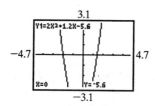

The TRACE shows that $y = 0$ when $x = -2$ and $x = 1.4$, and that $y = 5.6$ when $x = 0$.

The x-intercepts are $(-2, 0)$ and $(1.4, 0)$; the y-intercept is $(0, -5.6)$.

43. The answer is d because $x = 1.2$ is not an integer and therefore would not be displayed.

45. C

47. A

49. D

51. C

53. B

55. C

57.

x	$x+2$	$x-3$
-2	$-2+2 = 0$	$-2-3 = -5$
-1	$-1+2 = 1$	$-1-3 = -4$
0	$0+2 = 2$	$0-3 = -3$
1	$1+2 = 3$	$1-3 = -2$
2	$2+2 = 4$	$2-3 = -1$

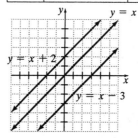

The basic equation is $y = x$.
$y = x + 2$ is obtained by moving $y = x$ up 2 units.
$y = x - 3$ is obtained by moving $y = x$ down 3 units.

59.

x	$-x^2$	$-x^2+1$	$-x^2-2$
-2	-4	-3	-6
-1	-1	0	-3
0	0	1	-2
1	-1	0	-3
2	-4	-3	-6

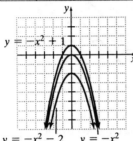

$y = -x^2$ is the basic equation.

$y = -x^2 + 1$ can be obtained by moving

$y = -x^2$ up 1 unit.

$y = -x^2 - 2$ can be obtained by moving

$y = -x^2$ down 2 units.

61. $x + y = 3$ or $y = 3 - x$

x	$3-x$	y
-1	$3-(-1)=4$	4
0	$3-0=3$	3
1	$3-1=2$	2

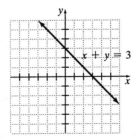

63.

x	$4x$	y
-1	$4(-1)=-4$	-4
0	$4(0)=0$	0
1	$4(1)=4$	4

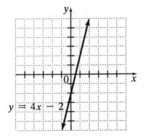

65.

x	$4x-2$	y
-1	$4(-1)-2=-6$	-6
0	$4(0)-2=-2$	-2
1	$4(1)-2=2$	2

67.

x	$|x|+3$	y
-3	$|-3|+3=3+3=6$	6
-2	$|-2|+3=2+3=5$	5
-1	$|-1|+3=1+3=4$	4
0	$|0|+3=0+3=3$	3
1	$|1|+3=1+3=4$	4
2	$|2|+3=2+3=5$	5
3	$|3|+3=3+3=6$	6

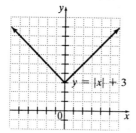

69. $2x-y=-5$ or $y=2x-5$

x	$2x-5$	y
0	$2(0)-5=-5$	-5
1	$2(1)-5=-3$	-3
2	$2(2)-5=-1$	-1

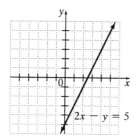

71.

x	$2x^2$	y
-3	$2(-3)^2=18$	18
-2	$2(-2)^2=8$	8
-1	$2(-1)^2=2$	2
0	$2(0)^2=0$	0
1	$2(1)^2=2$	2
2	$2(2)^2=8$	8
3	$2(3)^2=18$	18

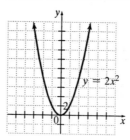

73.

x	$x^2 - 3$	y
-3	$(-3)^2 - 3 = 6$	6
-2	$(-2)^2 - 3 = 1$	1
-1	$(-1)^2 - 3 = -2$	-2
0	$(0)^2 - 3 = -3$	-3
1	$(1)^2 - 3 = -2$	-2
2	$(2)^2 - 3 = 1$	1
3	$(3)^2 - 3 = 6$	6

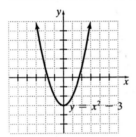

75.

x	$-2x$	y
-1	$-2(-1) = 2$	2
0	$-2(0) = 0$	0
1	$-2(1) = -2$	-2

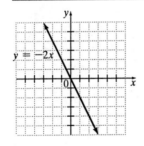

77.

x	$-2x + 3$	y
-1	$-2(-1) + 3 = 5$	5
0	$-2(0) + 3 = 3$	3
1	$-2(1) + 3 = 1$	1

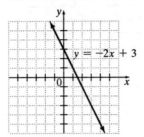

79.

x	$|x + 2|$	y
-4	$|-4 + 2| = 2$	2
-3	$|-3 + 2| = 1$	1
-2	$|-2 + 2| = 0$	0
-1	$|-1 + 2| = 1$	1
0	$|0 + 2| = 2$	2
1	$|1 + 2| = 3$	3

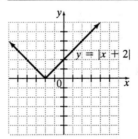

81.

x	x^3	y
-3	$(-3)^3 = -27$	-27
-2	$(-2)^3 = -8$	-8
-1	$(-1)^3 = -1$	-1
0	$0^3 = 0$	0
1	$1^3 = 1$	1
2	$2^3 = 8$	8

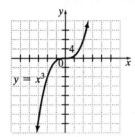

83.

x	$-|x|$	y
-3	$-|-3| = -3$	-3
-2	$-|-2| = -2$	-2
-1	$-|-1| = -1$	-1
0	$-|0| = -0$	0
1	$-|1| = -1$	-1
2	$-|2| = -2$	-2
3	$-|3| = -3$	-3

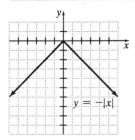

85.

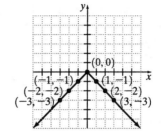

An equation is $y = -|x|$.

87. a. parabola
 b. line

89. line

91. a.

x	$2x+6$	y
0	$2(0)+6 = 6$	6
1	$2(1)+6 = 8$	8
2	$2(2)+6 = 10$	10

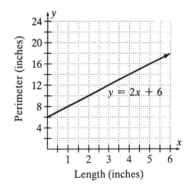

 b. At $x = 4$, $y = 14$. Therefore, the perimeter is 14 inches.

93. a. In April of 1991, it rose to $4.25.

 b. In September of 1997, it rose to $5.15.

 c. Answers may vary.

 d. $6.15 - 0.25 = 5.90$
 The minimum wage will have increased by $5.90 in 2002.

95. B; The graph shows a value of 40 around August and September, then a break until it shows a value of 60 around January and February.

97. C; The graph shows a line fluctuating between 10 and 30.

99.
$$3(x-2)+5x = 6x-16$$
$$3x-6+5x = 6x-16$$
$$8x-6 = 6x-16$$
$$2x = -10$$
$$x = -5$$
The solution is –5.

101.
$$3x+\frac{2}{5} = \frac{1}{10}$$
$$30x+4 = 1$$
$$30x = -3$$
$$x = -\frac{1}{10}$$

The solution is $-\frac{1}{10}$.

Exercise Set 2.2

1. Domain: {–1, 0, –2, 5}
Range: {7, 6, 2}
The relation is a function.

3. Domain:{–2, 6, –7}
Range: {4, –3, –8}
The relation is not a function since –2 is paired with both 4 and –3.

5. Domain: {1}
Range: {1, 2, 3, 4}
The relation is not a function since 1 is paired with both 1 and 2, for example.

7. Domain: $\left\{\frac{3}{2}, 0\right\}$

Range: $\left\{\frac{1}{2}, -7, \frac{4}{5}\right\}$

The relation is not a function since $\frac{3}{2}$ is paired with both $\frac{1}{2}$ and –7.

9. Domain: {–3, 0, 3}
Range: {–3, 0, 3}
The relation is a function.

11. Domain: {–1, 1, 2, 3}
Range: {2, 1}
The relation is a function.

13. Domain: {Colorado, Alaska, Delaware, Illinois, Connecticut, Texas}
Range: {6, 1, 20, 30}
The relation is a function.

15. Domain: {32°, 104°, 212°, 50°}
Range: {0°, 40°, 10°, 100°}
The relation is a function.

17. Domain: {2, –1, 5, 100}
Range: {0}
The relation is a function.

19. The relation is a function since each student will have only one grade average.

21. Answers may vary.

23. Function (passes the vertical line test)

25. Not a function (fails the vertical line test)

27. Function (passes the vertical line test)

29. Function (passes the vertical line test)

31. Not a function (fails the vertical line test)

33. Domain: $[0, \infty)$
Range: $(-\infty, \infty)$
The relation is not a function since it fails the vertical line test (try $x = 1$).

35. Domain: $[-1, 1]$
Range: $(-\infty, \infty)$
The relation is not a function since it fails the vertical line test (try $x = 0$).

37. Domain: $(-\infty, \infty)$
Range: $(-\infty, -3] \cup [3, \infty)$
The relation is not a function since it fails the vertical line test (try $x = 0$).

39. Domain: [2, 7]
Range: [1, 6]
The relation is not a function since it fails
the vertical line test (try $x = 3$).

41. Domain: {−2}
Range: $(-\infty, \infty)$
The relation is not a function since it fails
the vertical line test (try $x = -2$).

43. Domain: $(-\infty, \infty)$
Range: $(-\infty, 3]$
The relation is a function since it passes
the vertical line test.

45. Answers may vary.

47. Yes, $y = x + 1$ is a function. For each
x-value substituted into the equation, the
addition performed on each gives a single
result.

49. No, $x = 2y^2$ is not a function. If $y = -1$,
then $x = 2$ and if $y = 1$, the $x = 2$. The
x-value 2 corresponds to two y-values.

51. Yes, $y - x = 7$ or $y = x + 7$ is a function.
For each x-value substituted into the
equation, the addition performed on each
gives a single result.

53. Yes, $y = \dfrac{1}{x}$ is a function. For each x-value
substituted into the equation, the division
performed on each gives a single result.

55. Yes, $y = 5x - 12$ is a function. For each
x-value substituted into the equation, the
multiplication and subtraction performed
on each give a single result.

57. No, $x = y^2$ is not a function. If $y = -1$,
then $x = 1$ and if $y = 1$, then $x = 1$. The
x-value 1 corresponds to two y-values.

59. $f(x) = 3x + 3$
$f(4) = 3(4) + 3 = 12 + 3 = 15$

61. $h(x) = 5x^2 - 7$
$$\begin{aligned}
h(-3) &= 5(-3)^2 - 7 \\
&= 5(9) - 7 \\
&= 45 - 7 \\
&= 38
\end{aligned}$$

63. $g(x) = 4x^2 - 6x + 3$
$$\begin{aligned}
g(2) &= 4(2)^2 - 6(2) + 3 \\
&= 4(4) - 12 + 3 \\
&= 16 - 12 + 3 \\
&= 7
\end{aligned}$$

65. $g(x) = 4x^2 - 6x + 3$
$$\begin{aligned}
g(0) &= 4(0)^2 - 6(0) + 3 \\
&= 4(0) - 0 + 3 \\
&= 0 - 0 + 3 \\
&= 3
\end{aligned}$$

67. $f(x) = \dfrac{1}{2}x$

   a. $f(0) = \dfrac{1}{2}(0) = 0$

   b. $f(2) = \dfrac{1}{2}(2) = 1$

   c. $f(-2) = \dfrac{1}{2}(-2) = -1$

Chapter 2: *Graphs and Functions*   *SSM:* Intermediate Algebra

69. $g(x) = 2x^2 + 4$

a. $g(-11) = 2(-11)^2 + 4$
$= 2(121) + 4$
$= 242 + 4$
$= 246$

b. $g(-1) = 2(-1)^2 + 4$
$= 2(1) + 4$
$= 2 + 4$
$= 6$

c. $g\left(\dfrac{1}{2}\right) = 2\left(\dfrac{1}{2}\right)^2 + 4$
$= 2\left(\dfrac{1}{4}\right) + 4$
$= \dfrac{1}{2} + \dfrac{8}{2}$
$= \dfrac{9}{2}$

71. $f(x) = 1.3x^2 - 2.6x + 5.1$

Use a calculator for computations.

a. $f(2) = 5.1$

b. $f(-2) = 15.5$

c. $f(3.1) = 9.533$

73. a. For $x = -5$, $y_1 = 23$.
Therefore, $f(-5) = 23$.

b. For $x = 10$, $y_1 = 23$.
Therefore, $f(10) = 23$.

c. For $x = 15$, $y_1 = 33$.
Therefore, $f(15) = 33$.

75. For $x = 2$, $y = 7$.
Therefore, $f(2) = 7$.

77. Using the calculator, we evaluate
$C(t) = 10 + 35t$ for the given values of t.

```
1→T:20+35T
               55
2→T:20+35T
               90
3→T:20+35T
              125
```

```
4→T:20+35T
              160
5→T:20+35T
              195
■
```

Completing the table, we have

t	1	2	3	4	5
$C(t)$	55	90	125	160	195

79. Graph the function $y_1 = -16x^2 + 80x$.
Move the cursor to $x = 0.5$, $x = 1$, $x = 1.5$,
$x = 2$, $x = 2.5$, and $x = 5$ to find the
heights.

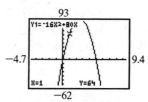

a. $h(0.5) = 36$

b. $h(1) = 64$

c. $h(1.5) = 84$

d. $h(2) = 96$

e. $h(2.5) = 100$

f. $h(5) = 0$

48

81. $f(1) = -10$ means $x = 1$ and $y = -10$.
The ordered pair is $(1, -10)$.

83. At $x = -1$, $y - 2$.
Therefore, $f(-1) = -2$.

85. $f(-4) = -5$ and $f(0) = -5$.
The values of x such that $f(x) = -5$
are -4 and 0.

87. Infinite number; Answers may vary.

89. a. On the graph, 1 corresponds to 1994.
In 1994, approximately \$13.4 billion
were spent.

   b. $f(x) = 1.882x + 11.79$
$f(1) = 1.882(1) + 11.79 = 13.672$
Using the function, the approximation
is \$13.672 billion.

91. $x = 12$ corresponds to 2005.
$f(x) = 1.882x + 11.79$
$f(12) = 1.882(12) + 11.79 = 34.374$
The prediction is that \$34.374 billion will
be spent in 2005.

93. $f(x) = x + 7$

95. $A(r) = \pi r^2$
$A(5) = \pi \left(5^2\right) = 25\pi$
The area is 25π square centimeters.

97. $V(x) = x^3$
$V(14) = (14)^3 = 2744$
The volume is 2744 cubic inches.

99. $H(f) = 2.59f + 47.24$
$H(46) = 2.59(46) + 47.24 = 166.38$
Her height is 166.38 centimeters.

101. $D(x) = \dfrac{136}{25}x$

$D(30) = \dfrac{136}{25}(30) = 163.2$
The proper dosage is 163.2 milligrams.

103. a. $C(x) = 1.7x + 88$
$C(2) = 1.7(2) + 88 = 91.4$
$1995 + 2 = 1997$
The per capita consumption of
poultry was 91.4 pounds in 1997.

   b. $x = 2006 - 1995 = 11$
$C(11) = 1.7(11) + 88 = 106.7$
The per capita consumption of
poultry is predicted to be 106.7
pounds in 2006.

105. $x - y = -5$

x	0	−5	1
y	5	0	6

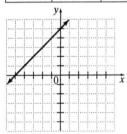

107. $7x + 4y = 8$

x	0	$\frac{8}{7}$	$\frac{12}{7}$
y	2	0	−1

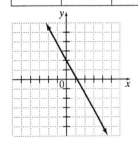

109. $y = 6x$

x	0	0	-1
y	0	0	-6

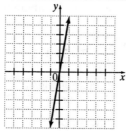

111. Yes; The two sides opposite and parallel to the 45-meter side also add up to 45 meters. The two sides opposite and parallel to the 40-meter side also add up to 40 meters. Therefore, the 6 sides add up to $2(45) + 2(40) = 170$.

The perimeter is 170 meters.

113. $g(x) = -3x + 12$
$g(s) = -3s + 12$
$g(r) = -3r + 12$

115. $f(x) = x^2 - 12$
$f(12) = 12^2 - 12 = 132$
$f(a) = a^2 - 12$

Exercise Set 2.3

1. $f(x) = -2x$
Plot the points to obtain the graph.

x	y
-1	2
0	0
1	-2

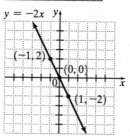

3. $f(x) = -2x + 3$
Plot the points to obtain the graph.

x	y
-1	5
0	3
1	1

Notice this the graph of $f(x) = -2x$ shifted upward 3 units.

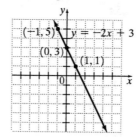

5. $f(x) = \dfrac{1}{2}x$

Plot the points to obtain the graph.

x	y
0	0
2	1
4	2

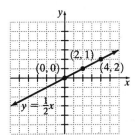

7. $f(x) = \dfrac{1}{2}x - 4$

Plot the points to obtain the graph.

x	y
0	-4
2	-3
4	-2

Notice this is the graph of $f(x) = \dfrac{1}{2}x$

shifted downward 4 units.

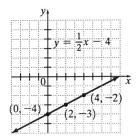

9. $f(x) = 5x - 6$ is C.

Notice it is the graph of $f(x) = 5x$ moved downward 6 units.

11. $f(x) = 5x + 7$ is D.

Notice it is the graph of $f(x) = 5x$ moved upward 7 units.

13. $x - y = 3$

Let $x = 0$	Let $y = 0$	Let $x = 2$
$0 - y = 3$	$x - 0 = 3$	$2 - y = 3$
$y = -3$	$x = 3$	$y = -1$

Intercepts occur at $(0, -3)$ and $(3, 0)$.

Use a third ordered pair solution, as in $(2, -1)$, to check your work.

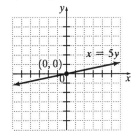

15. $x = 5y$

Let $x = 0$	Let $x = 5$	Let $x = -5$
$0 = 5y$	$5 = 5y$	$-5 = 5y$
$y = 0$	$y = 1$	$y = -1$

Intercept occurs at $(0, 0)$. Use two more ordered pair solutions, as in $(5, 1)$ and $(-5, -1)$, to graph.

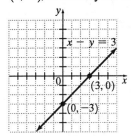

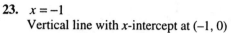
17. $-x + 2y = 6$

Let $x = 0$ Let $y = 0$
$-0 + 2y = 6$ $-x + 2(0) = 6$
     $y = 3$ $x = -6$

Let $x = -4$
$-(-4) + 2y = 6$
      $y = 1$

Intercepts occur at $(0, 3)$ and $(-6, 0)$. Use a third ordered pair solution, as in $(-4, 1)$, to check your work.

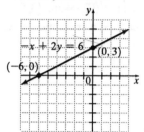

19. $2x - 4y = 8$

Let $x = 0$ Let $y = 0$
$2(0) - 4y = 8$ $2x - 4(0) = 8$
      $y = -2$ $x = 4$

Let $x = 2$
$2(2) - 4y = 8$
     $y = -1$

Intercepts occur at $(0, -2)$ and $(4, 0)$. Use a third ordered pair solution, as in $(2, -1)$, to check your work.

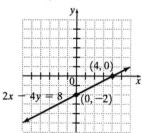

21. Answers may vary.

23. $x = -1$
Vertical line with x-intercept at $(-1, 0)$

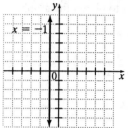

25. $y = 0$
Horizontal line with y-intercept at $(0, 0)$

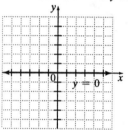

27. $y + 7 = 0$ or $y = -7$
Horizontal line with y-intercept at $(0, -7)$

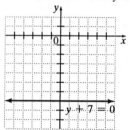

29. $y = 2$ matches graph C. Notice it is a horizontal line with y-intercept at $(0, 2)$.

31. $x - 2 = 0$ or $x = 2$ matches graph A. Notice it is a vertical line with x-intercept at $(2, 0)$.

33. The vertical line $x = 0$ has infinitely many y-intercepts.

For Exercises 35–57, find ordered pair solutions, or find the *x*-and *y*-intercepts, or find the *y*-intercept with the equation in the form $y = mx + b$. Check graphs using a graphing utility.

35. $x + 2y = 8$

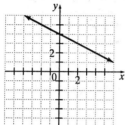

37. $f(x) = \frac{3}{4}x + 2$ or $y = \frac{3}{4}x + 2$

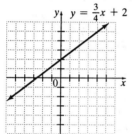

39. $x = -3$

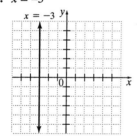

41. $3x + 5y = 7$

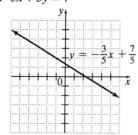

43. $f(x) = x$ or $y = x$

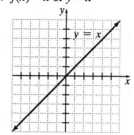

45. $x + 8y = 8$

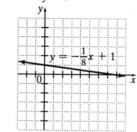

47. $5 = 6x - y$

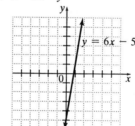

49. $-x + 10y = 11$

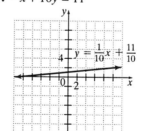

51. $y = 1$

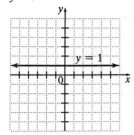

53. $f(x) = \dfrac{1}{2}x$ or $y = \dfrac{1}{2}x$

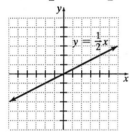

55. $x + 3 = 0$ or $x = -3$

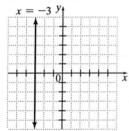

57. $f(x) = 4x - \dfrac{1}{3}$ or $y = 4x - \dfrac{1}{3}$

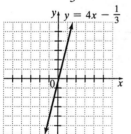

59. The point on the graph is $x = 0$, $y = 7$ or $(0, 7)$. Therefore $f(0) = 7$.
b and d are true.

61. $2x + 3y = 1500$

a. $2(0) + 3y = 1500$
$$3y = 1500$$
$$y = 500$$
(0, 500); If no tables are produced, 500 chairs can be produced.

b. $2x + 3(0) = 1500$
$$2x = 1500$$
$$x = 750$$
(750, 0); If no chairs are produced, 750 tables can be produced.

c. $2(50) + 3y = 1500$
$$100 + 3y = 1500$$
$$3y = 1400$$
$$y \approx 466.7$$
466 chairs is the greatest number of chairs the company can make if 50 tables are produced.

63. $C(x) = 0.2x + 24$

a. $C(200) = 0.2(200) + 24$
$$= 40 + 24$$
$$= 64$$
The cost of driving the car 200 miles is $64.

b.

c. The line moves upward from left to right.

65. $f(x) = 72.9x + 785.2$

   a. $f(20) = 72.9(20) + 785.2$
$$= 1458 + 785.2$$
$$= 2243.2$$
      The predicted cost is $2243.20.

   b. $2000 = 72.9x + 785.2$
$$1214.8 = 72.9x$$
$$16.66 \approx x$$
$$1990 + 17 = 2007$$
      The total cost is predicted to exceed
      $2000 in 2007.

   c. Answers may vary.

67. Graph the function defining
$y_1 = 15.75(40) + 1.5(15.75)x.$

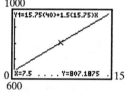

    Find the values in the table by calculating
    the function values on the graph.

x	2	3.25	7.5	9.75
$S(x)$	677.25	706.78	807.19	860.34

69. $\dfrac{-6-3}{2-8} = \dfrac{-9}{-6} = \dfrac{3}{2}$

71. $\dfrac{-8-(-2)}{-3-(-2)} = \dfrac{-8+2}{-3+2} = \dfrac{-6}{-1} = 6$

73. $\dfrac{0-6}{5-0} = -\dfrac{6}{5}$

Section 2.4

Mental Math

1. $m = \dfrac{7}{6}$ slants upward.

2. $m = -3$ slants downward.

3. $m = 0$ slants horizontally.

4. m is undefined, slants vertically.

Exercise Set 2.4

1. $m = \dfrac{y_2 - y_1}{x_2 - x_1} = \dfrac{11-2}{8-3} = \dfrac{9}{5}$

3. $m = \dfrac{y_2 - y_1}{x_2 - x_1} = \dfrac{3-8}{4-(-2)} = -\dfrac{5}{6}$

5. $m = \dfrac{y_2 - y_1}{x_2 - x_1} = \dfrac{-4-(-6)}{4-(-2)} = \dfrac{2}{6} = \dfrac{1}{3}$

7. $m = \dfrac{y_2 - y_1}{x_2 - x_1} = \dfrac{5-5}{3-(-2)} = \dfrac{0}{5} = 0$

9. $m = \dfrac{y_2 - y_1}{x_2 - x_1} = \dfrac{-5-1}{-1-(-1)} = -\dfrac{6}{0}$
Undefined slope

11. $m = \dfrac{y_2 - y_1}{x_2 - x_1} = \dfrac{4-2}{-3-(-1)} = \dfrac{2}{-3+1} = \dfrac{2}{-2} = -1$

13. $(3, -5), (15, -13)$
$$m = \dfrac{y_2 - y_1}{x_2 - x_1} = \dfrac{-13-(-5)}{15-3} = \dfrac{-8}{12} = -\dfrac{2}{3}$$

15. $(0, 5), (2, 8)$
$$m = \dfrac{y_2 - y_1}{x_2 - x_1} = \dfrac{8-5}{2-0} = \dfrac{3}{2} = 1.5$$

17. $(0, -8), (10, -6)$
$$m = \dfrac{y_2 - y_1}{x_2 - x_1} = \dfrac{-6-(-8)}{10-0} = \dfrac{2}{10} = 0.2$$

19. l_1 has a negative slope and l_2 has a
positive slope. Therefore, l_2 has the
greater slope.

21. l_1 has a negative slope and l_2 has a
slope of 0. Therefore, l_2 has the greater
slope.

23. l_1 and l_2 have positive slopes, but l_2 has
a greater slope because it is steeper.

25. **a.** m for $l_1 = \dfrac{-2-4}{2-(-1)} = \dfrac{-6}{3} = -2$

m for $l_2 = \dfrac{2-6}{-4-(-8)} = \dfrac{-4}{4} = -1$

m for $l_3 = \dfrac{-4-0}{0-(-6)} = \dfrac{-4}{6} = -\dfrac{2}{3}$

b. lesser

27. $f(x) = -2x + 6$ or $y = -2x + 6$
$m = -2$, $b = 6$

29. $-5x + y = 10$
$y = 5x + 10$
$m = 5$, $b = 10$

31. $-3x - 4y = 6$
$4y = -3x - 6$
$y = -\dfrac{3}{4}x - \dfrac{3}{2}$
$m = -\dfrac{3}{4}$, $b = -\dfrac{3}{2}$

33. $f(x) = -\dfrac{1}{4}x$ or $y = -\dfrac{1}{4}x + 0$
$m = -\dfrac{1}{4}$, $b = 0$

35. D (positive slope, y-intercept of $(0, -3)$)

37. C (negative slope, y-intercept of $(0, 3)$)

39. $y = -2$
$m = 0$

41. $x = 4$
m is undefined.

43. $y - 7 = 0$
$y = 7$
$m = 0$

45. Answers may vary.

47. $f(x) = x + 2$ or $y = x + 2$
$m = 1$, $b = 2$

49. $4x - 7y = 28$
$7y = 4x - 28$
$y = \dfrac{4}{7}x - 4$
$m = \dfrac{4}{7}$, $b = -4$

51. $2y - 7 = x$
$2y = x + 7$
$y = \dfrac{1}{2}x + \dfrac{7}{2}$
$m = \dfrac{1}{2}$, $b = \dfrac{7}{2}$

53. $x = 7$
Slope is undefined.
There is no y-intercept.

55. $f(x) = \dfrac{1}{7}x$ or $y = \dfrac{1}{7}x + 0$
$m = \dfrac{1}{7}$, $b = 0$

57. $x - 7 = 0$
$x = 7$
Slope is undefined.
There is no y-intercept.

59. $2y + 4 = -7$
$2y = -11$
$y = -\dfrac{11}{2}$
$m = 0$, $b = -\dfrac{11}{2}$

61. $f(x) = 5x - 6$ $g(x) = 5x + 2$
$m = 5$ $m = 5$
Since they have the same slope, the lines are parallel.

63. $2x - y = -10$ $2x + 4y = 2$
$y = 2x + 10$ $y = -\dfrac{1}{2}x + \dfrac{1}{2}$

$m = 2$ $m = -\dfrac{1}{2}$

Since the slopes are negative reciprocals of one another, the lines are perpendicular.

65.

$$x + 4y = 7 \qquad\qquad 2x - 5y = 0$$

$$y = -\frac{1}{4}x + \frac{7}{4} \qquad\qquad y = \frac{2}{5}x$$

$$m = -\frac{1}{4} \qquad\qquad m = \frac{2}{5}$$

Since their slopes are not equal, nor are they negative reciprocals of one another, the lines are neither parallel nor perpendicular.

67. Answers may vary.

69. Two points on the line: (1, 0), (0, 3)

$$m = \frac{3 - 0}{0 - 1} = \frac{3}{-1} = -3$$

71. Two points on the line: (−3, −1), (2, 4)

$$m = \frac{4 - (-1)}{2 - (-3)} = \frac{4 + 1}{2 + 3} = \frac{5}{5} = 1$$

73. $m = \dfrac{3}{25}$ since slope is change in vertical direction over change in horizontal direction.

75. $m = \dfrac{15}{100} = \dfrac{3}{20}$ since slope is change in vertical direction over change in horizontal direction.

77. $y = 1054.7x + 23{,}285.9$

   a. $x = 1996 - 1991 = 5$
      $y = 1054.7(5) + 23{,}285.9$
      $y = 28{,}559.4$
      The income is \$28,559.40.

   b. $m = 1054.7$; The annual average income increases by \$1054.70 for every one year.

   c. $b = 23{,}285.9$; At year $x = 0$, or 1991, the annual average income was \$23,285.90.

79.

$$-76x + 10y = 1130$$
$$10y = 76x + 1130$$
$$y = 7.6x + 113$$

   a. $m = 7.6, b = 113$

   b. The number of people employed as paralegals increases by 7.6 thousand for every one year.

   c. There were 113 thousand paralegals employed in 1996.

81. $f(x) = 72.9x + 785.2$

   a. $m = 72.9$; The yearly cost of tuition increases by \$72.90 every one year.

   b. $b = 785.2$; The yearly cost of tuition in 1990 was \$785.20.

83. $f(x) = x$
    $m = 1$
    The slope of a parallel line is 1.

85. $f(x) = x$
    $m = 1$
    The slope of a perpendicular line is −1.

87. $-3x + 4y = 10$

$$y = \frac{3}{4}x + \frac{10}{4}$$

$$m = \frac{3}{4}$$

The slope of a parallel line is $\dfrac{3}{4}$.

89. **a.** (6, 20)

   b. (10, 13)

   c. $m = \dfrac{13 - 20}{10 - 6} = -\dfrac{7}{4}$, or −1.75

      The rate of change is $-\dfrac{7}{4}$, or −1.75 yards per second.

   d. $m = \dfrac{8 - 2}{26 - 22} = \dfrac{6}{4} = \dfrac{3}{2}$, or 1.5

      The rate of change is $\dfrac{3}{2}$ or 1.5 yards per second.

91. $-4x + 2y = 5$
$2x - y = 7$

or $y_1 = 2x + \dfrac{5}{2}$

$y_2 = 2x - 7$

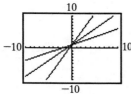

93. a. $y = \dfrac{1}{2}x + 1$
$y = x + 1$
$y = 2x + 1$

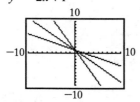

b. $y = -\dfrac{1}{2}x + 1$
$y = -x + 1$
$y = -2x + 1$

c. True

95. $P(B) = \dfrac{2}{11}$

97. $P(I \text{ or } T) = \dfrac{3}{11}$

99. $P(\text{vowel}) = \dfrac{4}{11}$

101. $y - 0 = -3[x - (-10)]$
$y = -3(x + 10)$
$y = -3x - 30$

103. $y - 9 = -8[x - (-4)]$
$y - 9 = -8(x + 4)$
$y - 9 = -8x - 32$
$y = -8x - 23$

Section 2.5

Mental Math

1. $m = -4, b = 12$

2. $m = \dfrac{2}{3}, b = -\dfrac{7}{2}$

3. $m = 5, b = 0$

4. $m = -1, b = 0$

5. $m = \dfrac{1}{2}, b = 6$

6. $m = -\dfrac{2}{3}, b = 5$

7. Parallel

8. Parallel

9. Neither

10. Neither

Exercise Set 2.5

1. $m = -1, \; b = 1$
$y = mx + b$
$y = -x + 1$

3. $m = 2, \; b = \dfrac{3}{4}$
$y = mx + b$
$y = 2x + \dfrac{3}{4}$

5. $m = \dfrac{2}{7},\ b = 0$

$y = mx + b$

$y = \dfrac{2}{7}x$

7. $y = 5x$

Possible points: $(0, 0),\ (1, 5)$

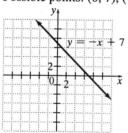

9. $x + y = 7$

$y = -x + 7$

Possible points: $(0, 7),\ (1, 6)$

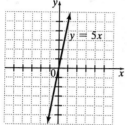

11. $-3x + 2y = 3$

$2y = 3x + 3$

$y = \dfrac{3}{2}x + \dfrac{3}{2}$

Possible points: $\left(0, \dfrac{3}{2}\right), \left(2, \dfrac{9}{2}\right)$

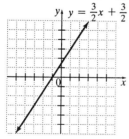

13. $y - y_1 = m(x - x_1)$

$y - 2 = 3(x - 1)$

$y - 2 = 3x - 3$

$y = 3x - 1$

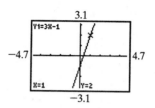

15. $y - y_1 = m(x - x_1)$

$y - (-3) = -2(x - 1)$

$y + 3 = -2x + 2$

$y = -2x - 1$

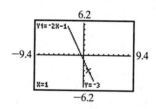

17. $y - y_1 = m(x - x_1)$

$y - 2 = \dfrac{1}{2}[x - (-6)]$

$y - 2 = \dfrac{1}{2}(x + 6)$

$y - 2 = \dfrac{1}{2}x + 3$

$y = \dfrac{1}{2}x + 5$

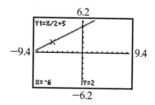

19. $y - y_1 = m(x - x_1)$

$y - 0 = -\dfrac{9}{10}[x - (-3)]$

$y = -\dfrac{9}{10}(x + 3)$

$y = -\dfrac{9}{10}x - \dfrac{27}{10}$

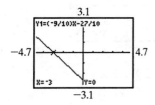

21. $(0, 3), (1, 1)$

$m = \dfrac{1-3}{1-0} = -\dfrac{2}{1} = -2$

$b = 3$

$y = -2x + 3$

$2x + y = 3$

23. $(-2, 1), (4, 5)$

$m = \dfrac{5-1}{4-(-2)} = \dfrac{4}{6} = \dfrac{2}{3}$

$y - 1 = \dfrac{2}{3}(x + 2)$

$3y - 3 = 2x + 4$

$2x - 3y = -7$

25. $m = \dfrac{y_2 - y_1}{x_2 - x_1} = \dfrac{6-0}{4-2} = \dfrac{6}{2} = 3$

$y - y_1 = m(x - x_1)$

$y - 0 = 3(x - 2)$

$y = 3x - 6$

$f(x) = 3x - 6$

27. $m = \dfrac{y_2 - y_1}{x_2 - x_1} = \dfrac{13-5}{-6-(-2)} = \dfrac{8}{-4} = -2$

$y - y_1 = m(x - x_1)$

$y - 5 = -2[x - (-2)]$

$y - 5 = -2(x + 2)$

$y - 5 = -2x - 4$

$y = -2x + 1$

$f(x) = -2x + 1$

29. $m = \dfrac{y_2 - y_1}{x_2 - x_1} = \dfrac{-3-(-4)}{-4-(-2)} = \dfrac{1}{-2} = -\dfrac{1}{2}$

$y - y_1 = m(x - x_1)$

$y - (-4) = -\dfrac{1}{2}[x - (-2)]$

$2(y + 4) = -(x + 2)$

$2y + 8 = -x - 2$

$y = -\dfrac{1}{2}x - \dfrac{10}{2}$

$f(x) = -\dfrac{1}{2}x - 5$

31. $m = \dfrac{y_2 - y_1}{x_2 - x_1} = \dfrac{-9-(-8)}{-6-(-3)} = \dfrac{-1}{-3} = \dfrac{1}{3}$

$y - y_1 = m(x - x_1)$

$y - (-8) = \dfrac{1}{3}[x - (-3)]$

$3(y + 8) = x + 3$

$3y + 24 = x + 3$

$3y = x - 21$

$f(x) = \dfrac{1}{3}x - 7$

33. Answers may vary.

35. $(0, -2); f(0) = -2$

37. $(2, 2); f(2) = 2$

39. $(-2, -6); f(-2) = -6; x = -2$

41. Since $m = 0$, the line is horizontal. A horizontal line has an equation of the form $y = b$. Since the line contains the point $(-2, -4)$, the equation is $y = -4$.

43. A vertical line has an equation of the form $x = c$. Since the line contains the point $(4, 7)$, the equation is $x = 4$.

45. A horizontal line has an equation of the form $y = b$. Since the line contains the point $(0, 5)$, the equation is $y = 5$.

47. $y = 4x - 2$, so $m = 4$.
$$y - y_1 = m(x - x_1)$$
$$y - 8 = 4(x - 3)$$
$$y - 8 = 4x - 12$$
$$y = 4x - 4$$
$$f(x) = 4x - 4$$

49. Note that the slope of the line $3y = x - 6$ is $\frac{1}{3}$ since $y = \frac{1}{3}x - 2$. So, the desired line must have a slope of -3.
$$y - y_1 = m(x - x_1)$$
$$y - (-5) = -3(x - 2)$$
$$y + 5 = -3x + 6$$
$$y = -3x + 1$$
$$f(x) = -3x + 1$$

51. Note that the slope of the line $3x + 2y = 5$ is $-\frac{3}{2}$ since $y = -\frac{3}{2}x + \frac{5}{2}$.
$$y - y_1 = m(x - x_1)$$
$$y - (-3) = -\frac{3}{2}[x - (-2)]$$
$$2(y + 3) = -3(x + 2)$$
$$2y + 6 = -3x - 6$$
$$y = -\frac{3}{2}x - 6$$
$$f(x) = -\frac{3}{2}x - 6$$

53. $$y - y_1 = m(x - x_1)$$
$$y - 3 = 2[x - (-2)]$$
$$y - 3 = 2(x + 2)$$
$$y - 3 = 2x + 4$$
$$2x - y = -7$$

55. $m = \dfrac{y_2 - y_1}{x_2 - x_1} = \dfrac{6 - 2}{1 - 5} = \dfrac{4}{-4} = -1$
$$y - y_1 = m(x - x_1)$$
$$y - 6 = -1(x - 1)$$
$$y - 6 = -x + 1$$
$$y = -x + 7$$
$$f(x) = -x + 7$$

57. $$y = mx + b$$
$$y = -\frac{1}{2}x + 11$$
$$2y = -x + 22$$
$$x + 2y = 22$$

59. $m = \dfrac{y_2 - y_1}{x_2 - x_1} = \dfrac{-6 - (-4)}{0 - (-7)} = -\dfrac{2}{7}$
$$y - y_1 = m(x - x_1)$$
$$y = -\frac{2}{7}x - 6$$
$$7y = -2x - 42$$
$$2x + 7y = -42$$

61. $$y - y_1 = m(x - x_1)$$
$$y - 0 = -\frac{4}{3}[x - (-5)]$$
$$3y = -4(x + 5)$$
$$3y = -4x - 20$$
$$4x + 3y = -20$$

63. $x = -2$

65. Note that the slope of the line $2x + 4y = 9$, or $y = -\frac{1}{2}x + \frac{9}{4}$, is $-\frac{1}{2}$.
$$y - y_1 = m(x - x_1)$$
$$y - (-2) = -\frac{1}{2}(x - 6)$$
$$2(y + 2) = -x + 6$$
$$2y + 4 = -x + 6$$
$$x + 2y = 2$$

67. $y = 12$

69. Note that the slope of the line $8x - y = 9$, or $y = 8x - 9$, is 8.
$$y - y_1 = m(x - x_1)$$
$$y - 1 = 8(x - 6)$$
$$y - 1 = 8x - 48$$
$$47 = 8x - y$$
$$8x - y = 47$$

71. $x = 5$

73. $m = \dfrac{y_2 - y_1}{x_2 - x_1} = \dfrac{-5 - (-8)}{-6 - 2} = \dfrac{3}{-8} = -\dfrac{3}{8}$

$y - y_1 = m(x - x_1)$

$y - (-8) = -\dfrac{3}{8}(x - 2)$

$8(y + 8) = -3(x - 2)$

$8y + 64 = -3x + 6$

$y = -\dfrac{3}{8}x - \dfrac{29}{4}$

$f(x) = -\dfrac{3}{8}x - \dfrac{29}{4}$

75. a. (1, 30,000), (4, 66,000)

$m = \dfrac{y_2 - y_1}{x_2 - x_1}$

$= \dfrac{66,000 - 30,000}{4 - 1}$

$= 12,000$

$y - y_1 = m(x - x_1)$

$y - 30,000 = 12,000(x - 1)$

$y = 12,000x + 18,000$

$P(x) = 12,000x + 18,000$

b. $P(7) = 12,000(7) + 18,000 = 102,000$
The predicted profit is $102,000.

c. $126,000 = 12,000x + 18,000$

$x = \dfrac{126,000 - 18,000}{12,000} = 9$

The profit should reach $126,000
in 9 years.

77. a. (3, 10,000), (5, 8000)

$m = \dfrac{y_2 - y_1}{x_2 - x_1} = \dfrac{8000 - 10,000}{5 - 3} = -1000$

$y - y_1 = m(x - x_1)$

$y - 10,000 = -1000(x - 3)$

$y = -1000x + 13,000$

b. $y = -1000(3.5) + 13,000$
$y = 9500$
If the price is $3.50, the daily sales
is 9500 Fun Noodles.

79. a. (0, 109,900), (4, 128,400)

$m = \dfrac{y_2 - y_1}{x_2 - x_1}$

$= \dfrac{128,400 - 109,900}{4 - 0}$

$= \dfrac{18,500}{4}$

$= 4625$

$y - y_1 = m(x - x_1)$

$y - 109,900 = 4625(x - 0)$

$y = 4625x + 109,900$

b. $x = 2008 - 1994 = 14$
$y = 4625(14) + 109,900$
$y = 174,650$
In 2008, the median existing home
price is predicted to be $174,650.

81. a. (0, 225), (10, 391)

$m = \dfrac{y_2 - y_1}{x_2 - x_1} = \dfrac{391 - 225}{10 - 0} = 16.6$

$y - y_1 = m(x - x_1)$

$y - 225 = 16.6(x - 0)$

$y = 16.6x + 225$

b. $x = 2004 - 1996 = 8$
$y = 16.6(8) + 225 = 357.8$
In 2004, the estimate is
357,800 people.

83. $f(x) = -x + 7$ or $y_1 = -x + 7$

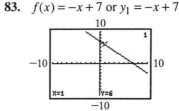

$(x, y) = (1, \ 6)$

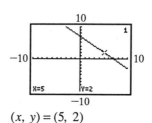

$(x, y) = (5, \ 2)$

85. $4x + 3y = -20$ or $y_1 = (-4x - 20)/3$

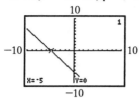

$(x, y) = (-5, 0)$

87.
$$2x - 7 = 21$$
$$2x - 7 + 7 = 21 + 7$$
$$2x = 28$$
$$\frac{2x}{2} = \frac{28}{2}$$
$$x = 14$$
The solution is 14.

89.
$$5(x - 2) = 3(x - 1)$$
$$5x - 10 = 3x - 3$$
$$5x - 3x = 10 - 3$$
$$2x = 7$$
$$\frac{2x}{2} = \frac{7}{2}$$
$$x = \frac{7}{2}$$
The solution is $\frac{7}{2}$.

91.
$$\frac{x}{2} + \frac{1}{4} = \frac{1}{8}$$
$$8\left(\frac{x}{2} + \frac{1}{4}\right) = 8\left(\frac{1}{8}\right)$$
$$4x + 2 = 1$$
$$4x = -1$$
$$x = -\frac{1}{4}$$
The solution is $-\frac{1}{4}$.

93. First, the midpoint of the segment with endpoints $(3, -1)$ and $(-5, 1)$ is $(-1, 0)$. The slope of this segment is
$$m = \frac{y_2 - y_1}{x_2 - x_1}$$
$$= \frac{1 - (-1)}{-5 - 3}$$
$$= \frac{1 + 1}{-5 - 3}$$
$$= \frac{2}{-8}$$
$$= -\frac{1}{4}$$
A line perpendicular to this segment must have slope 4. Finally, the equation of the perpendicular bisector is given by
$$y - y_1 = m(x - x_1)$$
$$y - 0 = 4[x - (-1)]$$
$$y = 4(x + 1)$$
$$y = 4x + 4$$
$$-4x + y = 4$$
$$4x - y = -4$$

95. First, the midpoint of the segment with endpoints $(-2, 6)$ and $(-22, -4)$ is $(-12, 1)$. The slope of this segment is
$$m = \frac{y_2 - y_1}{x_2 - x_1}$$
$$= \frac{-4 - 6}{-22 - (-2)}$$
$$= \frac{-4 - 6}{-22 + 2}$$
$$= \frac{-10}{-20}$$
$$= \frac{1}{2}$$
A line perpendicular to this segment must have slope -2. Finally, the equation of the perpendicular bisector is given by
$$y - y_1 = m(x - x_1)$$
$$y - 1 = -2[x - (-12)]$$
$$y - 1 = -2(x + 12)$$
$$y - 1 = -2x - 24$$
$$2x + y = -23$$

97. First, the midpoint of the segment with endpoints (2, 3) and (–4, 7) is (–1, 5). The slope of this segment is

$$m = \frac{y_2 - y_1}{x_2 - x_1} = \frac{7 - 3}{-4 - 2} = \frac{4}{-6} = -\frac{2}{3}$$

A line perpendicular to this segment must have slope $\frac{3}{2}$. Finally, the equation of the perpendicular bisector is given by

$$y - y_1 = m(x - x_1)$$

$$y - 5 = \frac{3}{2}[x - (-1)]$$

$$2y - 10 = 3(x + 1)$$

$$2y - 10 = 3x + 3$$

$$3x - 2y = -13$$

Exercise Set 2.6

1. $x = 2002 - 1993 = 9$
$f(x) = 1.882x + 11.79$
$f(9) = 1.882(9) + 11.79 = 28.728$
The amount spent is predicted to be $28.728 billion.

3. $f(x) = 1.505x + 32.56$
$40 = 1.505x + 32.56$
$7.44 = 1.505x$
$4.9 \approx x$
$1990 + 5 = 1995$
In 1995, the price was approximately $40.

5. a.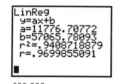

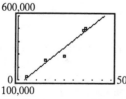

The linear regression equation is
$y = 11,776.708x + 57,065.781$.

b. $x = 2005 - 1960 = 45$

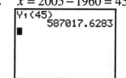

In 2005, the prediction is 587,018,000 visits.

c. From the regression equation, the slope is 11,776.708. Therefore, the number of visits is increasing at the rate of 11,777,000 visits per year.

7. a.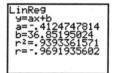

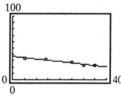

The linear regression equation is
$y = -0.412x + 36.852$.

b. $x = 2005 - 1960 = 45$

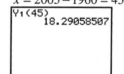

In 2005, the prediction is 18.3%.

c. From the regression equation, the slope is −0.412. Therefore, the percent of male smokers is decreasing at the rate of 0.412% per year.

9. a.

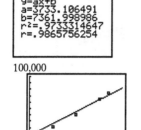

The linear regression equation is
$y = 3733.106x + 7361.999$.

b. $x = 2006 - 1980 = 26$

Y₁(26)
    104422.7677

In 2006, the prediction is
104,423 female prisoners.

c. From the regression equation, the
slope is 3733.106. Therefore, the
number of female prisoners is
increasing at the rate of 3733
prisoners per year.

11.

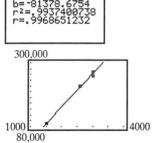

The linear regression equation is
$y = 114.453x - 81,378.675$.

13.

Y₁(2200)
    170418.0229

The selling price of a home with 2200
square feet is $170,418.

15.

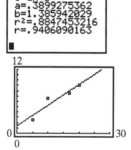

The linear regression equation is
$y = 0.08x + 700$.

17. a. Let $x = 0$ represent the year 1980.

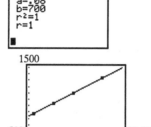

The linear regression equation is
$y = 0.390x + 1.386$.

b. $x = 2010 - 1980 = 30$

Y₁(30)
    13.08376812

In 2010, the predicted revenue is
$13.084 billion.

c. From the regression equation, the
slope is 0.390. Therefore, the rental
revenue is increasing at the rate of
$0.390 billion per year.

19. a.

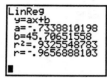

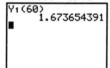

The linear regression equation is
$y = -0.734x + 45.707$.

b. $x = 2000 - 1940 = 60$

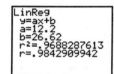

In 2000, the prediction is 1.674 per
1000 live births.

21. a.

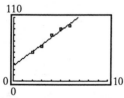

The linear regression equation is
$y = 12.2x + 26.62$.

b. $x = 2005 - 1990 = 15$

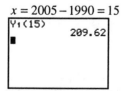

In 2005, the prediction is
209.62 million members.

23. a.

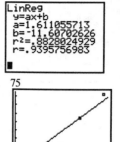

The linear regression equation is
$y = 1.611x - 11.607$.

b. $x = 2004 - 1950 = 54$

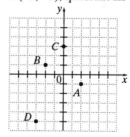

In 2004, the prediction is
$75.39 per ticket.

c.   From the regression equation, the
slope is 1.611. Therefore, the cost is
rising at a rate of $1.61 per year.

Chapter 2 Review

1. $A(2, -1)$, quadrant IV
$B(-2, 1)$, quadrant II
$C(0, 3)$, y-axis
$D(-3, -5)$, quadrant III

2. $A(-3, 4)$, quadrant II
$B(4, -3)$, quadrant IV
$C(-2, 0)$, x-axis
$D(-4, 1)$, quadrant II

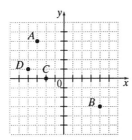

3. $7x - 8y = 56$
$(0, 56)$
$7(0) - 8(56) = 56$
$\qquad -448 = 56$ False
$(0, 56)$ is not a solution.
$(8, 0)$
$7(8) - 8(0) = 56$
$\qquad 56 = 56$ True
$(8, 0)$ is a solution.

4. $-2x + 5y = 10$
$(-5, 0)$
$-2(-5) + 5(0) = 10$
$\qquad\qquad 10 = 10$ True
$(-5, 0)$ is a solution.
$(1, 1)$
$-2(1) + 5(1) = 10$
$\qquad -2 + 5 = 10$
$\qquad\qquad 3 = 10$ False
$(1, 1)$ is not a solution.

5. $x = 13$
$(13, 5)$
$13 = 13$ True
$(13, 5)$ is a solution.
$(13, 13)$
$13 = 13$ True
$(13, 13)$ is a solution.

6. $y = 2$
$(7, 2)$
$2 = 2$ True
$(7, 2)$ is a solution.
$2 = 2$ True
$(2, 7)$
$7 = 2$ False
$(2, 7)$ is not a solution.

7. The equation is linear and the shape of its graph is a line.

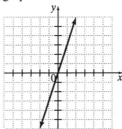

8. The equation is linear and the shape of its graph is a line.

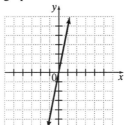

9. $3x - y = 4$
The equation is linear and the shape of its graph is a line. Find three ordered pair solutions, or find x- and y-intercepts, or find m and b.

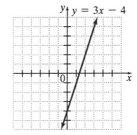

10. $x - 3y = 2$
The equation is linear and the shape of its graph is a line. Find three ordered pair solutions, or find x- and y-intercepts, or find m and b.

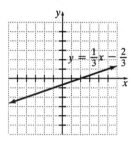

11. $y = |x| + 4$
The equation is nonlinear and the shape of its graph is V-shaped.

x	y
-3	7
-2	6
-1	5
0	4
1	5
2	6
3	7

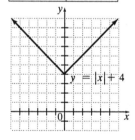

12. $y = x^2 + 4$
The equation is nonlinear and its graph is shaped like a parabola.

x	y
-3	13
-2	8
-1	5
0	4
1	5
2	8
3	13

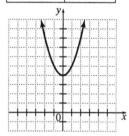

13. $y = -\dfrac{1}{2}x + 2$

The equation is linear and the shape of its graph is a line. Find three ordered pair solutions, or find x- and y-intercepts, or find m and b.

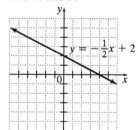

68

14. $y = -x + 5$

The equation is linear and the shape of its graph is a line. Find three ordered pair solutions, or find x- and y-intercepts, or find m and b.

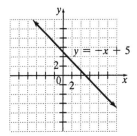

15. D; $y = x^2 + 2$; parabolic with y-intercept $(0, 2)$.

16. A; $y = x^2 - 4$; parabolic with y-intercept $(0, -4)$.

17. C; $y = |x| + 2$; V-shaped up with y-intercept $(0, 2)$.

18. B; $y = -|x| + 2$; V-shaped down with y-intercept $(0,2)$.

19. Domain: $\left\{ -\dfrac{1}{2}, \ 6, \ 0, \ 25 \right\}$

Range: $\left\{ \dfrac{3}{4}, \ -12, \ 25 \right\}$

The relation is a function.

20. Domain: $\left\{ \dfrac{3}{4}, \ -12, \ 25 \right\}$

Range: $\left\{ -\dfrac{1}{2}, \ 6, \ 0, \ 25 \right\}$

The relation is not a function since

$\dfrac{3}{4} = 0.75$ and it is paired with two different y-values.

21. Domain: $\{2, 4, 6, 8\}$
Range: $\{2, 4, 5, 6\}$
The relation is not a function since 2 is paired with both 2 and 4.

22. Domain: $\{$Triangle, Square, Rectangle, Parallelogram$\}$
Range: $\{3, 4\}$
The relation is a function.

23. Domain: $(-\infty, \infty)$
Range: $(-\infty, -1] \cup [1, \infty)$
The relation is not a function since it fails the vertical line test.

24. Domain: $\{-3\}$
Range: $(-\infty, \infty)$
The relation is not a function since it fails the vertical line test.

25. Domain: $(-\infty, \infty)$
Range: $\{4\}$
The relation is a function since it passes the vertical line test.

26. Domain: $[-1, 1]$
Range: $[-1, 1]$
The relation is not a function since it fails the vertical line test.

27. $f(x) = x - 5$
$f(2) = 2 - 5 = -3$

28. $g(x) = -3x$
$g(0) = -3(0) = 0$

29. $g(x) = -3x$
$g(-6) = -3(-6) = 18$

30. $h(x) = 2x^2 - 6x + 1$
$\begin{aligned} h(-1) &= 2(-1)^2 - 6(-1) + 1 \\ &= 2(1) + 6 + 1 \\ &= 9 \end{aligned}$

31. $h(x) = 2x^2 - 6x + 1$
$h(1) = 2(1)^2 - 6(1) + 1 = 2 - 6 + 1 = -3$

32. $f(x) = x - 5$
$f(5) = 5 - 5 = 0$

33. $J(x) = 2.54x$
$J(150) = 2.54(150) = 381$
The person would weigh 381 pounds on Jupiter.

34. $J(x) = 2.54x$
$J(2000) = 2.54(2000) = 5080$
The probe would weigh 5080 pounds on Jupiter.

35. $(-1, 0)$; $f(-1) = 0$

36. $(1, -2)$; $f(1) = -2$

37. $f(x) = 1$
$f(-2) = f(4) = 1$
The values of x are $-2, 4$.

38. $f(x) = -1$
$f(0) = f(2) = -1$
The values of x are $0, 2$.

39. $f(x) = x$ or $y = x$
$m = 1, b = 0$

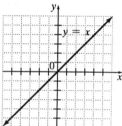

40. $f(x) = -\dfrac{1}{3}x$ or $y = -\dfrac{1}{3}x$

$m = -\dfrac{1}{3}, b = 0$

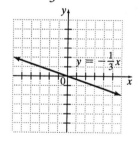

41. $g(x) = 4x - 1$ or $y = 4x - 1$
$m = 4, b = -1$

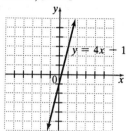

42. C; $m = 3, b = 1$

43. A; $m = 3, b = -2$

44. B; $m = 3, b = 2$

45. D; $m = 3, b = -5$

46. $4x + 5y = 20$

Let $x = 0$	Let $y = 0$
$4(0) + 5y = 20$	$4x + 5(0) = 20$
$y = 4$	$x = 5$

The intercepts are $(0, 4)$ and $(5, 0)$.

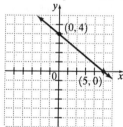

47. $3x - 2y = -9$

        Let $x = 0$ Let $y = 0$

      $3(0) - 2y = -9$ $3x - 2(0) = -9$

           $y = \dfrac{9}{2}$ $x = -3$

The intercepts are $\left(0, \ \dfrac{9}{2}\right)$ and $(-3, 0)$.

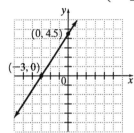

48. $4x - y = 3$

        Let $x = 0$ Let $y = 0$

      $4(0) - y = 3$ $4x - 0 = 3$

          $y = -3$

                     $x = \dfrac{3}{4}$

The intercepts are $(0, -3)$ and $\left(\dfrac{3}{4}, \ 0\right)$.

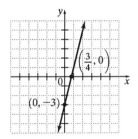

49. $2x + 6y = 9$

        Let $x = 0$ Let $y = 0$

      $2(0) + 6y = 9$ $2x + 6(0) = 9$

           $y = \dfrac{3}{2}$ $x = \dfrac{9}{2}$

The intercepts are $\left(0, \ \dfrac{3}{2}\right)$ and $\left(\dfrac{9}{2}, \ 0\right)$.

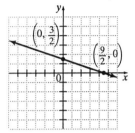

50. $y = 5$

Horizontal line with y-intercept $(0, 5)$

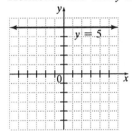

51. $x = -2$

Vertical line with x-intercept $(-2, 0)$

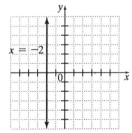

52. $x - 2 = 0$ or $x = 2$

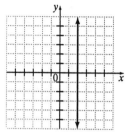

53. $y + 3 = 0$ or $y = -3$

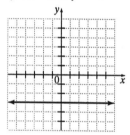

54. $C(x) = 0.3x + 42$

a. $C(150) = 0.3(150) + 42$
$$= 45 + 42$$
$$= 87$$
The cost is $87.

b. $m = 0.3$, $b = 42$

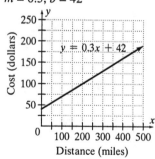

55. $m = \dfrac{y_2 - y_1}{x_2 - x_1} = \dfrac{-4 - 8}{6 - 2} = \dfrac{-12}{4} = -3$

56. $m = \dfrac{y_2 - y_1}{x_2 - x_1} = \dfrac{13 - 9}{5 - (-3)} = \dfrac{4}{8} = \dfrac{1}{2}$

57. $m = \dfrac{y_2 - y_1}{x_2 - x_1} = \dfrac{6 - (-4)}{-3 - (-7)} = \dfrac{10}{4} = \dfrac{5}{2}$

58. $m = \dfrac{y_2 - y_1}{x_2 - x_1} = \dfrac{7 - (-2)}{-5 - 7} = \dfrac{9}{-12} = -\dfrac{3}{4}$

59. $6x - 15y = 20$
$$6x - 20 = 15y$$
$$y = \dfrac{2}{5}x - \dfrac{4}{3}$$
$$m = \dfrac{2}{5}, \ b = -\dfrac{4}{3}$$

60. $4x + 14y = 21$
$$14y = -4x + 21$$
$$y = -\dfrac{2}{7}x + \dfrac{3}{2}$$
$$m = -\dfrac{2}{7}, \ b = \dfrac{3}{2}$$

61. $y - 3 = 0$
$$y = 3$$
Slope = 0

62. $x = -5$
Vertical line
Slope is undefined.

63. l_2 has the greater slope because the slope of l_2 is positive and the slope of l_1 is negative.

64. l_2 has the greater slope because the slope of l_2 is positive and the slope of l_1 is 0.

65. l_2 has the greater slope because it is steeper than l_1.

66. l_1 has the greater slope because the slope of l_1 is 0 and the slope of l_2 is negative.

67. $y = 0.3x + 42$

a. $m = 0.3$; The cost increases by $0.30 for each additional mile driven.

b. $b = 42$; The cost for 0 miles driven is $42.

68. $f(x) = -2x + 6$ $g(x) = 2x - 1$
$m = -2$ $m = 2$
Neither; The slopes are not the same, nor are they negative reciprocals of one another.

69. $-x + 3y = 2$ $6x - 18y = 3$
       $y = \dfrac{1}{3}x + \dfrac{2}{3}$ $y = \dfrac{1}{3}x - \dfrac{1}{6}$
$m = \dfrac{1}{3}, \; b = \dfrac{2}{3}$ $m = \dfrac{1}{3}, \; b = -\dfrac{1}{6}$
The lines are parallel since their slopes are equal and their y-intercepts are different.

70. $y = -x + 1$
$m = -1, \; b = 1, \;$ y-intercept $(0, 1)$

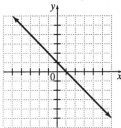

71. $y = 4x - 3$
$m = 4, \; b = -3, \;$ y-intercept $(0, -3)$

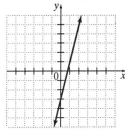

72. $3x - y = 6$
     $y = 3x - 6$
$m = 3, \; b = -6, \;$ y-intercept $(0, -6)$

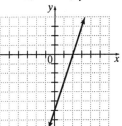

73. $y = -5x$
$m = -5, \; b = 0, \;$ y-intercept $(0, 0)$

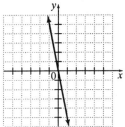

74. Horizontal lines have slope $= 0$.
The y-intercept is $(0 -1)$.
$y = -1$

75. Vertical lines have undefined slope.
The x-intercept is $(-2, 0)$.
$x = -2$

76. The slope is undefined.
The x-intercept is $(-4, 0)$.
$x = -4$

77. Horizontal line with y-intercept $(0, 5)$;
$y = 5$

78. $y - y_1 = m(x - x_1)$
   $y - 5 = 3[x - (-3)]$
   $y - 5 = 3(x + 3)$
   $y - 5 = 3x + 9$
   $3x - y = -14$

79. $y - y_1 = m(x - x_1)$
   $y - (-2) = 2(x - 5)$
     $y + 2 = 2x - 10$
     $2x - y = 12$

80. $m = \dfrac{y_2 - y_1}{x_2 - x_1} = \dfrac{-2 - (-1)}{-4 - (-6)} = -\dfrac{1}{2}$

$y - y_1 = m(x - x_1)$

$y - (-1) = -\dfrac{1}{2}[x - (-6)]$

$2(y + 1) = -(x + 6)$

$2y + 2 = -x - 6$

$x + 2y = -8$

81. $m = \dfrac{y_2 - y_1}{x_2 - x_1} = \dfrac{-8 - 3}{-4 - (-5)} = \dfrac{-11}{1} = -11$

$y - y_1 = m(x - x_1)$

$y - 3 = -11[x - (-5)]$

$y - 3 = -11x - 55$

$11x + y = -52$

82. $x = 4$ has undefined slope.
A line perpendicular to $x = 4$ has slope $= 0$.
y-intercept at $(0, 3)$
$y = 3$

83. Slope $= 0$, y-intercept $(0, -5)$
$y = -5$

84. $\quad y = mx + b$

$y = -\dfrac{2}{3}x + 4$

$f(x) = -\dfrac{2}{3}x + 4$

85. $\quad y = mx + b$
$y = -x - 2$
$f(x) = -x - 2$

86. Note that the slope of the line $6x + 3y = 5$
is -2 since $y = -2x + \dfrac{5}{3}$.

$y - y_1 = m(x - x_1)$

$y - (-6) = -2(x - 2)$

$y + 6 = -2x + 4$

$y = -2x - 2$

$f(x) = -2x - 2$

87. Note that the slope of the line $3x + 2y = 8$
is $-\dfrac{3}{2}$ since $y = -\dfrac{3}{2}x + 4$.

$y - y_1 = m(x - x_1)$

$y - (-2) = -\dfrac{3}{2}[x - (-4)]$

$y + 2 = -\dfrac{3}{2}x - 6$

$y = -\dfrac{3}{2}x - 8$

$f(x) = -\dfrac{3}{2}x - 8$

88. Note that the slope of the line $4x + 3y = 5$
is $-\dfrac{4}{3}$ since $y = -\dfrac{4}{3}x + \dfrac{5}{3}$. So, the

desired line must have a slope of $\dfrac{3}{4}$.

$y - y_1 = m(x - x_1)$

$y - (-1) = \dfrac{3}{4}[x - (-6)]$

$4(y + 1) = 3(x + 6)$

$4y + 4 = 3x + 18$

$y = \dfrac{3}{4}x + \dfrac{7}{2}$

$f(x) = \dfrac{3}{4}x + \dfrac{7}{2}$

89. Note that the slope of the line $2x - 3y = 6$
is $\dfrac{2}{3}$ since $y = \dfrac{2}{3}x - 2$. So, the desired

line must have a slope of $-\dfrac{3}{2}$.

$y - y_1 = m(x - x_1)$

$y - 5 = -\dfrac{3}{2}[x - (-4)]$

$y - 5 = -\dfrac{3}{2}x - 6$

$y = -\dfrac{3}{2}x - 1$

$f(x) = -\dfrac{3}{2}x - 1$

90. a. Use ordered pairs (0, 42) and (3, 58).

$$m = \frac{y_2 - y_1}{x_2 - x_1} = \frac{58 - 42}{3 - 0} = \frac{16}{3}$$

$$y - y_1 = m(x - x_1)$$

$$y - 42 = \frac{16}{3}(x - 0)$$

$$y = \frac{16}{3}x + 42$$

b. $x = 2007 - 1996 = 11$

$$y = \frac{16}{3}(11) + 42 = 100.7$$

There will be about 101 million subscribers.

91. a. Use ordered pairs (0, 43) and (22, 60).

$$m = \frac{y_2 - y_1}{x_2 - x_1} = \frac{60 - 43}{22 - 0} = \frac{17}{22}$$

$$y - y_1 = m(x - x_1)$$

$$y - 43 = \frac{17}{22}(x - 0)$$

$$y = \frac{17}{22}x + 43$$

b. $x = 2010 - 1998 = 12$

$$y = \frac{17}{22}(12) + 43 = 52.3$$

There will be about 52 million people reporting arthritis.

92. a.

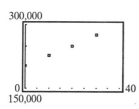

b.

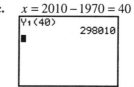

The linear regression equation is
$y = 2381.74x + 202,740.4$.

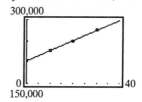

c. $x = 2010 - 1970 = 40$

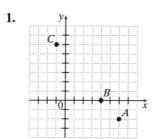

In 2010, the predicted population in the United States is 298,010 thousand.

Chapter 2 Test

1.

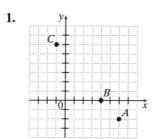

A is in quadrant IV.
B is on the x-axis, no quadrant.
C is in quadrant II.

2. $2y - 3x = 12$

$x = -6$:

$$2y - 3(-6) = 12$$

$$2y + 18 = 12$$

$$2y = -6$$

$$y = -3$$

$(-6, -3)$

3. $2x - 3y = -6$

$$-3y = -2x - 6$$

$$y = \frac{2}{3}x + 2$$

$$m = \frac{2}{3}, \; b = 2$$

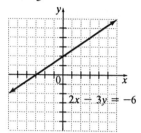

4. $4x + 6y = 7$

$$6y = -4x + 7$$

$$y = -\frac{2}{3}x + \frac{7}{6}$$

$$m = -\frac{2}{3}, \; b = \frac{7}{6}$$

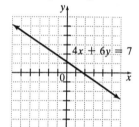

5. $f(x) = \frac{2}{3}x$ or $y = \frac{2}{3}x$

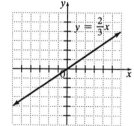

6. $y = -3$

Horizontal line with y-intercept at $(0, -3)$

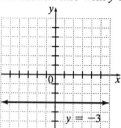

7. $m = \dfrac{y_2 - y_1}{x_2 - x_1} = \dfrac{10 - (-8)}{-7 - 5} = \dfrac{18}{-12} = -\dfrac{3}{2}$

8. $3x + 12y = 8$

$$12y = -3x + 8$$

$$y = -\frac{1}{4}x + \frac{2}{3}$$

$$m = -\frac{1}{4} \text{ and } b = \frac{2}{3}$$

9. A horizontal line has an equation of the form $y = b$. Since the line contains the point $(2, -8)$, the equation is $y = -8$.

10. A vertical line has an equation of the form $x = c$. Since the line contains the point $(-4, -3)$, the equation is $x = -4$.

11. Since the line $x = 5$ is vertical, a line perpendicular to it is horizontal. Horizontal lines have equations of the form $y = b$. Since the line contains the point $(3, -2)$, the equation is $y = -2$.

12. $\quad y - y_1 = m(x - x_1)$

$$y - (-1) = -3(x - 4)$$

$$y + 1 = -3x + 12$$

$$3x + y = 11$$

13. $\qquad y = mx + b$

$$y = 5x + (-2)$$

$$5x - y = 2$$

14. $m = \dfrac{y_2 - y_1}{x_2 - x_1} = \dfrac{-3 - (-2)}{6 - 4} = -\dfrac{1}{2}$

$y - y_1 = m(x - x_1)$

$y - (-2) = -\dfrac{1}{2}(x - 4)$

$2(y + 2) = -(x - 4)$

$2y + 4 = -x + 4$

$y = -\dfrac{1}{2}x$

$f(x) = -\dfrac{1}{2}x$

15. Note that the slope of the line $3x - y = 4$ is 3 since $y = 3x - 4$. So, the desired line must have a slope of $-\dfrac{1}{3}$.

$y - y_1 = m(x - x_1)$

$y - 2 = -\dfrac{1}{3}[x - (-1)]$

$3(y - 2) = -(x + 1)$

$3y - 6 = -x - 1$

$y = -\dfrac{1}{3}x + \dfrac{5}{3}$

$f(x) = -\dfrac{1}{3}x + \dfrac{5}{3}$

16. Note that the slope of the line $2y + x = 3$ is $-\dfrac{1}{2}$ since $y = -\dfrac{1}{2}x + \dfrac{3}{2}$.

$y - y_1 = m(x - x_1)$

$y - (-2) = -\dfrac{1}{2}(x - 3)$

$2(y + 2) = -(x - 3)$

$2y + 4 = -x + 3$

$y = -\dfrac{1}{2}x - \dfrac{1}{2}$

$f(x) = -\dfrac{1}{2}x - \dfrac{1}{2}$

17. $2x - 5y = 8$

$5y = 2x - 8$

$y = \dfrac{2}{5}x - \dfrac{8}{5}$,

so $m_1 = \dfrac{2}{5}$,

$m_2 = \dfrac{y_2 - y_1}{x_2 - x_1} = \dfrac{-1 - 4}{-1 - 1} = \dfrac{-5}{-2} = \dfrac{5}{2}$

Therefore, lines L_1 and L_2 are neither parallel nor perpendicular.

18. B; $y = x^2 + 2x + 3$; parabola; y-intercept $(0, 3)$

19. A; $y = 2|x - 1| + 3$; V-shaped; y-intercept $(0, 5)$

20. D; $y = 2x + 3$; linear; y-intercept $(0, 3)$

21. C; $y = 2(x - 1)^3 + 3$; y-intercept $(0, 1)$

22. Domain: $(-\infty, \infty)$
    Range: $\{5\}$
    The relation is a function.

23. Domain: $\{-2\}$
    Range: $(-\infty, \infty)$
    The relation is not a function since it fails the vertical line test.

24. Domain: $(-\infty, \infty)$
    Range: $[0, \infty)$
    The relation is a function since it passes the vertical line test.

25. Domain: $(-\infty, \infty)$
    Range: $(-\infty, \infty)$
    The relation is a function since it passes the vertical line test.

26. $f(x) = 732x + 21,428$

    a. $f(2) = 732(2) + 21,428 = 22,892$
    The average earnings were \$22,892.

    b. $x = 2005 - 1996 = 9$
        $f(9) = 732(9) + 21,428 = 28,016$
    The average earnings will be \$28,016.

    c.
$$f(x) > 30,000$$
$$732x + 21,428 > 30,000$$
$$732x > 8572$$
$$x > \frac{8572}{732} \approx 11.7$$

    The average yearly earnings for high school graduates will be greater than \$30,000 the twelfth year after 1996, or in 2008.

    d. $m = 732$; The average yearly earnings for high school graduates increase by \$732 per year.

    e. $b = 21,428$; The average yearly earnings for a high school graduate in 1996 were \$21,428.

27. a. Let $x = 0$ represent 1970.

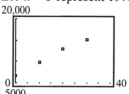

    b.

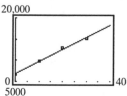

    The linear regression equation is $y = 285.18x + 6899.3$.

    c. $x = 2010 - 1970 = 40$

      Y₁(40)   18306.5

    In 2010, the predicted population of Florida is 18,306.5 thousand.

Chapter 3

Section 3.1

Mental Math

1. $x = 6$; The solution is 6.

2. $x = -2$; The solution is -2.

3. $x = 5$; The solution is 5.

4. $x = 1$; The solution is 1.

5. $x = 0$; The solution is 0.

6. $x = 7$; The solution is 7.

Exercise Set 3.1

1. Algebraic solution:
$$5x + 2 = 3x + 6$$
$$2x + 2 = 6$$
$$2x = 4$$
$$x = 2$$
Graphical solution:

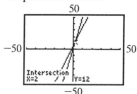

The solution is 2.

3. Algebraic solution:
$$9 - x = 2x + 12$$
$$9 - 3x = 12$$
$$-3x = 3$$
$$x = -1$$
Graphical solution:

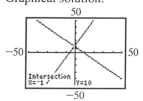

The solution is -1.

5. Algebraic solution:
$$8 - (2x - 1) = 13$$
$$8 - 2x + 1 = 13$$
$$9 - 2x = 13$$
$$-2x = 4$$
$$x = -2$$
Graphical solution:

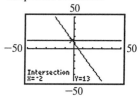

The solution is -2.

7. $3x - 13 = 2(x - 5) + 3$
The solution is 6.

9. $-(2x + 3) + 2 = 5x - 1 - 7x$
The lines are the same. The solution set is the set of all real numbers.

11. Algebraic solution:
$$7(x - 6) = 5(x + 2) + 2x$$
$$7x - 42 = 5x + 10 + 2x$$
$$7x - 42 = 7x + 10$$
$$7x - 42 - 7x = 7x + 10 - 7x$$
$$-42 = 10$$
This equation is a false statement no matter what value the variable x might have.
Graphical solution:

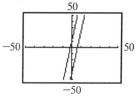

The lines are parallel. There is no solution. The solution set is $\varnothing$.

13. Algebraic solution:
$$3x - (6x + 2) = -(3x + 2)$$
$$3x - 6x - 2 = -3x - 2$$
$$-3x - 2 = -3x - 2$$
Since both sides are the same, we see that replacing x with any real number will result in a true statement.
Graphical solution:

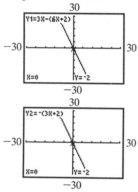

The lines are the same. The solution set is the set of all real numbers.

15. Algebraic solution:
$$5(x + 1) - 3(x - 7) = 2(x + 4) - 3$$
$$5x + 5 - 3x + 21 = 2x + 8 - 3$$
$$2x + 26 = 2x + 5$$
$$2x + 26 - 2x = 2x + 5 - 2x$$
$$26 = 5$$
This equation is a false statement no matter what value the variable x might have.
Graphical solution:

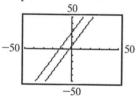

The lines are parallel. There is no solution. The solution set is $\varnothing$.

17. Algebraic solution:
$$3(x + 2) - 6(x - 5) = 36 - 3x$$
$$3x + 6 - 6x + 30 = 36 - 3x$$
$$-3x + 36 = -3x + 36$$
Since both sides are the same, we see that replacing x with any real number will result in a true statment.
Graphical solution:

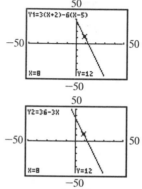

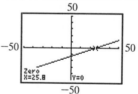

The lines are the same. The solution set is the set of all real numbers.

19.
$$x + 2.1 - (0.5x + 3) = 12$$
$$x + 2.1 - (0.5x + 3) - 12 = 0$$
Define $y_1 = x + 2.1 - (0.5x + 3) - 12$ and graph in an integer window. Choose the root, or zero, feature to solve.

$$x = 25.8$$
The solution is 25.8.

21. $5(a-12)+2(a+15)=a-9$

$5(a-12)+2(a+15)-a+9=0$

Define $y_1 = 5(x-12)+2(x+15)-x+9$ and graph in a standard window. Choose the root, or zero, feature to solve.

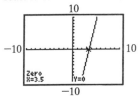

$x=3.5$ or $a=3.5$

The solution is 3.5.

23. $8(p-4)-5(2p+3)=3.5(2p-5)$

$8(p-4)-5(2p+3)-3.5(2p-5)=0$

Define $y_1 = 8(x-4)-5(2x+3)-3.5(2x-5)$ and graph in a standard window. Choose the root, or zero, feature to solve.

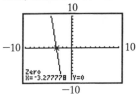

$x \approx -3.28$ or $p \approx -3.28$

The solution is approximately -3.28.

25. $5(x-2)+2x=7(x+4)$

$5(x-2)+2x-7(x+4)=0$

Define $y_1 = 5(x-2)+2x-7(x+4)$ and graph in an integer window. Choose the root, or zero, feature to solve.

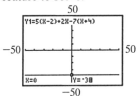

The graph is a horizontal line. There is no solution. The solution set is $\varnothing$.

27.
$$y + 0.2 = 0.6(y + 3)$$
$$y + 0.2 - 0.6(y + 3) = 0$$
Define $y_1 = x + 0.2 - 0.6(x + 3)$ and graph in a standard window. Choose the root, or zero, feature to solve.

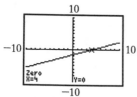

$x = 4$ or $y = 4$
The solution is 4.

29.
$$2y + 5(y - 4) = 4y - 2(y - 10)$$
$$2y + 5(y - 4) - 4y + 2(y - 10) = 0$$
Define $y_1 = 2x + 5(x - 4) - 4x + 2(x - 10)$ and graph in an integer window. Choose the root, or zero, feature to solve.

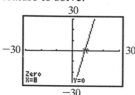

$x = 8$ or $y = 8$
The solution is 8.

31.
$$2(x - 8) + x = 3(x - 6) + 2$$
$$2(x - 8) + x - 3(x - 6) - 2 = 0$$
Define $y_1 = 2(x - 8) + x - 3(x - 6) - 2$ and graph in an integer window. Choose the root, or zero, feature to solve.

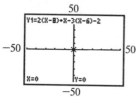

The graph is the x-axis. The solution set is the set of all real numbers.

33.
$$\frac{5x-1}{6} - 3x = \frac{1}{3} + \frac{4x+3}{9}$$
$$\frac{5x-1}{6} - 3x - \frac{1}{3} - \frac{4x+3}{9} = 0$$

Define $y_1 = \frac{5x-1}{6} - 3x - \frac{1}{3} - \frac{4x+3}{9}$ and graph in a standard window. Choose the root, or zero, feature to solve.

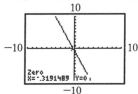

$x \approx -0.32$

The solution is approximately -0.32.

35.
$$-2(b-4) - (3b-1) = 5b+3$$
$$-2(b-4) - (3b-1) - 5b - 3 = 0$$

Define $y_1 = -2(x-4) - (3x-1) - 5x - 3$ and graph in a standard window. Choose the root, or zero, feature to solve.

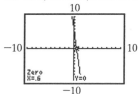

$x = 0.6$ or $b = 0.6$

The solution is 0.6.

37.
$$1.5(4-x) = 1.3(2-x)$$
$$1.5(4-x) - 1.3(2-x) = 0$$

Define $y_1 = 1.5(4-x) - 1.3(2-x)$ and graph in an integer window. Choose the root, or zero, feature to solve.

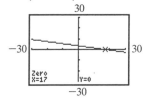

$x = 17$

The solution is 17.

$6 - 1.5x - 2.6 + 1.3x$

$3.4 - .2x = 0$
$3.4 - .2x + .2x = .2x$
$\dfrac{3.4}{2} = \dfrac{.2x}{2}$
$1.7 = x$

39.
$$\frac{1}{4}(a+2) = \frac{1}{6}(5-a)$$
$$\frac{1}{4}(a+2) - \frac{1}{6}(5-a) = 0$$

Define $y_1 = \frac{1}{4}(x+2) - \frac{1}{6}(5-x)$ and graph in a standard window. Choose the root, or zero, feature to solve.

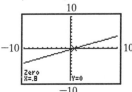

$x = 0.8$ or $a = 0.8$

The solution is 0.8.

41. a. $C_1(2) = 30 + 20(2) = 70$
$C_2(2) = 25(2) = 50$
The second consultant costs less.
($50 < $70)

    b. $C_1(8) = 30 + 20(8) = 190$
$C_2(8) = 25(8) = 200$
The first consultant costs less.
($190 < $200)

    c. $C_1(x) = C_2(x)$
$30 + 20x = 25x$
$30 = 5x$
$x = 6$
The cost is the same for a 6 hour job.

43. a. $R_1(50) = 25 + 0.30(50) = 40.00$
$R_2(50) = 28 + 0.25(50) = 40.50$
The first agency's cost is lower.
($40 < $40.50)

    b. $R_1(100) = 25 + 0.30(100) = 55.00$
$R_2(100) = 28 + 0.25(100) = 53.00$
The second agency's cost is lower.
($53 < $55)

    c. $R_1(x) = R_2(x)$
$25 + 0.30x = 28 + 0.25x$
$0.05x = 3$
$x = 60$
The cost is the same for 60 miles.

45. The screen shows the graphs intersect at $x = 12$, $y = 22$. Therefore, the ordered pair for the graphs of both y_1 and y_2 is (12, 22).

47. Since the graph of y_1 is above the graph of y_2 for x greater than 12, y_1 is greater than y_2 when x is greater than 12.

49. Define $y_1 = 1.75x - 2.5$ and $y_2 = 0$ and graph in a standard window.

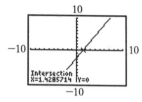

The solution, rounded to the nearest hundredth, is 1.43.

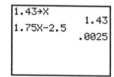

The solution is approximately 1.43.

51. Define $y_1 = 2.5x + 3$ and $y_2 = 7.8x - 5$ and graph in a standard window.

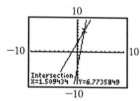

The solution, rounded to the nearest hundredth, is 1.51.

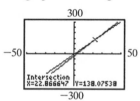

The solution is approximately 1.51.

53. Define $y_1 = 3x + \sqrt{5}$ and $y_2 = 7x - \sqrt{2}$ and graph in a standard window.

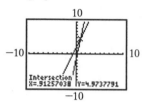

The solution, rounded to the nearest hundredth, is 0.91.

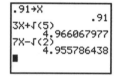

The solution is approximately 0.91.

55. Define $y_1 = 2\pi x - 5.6$ and $y_2 = 7(x - \pi)$ and graph in an integer window.

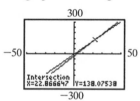

The solution, rounded to the nearest hundredth, is 22.87.

The solution is approximately 22.87.

57. The equation has no solution.

59. $\{-3, -2, -1\}$

61. $x + 5 \le 6$
$x \le 1$
$\{-3, -2, -1, 0, 1\}$

63. Answers may vary.

Section 3.2

Mental Math

1. $x - 2 < 4$
   $x < 6$
   The solution set is $\{x | x < 6\}$.

2. $x - 1 > 6$
   $x > 7$
   The solution set is $\{x | x > 7\}$.

3. $x + 5 \geq 15$
   $x \geq 10$
   The solution set is $\{x | x \geq 10\}$.

4. $x + 1 \leq 8$
   $x \leq 7$
   The solution set is $\{x | x \leq 7\}$.

5. $3x > 12$
   $x > 4$
   The solution set is $\{x | x > 4\}$.

6. $5x < 20$
   $x < 4$
   The solution set is $\{x | x < 4\}$.

7. $\dfrac{x}{2} \leq 1$
   $x \leq 2$
   The solution set is $\{x | x \leq 2\}$.

8. $\dfrac{x}{4} \geq 2$
   $x \geq 8$
   The solution set is $\{x | x \geq 8\}$.

Exercise Set 3.2

1. The solution set written in interval notation is $(-\infty, -3)$.

3. The solution set written in interval notation is $[0.3, \infty)$.

5. The solution set written in interval notation is $(5, \infty)$.

7. The solution set written in interval notation is $(-2, 5)$.

9. The solution set written in interval notation is $(-1, 5)$.

11. Answers may vary.

13. The point of intersection is $(4, 10)$. The graph of y_1 is below the graph of y_2 for all x-values less than 4. The solution set is $(-\infty, 4)$.

15. The point of intersection is $(-3, -5)$. The graph of y_1 is equal to or above the graph of y_2 for all x-values greater than or equal to -3. The solution set is $[-3, \infty)$.

17. The graph of y_1, is never above the graph of y_2. This means that no x-values satisfy the inequality $y_1 > y_2$. The solution set is $\varnothing$.

19. $\qquad 7x < 6x + 1$
    $\quad 7x - 6x < 6x + 1 - 6x$
    $\qquad\qquad x < 1$
    The solution set written in interval notation is $(-\infty, 1)$.

21.
$$8x - 7 \le 7x - 5$$
$$8x - 7 - 7x \le 7x - 5 - 7x$$
$$x - 7 \le -5$$
$$x - 7 + 7 \le -5 + 7$$
$$x \le 2$$

The solution set written in interval notation is $(-\infty, 2]$.

23.
$$2 + 4x > 5x + 6$$
$$2 + 4x - 4x > 5x + 6 - 4x$$
$$2 > x + 6$$
$$2 - 6 > x + 6 - 6$$
$$-4 > x \quad \text{or} \quad x < -4$$

The solution set written in interval notation is $(-\infty, -4)$.

25.
$$\frac{3}{4}x \ge 2$$
$$\frac{4}{3} \cdot \frac{3}{4}x \ge \frac{4}{3} \cdot 2$$
$$x \ge \frac{8}{3}$$

The solution set written in interval notation is $\left[\frac{8}{3}, \infty\right)$.

27. $5x < -23.5$
$$\frac{5x}{5} < \frac{-23.5}{5}$$
$$x < -4.7$$

The solution set written in interval notation is $(-\infty, -4.7)$.

29. $-3x \ge 9$
$$\frac{-3x}{-3} \le \frac{9}{-3}$$
$$x \le -3$$

The solution set written in interval notation is $(-\infty, -3]$.

31. $-x < -4$
$$\frac{-x}{-1} > \frac{-4}{-1}$$
$$x > 4$$

The solution set written in interval notation is $(4, \infty)$.

33. $-2x + 7 \ge 9$
$$-2x \ge 2$$
$$\frac{-2x}{-2} \le \frac{2}{-2}$$
$$x \le -1$$

The solution set written in interval notation is $(-\infty, -1]$.

35. $15 + 2x \ge 4x - 7$
$$15 \ge 2x - 7$$
$$22 \ge 2x$$
$$\frac{22}{2} \ge \frac{2}{2}$$
$$11 \ge x \quad \text{or} \quad x \le 11$$

The solution set written in interval notation is $(-\infty, 11]$.

37. $3(x - 5) < 2(2x - 1)$
$$3x - 15 < 4x - 2$$
$$-15 < x - 2$$
$$-13 < x \quad \text{or} \quad x > -13$$

The solution set written in interval notation is $(-13, \infty)$.

39.
$$\frac{1}{2}+\frac{2}{3}\geq\frac{x}{6}$$
$$6\left(\frac{1}{2}+\frac{2}{3}\right)\geq 6\left(\frac{x}{6}\right)$$
$$3+4\geq x$$
$$7\geq x \quad \text{or} \quad x\leq 7$$
The solution set written in interval notation is $(-\infty, 7]$.

41.
$$4(x-1)\geq 4x-8$$
$$4x-4\geq 4x-8$$
$$4x-4-4x\geq 4x-8-4x$$
$$-4\geq -8$$
The statement is true for all real numbers. The solution set written in interval notation is $(-\infty, \infty)$.

43.
$$7x < 7(x-2)$$
$$7x < 7x-14$$
$$7x-7x < 7x-14-7x$$
$$0 < -14$$
The statement is false for all values of x. The solution set is $\varnothing$.

45.
$$4(2x+1) > 4$$
$$8x+4 > 4$$
$$8x > 0$$
$$\frac{8x}{8} > \frac{0}{8}$$
$$x > 0$$
The solution set written in interval notation is $(0, \infty)$.

47.
$$\frac{x+7}{5} > 1$$
$$5\left(\frac{x+7}{5}\right) > 5(1)$$
$$x+7 > 5$$
$$x > -2$$
The solution set written in interval notation is $(-2, \infty)$.

49.
$$\frac{-5x+11}{2}\leq 7$$
$$2\left(\frac{-5x+11}{2}\right)\leq 2(7)$$
$$-5x+11\leq 14$$
$$-5x\leq 3$$
$$\frac{-5x}{-5}\geq \frac{3}{-5}$$
$$x\geq -\frac{3}{5}$$
The solution set written in interval notation is $\left[-\frac{3}{5}, \infty\right)$.

51.
$$8x-16.4\leq 10x+2.8$$
$$-16.4\leq 2x+2.8$$
$$-19.2\leq 2x$$
$$\frac{-19.2}{2}\leq \frac{2x}{2}$$
$$-9.6\leq x \quad \text{or} \quad x\geq -9.6$$
The solution set written in interval notation is $[-9.6, \infty)$.

53.
$$2(x-3) > 70$$
$$2x-6 > 70$$
$$2x > 76$$
$$\frac{2x}{2} > \frac{76}{2}$$
$$x > 38$$
The solution set written in interval notation is $(38, \infty)$.

55. Answers may vary.

57.
$$-5x+4\leq -4(x-1)$$
$$-5x+4\leq -4x+4$$
$$4\leq x+4$$
$$0\leq x \quad \text{or} \quad x\geq 0$$
The solution set written in interval notation is $[0, \infty)$.

59. $\frac{1}{4}(x-7) \geq x+2$

$$4\left[\frac{1}{4}(x-7)\right] \geq 4(x+2)$$

$$x-7 \geq 4x+8$$

$$-7 \geq 3x+8$$

$$-15 \geq 3x$$

$$\frac{-15}{3} \geq \frac{3x}{3}$$

$$-5 \geq x \quad \text{or} \quad x \leq -5$$

The solution set written in interval notation is $(-\infty, -5]$.

61. $\frac{2}{3}(x+2) < \frac{1}{5}(2x+7)$

$$15\left[\frac{2}{3}(x+2)\right] < 15\left[\frac{1}{5}(2x+7)\right]$$

$$10x+20 < 6x+21$$

$$4x+20 < 21$$

$$4x < 1$$

$$\frac{4x}{4} < \frac{1}{4}$$

$$x < \frac{1}{4}$$

The solution set written in interval notation is $\left(-\infty, \frac{1}{4}\right)$.

63. $4(x-6)+2x-4 \geq 3(x-7)+10x$

$$4x-24+2x-4 \geq 3x-21+10x$$

$$6x-28 \geq 13x-21$$

$$-28 \geq 7x-21$$

$$-7 \geq 7x$$

$$\frac{-7}{7} \geq \frac{7x}{7}$$

$$-1 \geq x \quad \text{or} \quad x \leq -1$$

The solution set written in interval notation is $(-\infty, -1]$.

65. $\frac{5x+1}{7} - \frac{2x-6}{4} \geq -4$

$$28\left(\frac{5x+1}{7} - \frac{2x-6}{4}\right) \geq 28(-4)$$

$$4(5x+1)-7(2x-6) \geq -112$$

$$20x+4-14x+42 \geq -112$$

$$6x+46 \geq -112$$

$$6x \geq -158$$

$$\frac{6x}{6} \geq \frac{-158}{6}$$

$$x \geq -\frac{79}{3}$$

The solution set written in interval notation is $\left[-\frac{79}{3}, \infty\right)$.

67. $\frac{-x+2}{2} - \frac{1-5x}{8} < -1$

$$8\left(\frac{-x+2}{2} - \frac{1-5x}{8}\right) < 8(-1)$$

$$4(-x+2)-(1-5x) < -8$$

$$-4x+8-1+5x < -8$$

$$x+7 < -8$$

$$x < -15$$

The solution set written in interval notation is $(-\infty, -15)$.

69. $0.8x+0.6x \geq 4.2$

$$1.4x \geq 4.2$$

$$\frac{1.4x}{1.4} \geq \frac{4.2}{1.4}$$

$$x \geq 3$$

The solution set written in interval notation is $[3, \infty)$.

71.
$$\frac{x+5}{5} - \frac{3+x}{8} \geq -\frac{3}{10}$$
$$40\left(\frac{x+5}{5} - \frac{3+x}{8}\right) \geq 40\left(-\frac{3}{10}\right)$$
$$8(x+5) - 5(3+x) \geq 4(-3)$$
$$8x + 40 - 15 - 5x \geq -12$$
$$3x + 25 \geq -12$$
$$3x \geq -37$$
$$\frac{3x}{3} \geq \frac{-37}{3}$$
$$x \geq -\frac{37}{3}$$

The solution set written in interval

notation is $\left[-\frac{37}{3}, \infty\right)$.

73.
$$\frac{x+3}{12} + \frac{x-5}{15} < \frac{2}{3}$$
$$60\left(\frac{x+3}{12} + \frac{x-5}{15}\right) < 60\left(\frac{2}{3}\right)$$
$$5(x+3) + 4(x-5) < 20(2)$$
$$5x + 15 + 4x - 20 < 40$$
$$9x - 5 < 40$$
$$9x < 45$$
$$\frac{9x}{9} < \frac{45}{9}$$
$$x < 5$$

The solution set written in interval notation is $(-\infty, 5)$.

75. Let $x =$ her score on the final.
$$\frac{72 + 67 + 82 + 79 + 2x}{6} \geq 60$$
$$72 + 67 + 82 + 79 + 2x \geq 360$$
$$300 + 2x \geq 360$$
$$2x \geq 60$$
$$x \geq 30$$

Therefore, she must score at least 30 on the final exam.

77. Let $x =$ the weight of the luggage and cargo.
$$6(160) + x \leq 2000$$
$$960 + x \leq 2000$$
$$x \leq 1040$$
The plane can carry a maximum of 1040 pounds of luggage and cargo.

79. Let $x =$ the number of ounces.
$$0.33 + 0.22(x-1) \leq 4.00$$
$$100[0.33 + 0.22(x-1)] \leq 100(4.00)$$
$$33 + 22(x-1) \leq 400$$
$$33 + 22x - 22 \leq 400$$
$$22x + 11 \leq 400$$
$$22x \leq 389$$
$$x \leq 17.68$$
Thus, at most 17 whole ounces can be mailed for \$4.00.

81. Let $n =$ the number of calls made in a given month.
$$25 < 13 + 0.06n$$
$$12 < 0.06n$$
$$200 < n \quad \text{or} \quad n > 200$$
Therefore, Plan 1 is more economical than Plan 2 when more than 200 calls are made.

83. Given that $C \geq 500$, we use substitution in the following:
$$F \geq \frac{9}{5}C + 32$$
$$F \geq \frac{9}{5}(500) + 32$$
$$F \geq 900 + 32$$
$$F \geq 932$$
Glass is a liquid at temperatures of $932°F$ or higher.

85. a. Let $s = 50,000$.
$$50,000 < 2806.6t + 32,558$$
$$17,442 < 2806.6t$$
$$6.21 < t$$
Beginning salaries will be greater than \$50,000 in
$$1995 + 6.21 = 2001.21, \text{ or in } 2001.$$

b. Answers may vary.

87. a. The consumption of whole milk is decreasing. The graph of the line is going down over time.

b. The consumption of skim milk is increasing. The graph of the line is going up over time.

c. $t = 2005 - 1994 = 11$
$w = -0.18t + 8.72$
$w = -0.18(11) + 8.72$
$w = 6.74$
The consumption of whole milk will be about 6.74 gallons per person per year in 2005.

d. $t = 2005 - 1994 = 11$
$s = 0.26t + 5.86$
$s = 0.26(11) + 5.86$
$s = 8.72$
The consumption of skim milk will be about 8.72 gallons per person per year in 2005.

89. Define $y_1 = -0.18x + 8.72$ and
$y_2 = 0.26x + 5.86$ and graph in a standard window.

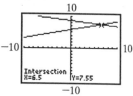

The point of intersection is (6.5, 7.55). The consumption of whole milk will be the same as the consumption of skim milk in $1994 + 7 = 2001$.

91. a.
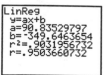

The linear regression equation is $y = 90.835x - 349.646$.

b. $x = 2005 - 1980 = 25$

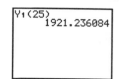

Approximately 1,921,000 students will be home educated in 2005.

c. From the regression equation, the slope is 90.835. Therefore, the number of students is increasing at the rate of 90,835 per year.

93. $x \geq 0$ and $x \leq 7$ describes the integers less than or equal to 7 and greater than or equal to 0. The integers are 0, 1, 2, 3, 4, 5, 6, 7 or {0, 1, 2, 3, 4, 5, 6, 7}.

95. $x < 6$ and $x < -5$ describes the integers less than 6 and less than −5. The integers are −6, −7, −8, −9,... or {..., −9, −8, −7, −6}.

97. The solution set written in interval notation is (−7, 1].

-7 1

99. The solution set written in interval notation is [−2.5, 5.3).

-2.5 5.3

Exercise Set 3.3

1. The numbers that are in either set or both sets are {2, 3, 4, 5, 6, 7}.

3. The numbers 4 and 6 are in both sets. The intersection is {4, 6}.

5. The numbers that are in either set or both sets are {..., −2, −1, 0, 1,...}.

7. The numbers 5 and 7 are in both sets. The intersection is {5, 7}.

9. The numbers that are in either set or both sets are
$\{x | x$ is an odd integer or $x = 2$ or $x = 4\}$.

11. The numbers 2 and 4 are in both sets. The intersection is {2, 4}.

13. Graph the two intervals and find their intersection.

$\{x|x < 5\}$

$\{x|x > -2\}$

$\{x|x < 5 \text{ and } x > -2\}$

The solution set is $(-2, 5)$.

15. $x + 1 \geq 7$ and $3x - 1 \geq 5$

$\quad\quad x \geq 6$ and $\quad\quad 3 \geq 6$

$\quad\quad x \geq 6$ and $\quad\quad x \geq 2$

Graph the two intervals and find their intersection.

$\{x|x \geq 6\}$

$\{x|x \geq 2\}$

$\{x|x \geq 6 \text{ and } x \geq 2\}$

$= \{x|x \geq 6\}$

The solution set is $[6, \infty)$.

17. $4x + 2 \leq -10$ and $2x \leq 0$

$\quad\quad 4x \leq -12$ and $\quad x \leq 0$

$\quad\quad\; x \leq -3$ and $\quad x \leq 0$

Graph the two intervals and find their intersection.

$\{x|x \leq -3\}$

$\{x|x \leq 0\}$

$\{x|x \leq -3 \text{ and } x \leq 0\}$

$= \{x|x \leq -3\}$

The solution set is $(-\infty, -3]$.

19. $\quad 5 < x - 6 < 11$

$\quad 5 + 6 < x - 6 + 6 < 11 + 6$

$\quad\quad 11 < x < 17$

The solution set written in interval notation is $(11, 17)$.

21. $\quad -2 \leq 3x - 5 \leq 7$

$\quad -2 + 5 \leq 3x - 5 + 5 \leq 7 + 5$

$\quad\quad\quad 3 \leq 3x \leq 12$

$\quad\quad\quad \dfrac{3}{3} \leq \dfrac{3x}{3} \leq \dfrac{12}{3}$

$\quad\quad\quad 1 \leq x \leq 4$

The solution set written in interval notation is $[1, 4]$.

23. $\quad 1 \leq \dfrac{2}{3}x + 3 \leq 4$

$\quad 3(1) \leq 3\left(\dfrac{2}{3}x + 3\right) \leq 3(4)$

$\quad\quad 3 \leq 2x + 9 \leq 12$

$\quad 3 - 9 \leq 2x + 9 - 9 \leq 12 - 9$

$\quad\quad -6 \leq 2x \leq 3$

$\quad\quad \dfrac{-6}{2} \leq \dfrac{2x}{2} \leq \dfrac{3}{2}$

$\quad\quad -3 \leq x \leq \dfrac{3}{2}$

The solution set written in interval notation is $\left[-3, \dfrac{3}{2}\right]$.

25. $\quad -5 \leq \dfrac{x+1}{4} \leq -2$

$\quad 4(-5) \leq 4\left(\dfrac{x+1}{4}\right) \leq 4(-2)$

$\quad\quad -20 \leq x + 1 \leq -8$

$\quad -20 - 1 \leq x + 1 - 1 \leq -8 - 1$

$\quad\quad -21 \leq x \leq -9$

The solution set written in interval notation is $[-21, -9]$.

27. Graph the two intervals and find their union.

$\{x|x < -1\}$

$\{x|x > 0\}$

$\{x|x < -1 \text{ or } x > 0\}$

The solution set is $(-\infty, -1) \cup (0, \infty)$.

29. $-2x \le -4$ or $5x - 20 \ge 5$

$\qquad x \ge 2 \quad$ or $\qquad 5x \ge 25$

$\qquad x \ge 2 \quad$ or $\qquad x \ge 5$

Graph the two intervals and find their union.

$\{x|x \ge 2\}$

$\{x|x \ge 5\}$

$\{x|x \ge 2 \text{ or } x \ge 5\}$

The solution set is $[2, \infty)$.

31. $3(x-1) < 12$ or $x + 7 > 10$

$\qquad x - 1 < 4 \quad$ or $\qquad x > 3$

$\qquad x < 5 \quad$ or $\qquad x > 3$

Graph the two intervals and find their union.

$\{x|x < 5\}$

$\{x|x > 3\}$

$\{x|x < 5 \text{ or } x > 3\}$

The solution set is $(-\infty, \infty)$.

33. Answers may vary.

35. Graph the two intervals and find their intersection.

$\{x|x < 2\}$

$\{x|x > -1\}$

$\{x|x < 2 \text{ and } x > -1\}$

The solution set is $(-1, 2)$.

37. Graph the two intervals and find their union.

$\{x|x < 2\}$

$\{x|x > -1\}$

$\{x|x < 2 \text{ or } x > -1\}$

The solution set is $(-\infty, \infty)$.

39. Graph the two intervals and find their intersection.

$\{x|x \ge -5\}$

$\{x|x \ge -1\}$

$\{x|x \ge -5 \text{ and } x \ge -1\}$

The solution set is $[-1, \infty)$.

41. Graph the two intervals and find their union.

$\{x|x \ge -5\}$

$\{x|x \ge -1\}$

$\{x|x \ge -5 \text{ or } x \ge -1\}$

The solution set is $[-5, \infty)$.

43. $\qquad 0 \le 2x - 3 \le 9$

$0 + 3 \le 2x - 3 + 3 \le 9 + 3$

$\qquad 3 \le 2x \le 12$

$\qquad \dfrac{3}{2} \le \dfrac{2x}{2} \le \dfrac{12}{2}$

$\qquad \dfrac{3}{2} \le x \le 6$

The solution set written in interval notation is $\left[\dfrac{3}{2},\ 6\right]$.

45. $\quad \dfrac{1}{2} < x - \dfrac{3}{4} < 2$

$$\dfrac{1}{2} + \dfrac{3}{4} < x - \dfrac{3}{4} + \dfrac{3}{4} < 2 + \dfrac{3}{4}$$

$$\dfrac{5}{4} < x < \dfrac{11}{4}$$

The solution set written in interval

notation is $\left(\dfrac{5}{4}, \ \dfrac{11}{4} \right)$.

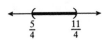

$\qquad \dfrac{5}{4} \qquad \dfrac{11}{4}$

47. $\quad x + 3 \geq 3 \quad$ and $\quad x + 3 \leq 2$

$\qquad x \geq 0 \quad$ and $\qquad x \leq -1$

Graph the two intervals and find their intersection.

$\left\{ x \mid x \geq 0 \right\}$

$\left\{ x \mid x \leq -1 \right\}$

$\left\{ x \mid x \geq 0 \text{ and } x \leq -1 \right\}$

The solution set is $\varnothing$.

49. $\quad 3x \geq 5 \quad$ or $\quad -x - 6 < 1$

$\qquad x \geq \dfrac{5}{3} \quad$ or $\qquad -x < 7$

$\qquad x \geq \dfrac{5}{3} \quad$ or $\qquad x > -7$

Graph the two intervals and find their union

$\left\{ x \mid x \geq \dfrac{5}{3} \right\}$

$\left\{ x \mid x > -7 \right\}$

$\left\{ x \mid x \geq \dfrac{5}{3} \text{ or } x > -7 \right\}$

The solution set is $(-7, \infty)$.

51. $\quad 0 < \dfrac{5 - 2x}{3} < 5$

$$3(0) < 3\left(\dfrac{5 - 2x}{3} \right) < 3(5)$$

$$0 < 5 - 2x < 15$$

$$0 - 5 < 5 - 2x - 5 < 15 - 5$$

$$-5 < -2x < 10$$

$$\dfrac{-5}{-2} > \dfrac{-2x}{-2} > \dfrac{10}{-2}$$

$$\dfrac{5}{2} > x > -5$$

$$-5 < x < \dfrac{5}{2}$$

The solution set written in interval

notation is $\left(-5, \dfrac{5}{2} \right)$.

$\qquad -5 \qquad \dfrac{5}{2}$

53. $\quad -6 < 3(x - 2) \leq 8$

$$-6 < 3x - 6 < 8$$

$$-6 + 6 < 3x - 6 + 6 \leq 8 + 6$$

$$0 < 3x \leq 14$$

$$\dfrac{0}{3} < \dfrac{3x}{3} \leq \dfrac{14}{3}$$

$$0 < x \leq \dfrac{14}{3}$$

The solution set written in interval

notation is $\left(0, \ \dfrac{14}{3} \right]$.

$\qquad 0 \qquad \dfrac{14}{3}$

55. $\quad -x + 5 > 6 \quad$ and $\quad 1 + 2x \leq -5$

$\qquad -x > 1 \quad$ and $\qquad 2x \leq -6$

$\qquad x < -1 \quad$ and $\qquad x \leq -3$

Graph the two intervals and find their intersection.

$\left\{ x \mid x < -1 \right\}$

$\left\{ x \mid x \leq -3 \right\}$

$\left\{ x \mid x < -1 \text{ and } x \leq -3 \right\}$

The solution set is $\left(-\infty, -3 \right]$.

57. $3x + 2 \leq 5$ or $7x > 29$

$$3x \leq 3 \quad \text{or} \quad x > \frac{29}{7}$$

$$x \leq 1 \quad \text{or} \quad x > \frac{29}{7}$$

Graph the two intervals and find their union.

$\{x | x \leq 1\}$

$\left\{ x \middle| x > \frac{29}{7} \right\}$

$\left\{ x \middle| x \leq 1 \text{ or } x > \frac{29}{7} \right\}$

The solution set is $\left(-\infty, 1 \right] \cup \left(\frac{29}{7}, \infty \right)$.

59.

$$-\frac{1}{2} \leq \frac{4x-1}{6} < \frac{5}{6}$$

$$6\left(-\frac{1}{2}\right) \leq 6\left(\frac{4x-1}{6}\right) < 6\left(\frac{5}{6}\right)$$

$$-3 \leq 4x - 1 < 5$$

$$-3 + 1 \leq 4x - 1 + 1 < 5 + 1$$

$$-2 \leq 4x < 6$$

$$\frac{-2}{4} \leq \frac{4x}{4} < \frac{6}{4}$$

$$-\frac{1}{2} \leq x < \frac{3}{2}$$

The solution set written in interval notation is $\left[-\frac{1}{2}, \frac{3}{2} \right)$.

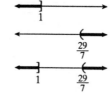

61.

$$0.3 < 0.2x - 0.9 < 1.5$$

$$0.3 + 0.9 < 0.2x - 0.9 + 0.9 < 1.5 + 0.9$$

$$1.2 < 0.2x < 2.4$$

$$\frac{1.2}{0.2} < \frac{0.2x}{0.2} < \frac{2.4}{0.2}$$

$$6 < x < 12$$

The solution set written in interval notation is $(6, 12)$.

63. The intersection point of y_1 and y_2 is $(2.5, 0)$. The intersection point of y_2 and y_3 is $(-5, 5)$.

   a. $y_1 < y_2 < y_3$ is true for x-values between -5 and 2.5, or in interval notation $(-5, 2.5)$.

   b. $y_2 < y_1$ or $y_2 > y_3$ is true for x-values less than -5 or greater than 2.5, or in interval notation $(-\infty, -5) \cup (2.5, \infty)$.

65. The intersection point of y_1 and y_2 is $(2, -6)$. The intersection point of y_2 and y_3 is $(9, 8)$.

   a. $y_1 \leq y_2 \leq y_3$ is true for x-values equal to and between 2 and 9, or in interval notation $[2, 9]$.

   b. $y_2 \leq y_1$ or $y_2 \geq y_3$ is true for x-values less than or equal to 2 or greater than or equal to 9, or in interval notation $(-\infty, 2] \cup [9, \infty)$.

67.

$$-29 \leq C \leq 35$$

$$-29 \leq \frac{5}{9}(F - 32) \leq 35$$

$$9(-29) \leq 9\left[\frac{5}{9}(F - 32)\right] \leq 9(35)$$

$$-261 \leq 5F - 160 \leq 315$$

$$-261 + 160 \leq 5F - 160 + 160 \leq 315 + 160$$

$$-101 \leq 5F \leq 475$$

$$\frac{101}{5} \leq \frac{5F}{5} \leq \frac{475}{5}$$

$$-20.2 \leq F \leq 95$$

The temperatures ranged from $-20.2°$ to $95°$ Fahrenheit.

69. Let x = Christian's score on the final exam.

$$70 \le \frac{68 + 65 + 75 + 78 + 2x}{6} \le 79$$

$$420 \le 286 + 2x \le 474$$
$$134 \le 2x \le 188$$
$$67 \le x \le 94$$

If Christian scores between 67 and 94 inclusive on his final exam, he will receive a C in the course.

71. The years that the consumption of pork was greater than 48 pounds per person were 1992, 1993, 1994, and 1995. The years that the consumption of chicken was greater than 48 pounds per person were 1993, 1994, 1995, 1996, and 1997. The years in common are 1993, 1994, and 1995.

73. $|-7| - |19| = 7 - 19 = -12$

75. $-(-6) - |-10| = 6 - 10 = -4$

77. $|x| = 7$
$x = -7, 7$
The solutions are -7 and 7.

79. $|x| = 0$
$x = 0$
The solution is 0.

81. $2x - 3 < 3x + 1$　and　$3x + 1 < 4x - 5$
$　　-3 < x + 1$　and　$　　1 < x - 5$
$　　-4 < x$　　and　$　　6 < x$
$　　x > -4$　and　$　　x > 6$

$x > -4$

$x > 6$

$x > 6$　or　$(6, \infty)$

83. $-3(x - 2) \le 3 - 2x$　and　$3 - 2x \le 10 - 3x$
$-3x + 6 \le 3 - 2x$　and　$3 + x \le 10$
$　　6 \le 3 + x$　and　$　　x \le 7$
$　　3 \le x$　　and　$　　x \le 7$
$　　x \ge 3$　　and　$　　x \le 7$

$x \ge 3$

$x \le 7$

$3 \le x \le 7$　or　$[3, 7]$

85. $5x - 8 < 2(2 + x)$　and　$2(2 + x) < -2(1 + 2x)$
$5x - 8 < 4 + 2x$　and　$4 + 2x < -2 - 4x$
$3x - 8 < 4$　　and　$4 + 6x < -2$
$　　3x < 12$　and　$　　6x < -6$
$　　x < 4$　　and　$　　x < -1$

$x < 4$

$x < -1$

$x < -1$　or　$(-\infty, -1)$

Section 3.4

Mental Math

1. $|-7| = 7$

2. $|-8| = 8$

3. $-|5| = -5$

4. $-|10| = -10$

5. $-|-6| = -6$

6. $-|-3| = -3$

7. $|-3| + |-2| + |-7| = 3 + 2 + 7 = 12$

8. $|-1| + |-6| + |-8| = 1 + 6 + 8 = 15$

Exercise Set 3.4

1. $|x| = 7$
$x = 7$ or $x = -7$
The solutions are 7 and –7.

3. $|3x| = 12.6$
$3x = 12.6$ or $3x = -12.6$
$x = 4.2$ or $x = -4.2$
The solutions are 4.2 and –4.2.

5. $|2x - 5| = 9$
$2x - 5 = 9$ or $2x - 5 = -9$
$2x = 14$ or $2x = -4$
$x = 7$ or $x = -2$
The solutions are 7 and –2.

7. $\left|\dfrac{x}{2} - 3\right| = 1$

$\dfrac{x}{2} - 3 = 1$ or $\dfrac{x}{2} - 3 = -1$

$2\left(\dfrac{x}{2} - 3\right) = 2(1)$ or $2\left(\dfrac{x}{2} - 3\right) = 2(-1)$

$x - 6 = 2$ or $x - 6 = -2$
$x = 8$ or $x = 4$
The solutions are 8 and 4.

9. $|z| + 4 = 9$
$|z| = 5$
$z = 5$ or $z = -5$
The solutions are 5 and –5.

11. $|3x| + 5 = 14$
$|3x| = 9$
$3x = 9$ or $3x = -9$
$x = 3$ or $x = -3$
The solutions are 3 and –3.

13. $|2x| = 0$
$2x = 0$
$x = 0$
The solution is 0.

15. $|4n + 1| + 10 = 4$
$|4n + 1| = -6$
The absolute value of any expression is never negative, so no solution exists. The solution set is $\varnothing$.

17. $|5x - 1| = 0$
$5x - 1 = 0$
$5x = 1$
$x = \dfrac{1}{5}$
The solution is $\dfrac{1}{5}$.

19. Translating directly, we get the equation $|x| = 5$.

21. $|5x - 7| = |3x + 11|$
$5x - 7 = 3x + 11$ or $5x - 7 = -(3x + 11)$
$2x - 7 = 11$ or $5x - 7 = -3x - 11$
$2x = 18$ or $8x - 7 = -11$
$x = 9$ or $8x = -4$
$x = 9$ or $x = -\dfrac{1}{2}$

The solutions are 9 and $-\dfrac{1}{2}$.

23. $|z + 8| = |z - 3|$
$z + 8 = z - 3$ or $z + 8 = -(z - 3)$
$z + 8 - z = z - 3 - z$ or $z + 8 = -z + 3$
$8 = -3$ or $2z + 8 = 3$
false   or $2z = -5$
$z = -\dfrac{5}{2}$

The only solution is $-\dfrac{5}{2}$.

25. Answers may vary.

27. The points of intersection are (–1, 5) and (4, 5). Therefore, the solution set is $\{-1, 4\}$.

29. The point of intersection is (2.5, 1.5). Therefore, the solution set is $\{2.5\}$.

31. $|x| = 4$

$x = 4$ or $x = -4$

The solutions are 4 and –4.

33. The absolute value of a number is never negative, so no solution exists. The solution set is $\varnothing$.

35. $|7 - 3x| = 7$

$7 - 3x = 7$ or $7 - 3x = -7$

$-3x = 0$ or $-3x = -14$

$x = 0$ or $x = \dfrac{14}{3}$

The solutions are 0 and $\dfrac{14}{3}$.

37. $|6x| - 1 = 11$

$|6x| = 12$

$6x = 12$ or $6x = -12$

$x = 2$ or $x = -2$

The solutions are 2 and –2.

39. $|x - 3| + 3 = 7$

$|x - 3| = 4$

$x - 3 = 4$ or $x - 3 = -4$

$x = 7$ or $x = -1$

The solutions are 7 and –1.

41. The absolute value of any expression is never negative, so no solution exists. The solution set is $\varnothing$.

43. The absolute value of any expression is never negative, so no solution exists. The solution set is $\varnothing$.

45. $|8n + 1| = 0$

$8n + 1 = 0$

$8n = -1$

$n = -\dfrac{1}{8}$

The solution is $-\dfrac{1}{8}$.

47. $|1 + 6c| - 7 = -3$

$|1 + 6c| = 4$

$1 + 6c = 4$ or $1 + 6c = -4$

$6c = 3$ or $6c = -5$

$c = \dfrac{1}{2}$ or $c = -\dfrac{5}{6}$

The solutions are $\dfrac{1}{2}$ and $-\dfrac{5}{6}$.

49. $|5x + 1| = 11$

$5x + 1 = 11$ or $5x + 1 = -11$

$5x = 10$ or $5x = -12$

$x = 2$ or $x = -\dfrac{12}{5}$

The solutions are 2 and $-\dfrac{12}{5}$.

51. $|4x - 2| = |-10|$

$|4x - 2| = 10$

$4x - 2 = 10$ or $4x - 2 = -10$

$4x = 12$ or $4x = -8$

$x = 3$ or $x = -2$

The solutions are 3 and –2.

53. $|5x + 1| = |4x - 7|$

$5x + 1 = 4x - 7$ or $5x + 1 = -(4x - 7)$

$x + 1 = -7$ or $5x + 1 = -4x + 7$

$x = -8$ or $9x + 1 = 7$

$x = -8$ or $9x = 6$

$x = -8$ or $x = \dfrac{2}{3}$

The solutions are –8 and $\dfrac{2}{3}$.

55. $|6 + 2x| = -|-7|$

$|6 + 2x| = -7$

The absolute value of any expression is never negative, so no solution exists. The solution set is $\varnothing$.

57. $|2x - 6| = |10 - 2x|$

$2x - 6 = 10 - 2x$ or $\quad 2x - 6 = -(10 - 2x)$

$4x - 6 = 10$ or $\quad 2x - 6 = -10 + 2x$

$\qquad 4x = 16$ or $2x - 6 - 2x = -10 + 2x - 2x$

$\qquad x = 4$ or $\qquad -6 = -10$

$\qquad\qquad\qquad\qquad\qquad$ false

The only solution is 4.

59. $\left|\dfrac{2x - 5}{3}\right| = 7$

$\dfrac{2x - 5}{3} = 7$ or $\dfrac{2x - 5}{3} = -7$

$2x - 5 = 21$ or $2x - 5 = -21$

$\quad 2x = 26$ or $\quad 2x = -16$

$\qquad x = 13$ or $\qquad x = -8$

The solutions are 13 and –8.

61. $2 + |5n| = 17$

$\quad |5n| = 15$

$5n = 15$ or $5n = -15$

$n = 3$ or $\quad n = -3$

The solutions are 3 and –3.

63. $\left|\dfrac{2x - 1}{3}\right| = |-5|$

$\left|\dfrac{2x - 1}{3}\right| = 5$

$\dfrac{2x - 1}{3} = 5$ or $\dfrac{2x - 1}{3} = -5$

$2x - 1 = 15$ or $2x - 1 = -15$

$\quad 2x = 16$ or $\quad 2x = -14$

$\qquad x = 8$ or $\qquad x = -7$

The solutions are 8 and –7.

65. $|2y - 3| = |9 - 4y|$

$2y - 3 = 9 - 4y$ or $2y - 3 = -(9 - 4y)$

$6y - 3 = 9$ or $2y - 3 = -9 + 4y$

$\quad 6y = 12$ or $-2y - 3 = -9$

$\qquad y = 2$ or $\quad -2y = -6$

$\qquad y = 2$ or $\qquad y = 3$

The solutions are 2 and 3.

67. $\left|\dfrac{3n + 2}{8}\right| = |-1|$

$\left|\dfrac{3n + 2}{8}\right| = 1$

$\dfrac{3n + 2}{8} = 1$ or $\dfrac{3n + 2}{8} = -1$

$3n + 2 = 8$ or $3n + 2 = -8$

$\quad 3n = 6$ or $\quad 3n = -10$

$\qquad n = 2$ or $\qquad n = -\dfrac{10}{3}$

The solutions are 2 and $-\dfrac{10}{3}$.

69. $|x + 4| = |7 - x|$

$x + 4 = 7 - x$ or $\quad x + 4 = -(7 - x)$

$2x + 4 = 7$ or $\quad x + 4 = -7 + x$

$\quad 2x = 3$ or $x + 4 - x = -7 + x - x$

$\qquad x = \dfrac{3}{2}$ or $\qquad 4 = -7$

$\qquad\qquad\qquad\qquad\qquad$ false

The only solution is $\dfrac{3}{2}$.

71. $\left|\dfrac{8c - 7}{3}\right| = -|-5|$

$\left|\dfrac{8c - 7}{3}\right| = -5$

The absolute value of any expression is never negative, so no solution exists. The solution set is $\varnothing$.

73. $|2.3x - 1.5| = 5$

$y_1 = |2.3x - 1.5|$

$y_2 = 5$

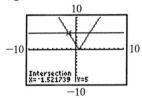

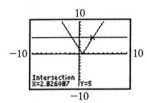

The graphs intersect when $x \approx -1.52$ and $x \approx 2.83$. The solutions are -1.52 and 2.83.

75. $3.6 - |4.1x - 2.6| = |x - 1.4|$

$y_1 = 3.6 - |4.1x - 2.6|$

$y_2 = |x - 1.4|$

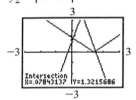

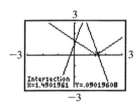

The graphs intersect when $x \approx 0.08$ and $x \approx 1.49$. The solutions are 0.08 and 1.49.

77. Answers may vary.

79. 13% of Disney's operating income came from consumer products.

81. 32%(3.4 billion) = 0.32(3.4 billion)
$$= 1.088 \text{ billion}$$
$1.088 billion is expected from the media networks segment.

83. Answers may vary.

$|x| \geq -2$

$-2, -1, 0, 1, 2$, for example

85. $|y| < 0$

There is no solution.

Exercise Set 3.5

1. $|x| \leq 4$

$-4 \leq x \leq 4$

The solution set is $[-4, 4]$.

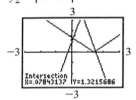

3. $|x - 3| < 2$

$-2 < x - 3 < 2$

$-2 + 3 < x - 3 + 3 < 2 + 3$

$1 < x < 5$

The solution set is $(1, 5)$.

5. $|x + 3| < 2$

$-2 < x + 3 < 2$

$-2 - 3 < x + 3 - 3 < 2 - 3$

$-5 < x < -1$

The solution set is $(-5, -1)$.

7. $|2x + 7| \leq 13$

$-13 \leq 2x + 7 \leq 13$

$-13 - 7 \leq 2x + 7 - 7 \leq 13 - 7$

$-20 \leq 2x \leq 6$

$-10 \leq x \leq 3$

The solution set is $[-10, 3]$.

9. $|x| + 7 \leq 12$

$|x| \leq 5$

$-5 \leq x \leq 5$

The solution set is $[-5, 5]$.

11. $|3x-1| < -5$

The absolute value of an expression is never negative, so no solution exists. The solution set is $\varnothing$.

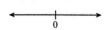

13. $|x-6|-7 \le -1$

$|x-6| \le 6$

$-6 \le x-6 \le 6$

$-6+6 \le x-6+6 \le 6+6$

$0 \le x \le 12$

The solution set is $[0, 12]$.

15. $|x| > 3$

$x < -3$ or $x > 3$

The solution set is $(-\infty, -3) \cup (3, \infty)$.

17. $|x+10| \ge 14$

$x+10 \le -14$ or $x+10 \ge 14$

$x+10-10 \le -14-10$ or $x+10-10 \ge 14-10$

$x \le -24$ or $x \ge 4$

The solution set is $(-\infty, -24] \cup [4, \infty)$.

19. $|x|+2 > 6$

$|x| > 4$

$x < -4$ or $x > 4$

The solution set is $(-\infty, -4) \cup (4, \infty)$.

21. $|5x| > -4$

An absolute value is always greater than a negative number. The solution set is $(-\infty, \infty)$.

23. $|6x-8|+3 > 7$

$|6x-8| > 4$

$6x-8 < -4$ or $6x-8 > 4$

$6x-8+8 < -4+8$ or $6x-8+8 > 4+8$

$6x < 4$ or $6x > 12$

$x < \dfrac{2}{3}$ or $x > 2$

The solution set is $\left(-\infty, \dfrac{2}{3}\right) \cup (2, \infty)$.

25. $|x| \le 0$

$|x| = 0$

$x = 0$

The solution set is $\{0\}$.

27. $|8x+3| > 0$ only excludes $|8x+3| = 0$

$8x+3 = 0$

$8x = -3$

$x = -\dfrac{3}{8}$

The solution is all real numbers except $-\dfrac{3}{8}$.

The solution set is $\left(-\infty, -\dfrac{3}{8}\right) \cup \left(-\dfrac{3}{8}, \infty\right)$.

29. Translating directly, we get the inequality $|x| < 7$.

31. From the property, we have $|x| \le 5$.

33. $|x| \le 2$

$-2 \le x \le 2$

The solution set is $[-2, 2]$.

35. $|y| > 1$

$y < -1$ or $y > 1$
The solution set is $(-\infty, -1) \cup (1, \infty)$.

37. $|x - 3| < 8$

$-8 < x - 3 < 8$

$-8 + 3 < x - 3 + 3 < 8 + 3$

$-5 < x < 11$
The solution set is $(-5, 11)$.

39. $|0.6x - 3| > 0.6$

$0.6x - 3 < -0.6$ or $0.6x - 3 > 0.6$

$0.6x - 3 + 3 < -0.6 + 3$ or $0.6x - 3 + 3 > 0.6 + 3$

$0.6x < 2.4$ or $0.6x > 3.6$

$x < 4$ or $x > 6$
The solution set is $(-\infty, 4) \cup (6, \infty)$.

41. $5 + |x| \le 2$

$|x| \le -3$

An absolute value is never negative, so no solution exists. The solution set is $\varnothing$.

43. $|x| > -4$

An absolute value is always greater than or equal to 0. The solution set is $(-\infty, \infty)$.

45. $|2x - 7| \le 11$

$-11 \le 2x - 7 \le 11$

$-11 + 7 \le 2x - 7 + 7 \le 11 + 7$

$-4 \le 2x \le 18$

$-2 \le x \le 9$
The solution set is $[-2, 9]$.

47. $|x+5|+2 \geq 8$

$\qquad |x+5| \geq 6$

$\qquad x+5 \leq -6 \qquad$ or $\qquad x+5 \geq 6$

$\qquad x+5-5 \leq -6-5 \quad$ or $\quad x+5-5 \geq 6-5$

$\qquad\quad x \leq -11 \qquad$ or $\qquad\quad x \geq 1$

The solution set is $(-\infty, -11] \cup [1, \infty)$.

49. $|x| > 0$ excludes only $|x| = 0$, or $x = 0$. The solution is all real numbers except $x = 0$. The solution set is $(-\infty, 0) \cup (0, \infty)$.

51. $9 + |x| > 7$

$\qquad |x| > -2$

An absolute value is always greater than or equal to 0. The solution set is $(-\infty, \infty)$.

53. $6 + |4x - 1| \leq 9$

$\qquad |4x - 1| \leq 3$

$\qquad -3 \leq 4x - 1 \leq 3$

$\qquad -3 + 1 \leq 4x - 1 + 1 \leq 3 + 1$

$\qquad\quad -2 \leq 4x \leq 4$

$\qquad\quad -\dfrac{1}{2} \leq x \leq 1$

The solution set is $\left[-\dfrac{1}{2}, \ 1\right]$.

55. $\left|\dfrac{2}{3}x + 1\right| > 1$

$\qquad \dfrac{2}{3}x + 1 < -1 \qquad$ or $\qquad \dfrac{2}{3}x + 1 > 1$

$\qquad 3\left(\dfrac{2}{3}x + 1\right) < 3(-1) \quad$ or $\quad 3\left(\dfrac{2}{3}x + 1\right) > 3(1)$

$\qquad\quad 2x + 3 < -3 \qquad$ or $\qquad 2x + 3 > 3$

$\qquad 2x + 3 - 3 < -3 - 3 \quad$ or $\quad 2x + 3 - 3 > 3 - 3$

$\qquad\qquad 2x < -6 \qquad$ or $\qquad\qquad 2x > 0$

$\qquad\qquad x < -3 \qquad$ or $\qquad\qquad x > 0$

The solution set is $(-\infty, -3) \cup (0, \infty)$.

57. $|5x + 3| < -6$

The absolute value of an expression is never negative, so no solution exists. The solution set is $\varnothing$.

59. $|8x + 3| \geq 0$

An absolute value is always greater than or equal to 0. The solution set is $(-\infty, \infty)$.

61. $|1 + 3x| + 4 < 5$

$\qquad |1 + 3x| < 1$

$\qquad -1 < 1 + 3x < 1$

$\qquad -1 - 1 < 1 + 3x - 1 < 1 - 1$

$\qquad\quad -2 < 3x < 0$

$\qquad\quad -\dfrac{2}{3} < x < 0$

The solution set is $\left(-\dfrac{2}{3}, \ 0\right)$.

63. $|x| - 3 \geq -3$

$\qquad |x| \geq 0$

An absolute value is always greater than or equal to 0. The solution set is $(-\infty, \infty)$.

65. $|8x| - 10 > -2$

$\qquad |8x| > 8$

$8x < -8 \quad$ or $\quad 8x > 8$

$x < -1 \quad$ or $\quad x > 1$

The solution set is $(-\infty, -1) \cup (1, \infty)$.

67. $\left| \dfrac{x+6}{3} \right| > 2$

$\dfrac{x+6}{3} < -2 \qquad$ or $\qquad \dfrac{x+6}{3} > 2$

$x + 6 < -6 \qquad$ or $\qquad x + 6 > 6$

$x + 6 - 6 < -6 - 6 \quad$ or $\quad x + 6 - 6 > 6 - 6$

$\qquad x < -12 \qquad$ or $\qquad x > 0$

The solution set is $(-\infty, -12) \cup (0, \infty)$.

69. $-15 + |2x - 7| \le -6$

$\qquad |2x - 7| \le 9$

$-9 \le 2x - 7 \le 9$

$-9 + 7 \le 2x - 7 + 7 \le 9 + 7$

$-2 \le 2x \le 16$

$-1 \le x \le 8$

The solution set is $[-1, 8]$.

71. $\left| 2x + \dfrac{3}{4} \right| - 7 \le -2$

$\left| 2x + \dfrac{3}{4} \right| \le 5$

$-5 \le 2x + \dfrac{3}{4} \le 5$

$4(-5) \le 4\left(2x + \dfrac{3}{4} \right) \le 4(5)$

$-20 \le 8x + 3 \le 20$

$-20 - 3 \le 8x + 3 - 3 \le 20 - 3$

$-23 \le 8x \le 17$

$-\dfrac{23}{8} \le x \le \dfrac{17}{8}$

The solution set is $\left[-\dfrac{23}{8}, \ \dfrac{17}{8} \right]$.

73. a. The points of intersection of y_1 and y_2 are $(-5, 6)$ and $(11, 6)$. The solution set consists of the *x*-values of the intersection points, or $\{-5, 11\}$.

b. The graph of y_1 is below the graph of y_2 for *x*-values between -5 and 11. The solution set is $(-5, 11)$.

c. The graph of y_1 is on or above the graph of y_2 for *x*-values less than or equal to -5 or greater than or equal to 11. The solution set is $(-\infty, -5] \cup [11, \infty)$.

75. a. The points of intersection of y_1 and y_2 are $(-8, -4)$ and $(4, -4)$. The solution set consists of the *x*-values of the intersection points, or $\{-8, 4\}$.

b. The graph of y_1 is on or below the graph of y_2 for *x*-values between, and including, -8 and 4. The solution set is $[-8, 4]$.

c. The graph of y_1 is above the graph of y_2 for *x*-values less than -8 or greater than 4. The solution set is $(-\infty, -8) \cup (4, \infty)$.

77. $|2x - 3| < 7$

$-7 < 2x - 3 < 7$

$-7 + 3 < 2x - 3 + 3 < 7 + 3$

$-4 < 2x < 10$

$-2 < x < 5$

The solution set is $(-2, 5)$.

79. $|2x - 3| = 7$

$2x - 3 = 7 \quad$ or $\quad 2x - 3 = -7$

$2x = 10 \quad$ or $\qquad 2x = -4$

$x = 5 \qquad$ or $\qquad x = -2$

The solutions are 5 and -2.

81. $|x - 5| \geq 12$

$$x - 5 \leq -12 \quad \text{or} \quad x - 5 \geq 12$$
$$x - 5 + 5 \leq -12 + 5 \quad \text{or} \quad x - 5 + 5 \geq 12 + 5$$
$$x \leq -7 \quad \text{or} \quad x \geq 17$$

The solution set is $(-\infty, -7] \cup [17, \infty)$.

83. $|9 + 4x| = 0$

$$9 + 4x = 0$$
$$4x = -9$$
$$x = -\frac{9}{4}$$

The solution is $-\frac{9}{4}$.

85. $|2x + 1| + 4 < 7$

$$|2x + 1| < 3$$
$$-3 < 2x + 1 < 3$$
$$-3 - 1 < 2x + 1 - 1 < 3 - 1$$
$$-4 < 2x < 2$$
$$-2 < x < 1$$

The solution set is $(-2, 1)$.

87. $|3x - 5| + 4 = 5$

$$|3x - 5| = 1$$
$$3x - 5 = -1 \quad \text{or} \quad 3x - 5 = 1$$
$$3x = 4 \quad \text{or} \quad 3x = 6$$
$$x = \frac{4}{3} \quad \text{or} \quad x = 2$$

The solutions are $\frac{4}{3}$ and 2.

89. $|x + 11| = -1$

The absolute value of any expression is never negative, so no solution exists. The solution set is $\varnothing$.

91. $\left|\dfrac{2x - 1}{3}\right| = 6$

$$\frac{2x - 1}{3} = -6 \quad \text{or} \quad \frac{2x - 1}{3} = 6$$
$$2x - 1 = -18 \quad \text{or} \quad 2x - 1 = 18$$
$$2x = -17 \quad \text{or} \quad 2x = 19$$
$$x = -\frac{17}{2} \quad \text{or} \quad x = \frac{19}{2}$$

The solutions are $-\dfrac{17}{2}$ and $\dfrac{19}{2}$.

93. $\left|\dfrac{3x - 5}{6}\right| > 5$

$$\frac{3x - 5}{6} < -5 \quad \text{or} \quad \frac{3x - 5}{6} > 5$$
$$3x - 5 + 5 < -30 + 5 \quad \text{or} \quad 3x - 5 + 5 > 30 + 5$$
$$3x - 5 < -30 \quad \text{or} \quad 3x - 5 > 30$$
$$3x < -25 \quad \text{or} \quad 3x > 35$$
$$x < -\frac{25}{3} \quad \text{or} \quad x > \frac{35}{3}$$

The solution set is $\left(-\infty, -\dfrac{25}{3}\right) \cup \left(\dfrac{35}{3}, \infty\right)$.

95. Answers may vary.

97. $|3.5 - x| < 0.05$

$$-0.05 < 3.5 - x < 0.05$$
$$-0.05 - 3.5 < 3.5 - x - 3.5 < 0.05 - 3.5$$
$$-3.55 < -x < -3.45$$
$$3.55 > x > 3.45$$
$$3.45 < x < 3.55$$

99. $P(\text{rolling a } 2) = \dfrac{1 \text{ way to occur}}{6 \text{ possible outcomes}} = \dfrac{1}{6}$

101. $P(\text{rolling a } 7) = \dfrac{0 \text{ ways to occur}}{6 \text{ possible outcomes}} = 0$

103. $P(\text{rolling a 1 or 3}) = \dfrac{2 \text{ ways to occur}}{6 \text{ possible outcomes}}$

$$= \frac{2}{6}$$
$$= \frac{1}{3}$$

105.
$$3x - 4y = 12$$
$$3(2) - 4y = 12$$
$$6 - 4y = 12$$
$$-4y = 6$$
$$y = -\frac{6}{4}$$
$$y = -\frac{3}{2} = -1.5$$

107.
$$3x - 4y = 12$$
$$3x - 4(-3) = 12$$
$$3x + 12 = 12$$
$$3x = 0$$
$$x = 0$$

Exercise Set 3.6

1. Graph $x = 2$ as a dashed line.
Test: (0, 0)
$0 < 2$ True
Shade the half-plane that contains (0, 0).

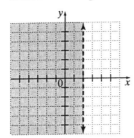

3. Graph $x - y = 7$ as a solid line.
Test: (0, 0)
$0 - 0 \geq 7$
   $0 \geq 7$ False
Shade the half-plane that does not contain (0, 0).

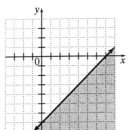

5. Graph $3x + y = 6$ as a dashed line.
Test: (0, 0)
$3(0) + 0 > 6$
   $0 > 6$ False
Shade the half-plane that does not contain (0, 0).

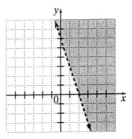

7. Graph $y = -2x$ as a solid line.
Test: (1, 1)
$1 \leq -2(1)$
$1 \leq -2$ False
Shade the half-plane that does not contain (1, 1).

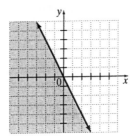

9. Graph $2x + 4y = 8$ as a solid line.
Test: (0, 0)
$2(0) + 4(0) \geq 8$
   $0 \geq 8$ False
Shade the half-plane that does not contain (0, 0).

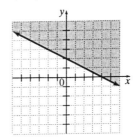

11. Graph $5x + 3y = -15$ as a dashed line.
Test: $(0, 0)$
$5(0) + 3(0) > -15$
$0 > -15$　True
Shade the half-plane that contains $(0, 0)$.

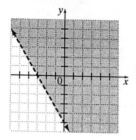

13. Answers may vary. A dashed boundary line should be used when the inequality contains a $<$ or $>$ symbol.

15. $x \geq 3$ and $y \leq -2$
Graph each inequality. The intersection of the two graphs is all points common to both regions, as shown by the heaviest shading in the graph.

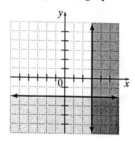

17. $x \leq -2$ or $y \geq 4$
Graph each inequality. The union of the two inequalities is both shaded regions, including the solid boundary lines shown in the graph.

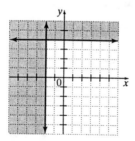

19. $x - y < 3$ and $x > 4$
Graph each inequality. The intersection of the two graphs is all points common to both regions, as shown by the heaviest shading in the graph.

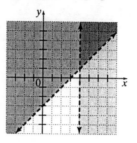

21. $x + y \leq 3$ or $x - y \geq 5$
Graph each inequality. The union of the two inequalities is both shaded regions, including the solid boundary lines shown in the graph.

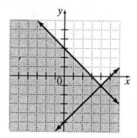

23. Graph $y = -2$ as a solid line.
Test: $(0, 0)$
$0 \geq -2$　True
Shade the half-plane that contains $(0, 0)$.

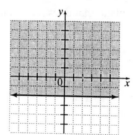

25. Graph $x - 6y = 12$ as a dashed line.
Test: $(0, 0)$
$0 - 6(0) < 12$
$\quad\quad 0 < 12 \quad$ True
Shade the half-plane that contains $(0, 0)$.

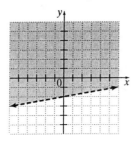

27. Graph $x = 5$ as a dashed line.
Test: $(0, 0)$
$0 > 5 \quad$ False
Shade the half-plane that does not contain $(0, 0)$.

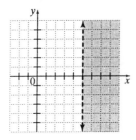

29. Graph $-2x + y = 4$ as a solid line.
Test: $(0, 0)$
$-2(0) + 0 \le 4$
$\quad\quad 0 \le 4 \quad$ True
Shade the half-plane that contains $(0, 0)$.

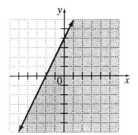

31. Graph $x - 3y = 0$ as a dashed line.
Test: $(0, 1)$
$0 - 3(1) < 0$
$\quad\quad -3 < 0 \quad$ True
Shade the half-plane that contains $(0, 1)$.

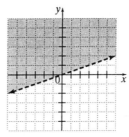

33. Graph $3x - 2y = 12$ as a solid line.
Test: $(0, 0)$
$3(0) - 2(0) \le 12$
$\quad\quad 0 \le 12 \quad$ True
Shade the half-plane that contains $(0, 0)$.

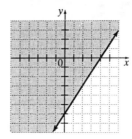

35. $x - y \ge 2$ or $y < 5$
Graph each inequality. The union of the two inequalities is both shaded regions, including the solid boundary line shown in the graph.

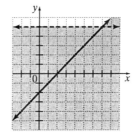

37. $x + y \le 1$ and $y \le -1$

Graph each inequallity. The intersection of the two graphs is all points common to both regions, as shown by the heaviest shading in the graph.

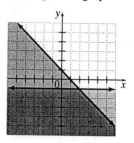

39. $2x + y > 4$ or $x \ge 1$

Graph each inequality. The union of the two inequalities is both shaded regions, including the solid boundary line shown in the graph.

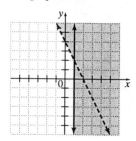

41. $x \ge -2$ and $x \le 1$

Graph each inequality. The intersection of the two graphs is all points common to both regions, as shown by the heaviest shading in the graph.

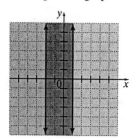

43. $x + y \le 0$ or $3x - 6y \ge 12$

Graph each inequality. The union of the two inequalities is both shaded regions, including the solid boundary lines shown in the graph.

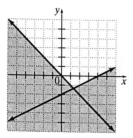

45. $2x - y > 3$ and $x \ge 0$

Graph each inequality. The intersection of the two graphs is all points common to both regions, as shown by the heaviest shading in the graph.

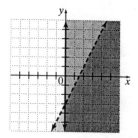

47. D; Solid line, half-plane shaded that contains $(0, 0)$.

49. A; Dashed line, half-plane shaded that does not contain $(0, 0)$.

51. Solid line at $x = 2$. Half-plane shaded that does not contain $(0, 0)$. Therefore, $x \ge 2$.

53. Solid line at $y = -3$. Half-plane shaded that does not contain $(0, 0)$. Therefore, $y \le -3$.

55. Dashed line at $y = 4$. Half-plane shaded that does not contain $(0, 0)$. Therefore, $y > 4$.

57. Dashed line at $x = 1$. Half-plane shaded that contains $(0, 0)$. Therefore, $x < 1$.

59. $x \leq 20$ and $y \geq 10$

Graph each inequality. The intersection of the two graphs is all points common to both regions, as shown by the heaviest shading in the graph.

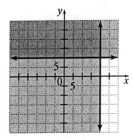

61. $\begin{cases} x \geq 0 \\ y \geq 0 \\ 2x + 4y \leq 40 \end{cases}$

Graph each inequality. The intersection of the three graphs is all points common to the three regions, as shown by the heaviest shading in the graph.

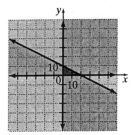

63. $3^2 = 3 \cdot 3 = 9$

65. $(-5)^2 = (-5)(-5) = 25$

67. $-2^4 = -2 \cdot 2 \cdot 2 \cdot 2 = -16$

69. $\left(\dfrac{2}{7}\right)^2 = \left(\dfrac{2}{7}\right)\left(\dfrac{2}{7}\right) = \dfrac{4}{49}$

71. Domain: $(-\infty, 2] \cup [2, \infty)$
Range: $(-\infty, \infty)$
This relation is not a function since it fails the vertical line test.

Chapter 3 Review

1. Algebraic solution:
$$4(x - 6) + 3 = 27$$
$$4x - 24 + 3 = 27$$
$$4x - 21 = 27$$
$$4x = 48$$
$$x = 12$$
Graphical solution:

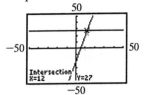

The solution is 12.

2. Algebraic solution:
$$15(x + 2) - 6 = 18$$
$$15x + 30 - 6 = 18$$
$$15x + 24 = 18$$
$$15x = -6$$
$$x = -\dfrac{2}{5}$$
Graphical solution:

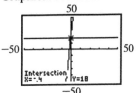

The solution is $-\dfrac{2}{5}$.

3. Algebraic solution:
$$5x + 15 = 3(x + 2) + 2(x - 3)$$
$$5x + 15 = 3x + 6 + 2x - 6$$
$$5x + 15 = 5x$$
$$5x + 15 - 5x = 5x - 5x$$
$$15 = 0$$
This equation is a false statement no matter what value the variable x might have.

Graphical solution:

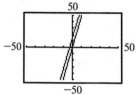

The lines are parallel. There is no solution. The solution set is ∅.

4. Algebraic solution:
$$2x - 5 + 3(x - 4) = 5(x + 2) - 27$$
$$2x - 5 + 3x - 12 = 5x + 10 - 27$$
$$5x - 17 = 5x - 17$$
Since both sides are the same, we see that replacing x with any real number will result in a true statement.

Graphical solution:

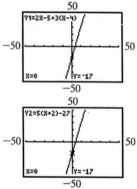

The lines are the same. The solution set is the set of all real numbers.

5. Algebraic solution:
$$14 - 2(x + 3) = 3(x - 9) + 18$$
$$14 - 2x - 6 = 3x - 27 + 18$$
$$8 - 2x = 3x - 9$$
$$8 - 5x = -9$$
$$-5x = -17$$
$$x = \frac{17}{5}$$

Graphical solution:

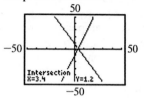

The solution is $\frac{17}{5}$.

6. Algebraic solution:
$$16 + 2(5 - x) = 19 - 3(x + 2)$$
$$16 + 10 - 2x = 19 - 3x - 6$$
$$26 - 2x = 13 - 3x$$
$$26 + x = 13$$
$$x = -13$$
Graphical solution:

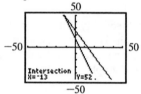

The solution is −13.

7. Define $y_1 = 0.4(x - 6)$ and $y_2 = \pi x + \sqrt{3}$ and graph in an integer window.

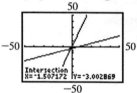

The solution is approximately −1.51.

8. Define $y_1 = 1.7x + \sqrt{7}$ and $y_2 = -0.4x - \sqrt{6}$ and graph in an integer window.

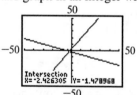

The solution is approximately −2.43.

9. $3(x-5) > -(x+3)$
$$3x - 15 > -x - 3$$
$$4x - 15 > -3$$
$$4x > 12$$
$$\frac{4x}{4} > \frac{12}{4}$$
$$x > 3$$
The solution set is $(3, \infty)$.

10. $-2(x+7) \geq 3(x+2)$
$$-2x - 14 \geq 3x + 6$$
$$-14 \geq 5x + 6$$
$$-20 \geq 5x$$
$$\frac{-20}{5} \geq \frac{5x}{5}$$
$$-4 \geq x \quad \text{or} \quad x \leq -4$$
The solution set is $(-\infty, -4]$.

11. $4x - (5+2x) < 3x - 1$
$$4x - 5 - 2x < 3x - 1$$
$$2x - 5 < 3x - 1$$
$$-5 < x - 1$$
$$-4 < x \quad \text{or} \quad x > -4$$
The solution set is $(-4, \infty)$.

12. $3(x-8) < 7x + 2(5-x)$
$$3x - 24 < 7x + 10 - 2x$$
$$3x - 24 < 5x + 10$$
$$-24 < 2x + 10$$
$$-34 < 2x$$
$$\frac{-34}{2} < \frac{2x}{2}$$
$$-17 < x \quad \text{or} \quad x > -17$$
The solution set is $(-17, \infty)$.

13. $24 \geq 6x - 2(3x-5) + 2x$
$$24 \geq 6x - 6x + 10 + 2x$$
$$24 \geq 10 + 2x$$
$$14 \geq 2x$$
$$\frac{14}{2} \geq \frac{2x}{2}$$
$$7 \geq x \quad \text{or} \quad x \leq 7$$
The solution set is $(-\infty, 7]$.

14. $48 + x \geq 5(2x+4) - 2x$
$$48 + x \geq 10x + 20 - 2x$$
$$48 + x \geq 8x + 20$$
$$48 \geq 7x + 20$$
$$28 \geq 7x$$
$$\frac{28}{7} \geq \frac{7x}{7}$$
$$4 \geq x \quad \text{or} \quad x \leq 4$$
The solution set is $(-\infty, 4]$.

15. $\dfrac{x}{3} + \dfrac{1}{2} > \dfrac{2}{3}$
$$6\left(\frac{x}{3} + \frac{1}{2}\right) > 6\left(\frac{2}{3}\right)$$
$$2x + 3 > 4$$
$$2x > 1$$
$$\frac{2x}{2} > \frac{1}{2}$$
$$x > \frac{1}{2}$$
The solution set is $\left(\dfrac{1}{2}, \infty\right)$.

16. $x + \dfrac{3}{4} < -\dfrac{x}{2} + \dfrac{9}{4}$
$$4\left(x + \frac{3}{4}\right) < 4\left(-\frac{x}{2} + \frac{9}{4}\right)$$
$$4x + 3 < -2x + 9$$
$$6x + 3 < 9$$
$$6x < 6$$
$$\frac{6x}{6} < \frac{6}{6}$$
$$x < 1$$
The solution set is $(-\infty, 1)$.

17.
$$\frac{x-5}{2} \leq \frac{3}{8}(2x+6)$$
$$8\left(\frac{x-5}{2}\right) \leq 8\left[\frac{3}{8}(2x+6)\right]$$
$$4(x-5) \leq 3(2x+6)$$
$$4x-20 \leq 6x+18$$
$$-20 \leq 2x+18$$
$$-38 \leq 2x$$
$$\frac{-38}{2} \leq \frac{2x}{2}$$
$$-19 \leq x \quad \text{or} \quad x \geq -19$$

The solution set is $[-19, \infty)$.

18.
$$\frac{3(x-2)}{5} > \frac{-5(x-2)}{3}$$
$$15\left[\frac{3(x-2)}{5}\right] > 15\left[\frac{-5(x-2)}{3}\right]$$
$$9(x-2) > -25(x-2)$$
$$9x-18 > -25x+50$$
$$34x-18 > 50$$
$$34x > 68$$
$$\frac{34x}{34} > \frac{68}{34}$$
$$x > 2$$

The solution set is $(2, \infty)$.

19. Let $n = $ the number of pounds of laundry.
$$25 < 0.9(10) + 0.8(n-10)$$
$$25 < 9 + 0.8n - 8$$
$$25 < 1 + 0.8n$$
$$24 < 0.8n$$
$$\frac{24}{0.8} < \frac{0.8n}{0.8}$$
$$30 < n \quad \text{or} \quad n > 30$$
It is more economical to use the housekeeper for more than 30 pounds of laundry per week.

20. $500 \leq F \leq 1000$
$$500 \leq \frac{9}{5}C + 32 \leq 1000$$
$$468 \leq \frac{9}{5}C \leq 968$$
$$260 \leq C \leq \frac{4840}{9}$$
So, rounded to the nearest degree, firing temperatures range from 260°C to 538°C.

21. Let $x = $ the minimum score that the last judge can give.
$$\frac{9.5 + 9.7 + 9.9 + 9.7 + 9.7 + 9.6 + 9.5 + x}{8} \geq 9.65$$
$$\frac{67.6 + x}{8} \geq 9.65$$
$$67.6 + x \geq 77.2$$
$$x \geq 9.6$$
The last judge must give Nana at least a 9.6 for her to win a silver medal.

22. Let $x = $ the amount she saves each summer.
$$4000 \leq 2x + 500 \leq 8000$$
$$3500 \leq 2x \leq 7500$$
$$\frac{3500}{2} \leq \frac{2x}{2} \leq \frac{7500}{2}$$
$$1750 \leq x \leq 3750$$
She must save between $1750 and $3750 each summer.

23.
$$1 \leq 4x - 7 \leq 3$$
$$1+7 \leq 4x - 7 + 7 \leq 3 + 7$$
$$8 \leq 4x \leq 10$$
$$\frac{8}{4} \leq \frac{4x}{4} \leq \frac{10}{4}$$
$$2 \leq x \leq \frac{5}{2}$$

The solution set is $\left[2, \frac{5}{2}\right]$.

24. $-2 \le 8 + 5x < -1$

$-2 - 8 \le 8 + 5x - 8 < -1 - 8$

$-10 \le 5x < -9$

$\dfrac{-10}{5} \le \dfrac{5x}{5} < \dfrac{-9}{5}$

$-2 \le x < -\dfrac{9}{5}$

The solution set is $\left[-2, \ -\dfrac{9}{5}\right)$.

25. $-3 < 4(2x - 1) < 12$

$-3 < 8x - 4 < 12$

$-3 + 4 < 8x - 4 + 4 < 12 + 4$

$1 < 8x < 16$

$\dfrac{1}{8} < \dfrac{8x}{8} < \dfrac{16}{8}$

$\dfrac{1}{8} < x < 2$

The solution set is $\left(\dfrac{1}{8}, \ 2\right)$.

26. $-6 < x - (3 - 4x) < -3$

$-6 < x - 3 + 4x < -3$

$-6 < 5x - 3 < -3$

$-6 + 3 < 5x - 3 + 3 < -3 + 3$

$-3 < 5x < 0$

$\dfrac{-3}{5} < \dfrac{5x}{5} < \dfrac{0}{5}$

$-\dfrac{3}{5} < x < 0$

The solution set is $\left(-\dfrac{3}{5}, \ 0\right)$.

27. $\dfrac{1}{6} < \dfrac{4x - 3}{3} \le \dfrac{4}{5}$

$30\left(\dfrac{1}{6}\right) < 30\left(\dfrac{4x - 3}{3}\right) \le 30\left(\dfrac{4}{5}\right)$

$5 < 10(4x - 3) \le 24$

$5 < 40x - 30 \le 24$

$5 + 30 < 40x - 30 + 30 \le 24 + 30$

$35 < 40x \le 54$

$\dfrac{35}{40} < \dfrac{40x}{40} \le \dfrac{54}{40}$

$\dfrac{7}{8} < x \le \dfrac{27}{20}$

The solution set is $\left(\dfrac{7}{8}, \ \dfrac{27}{20}\right]$.

28. $0 \le \dfrac{2(3x + 4)}{5} \le 3$

$5(0) \le 5\left[\dfrac{2(3x + 4)}{5}\right] \le 5(3)$

$0 \le 6x + 8 \le 15$

$0 - 8 \le 6x + 8 - 8 \le 15 - 8$

$-8 \le 6x \le 7$

$\dfrac{-8}{6} \le \dfrac{6x}{6} \le \dfrac{7}{6}$

$-\dfrac{4}{3} \le x \le \dfrac{7}{6}$

The solution set is $\left[-\dfrac{4}{3}, \ \dfrac{7}{6}\right]$.

29. Graph the two intervals and find their intersection.

$\{x \mid x \le 2\}$

$\{x \mid x > -5\}$

$\{x \mid x \le 2 \text{ and } x > -5\}$

The solution set is $(-5, 2]$.

30. Graph the two intervals and find their union.

$\{x \mid x \le 2\}$

$\{x \mid x > -5\}$

$\{x \mid x \le 2 \text{ or } x > -5\}$

The solution set is $(-\infty, \infty)$.

31. $3x - 5 > 6$ or $-x < -5$

$3x > 11$ or $x > 5$

$x > \dfrac{11}{3}$ or $x > 5$

Graph the two intervals and find their union.

$\left\{ x \middle| x > \dfrac{11}{3} \right\}$

$\left\{ x \middle| x > 5 \right\}$

$\left\{ x \middle| x > \dfrac{11}{3} \text{ or } x > 5 \right\}$

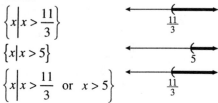

The solution set is $\left(\dfrac{11}{3}, \infty \right)$.

32. $-2x \le 6$ and $-2x + 3 < -7$

$x \ge -3$ and $-2x < -10$

$x \ge -3$ and $x > 5$

Graph the two intervals and find their intersection.

$\left\{ x \middle| x \ge -3 \right\}$

$\left\{ x \middle| x > 5 \right\}$

$\left\{ x \middle| x \ge -3 \text{ and } x > 5 \right\}$

The solution set is $(5, \infty)$.

33. $|x - 7| = 9$

$x - 7 = 9$ or $x - 7 = -9$

$x = 16$ or $x = -2$

The solutions are 16 and –2.

34. $|8 - x| = 3$

$8 - x = 3$ or $8 - x = -3$

$-x = -5$ or $-x = -11$

$x = 5$ or $x = 11$

The solutions are 5 and 11.

35. $|2x + 9| = 9$

$2x + 9 = 9$ or $2x + 9 = -9$

$2x = 0$ or $2x = -18$

$x = 0$ or $x = -9$

The solutions are 0 and –9.

36. $|-3x + 4| = 7$

$-3x + 4 = 7$ or $-3x + 4 = -7$

$-3x = 3$ or $-3x = -11$

$x = -1$ or $x = \dfrac{11}{3}$

The solutions are -1 and $\dfrac{11}{3}$.

37. $|3x - 2| + 6 = 10$

$|3x - 2| = 4$

$3x - 2 = 4$ or $3x - 2 = -4$

$3x = 6$ or $3x = -2$

$x = 2$ or $x = -\dfrac{2}{3}$

The solutions are 2 and $-\dfrac{2}{3}$.

38. $5 + |6x + 1| = 5$

$|6x + 1| = 0$

$6x + 1 = 0$

$6x = -1$

$x = -\dfrac{1}{6}$

The solution is $-\dfrac{1}{6}$.

39. The absolute value of any expression is never negative, so no solution exists. The solution set is $\varnothing$.

40. $|5 - 6x| + 8 = 3$

$|5 - 6x| = -5$

The absolute value of any expression is never negative, so no solution exists. The solution set is $\varnothing$.

41. $|7x| - 26 = -5$

$|7x| = 21$

$7x = 21$ or $7x = -21$

$x = 3$ or $x = -3$

The solutions are 3 and –3.

42. $-8 = |x - 3| - 10$
$2 = |x - 3|$
$x - 3 = 2$ or $x - 3 = -2$
$x = 5$ or $x = 1$
The solutions are 1 and 5.

43. $\left|\dfrac{3x - 7}{4}\right| = 2$

$\dfrac{3x - 7}{4} = 2$ or $\dfrac{3x - 7}{4} = -2$

$3x - 7 = 8$ or $3x - 7 = -8$

$3x = 15$ or $3x = -1$

$x = 5$ or $x = -\dfrac{1}{3}$

The solutions are 5 and $-\dfrac{1}{3}$.

44. The absolute value of any expression is never negative, so no solution exists. The solution set is $\varnothing$.

45. $|6x + 1| = |15 + 4x|$
$6x + 1 = 15 + 4x$ or $6x + 1 = -(15 + 4x)$
$2x + 1 = 15$ or $6x + 1 = -15 - 4x$
$2x = 14$ or $10x + 1 = -15$
$x = 7$ or $10x = -16$
$x = 7$ or $x = -\dfrac{8}{5}$

The solutions are 7 and $-\dfrac{8}{5}$.

46. $|x - 3| = |7 + 2x|$
$x - 3 = 7 + 2x$ or $x - 3 = -(7 + 2x)$
$-3 = 7 + x$ or $x - 3 = -7 - 2x$
$-10 = x$ or $3x - 3 = -7$
$x = -10$ or $3x = -4$
$x = -10$ or $x = -\dfrac{4}{3}$

The solutions are -10 and $-\dfrac{4}{3}$.

47. $|5x - 1| < 9$
$-9 < 5x - 1 < 9$
$-9 + 1 < 5x - 1 + 1 < 9 + 1$
$-8 < 5x < 10$
$-\dfrac{8}{5} < x < 2$

The solution set is $\left(-\dfrac{8}{5},\ 2\right)$.

48. $|6 + 4x| \geq 10$
$6 + 4x \leq -10$ or $6 + 4x \geq 10$
$6 + 4x - 6 \leq -10 - 6$ or $6 + 4x - 6 \geq 10 - 6$
$4x \leq -16$ or $4x \geq 4$
$x \leq -4$ or $x \geq 1$
The solution set is $(-\infty, -4] \cup [1, \infty)$.

49. $|3x| - 8 > 1$
$|3x| > 9$
$3x < -9$ or $3x > 9$
$x < -3$ or $x > 3$
The solution set is $(-\infty, -3) \cup (3, \infty)$.

50. $9 + |5x| < 24$
$|5x| < 15$
$-15 < 5x < 15$
$-3 < x < 3$
The solution set is $(-3, 3)$.

51. The absolute value of an expression is never negative, so no solution exists. The solution set is $\varnothing$.

52. $|6x - 5| \geq -1$

An absolute value is always greater than or equal to 0. Thus, the solution set is $(-\infty, \infty)$.

53. $\left|3x + \dfrac{2}{5}\right| \geq 4$

$3x + \dfrac{2}{5} \leq -4$ or $3x + \dfrac{2}{5} \geq 4$

$5\left(3x + \dfrac{2}{5}\right) \leq 5(-4)$ or $5\left(3x + \dfrac{2}{5}\right) \geq 5(4)$

$15x + 2 \leq -20$ or $15x + 2 \geq 20$

$15x + 2 - 2 \leq -20 - 2$ or $15x + 2 - 2 \geq 20 - 2$

$15x \leq -22$ or $15x \geq 18$

$x \leq -\dfrac{22}{15}$ or $x \geq \dfrac{6}{5}$

The solution set is $\left(-\infty, -\dfrac{22}{15}\right] \cup \left[\dfrac{6}{5}, \infty\right)$.

54. $\left|\dfrac{4x - 3}{5}\right| < 1$

$-1 < \dfrac{4x - 3}{5} < 1$

$-5 < 4x - 3 < 5$

$-5 + 3 < 4x - 3 + 3 < 5 + 3$

$-2 < 4x < 8$

$-\dfrac{1}{2} < x < 2$

The solution set is $\left(-\dfrac{1}{2}, \ 2\right)$.

55. $\left|\dfrac{x}{3} + 6\right| - 8 > -5$

$\left|\dfrac{x}{3} + 6\right| > 3$

$\dfrac{x}{3} + 6 < -3$ or $\dfrac{x}{3} + 6 > 3$

$3\left(\dfrac{x}{3} + 6\right) < 3(-3)$ or $3\left(\dfrac{x}{3} + 6\right) > 3(3)$

$x + 18 < -9$ or $x + 18 > 9$

$x + 18 - 18 < -9 - 18$ or $x + 18 - 18 > 9 - 18$

$x < -27$ or $x > -9$

The solution set is $(-\infty, \ -27) \cup (-9, \infty)$.

56. $\left|\dfrac{4(x - 1)}{7}\right| + 10 < 2$

$\left|\dfrac{4(x - 1)}{7}\right| < -8$

The absolute value of an expression is never negative, so no solution exists. The solution set is $\varnothing$.

57. Graph $3x + y = 4$ as a dashed line.

Test: $(0, 0)$

$3(0) + 0 > 4$

$0 > 4$ False

Shade the half-plane that does not contain $(0, 0)$.

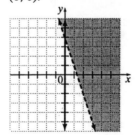

58. Graph $\frac{1}{2}x - y = 2$ as a dashed line.

Test: (0, 0)

$\frac{1}{2}(0) - 0 < 2$

$\qquad 0 < 0 \quad$ True

Shade the half-plane that contains (0, 0).

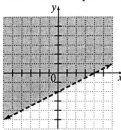

59. Graph $5x - 2y = 9$ as a solid line.

Test: (0, 0)

$5(0) - 2(0) \le 9$

$\qquad 0 \le 9 \quad$ True

Shade the half-plane that contains (0, 0).

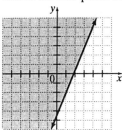

60. Graph $3y = x$ as a solid line.

Test: (0, 1)

$3(1) \ge 0$

$\qquad 3 \ge 0 \quad$ True

Shade the half-plane that contains (0, 1).

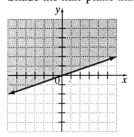

61. Graph $y = 1$ as a dashed line.

Test: (0, 0)

$0 < 1 \quad$ True

Shade the half-plane that contains (0, 0).

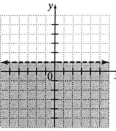

62. Graph $x = -2$ as a dashed line.

Test: (0, 0)

$0 > -2 \quad$ True

Shade the half-plane that contains (0, 0).

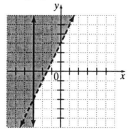

63. $y > 2x + 3$ or $x \le -3$

Graph each inequality. The union of the two inequalities is both shaded regions, including the solid boundary line shown in the graph.

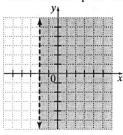

64. $2x < 3y + 8$ and $y \geq -2$
Graph each inequality. The intersection of the two graphs is all points common to both regions, as shown by the heaviest shading in the graph.

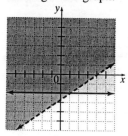

Chapter 3 Test

1. Algebraic solution:
$$15x + 26 = -2(x+1) - 1$$
$$15x + 26 = -2x - 2 - 1$$
$$15x + 26 = -2x - 3$$
$$17x + 26 = -3$$
$$17x = -29$$
$$x = -\frac{29}{17}$$

Graphical solution:

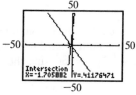

The solution is $-\dfrac{29}{17}$.

2. Define $y_1 = -3x - \sqrt{5}$ and $y_2 = \pi(x-1)$ and graph in an integer window.

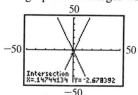

The solution is approximately 0.15.

3. $|6x - 5| = 1$
$$6x - 5 = -1 \quad \text{or} \quad 6x - 5 = 1$$
$$6x = 4 \quad \text{or} \quad 6x = 6$$
$$x = \frac{2}{3} \quad \text{or} \quad x = 1$$

The solutions are $\dfrac{2}{3}$ and 1.

4. $|8 - 2t| = -6$
The absolute value of an expression is never negative, so no solution exists. The solution set is $\varnothing$.

5. $3(2x - 7) - 4x > -(x + 6)$
$$6x - 21 - 4x > -x - 6$$
$$2x - 21 > -x - 6$$
$$3x - 21 > -6$$
$$3x > 15$$
$$\frac{3x}{3} > \frac{15}{3}$$
$$x > 5$$
The solution set is $(5, \infty)$.

6. $8 - \dfrac{x}{2} \leq 7$
$$2\left(8 - \frac{x}{2}\right) \leq 2(7)$$
$$16 - x \leq 14$$
$$-x \leq -2$$
$$\frac{-x}{-1} \geq \frac{-2}{-1}$$
$$x \geq 2$$
The solution set is $[2, \infty)$.

7. $-3 < 2(x - 3) \leq 4$
$$-3 < 2x - 6 \leq 4$$
$$-3 + 6 < 2x - 6 + 6 \leq 4 + 6$$
$$3 < 2x \leq 10$$
$$\frac{3}{2} < \frac{2x}{2} \leq \frac{10}{2}$$
$$\frac{3}{2} < x \leq 5$$
The solution set is $\left(\dfrac{3}{2},\ 5\right]$.

8. $|3x+1| > 5$

$3x+1 < -5$ or $3x+1 > 5$

$3x < -6$ or $\quad 3x > 4$

$x < -2$ or $\quad x > \dfrac{4}{3}$

The solution set is $(-\infty, -2) \cup \left(\dfrac{4}{3}, \infty\right)$.

9. Graph the two intervals and find their intersection.

$\{x \mid x \geq 5\}$

$\{x \mid x \geq 4\}$

$\{x \mid x \geq 5 \text{ and } x \geq 4\}$

The solution set is $[5, \infty)$.

10. Graph the two intervals and find their union.

$\{x \mid x \geq 5\}$

$\{x \mid x \geq 4\}$

$\{x \mid x \geq 5 \text{ or } x \geq 4\}$

The solution set is $[4, \infty)$.

11. $-x > 1 \quad$ and $\quad 3x+3 \geq x-3$

$x < -1$ and $2x+3 \geq -3$

$x < -1$ and $\quad 2x \geq -6$

$x < -1$ and $\quad\quad x \geq -3$

Graph the two intervals and find their intersection.

$\{x \mid x < -1\}$

$\{x \mid x \geq -3\}$

$\{x \mid x < -1 \text{ and } x \geq -3\}$

The solution set is $[-3, -1)$.

12. $6x+1 > 5x+4 \quad$ or $\quad 1-x > -4$

$x+1 > 4 \quad\quad$ or $\quad -x > -5$

$x > 3 \quad\quad$ or $\quad\quad x < 5$

Graph the two intervals and find their union.

$\{x \mid x > 3\}$

$\{x \mid x < 5\}$

$\{x \mid x > 3 \text{ or } x < 5\}$

The solution set is $(-\infty, \infty)$.

13. $\left|\dfrac{2x-6}{5}\right| = 4$

$\dfrac{2x-6}{5} = -4 \quad$ or $\quad \dfrac{2x-6}{5} = 4$

$2x-6 = -20$ or $\quad 2x-6 = 20$

$2x = -14$ or $\quad\quad 2x = 26$

$x = -7 \quad$ or $\quad\quad\quad x = 13$

The solutions are -7 and 13.

14. $\left|\dfrac{7x-1}{2}\right| \leq 3$

$-3 \leq \dfrac{7x-1}{2} \leq 3$

$-6 \leq 7x-1 \leq 6$

$-6+1 \leq 7x-1+1 \leq 6+1$

$-5 \leq 7x \leq 7$

$\dfrac{-5}{7} \leq \dfrac{7x}{7} \leq \dfrac{7}{7}$

$-\dfrac{5}{7} \leq x \leq 1$

The solution set is $\left[-\dfrac{5}{7}, 1\right]$.

15. Graph $x = -4$ as a solid line.
Test: (0, 0)
$0 \le -4$ False
Shade the half-plane that does not contain (0, 0).

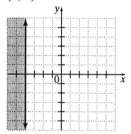

16. Graph $y = -2$ as a dashed line.
Test: (0, 0)
$0 > -2$ True
Shade the half-plane that contains (0, 0).

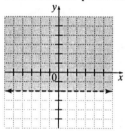

17. Graph $2x - y = 5$ as a dashed line.
Test: (0, 0)
$2(0) - 0 > 5$
$\qquad 0 > 5$ False
Shade the half-plane that does not contain (0, 0).

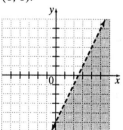

18. $2x + 4y < 6$ and $y \le -4$
Graph each inequality. The intersection of the two graphs is all points common to both regions, as shown by the heaviest shading in the graph.

19. The point of intersection is $(-3, 5)$. The graph of y_1 is below the graph of y_2 for x-values less than -3. The solution set is $(-\infty, -3)$.

20. The point of intersection is $(-3, 5)$. The graph of y_1 is above the graph of y_2 for x-values greater than -3. The solution set is $(-3, \infty)$.

Chapter 4

Section 4.1

Mental Math

1. B

2. C

3. A

4. D

Exercise Set 4.1

1. $\begin{cases} 2x - 3y = -9 \\ 4x + 2y = -2 \end{cases}$

   $2x - 3y = -9$

   $2(3) - 3(5) \stackrel{?}{=} -9$

   $6 - 15 \stackrel{?}{=} -9$

   $-9 = -9$ True

   $4x + 2y = -2$

   $4(3) + 2(5) \stackrel{?}{=} -2$

   $12 + 10 \stackrel{?}{=} -2$

   $22 = -2$ False

   No, (3, 5) is not a solution.

3. From the display, $x = 1.5$ and $y = 1$.

   $2x + 5y = 8$

   $2(1.5) + 5(1) = 3 + 5 = 8$

   $6x + y = 10$

   $6(1.5) + 1 = 9 + 1 = 10$

   The solution is (1.5, 1).

5. From the display, $x = -2$ and $y = 0.75$.

   $x - 4y = -5$

   $-2 - 4(0.75) = -2 - 3 = -5$

   $-3x - 8y = 0$

   $-3(-2) - 8(0.75) = 6 - 6 = 0$

   The solution is (-2, 0.75).

7. $\begin{cases} -3x + y = 13 \\ x + 2y = 5 \end{cases}$

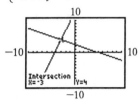

The solution is (-3, 4).

9. $\begin{cases} 2y - 4 = 0 \\ x + 2y = 5 \end{cases}$

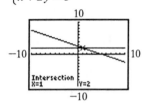

The solution is (1, 2).

11. $\begin{cases} 3x - y = 4 \\ 6x - 2y = 4 \end{cases}$

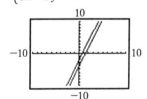

The solution set is $\varnothing$.

13. No; answers may vary.

15. $\begin{cases} x + y = 10 \\ y = 4x \end{cases}$

   Replace y with $4x$ in the first equation.

   $x + 4x = 10$

   $5x = 10$

   $x = 2$

   Replace x with 2 in the second equation.

   $y = 4(2)$

   $y = 8$

   The solution is (2, 8).

17. $\begin{cases} 4x - y = 9 \\ 2x + 3y = -27 \end{cases}$

Solve the first equation for y.
$4x - y = 9$
$\quad y = 4x - 9$
Replace y with $4x - 9$ in the second equation.
$2x + 3(4x - 9) = -27$
$2x + 12x - 27 = -27$
$\qquad\quad 14x = 0$
$\qquad\qquad x = 0$
Replace x with 0 in the first equation.
$4(0) - y = 9$
$\qquad\quad y = -9$
The solution is (0, –9).

19. $\begin{cases} \dfrac{1}{2}x + \dfrac{3}{4}y = -\dfrac{1}{4} \\ \dfrac{3}{4}x - \dfrac{1}{4}y = 1 \end{cases}$

Clear fractions by multiplying each equation by 4.
$\begin{cases} 2x + 3y = -1 \\ 3x - y = 4 \end{cases}$
Now solve the second equation for y.
$3x - y = 4$
$\quad y = 3x - 4$
Replace y with $3x - 4$ in the first equation.
$2x + 3(3x - 4) = -1$
$2x + 9x - 12 = -1$
$\qquad\quad 11x = 11$
$\qquad\qquad x = 1$
Replace x with 1 in the equation
$y = 3x - 4$.
$y = 3(1) - 4$
$y = -1$
The solution is (1, –1).

21. $\begin{cases} \dfrac{x}{3} + y = \dfrac{4}{3} \\ -x + 2y = 11 \end{cases}$

Clear fractions by multiplying the first equation by 3.
$\begin{cases} x + 3y = 4 \\ -x + 2y = 11 \end{cases}$
Solve the second equation for x.
$2y - 11 = x$
$\qquad x = 2y - 11$
Replace x with $2y - 11$ in the first equation.
$2y - 11 + 3y = 4$
$\qquad\quad 5y = 15$
$\qquad\quad y = 3$
Replace y with 3 in the equation
$x = 2y - 11$.
$x = 2(3) - 11$
$x = -5$
The solution is (–5, 3).

23. $\begin{cases} 2x - 4y = 0 \\ x + 2y = 5 \end{cases}$

Multiply the first equation by $\dfrac{1}{2}$.
$\begin{cases} x - 2y = 0 \\ x + 2y = 5 \end{cases}$
Add the equations.
$x - 2y = 0$
$\underline{x + 2y = 5}$
$2x \quad\;\; = 5$
$\qquad x = \dfrac{5}{2}$
Replace x with $\dfrac{5}{2}$ in the first equation.
$\dfrac{5}{2} - 2y = 0$
$\qquad \dfrac{5}{2} = 2y$
$\qquad\quad y = \dfrac{5}{4}$
The solution is $\left(\dfrac{5}{2},\ \dfrac{5}{4}\right)$.

25. $\begin{cases} 5x + 2y = 1 \\ x - 3y = 7 \end{cases}$

Multiply the second equation by –5.

$\begin{cases} 5x + 2y = 1 \\ -5x + 15y = -35 \end{cases}$

Add the equations.

$5x + 2y = 1$

$-5x + 15y = -35$

$\overline{\qquad 17y = -34}$

$y = -2$

Replace y with –2 in the second equation.

$x - 3(-2) = 7$

$x + 6 = 7$

$x = 1$

The solution is (1, –2).

27. $\begin{cases} 5x - 2y = 27 \\ -3x + 5y = 18 \end{cases}$

Multiply the first equation by 3 and the second equation by 5.

$\begin{cases} 15x - 6y = 81 \\ -15x + 25y = 90 \end{cases}$

Add the equations.

$15x - 6y = 81$

$-15x + 25y = 90$

$\overline{\qquad 19y = 171}$

$y = 9$

Replace y with 9 in the first equation.

$5x - 2(9) = 27$

$5x - 18 = 27$

$5x = 45$

$x = 9$

The solution is (9, 9).

29. $\begin{cases} 3x - 5y = 11 \\ 2x - 6y = 2 \end{cases}$

Multiply the first equation by –2 and the second equation by 3.

$\begin{cases} -6x + 10y = -22 \\ 6x - 18y = 6 \end{cases}$

Add the equations.

$-6x + 10y = -22$

$6x - 18y = 6$

$\overline{\qquad -8y = -16}$

$y = 2$

Replace y with 2 in the first equation.

$3x - 5(2) = 11$

$3x - 10 = 11$

$3x = 21$

$x = 7$

The solution is (7, 2).

31. $\begin{cases} x - 2y = 4 \\ 2x - 4y = 4 \end{cases}$

Multiply the first equation by –2.

$\begin{cases} -2x + 4y = -8 \\ 2x - 4y = 4 \end{cases}$

Add the equations.

$-2x + 4y = -8$

$2x - 4y = 4$

$\overline{\qquad 0 = -4 \quad \text{False}}$

Inconsistent system
The solution set is $\varnothing$.

33. $\begin{cases} 3x + y = 1 \\ 2y = 2 - 6x \end{cases}$

$\begin{cases} 3x + y = 1 \\ 6x + 2y = 2 \end{cases}$

Multiply the first equation by –2.

$\begin{cases} -6x - 2y = -2 \\ 6x + 2y = 2 \end{cases}$

Add the equations.

$-6x - 2y = -2$

$\underline{6x + 2y = 2}$

$\qquad 0 = 0 \quad$ True

Dependent system

The solution set is $\{(x,\ y)\,|\,3x + y = 1\}$.

35. Answers may vary.

One possibility: $\begin{cases} -2x + y = 1 \\ x - 2y = -8 \end{cases}$

37. $\begin{cases} 2x + 5y = 8 \\ 6x + y = 10 \end{cases}$

Multiply the first equation by –3.

$\begin{cases} -6x - 15y = -24 \\ 6x + y = 10 \end{cases}$

Add the equations.

$-6x - 15y = -24$

$\underline{6x + y = 10}$

$\qquad -14y = -14$

$\qquad\quad y = 1$

Replace y with 1 in the second equation.

$6x + 1 = 10$

$6x = 9$

$x = \dfrac{9}{6} = \dfrac{3}{2}$

The solution is $\left(\dfrac{3}{2},\ 1\right)$.

39. $\begin{cases} 0.7x - 0.2y = -1.6 \\ 0.2x - y = -1.4 \end{cases}$

Multiply both equations by 10 to clear decimals.

$\begin{cases} 7x - 2y = -16 \\ 2x - 10y = -14 \end{cases}$

Multiply the first equation by –5.

$\begin{cases} -35x + 10y = 80 \\ 2x - 10y = -14 \end{cases}$

Add the equations.

$-35x + 10y = 80$

$\underline{2x - 10y = -14}$

$-33x \qquad\quad = 66$

$\qquad\quad x = -2$

Replace x with –2 in the equation

$2x - 10y = -14$.

$2(-2) - 10y = -14$

$-4 - 10y = -14$

$-10y = -10$

$y = 1$

The solution is (–2, 1).

41. $\begin{cases} \dfrac{1}{3}x + y = \dfrac{4}{3} \\ -\dfrac{1}{4}x - \dfrac{1}{2}y = -\dfrac{1}{4} \end{cases}$

Clear fractions by multiplying the first equation by 3 and the second equation by 4.

$\begin{cases} x + 3y = 4 \\ -x - 2y = -1 \end{cases}$

Add the equations.

$x + 3y = 4$

$\underline{-x - 2y = -1}$

$\qquad\quad y = 3$

Replace y with 3 in the equation

$x + 3y = 4$.

$x + 3(3) = 4$

$x + 9 = 4$

$x = -5$

The solution is (–5, 3).

43. $\begin{cases} 2x + 6y = 8 \\ 3x + 9y = 12 \end{cases}$

Multiply the first equation by –3 and the second equation by 2.

$\begin{cases} -6x - 18y = -24 \\ 6x + 18y = 24 \end{cases}$

Add the equations.

$-6x - 18y = -24$

$\underline{6x + 18y = 24}$

$\qquad 0 = 0 \quad$ True

Dependent system

The solution set is $\{(x, y) \mid 3x + 9y = 12\}$.

45. $\begin{cases} 4x + 2y = 5 \\ 2x + y = -1 \end{cases}$

Multiply the second equation by –2.

$\begin{cases} 4x + 2y = 5 \\ -4x - 2y = 2 \end{cases}$

Add the equations.

$4x + 2y = 5$

$\underline{-4x - 2y = 2}$

$\qquad 0 = 7 \quad$ False

Inconsistent system

The solution set is $\varnothing$.

47. $\begin{cases} 10y - 2x = 1 \\ \quad 5y = 4 - 6x \end{cases}$

$\begin{cases} -2x + 10y = 1 \\ 6x + 5y = 4 \end{cases}$

Multiply the second equation by –2.

$\begin{cases} -2x + 10y = 1 \\ -12x - 10y = -8 \end{cases}$

Add the equations.

$-2x + 10y = 1$

$\underline{-12x - 10y = -8}$

$-14x \qquad = -7$

$\qquad x = \dfrac{-7}{-14} = \dfrac{1}{2}$

Replace x with $\dfrac{1}{2}$ in the first equation.

$10y - 2\left(\dfrac{1}{2}\right) = 1$

$\qquad 10y - 1 = 1$

$\qquad 10y = 2$

$\qquad y = \dfrac{2}{10} = \dfrac{1}{5}$

The solution is $\left(\dfrac{1}{2}, \dfrac{1}{5}\right)$.

49. $\begin{cases} \dfrac{3}{4}x + \dfrac{5}{2}y = 11 \\ \dfrac{1}{16}x - \dfrac{3}{4}y = -1 \end{cases}$

Clear fractions by multiplying the first equation by 4 and the second equation by 16.

$\begin{cases} 3x + 10y = 44 \\ x - 12y = -16 \end{cases}$

Multiply the second equation by –3.

$\begin{cases} 3x + 10y = 44 \\ -3x + 36y = 48 \end{cases}$

Add the equations.

$3x + 10y = 44$

$\underline{-3x + 36y = 48}$

$\qquad 46y = 92$

$\qquad y = 2$

Replace y with 2 in the equation

$x - 12y = -16$.

$x - 12(2) = -16$

$\quad x - 24 = -16$

$\qquad x = 8$

The solution is (8, 2).

51. $\begin{cases} x = 3y + 2 \\ 5x - 15y = 10 \end{cases}$

Replace x with $3y + 2$ in the second equation.

$5(3y + 2) - 15y = 10$
$15y + 10 - 15y = 10$
$ 10 = 10$ True

The system is dependent.

The solution set is $\{(x, y) \mid x = 3y + 2\}$.

53. $\begin{cases} 2x - y = -1 \\ y = -2x \end{cases}$

Replace y with $-2x$ in the first equation.

$2x - (-2x) = -1$
$ 4x = -1$
$ x = -\dfrac{1}{4}$

Replace x with $-\dfrac{1}{4}$ in the second equation.

$y = -2\left(-\dfrac{1}{4}\right) = \dfrac{1}{2}$

The solution is $\left(-\dfrac{1}{4}, \dfrac{1}{2}\right)$.

55. $\begin{cases} 2x = 6 \\ y = 5 - x \end{cases}$

The first equation yields $x = 3$.
Replace x with 3 in the second equation.
$y = 5 - 3 = 2$
The solution is $(3, 2)$.

57. $\begin{cases} \dfrac{x + 5}{2} = \dfrac{6 - 4y}{3} \\ \dfrac{3x}{5} = \dfrac{21 - 7y}{10} \end{cases}$

Multiply the first equation by 6 and the second equation by 10.

$\begin{cases} 3x + 15 = 12 - 8y \\ 6x = 21 - 7y \end{cases}$

$\begin{cases} 3x + 8y = -3 \\ 6x + 7y = 21 \end{cases}$

Multiply the first equation by -2.

$\begin{cases} -6x - 16y = 6 \\ 6x + 7y = 21 \end{cases}$

Add the equations.

$\begin{aligned} -6x - 16y &= 6 \\ \underline{6x + 7y} &= \underline{21} \\ -9y &= 27 \\ y &= -3 \end{aligned}$

Replace y with -3 in the equation $3x + 8y = -3$.

$3x + 8(-3) = -3$
$ 3x - 24 = -3$
$ 3x = 21$
$ x = 7$

The solution is $(7, -3)$.

59. $\begin{cases} 4x - 7y = 7 \\ 12x - 21y = 24 \end{cases}$

Multiply the first equation by -3.

$\begin{cases} -12x + 21y = -21 \\ 12x - 21y = 24 \end{cases}$

Add the equations.

$\begin{aligned} -12x + 21y &= -21 \\ \underline{12x - 21y} &= \underline{24} \\ 0 &= 3 \quad \text{False} \end{aligned}$

Inconsistent system
The solution set is $\varnothing$.

61.

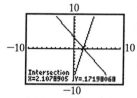

The solution is $(2.11, 0.17)$.

63.

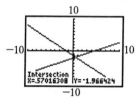

The solution is $(0.57, -1.97)$.

65.

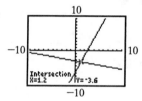

The solution is $(1.2, -3.6)$.

67. Supply equals demand at the intersection point, $(5, 21)$. Supply equals demand for 5000 ties and the price per tie is \$21.

69. Supply is greater than demand because the graph for supply is above the graph for demand for x-values greater than 6.

71. $\begin{cases} y = 2.5x \\ y = 0.9x + 3000 \end{cases}$

Replace y with $2.5x$ in the second equation.
$2.5x = 0.9x + 3000$
$1.6x = 3000$
$\quad x = 1875$
Replace x with the 1875 in the first equation.
$y = 2.5(1875) = 4687.5$
The point of intersection is $(1875, 4687.5)$.

73. The company makes money because revenue is greater than cost at $x = 2000$.

75. The company makes a profit for values of x greater than 1875 because the x-value at the intersection is 1875.

77. a. Consumption of red meat is decreasing while consumption of poultry is increasing.

    b. $\begin{cases} y = -0.6x + 121.2 \\ y = 1.7x + 88 \end{cases}$
Replace y with $-0.6x + 121.2$ in the second equation.
$-0.6x + 121.2 = 1.7x + 88$
$\qquad\qquad 33.2 = 2.3x$
$\qquad\qquad\; 14 \approx x$
Replace x with 14 in the first equation.
$y = -0.6(14) + 121.2$
$y \approx 113$
The solution is $(14, 113)$.

    c. $x = 14$, $1995 + 14 = 2009$
In the year 2009, red meat and poultry consumption will each be about 113 pounds per person.

79. $\begin{cases} 107x + y = 11,096 \\ 399x + y = 15,149 \end{cases}$
Multiply the first equation by -1.
$\begin{cases} -107x - y = -11,096 \\ \;\;\;399x + y = 15,149 \end{cases}$
Add the two new equations:
$292x = 4053$
$\quad x \approx 13.88$
$1980 + 13 = 1993$
The year is 1993.

81. $\quad x + 2y - z = 7$
$\quad 2 + 2(-3) - 3 = 7$
$\qquad\quad 2 - 6 - 3 = 7$
$\qquad\qquad\qquad -7 = 7$
False

83.
$$-4x + y - 8z = 4$$
$$-4(1) + 0 - 8(-1) = 4$$
$$-4 + 8 = 4$$
$$4 = 4$$
True

85.
$$\begin{array}{r} x + 4y - 5z = 20 \\ + \ \ 2x - 4y - 2z = -17 \\ \hline 3x \qquad - 7z = 3 \end{array}$$

87.
$$\begin{array}{r} -9x - 8y \ - z = 31 \\ + \ \ 9x + 4y \ - z = 12 \\ \hline -4y - 2z = 43 \end{array}$$

89.
$$\begin{cases} x + \dfrac{2}{y} = 7 \\ 3x + \dfrac{3}{y} = 6 \end{cases}$$

Replacing $\dfrac{1}{y}$ with a, we have the following:
$$\begin{cases} x + 2a = 7 \\ 3x + 3a = 6 \end{cases}$$
Multiply the first equation by -3.
$$\begin{cases} -3x - 6a = -21 \\ 3x + 3a = 6 \end{cases}$$
Add the two new equations:
$$-3a = -15$$
$$a = 5$$
Substitute.
$$x + 2a = 7$$
$$x + 2(5) = 7$$
$$x + 10 = 7$$
$$x = -3$$
Since $a = 5$, $y = \dfrac{1}{5}$. The solution is
$$\left(-3, \dfrac{1}{5} \right).$$

91.
$$\begin{cases} \dfrac{2}{x} + \dfrac{3}{y} = 5 \\ \dfrac{5}{x} - \dfrac{3}{y} = 2 \end{cases}$$

Replacing $\dfrac{1}{x}$ with a, and $\dfrac{1}{y}$ with b, we have the following:
$$\begin{cases} 2a + 3b = 5 \\ 5a - 3b = 2 \end{cases}$$
Add the two new equations:
$$7a = 7$$
$$a = 1$$
$$2a + 3b = 5$$
$$2 + 3b = 5$$
$$3b = 3$$
$$b = 1$$
Since $a = 1$, $x = \dfrac{1}{1} = 1$. Similarly, $y = 1$.
The solution is $(1, 1)$.

93.
$$\begin{cases} \dfrac{3}{x} - \dfrac{2}{y} = -18 \\ \dfrac{2}{x} + \dfrac{3}{y} = 1 \end{cases}$$

Replacing $\dfrac{1}{x}$ with a, and $\dfrac{1}{y}$ with b, we have the following:
$$\begin{cases} 3a - 2b = -18 \\ 2a + 3b = 1 \end{cases}$$
Multiplying the first equation by 3 and the second equation by 2 yields the following:
$$\begin{cases} 9a - 6b = -54 \\ 4a + 6b = 2 \end{cases}$$
Add the two new equations:
$$13a = -52$$
$$a = -4$$
$$3a - 2b = -18$$
$$-12 - 2b = -18$$
$$-2b = -6$$
$$b = 3$$
Since $a = -4$, $x = -\dfrac{1}{4}$. Similarly, $y = \dfrac{1}{3}$.
The solution is $\left(-\dfrac{1}{4}, \dfrac{1}{3} \right)$.

95. $\begin{cases} \dfrac{5}{x} + \dfrac{7}{y} = 1 \\ -\dfrac{10}{x} - \dfrac{14}{y} = 0 \end{cases}$

Replacing $\dfrac{1}{x}$ with a, and $\dfrac{1}{y}$ with b, we have the following:

$\begin{cases} 5a + 7b = 1 \\ -10a - 14b = 0 \end{cases}$

Multiplying the first equation by 2 yields the following:

$\begin{cases} 10a + 14b = 2 \\ -10a - 14b = 0 \end{cases}$

Add the two new equations:

$0 = 2$

This is a false statement. The solution set is $\varnothing$.

Exercise Set 4.2

1. $\begin{cases} x + y \quad\;\; = 3 \quad (1) \\ \quad\;\; 2y \quad\;\; = 10 \quad (2) \\ 3x + 2y - 3z = 1 \quad (3) \end{cases}$

Solve the second equation for y.

$y = 5$

Replace y with 5 in the first equation.

$x + 5 = 3$

$x = -2$

Replace x with -2 and y with 5 in the third equation.

$3(-2) + 2(5) - 3z = 1$

$-6 + 10 - 3z = 1$

$4 - 3z = 1$

$-3z = -3$

$z = 1$

The solution is $(-2, 5, 1)$.

3. $\begin{cases} 2x + 2y + z = 1 \quad (1) \\ -x + y + 2z = 3 \quad (2) \\ x + 2y + 4z = 0 \quad (3) \end{cases}$

Add equations (2) and (3).

$3y + 6z = 3$ or $y + 2z = 1$

Add twice equation (2) to equation (1).

$-2x + 2y + 4z = 6$

$\underline{2x + 2y + z = 1}$

$4y + 5z = 7$

Solve the new system:

$\begin{cases} y + 2z = 1 \\ 4y + 5z = 7 \end{cases}$

Multiply the first equation by -4.

$\begin{cases} -4y - 8z = -4 \\ 4y + 5z = 7 \end{cases}$

Add the equations.

$-4y - 8z = -4$

$\underline{4y + 5z = 7}$

$-3z = 3$

$z = -1$

Replace z with -1 in the equation $y + 2z = 1$.

$y + 2(-1) = 1$

$y - 2 = 1$

$y = 3$

Replace y with 3 and z with -1 in equation (3).

$x + 2(3) + 4(-1) = 0$

$x + 6 - 4 = 0$

$x + 2 = 0$

$x = -2$

The solution is $(-2, 3, -1)$.

5. $\begin{cases} x - 2y + z = -5 & (1) \\ -3x + 6y - 3z = 15 & (2) \\ 2x - 4y + 2z = -10 & (3) \end{cases}$

Multiply equation (2) by $-\dfrac{1}{3}$ and

equation (3) by $\dfrac{1}{2}$.

$\begin{cases} x - 2y + z = -5 \\ x - 2y + z = -5 \\ x - 2y + z = -5 \end{cases}$

All three equations are identical. There are infinitely many solutions. The solution set is $\{(x, y, z) \mid x - 2y + z = -5\}$.

7. $\begin{cases} 4x - y + 2z = 5 & (1) \\ 2y + z = 4 & (2) \\ 4x + y + 3z = 10 & (3) \end{cases}$

Multiply equation (1) by -1 and add to equation (3).

$\begin{array}{r} -4x + y - 2z = -5 \\ 4x + y + 3z = 10 \\ \hline 2y + z = 5 \quad (4) \end{array}$

Multiply equation (4) by -1 and add to equation (2).

$\begin{array}{r} -2y - z = -5 \\ 2y + z = 4 \\ \hline 0 = -1 \quad \text{False} \end{array}$

Inconsistent system

The solution set is $\varnothing$.

9. Answers may vary.

One possibility is: $\begin{cases} 3x = -3 \\ 2x + 4y = 6 \\ x - 3y + z = -11 \end{cases}$

11. $\begin{cases} x + 5z = 0 & (1) \\ 5x + y = 0 & (2) \\ y - 3z = 0 & (3) \end{cases}$

Multiply equation (3) by -1 and add to equation (2).

$\begin{array}{r} -y + 3z = 0 \\ 5x + y = 0 \\ \hline 5x + 3z = 0 \quad (4) \end{array}$

Multiply equation (1) by -5 and add to equation (4).

$\begin{array}{r} -5x - 25z = 0 \\ 5x + 3z = 0 \\ \hline -22z = 0 \\ z = 0 \end{array}$

Replace z with 0 in equation (4).

$5x + 3(0) = 0$
$5x = 0$
$x = 0$

Replace x with 0 in equation (2).

$5(0) + y = 0$
$y = 0$

The solution is $(0, 0, 0)$.

13. $\begin{cases} 6x - 5z = 17 & (1) \\ 5x - y + 3z = -1 & (2) \\ 2x + y = -41 & (3) \end{cases}$

Add equations (2) and (3).

$7x + 3z = -42 \quad (4)$

Multiply equation (4) by 5, multiply equation (1) by 3, and add.

$\begin{array}{r} 35x + 15z = -210 \\ 18x - 15z = 51 \\ \hline 53x = -159 \\ x = -3 \end{array}$

Replace x with -3 in equation (1).

$6(-3) - 5z = 17$
$-18 - 5z = 17$
$-5z = 35$
$z = -7$

Replace x with -3 in equation (3).

$2(-3) + y = -41$
$-6 + y = -41$
$y = -35$

The solution is $(-3, -35, -7)$.

15. $\begin{cases} x + y + z = 8 & (1) \\ 2x - y - z = 10 & (2) \\ x - 2y - 3z = 22 & (3) \end{cases}$

Add equations (1) and (2).
$3x = 18$ or $x = 6$
Add twice equation (1) to equation (3).
$2x + 2y + 2z = 16$

$\dfrac{x - 2y - 3z = 22}{3x \qquad - z = 38}$

Replace x with 6 in this equation.
$3(6) - z = 38$
$18 - z = 38$
$-z = 20$
$z = -20$
Replace x with 6 and z with -20 in equation (1).
$6 + y + (-20) = 8$
$y - 14 = 8$
$y = 22$
The solution is $(6, 22, -20)$.

17. $\begin{cases} x + 2y - z = 5 & (1) \\ 6x + y + z = 7 & (2) \\ 2x + 4y - 2z = 5 & (3) \end{cases}$

Add equations (1) and (2).
$7x + 3y = 12$ (4)
Add twice equation (2) to equation (3).
$12x + 2y + 2z = 14$

$\dfrac{2x + 4y - 2z = 5}{14x + 6y \qquad = 19} \quad (5)$

Multiply equation (4) by -2 and add to equation (5).
$-14x - 6y = -24$

$\dfrac{14x + 6y = 19}{0 = -5 \quad \text{False}}$

Inconsistent system
The solution set is $\varnothing$.

19. $\begin{cases} 2x - 3y + z = 2 & (1) \\ x - 5y + 5z = 3 & (2) \\ 3x + y - 3z = 5 & (3) \end{cases}$

Add -2 times equation (2) to equation (1).
$2x - 3y + z = 2$

$\dfrac{-2x + 10y - 10z = -6}{7y - 9z = -4}$

Add -3 times equation (2) to equation (3).
$-3x + 15y - 15z = -9$

$\dfrac{3x + y - 3z = 5}{16y - 18z = -4}$

We now have the system:
$\begin{cases} 7y - 9z = -4 & (4) \\ 16y - 18z = -4 & (5) \end{cases}$

Multiply equation (4) by -2 and add to equation (5).
$-14y + 18z = 8$

$\dfrac{16y - 18z = -4}{2y \qquad = 4}$

$y = 2$
Replace y with 2 in equation (4).
$7(2) - 9z = -4$
$-9z = -18$
$z = 2$
Replace y with 2 and z with 2 in equation (1).
$2x - 3(2) + 2 = 2$
$x = 3$
The solution is $(3, 2, 2)$.

21. $\begin{cases} -2x - 4y + 6z = -8 & (1) \\ x + 2y - 3z = 4 & (2) \\ 4x + 8y - 12z = 16 & (3) \end{cases}$

Add 2 times equation (2) to equation (1).

$2x + 4y - 6z = 8$

$\dfrac{-2x - 4y + 6z = -8}{0 = 0}$

Add –4 times equation (2) to equation (3).

$-4x - 8y + 12z = -16$

$\dfrac{4x + 8y - 12z = 16}{0 = 0}$

The system is dependent.
The solution set is
$\{(x, y, z) \mid x + 2y - 3z = 4\}$.

23. $\begin{cases} 2x + 2y - 3z = 1 & (1) \\ y + 2z = -14 & (2) \\ 3x - 2y = -1 & (3) \end{cases}$

Add equations (1) and (3).

$5x - 3z = 0$ (4)

Add twice equation (2) to equation (3).

$2y + 4z = -28$

$\dfrac{3x - 2y = -1}{3x + 4z = -29}$ (5)

Multiply equation (4) by 4, multiply
equation (5) by 3 and add.

$20x - 12z = 0$

$\dfrac{9x + 12z = -87}{29x = -87}$

$x = -3$

Replace x with –3 in equation (4).

$5(-3) - 3z = 0$

$3z = -15$

$z = -5$

Replace z with –5 in equation (2).

$y + 2(-5) = -14$

$y - 10 = -14$

$y = -4$

The solution is (–3, –4, –5).

25. $\begin{cases} \dfrac{3}{4}x - \dfrac{1}{3}y + \dfrac{1}{2}z = 9 & (1) \\[2mm] \dfrac{1}{6}x + \dfrac{1}{3}y - \dfrac{1}{2}z = 2 & (2) \\[2mm] \dfrac{1}{2}x - y + \dfrac{1}{2}z = 2 & (3) \end{cases}$

Multiply equation (1) by 12, multiply
equation (2) by 6, and multiply equation
(3) by 2.

$\begin{cases} 9x - 4y + 6z = 108 & (4) \\ x + 2y - 3z = 12 & (5) \\ x - 2y + z = 4 & (6) \end{cases}$

Add twice equation (5) to equation (4).

$2x + 4y - 6z = 24$

$\dfrac{9x - 4y + 6z = 108}{11x = 132}$

$x = 12$

Add equations (5) and (6).

$2x - 2z = 16$ or $x - z = 8$

Replace x with 12 in this equation.

$12 - z = 8$

$z = 4$

Replace x with 12 and z with 4 in
equation (6).

$12 - 2y + 4 = 4$

$12 - 2y = 0$

$12 = 2y$

$y = 6$

The solution is (12, 6, 4).

27. $\begin{cases} x + y + z = 1 & (1) \\ 2x - y + z = 0 & (2) \\ -x + 2y + 2z = -1 & (3) \end{cases}$

Multiply equation (3) by 2 and add to equation (2).

$2x - y + z = 0$

$-2x + 4y + 4z = -2$

$\overline{\qquad 3y + 5z = -2 \quad (4)}$

Multiply equation (1) by −2 and add to equation (2).

$-2x - 2y - 2z = -2$

$2x - y + z = 0$

$\overline{\qquad -3y - z = -2 \quad (5)}$

Add equations (4) and (5).

$3y + 5z = -2$

$-3y - z = -2$

$\overline{\qquad 4z = -4}$

$z = -1$

Replace z with −1 in equation (4).

$3y + 5(-1) = -2$

$3y = 3$

$y = 1$

Replace y with 1 and z with −1 in equation (1).

$x + 1 + (-1) = 1$

$x = 1$

The solution is $(1, 1, -1)$.

$\dfrac{1}{24} = \dfrac{x}{8} + \dfrac{y}{4} + \dfrac{z}{3}$

$\dfrac{1}{24} = \dfrac{1}{8} + \dfrac{1}{4} - \dfrac{1}{3}$

$\dfrac{1}{24} = \dfrac{3}{24} + \dfrac{6}{24} - \dfrac{8}{24}$

$\dfrac{1}{24} = \dfrac{1}{24}$ True

29. Let $x =$ the first number

$45 - x =$ the second number

$45 - x = 2x$

$45 = 3x$

$15 = x$

$45 - 15 = 30$

The numbers are 15 and 30.

31. $2(x - 1) - 3x = x - 12$

$2x - 2 - 3x = x - 12$

$-2 - x = x - 12$

$10 = 2x$

$5 = x$

The solution is 5.

33. $-y - 5(y + 5) = 3y - 10$

$-y - 5y - 25 = 3y - 10$

$-6y - 25 = 3y - 10$

$-15 = 9y$

$-\dfrac{15}{9} = y$

$y = -\dfrac{5}{3}$

The solution is $-\dfrac{5}{3}$.

35. $\begin{cases} x+y \quad\ -w=0 & (1) \\ \quad\ y+2z+w=3 & (2) \\ x \quad\quad -z \quad=1 & (3) \\ 2x-y \quad\quad -w=-1 & (4) \end{cases}$

Add equations (2) and (4).
$2x+2z=2 \quad (5)$
Add equation (5) to 2 times equation (3).
$2x-2z=2$

$\dfrac{2x+2z=2}{4x \quad\quad =4}$

$\qquad x=1$
Replace x with 1 in equation (3).
$1-z=1$
$\quad z=0$
Replace x with 1 in equation (1).
$1+y-w=0$
$\quad y-w=-1 \quad (6)$
Replace x with 1 in equation (4).
$2-y-w=-1$
$\quad -y-w=-3 \quad (7)$
Add equations (6) and (7).
$\quad y-w=-1$

$\dfrac{-y-w=-3}{\ -2w=-4}$

$\quad w=2$
Replace w with 2 in equation (6).
$y-2=-1$
$\quad y=1$
The solution is $(1, 1, 0, 2)$.

37. $\begin{cases} x+y+z+w=5 & (1) \\ 2x+y+z+w=6 & (2) \\ x+y+z \quad\ =2 & (3) \\ x+y \quad\quad =0 & (4) \end{cases}$

Multiply equation (1) by -1 and add to equation (2).
$-x-y-z-w=-5$

$\dfrac{2x+y+z+w=6}{x \quad\quad\quad =1}$

Replace x with 1 in equation (4).
$1+y=0$
$\quad y=-1$
Replace x with 1 and y with -1 in equation (3).
$1+(-1)+z=2$
$\quad\quad z=2$
Replace x with 1, y with -1, and z with 2 in equation (1).
$1+(-1)+2+w=5$
$\quad\quad 2+w=5$
$\quad\quad\quad w=3$
The solution is $(1, -1, 2, 3)$.

Exercise Set 4.3

1. Let $m=$ the first number
$\quad\quad n=$ the second number
$\begin{cases} m= \ n+2 \\ 2m=3n-4 \end{cases}$
Substitute $m=n+2$ in the second equation.
$2(n+2)=3n-4$
$\quad 2n+4=3n-4$
$\quad\quad\quad n=8$
Replace n with 8 in the first equation.
$m=8+2=10$
The numbers are 10 and 8.

3. Let p = the speed of the plane in still air
     w = the speed of the wind
$$\begin{cases} p + w = 560 \\ p - w = 480 \end{cases}$$
Add the equations.
$$2p = 1040$$
$$p = 520$$
Replace p with 520 in the first equation.
$$520 + w = 560$$
$$w = 40$$
The speed of the plane in still air is 520 miles per hour and the speed of the wind is 40 miles per hour.

5. Let x = the number of quarts of
     4% butterfat milk
     y = the number of quarts of
     1% butterfat milk
$$\begin{cases} x + y = 60 \\ 0.04x + 0.01y = 0.02(60) \end{cases}$$
Multiply the second equation by -100 and add to the first equation.
$$x + y = 60$$
$$\underline{-4x - y = -120}$$
$$-3x = -60$$
$$x = 20$$
Replace x with 20 in the first equation.
$$20 + y = 60$$
$$y = 40$$
20 quarts of 4% butterfat milk and 40 quarts of 1% butterfat milk should be used.

7. Let l = the number of large frames
     s = the number of small frames
$$\begin{cases} l + s = 22 \\ 15l + 8s = 239 \end{cases}$$
Multiply the first equation by -8 and add to the second equation.
$$-8l - 8s = -176$$
$$\underline{15l + 8s = 239}$$
$$7l = 63$$
$$l = 9$$
Replace l with 9 in the first equation.
$$9 + s = 22$$
$$s = 13$$
She bought 9 large frames and 13 small frames.

9. Let m = the first number
     n = the second number
$$\begin{cases} m = n - 2 \\ 2m = 3n + 4 \end{cases}$$
Substitute $m = n - 2$ in the second equation.
$$2(n - 2) = 3n + 4$$
$$2n - 4 = 3n + 4$$
$$n = -8$$
Replace n with -8 in the first equation.
$$m = -8 - 2 = -10$$
The numbers are -10 and -8.

11. Let x = the price of each tablet
     y = the price of each pen
$$\begin{cases} 7x + 4y = 6.40 \\ 2x + 19y = 5.40 \end{cases}$$
Multiply the first equation by 2 and the second equation by -7 and add.
$$14x + 8y = 12.80$$
$$\underline{-14x - 133y = -37.80}$$
$$-125y = -25$$
$$y = \frac{25}{125} = 0.20$$
Replace y with 0.20 in the first equation.
$$7x + 4(0.20) = 6.40$$
$$7x + 0.80 = 6.40$$
$$7x = 5.60$$
$$x = 0.80$$
Tablets cost \$0.80 each and pens cost \$0.20 each.

13. Let p = the speed of the plane in still air
w = the speed of the wind
First note:
$$\frac{2160 \text{ miles}}{3 \text{ hours}} = 720 \text{ miles per hour and}$$
$$\frac{2160 \text{ miles}}{4 \text{ hours}} = 540 \text{ miles per hour}$$
Now,
$$\begin{cases} p + w = 720 \\ p - w = 540 \end{cases}$$
Add the equations.
$2p = 1260$
$p = 630$
Replace p with 630 in the first equation.
$630 + w = 720$
$\quad\quad w = 90$
The speed of the plane in still air is 630 miles per hour and the speed of the wind is 90 miles per hour.

15. Let s = the length of the shortest side
l = the length of the longest side
m = the length of the other two sides
$$\begin{cases} s + 2m + l = 29 \\ \quad\quad l = 2s \\ \quad\quad m = s + 2 \end{cases}$$
Substitute $l = 2s$ and $m = s + 2$ in the first equation.
$s + 2(s + 2) + 2s = 29$
$s + 2s + 4 + 2s = 29$
$\quad\quad\quad 5s + 4 = 29$
$\quad\quad\quad\quad 5s = 25$
$\quad\quad\quad\quad\; s = 5$
Replace s with 5 in the second and third equations.
$l = 2(5) = 10$
$m = 5 + 2 = 7$
The shortest side is 5 inches, the longest side is 10 inches, and the other two sides are 7 inches.

17. Let x = the first number
y = the second number
z = the third number
$$\begin{cases} x + y + z = 40 \\ \quad\quad x = y + 5 \\ \quad\quad x = 2z \end{cases}$$
$$\begin{cases} x + y + z = 40 \\ \quad\quad y = x - 5 \\ \quad\quad z = \frac{1}{2}x \end{cases}$$
Substitute $y = x - 5$ and $z = \frac{1}{2}x$ in the first equation.
$$x + x - 5 + \frac{1}{2}x = 40$$
$$\frac{5}{2}x - 5 = 40$$
$$\frac{5}{2}x = 45$$
$$x = \frac{2}{5}(45) = 18$$
$y = 18 - 5 = 13$
$z = \frac{1}{2}(18) = 9$
The three numbers are 18, 13, and 9.

19. Let x = total weekly sales
y = total weekly salary
$$\begin{cases} y = 200 + 0.05x \\ y = 0.15x \end{cases}$$
Substitute $y = 0.15x$ in the first equation.
$200 + 0.05x = 0.15x$
$\quad\quad\quad 200 = 0.10x$
$$x = \frac{200}{0.10} = 2000$$
Jack's salary would be the same, regardless of pay arrangement, at $2000 worth of sales.

21. Let $x =$ the price for a template
$y =$ the price for a pencil
$z =$ the price for a pad of paper

$$\begin{cases} 3x + y = 6.45 \\ 2z + 4y = 7.50 \\ \quad z = 3y \end{cases}$$

Substitute $z = 3y$ in the second equation.
$$2(3y) + 4y = 7.50$$
$$6y + 4y = 7.50$$
$$10y = 7.50$$
$$y = 0.75$$

Replace y with 0.75 in the third equation.
$$3(0.75) = z$$
$$2.25 = z$$

Replace y with 0.75 in the first equation.
$$3x + 0.75 = 6.45$$
$$3x = 5.70$$
$$x = 1.90$$

The prices are \$1.90 for a template, \$0.75 for a pencil, and \$2.25 for a pad of paper.

23. $C(x) = 12x + 15,000$
$R(x) = 32x$
$$12x + 15,000 = 32x$$
$$15,000 = 20x$$
$$750 = x$$
750 units must be sold to break even.

25. $C(x) = 0.8x + 900$
$R(x) = 2x$
$$0.8x + 900 = 2x$$
$$900 = 1.2x$$
$$750 = x$$
750 units must be sold to break even.

27. $C(x) = 105x + 70,000$
$R(x) = 245x$
$$105x + 70,000 = 245x$$
$$70,000 = 140x$$
$$500 = x$$
500 units must be sold to break even.

29. a. $R(x) = 31x$

b. $C(x) = 15x + 500$

c. $R(x) = C(x)$
$$31x = 15x + 500$$
$$16x = 500$$
$$x = 31.25$$
32 baskets is the break-even point.

31. $$\begin{cases} x + 2y = 180 \\ 3x - 10 + y = 180 \end{cases}$$

$$\begin{cases} x = 180 - 2y \\ 3x + y = 190 \end{cases}$$

Substitute $x = 180 - 2y$ in the second equation.
$$3(180 - 2y) + y = 190$$
$$540 - 6y + y = 190$$
$$-5y = -350$$
$$y = 70$$

Replace y with 70 in the equation
$x = 180 - 2y$.
$$x = 180 - 2(70)$$
$$x = 40$$
$x = 40$ and $y = 70$

33. Let $x =$ the number of ounces of 20% acid
$y =$ the number of ounces of 50% acid.

$$\begin{cases} x + y = 60 & (1) \\ 0.2x + 0.5y = 0.3(60) & (2) \end{cases}$$

Multiply the second equation by 10.
$$x + y = 60$$
$$2x + 5y = 180$$

Multiply the first equation by -2 and add to the second.
$$-2x - 2y = -120$$
$$\underline{2x + 5y = 180}$$
$$3y = 60$$
$$y = 20$$

Replace y with 20 in the first equation.
$$x + 20 = 60$$
$$x = 40$$
40 ounces of 20% acid solution and 20 ounces of 50% acid solution should be mixed.

35.

Concentration	Amount	Solution
25%	$2x$	$0.25(2x) = 0.5x$
40%	x	$0.4x$
50%	y	$0.5y$
32%	200	$0.32(200) = 64$

$$\begin{cases} 2x + x + y = 200 \\ 0.5x + 0.4x + 0.5y = 64 \end{cases}$$

$$\begin{cases} 3x + \quad y = 200 \\ 0.9x + 0.5y = 64 \end{cases}$$

Solve the first equation for y and substitute into the second equation.
$$y = 200 - 3x$$
$$0.9x + 0.5(200 - 3x) = 64$$
$$0.9x + 100 - 1.5x = 64$$
$$-0.6x = -36$$
$$x = 60$$
$$2x = 120$$

Replace x with 60 in the equation $y = 200 - 3x$.
$$y = 200 - 3(60) = 20$$
120 liters of 25% solution,
60 liters of 40% solution, and
20 liters of 50% solution are used.

37. Let x = the number of free throws
 y = the number of two-point field goals
 z = the number of three-point field goals

$$\begin{cases} x + 2y + 3z = 26 \\ \quad\quad x = z + 2 \\ \quad\quad y = x + 4 \end{cases}$$

$$\begin{cases} x + 2y + 3z = 26 \quad (1) \\ x \quad\quad - z = 2 \quad (2) \\ -x + \ y \quad\quad = 4 \quad (3) \end{cases}$$

Multiply equation (2) by -1 and add to equation (1).
$$x + 2y + 3z = 26$$
$$\underline{-x + \quad\quad z = -2}$$
$$2y + 4z = 24 \quad (4)$$

Add equation (3) to equation (1).
$$x + 2y + 3z = 26$$
$$\underline{-x + \ y \quad\quad = 4}$$
$$3y + 3z = 30 \quad (5)$$

Multiply equation (5) by $-\dfrac{2}{3}$ and add to equation (4).
$$2y + 4z = 24$$
$$\underline{-2y - 2z = -20}$$
$$2z = 4$$
$$z = 2$$

Replace z with 2 in equation (2).
$$x - (2) = 2$$
$$x = 4$$

Replace x with 4 in equation (3).
$$-(4) + y = 4$$
$$y = 8$$

He scored 4 free throws, 8 two-point field goals, and 2 three-point field goals.

39. Answers may vary. No matter how much of each acid solution is used, the strength of the mixture should be somewhere between 25% and 60%. So, you should suspect an error in the calculation if the result is 14%.

41. $y = ax^2 + bx + c$

$(1,2):$ $2 = a + b + c$ (1)

$(2,3):$ $3 = 4a + 2b + c$ (2)

$(-1,6):$ $6 = a - b + c$ (3)

Add equations (1) and (3).

$2 = a + b + c$

$\underline{6 = a - b + c}$

$8 = 2a \quad + 2c$ (4)

Multiply equation (3) by 2 and add to equation (2).

$3 = 4a + 2b + c$

$\underline{12 = 2a - 2b + 2c}$

$15 = 6a \quad + 3c$ (5)

Multiply equation (4) by -3 and add to equation (5).

$-24 = -6a - 6c$

$\underline{15 = \ 6a + 3c}$

$-9 = \quad\quad -3c$

$c = 3$

Replace c with 3 in equation (4).

$8 = 2a + 2(3)$

$2 = 2a$

$a = 1$

Replace a with 1 and c with 3 in equation (1).

$2 = 1 + b + 3$

$b = -2$

The values are $a = 1$, $b = -2$, and $c = 3$.

43. $x = 180 - (z + 15)$

$x = 165 - z$

$y = 180 - (z - 13)$

$y = 193 - z$

$360 = x + y + z + 72$

$288 = x + y + z$

$$\begin{cases} z = 165 - x \\ z = 193 - y \\ x + y + z = 288 \end{cases}$$

Replace z with $165 - x$ in the third equation.

$x + y + 165 - x = 288$

$\quad\quad\quad\quad y = 123$

Replace z with $193 - y$ in the third equation.

$x + y + 193 - y = 288$

$\quad\quad\quad\quad x = 95$

Replace x with 95 in the first equation.

$z = 165 - 95 = 70$

The values are $x = 95$, $y = 123$, and $z = 70$.

45. $y = ax^2 + bx + c$

$(4, 2.47)$: $\quad 2.47 = 16a + 4b + c$

$(7, 0.6)$: $\quad 0.6 = 49a + 7b + c$

$(8, 1.1)$: $\quad 1.1 = 64a + 8b + c$

Multiply the second equation by -1 and add to the first equation.

$$2.47 = \quad 16a + 4b + c$$
$$\underline{-0.6 = -49a - 7b - c}$$
$$1.87 = -33a - 3b$$

Multiply the second equation by -1 and add to the third equation.

$$-0.6 = -49a - 7b - c$$
$$\underline{1.1 = \quad 64a + 8b + c}$$
$$0.5 = \quad 15a + \quad b$$

$$b = 0.5 - 15a$$

Replace b with $0.5 - 15a$.

$1.87 = -33a - 3(0.5 - 15a)$

$1.87 = -33a - 1.5 + 45a$

$3.37 = 12a$

$a = \dfrac{3.37}{12} \approx 0.28$

Replace a with $\dfrac{3.37}{12}$.

$b = 0.5 - 15\left(\dfrac{3.37}{12}\right)$

$b = \dfrac{-297}{80} \approx -3.71$

$c = 2.47 - 16a - 4b$

$c = 2.47 - 16\left(\dfrac{3.37}{12}\right) - 4\left(\dfrac{-297}{80}\right)$

$c \approx 12.83$

The values are $a \approx 0.28$, $b \approx -3.71$, $c \approx 12.83$. The model is

$y = 0.28x^2 - 3.71x + 12.83$.

For September, $x = 9$.

$y = 0.28(9)^2 - 3.71(9) + 12.83 = 2.12$

Portland should expect 2.12 inches of rain in September.

47. $\begin{cases} 2x + y + 3z = 7 & (1) \\ -4x + y + 2z = 4 & (2) \end{cases}$

Multiply equation (1) by 2 and add to equation (2).

$$4x + 2y + 6z = 14$$
$$\underline{-4x + \quad y + 2z = 4}$$
$$3y + 8z = 18$$

49. $\begin{cases} 2x - 3y + 2z = 5 & (1) \\ x - 9y + \quad z = -1 & (2) \end{cases}$

Multiply equation (1) by -3 and add to equation (2).

$$-6x + 9y - 6z = -15$$
$$\underline{x - 9y + \quad z = \quad -1}$$
$$-5x \qquad - 5z = -16$$

51. $P(\text{green}) = \dfrac{3}{8}$

53. $P(\text{red or blue}) = \dfrac{5}{8}$

Exercise Set 4.4

1. $\begin{cases} x + \quad y = 1 \\ x - 2y = 4 \end{cases}$

$\begin{bmatrix} 1 & 1 & \vdots & 1 \\ 1 & -2 & \vdots & 4 \end{bmatrix}$

Multiply row 1 by -1 and add to row 2.

$\begin{bmatrix} 1 & 1 & \vdots & 1 \\ 0 & -3 & \vdots & 3 \end{bmatrix}$

Divide row 2 by -3.

$\begin{bmatrix} 1 & 1 & \vdots & 1 \\ 0 & 1 & \vdots & -1 \end{bmatrix}$

This corresponds to:

$\begin{cases} x + y = \quad 1 \\ \quad\quad y = -1 \end{cases}$

$x + (-1) = 1$

$x - 1 = 1$

$x = 2$

The solution is $(2, -1)$.

3. $\begin{cases} x + 3y = 2 \\ x + 2y = 0 \end{cases}$

$\begin{bmatrix} 1 & 3 & | & 2 \\ 1 & 2 & | & 0 \end{bmatrix}$

Multiply row 1 by –1 and add to row 2.

$\begin{bmatrix} 1 & 3 & | & 2 \\ 0 & -1 & | & -2 \end{bmatrix}$

Multiply row 2 by –1.

$\begin{bmatrix} 1 & 3 & | & 2 \\ 0 & 1 & | & 2 \end{bmatrix}$

This corresponds to:

$\begin{cases} x + 3y = 2 \\ \quad\quad y = 2 \end{cases}$

$x + 3(2) = 2$

$x + 6 = 2$

$x = -4$

The solution is (–4, 2).

5. $\begin{cases} x - 2y = 4 \\ 2x - 4y = 4 \end{cases}$

$\begin{bmatrix} 1 & -2 & | & 4 \\ 2 & -4 & | & 4 \end{bmatrix}$

Multiply row 1 by –2 and add to row 2.

$\begin{bmatrix} 1 & -2 & | & 4 \\ 0 & 0 & | & -4 \end{bmatrix}$

This corresponds to:

$\begin{cases} x - 2y = 4 \\ \quad\quad 0 = -4 \end{cases}$

This is an inconsistent system.
The solution set is $\varnothing$.

7. $\begin{cases} 3x - 3y = 9 \\ 2x - 2y = 6 \end{cases}$

$\begin{bmatrix} 3 & -3 & | & 9 \\ 2 & -2 & | & 6 \end{bmatrix}$

Divide row 1 by 3.

$\begin{bmatrix} 1 & -1 & | & 3 \\ 2 & -2 & | & 6 \end{bmatrix}$

Multiply row 1 by –2 and add to row 2.

$\begin{bmatrix} 1 & -1 & | & 3 \\ 0 & 0 & | & 0 \end{bmatrix}$

This corresponds to:

$\begin{cases} x - y = 3 \\ \quad\; 0 = 0 \end{cases}$

This is a dependent system.
The solution set is $\{(x, y) \mid x - y = 3\}$.

9. $\begin{cases} x+ \ y \qquad = 3 \\ \quad \ 2y \qquad = 10 \\ 3x+2y-4z = 12 \end{cases}$

$$\begin{bmatrix} 1 & 1 & 0 & | & 3 \\ 0 & 2 & 0 & | & 10 \\ 3 & 2 & -4 & | & 12 \end{bmatrix}$$

Multiply row 1 by –3 and add to row 3.

$$\begin{bmatrix} 1 & 1 & 0 & | & 3 \\ 0 & 2 & 0 & | & 10 \\ 0 & -1 & -4 & | & 3 \end{bmatrix}$$

Divide row 2 by 2.

$$\begin{bmatrix} 1 & 1 & 0 & | & 3 \\ 0 & 1 & 0 & | & 5 \\ 0 & -1 & -4 & | & 3 \end{bmatrix}$$

Add row 2 to row 3.

$$\begin{bmatrix} 1 & 1 & 0 & | & 3 \\ 0 & 1 & 0 & | & 5 \\ 0 & 0 & -4 & | & 8 \end{bmatrix}$$

Divide row 3 by –4.

$$\begin{bmatrix} 1 & 1 & 0 & | & 3 \\ 0 & 1 & 0 & | & 5 \\ 0 & 0 & 1 & | & -2 \end{bmatrix}$$

This corresponds to:

$$\begin{cases} x+y = 3 \\ \quad y = 5 \\ \qquad z = -2 \end{cases}$$

$x+5 = 3$
$\quad x = -2$

The solution is $(-2, 5, -2)$.

11. $\begin{cases} \quad \ 2y- \ z = -7 \\ x+4y+ \ z = -4 \\ 5x-y+2z = 13 \end{cases}$

$$\begin{bmatrix} 0 & 2 & -1 & | & -7 \\ 1 & 4 & 1 & | & -4 \\ 5 & -1 & 2 & | & 13 \end{bmatrix}$$

Interchange rows 1 and 2.

$$\begin{bmatrix} 1 & 4 & 1 & | & -4 \\ 0 & 2 & -1 & | & -7 \\ 5 & -1 & 2 & | & 13 \end{bmatrix}$$

Multiply row 1 by –5 and add to row 3.

$$\begin{bmatrix} 1 & 4 & 1 & | & -4 \\ 0 & 2 & -1 & | & -7 \\ 0 & -21 & -3 & | & 33 \end{bmatrix}$$

Divide row 2 by 2.

$$\begin{bmatrix} 1 & 4 & 1 & | & -4 \\ 0 & 1 & -\frac{1}{2} & | & -\frac{7}{2} \\ 0 & -21 & -3 & | & 33 \end{bmatrix}$$

Multiply row 2 by 21 and add to row 3.

$$\begin{bmatrix} 1 & 4 & 1 & | & -4 \\ 0 & 1 & -\frac{1}{2} & | & -\frac{7}{2} \\ 0 & 0 & -\frac{27}{2} & | & -\frac{81}{2} \end{bmatrix}$$

Multiply row 3 by $-\dfrac{2}{27}$.

$$\begin{bmatrix} 1 & 4 & 1 & | & -4 \\ 0 & 1 & -\frac{1}{2} & | & -\frac{7}{2} \\ 0 & 0 & 1 & | & 3 \end{bmatrix}$$

This corresponds to:

$$\begin{cases} x+4y+ \ z = -4 \\ \quad y-\frac{1}{2}z = -\frac{7}{2} \\ \qquad z = 3 \end{cases}$$

$y-\dfrac{1}{2}(3) = -\dfrac{7}{2}$

$\quad y-\dfrac{3}{2} = -\dfrac{7}{2}$

$\qquad y = -2$

$x+4(-2)+3 = -4$

$\quad x-8+3 = -4$

$\qquad x = 1$

The solution is $(1, -2, 3)$.

13. $\begin{cases} x - 4 = 0 \\ x + y = 1 \end{cases}$ or $\begin{cases} x = 4 \\ x + y = 1 \end{cases}$

$$\begin{bmatrix} 1 & 0 & | & 4 \\ 1 & 1 & | & 1 \end{bmatrix}$$

Multiply row 1 by -1 and add to row 2.

$$\begin{bmatrix} 1 & 0 & | & 4 \\ 0 & 1 & | & -3 \end{bmatrix}$$

This corresponds to:

$$\begin{cases} x = 4 \\ y = -3 \end{cases}$$

The solution is $(4, -3)$.

15. $\begin{cases} x + y + z = 2 \\ 2x \quad - z = 5 \\ 3y + z = 2 \end{cases}$

$$\begin{bmatrix} 1 & 1 & 1 & | & 2 \\ 2 & 0 & -1 & | & 5 \\ 0 & 3 & 1 & | & 2 \end{bmatrix}$$

Multiply row 1 by -2 and add to row 2.

$$\begin{bmatrix} 1 & 1 & 1 & | & 2 \\ 0 & -2 & -3 & | & 1 \\ 0 & 3 & 1 & | & 2 \end{bmatrix}$$

Divide row 2 by -2.

$$\begin{bmatrix} 1 & 1 & 1 & | & 2 \\ 0 & 1 & \frac{3}{2} & | & -\frac{1}{2} \\ 0 & 3 & 1 & | & 2 \end{bmatrix}$$

Multiply row 2 by -3 and add to row 3.

$$\begin{bmatrix} 1 & 1 & 1 & | & 2 \\ 0 & 1 & \frac{3}{2} & | & -\frac{1}{2} \\ 0 & 0 & -\frac{7}{2} & | & \frac{7}{2} \end{bmatrix}$$

Multiply row 3 by $-\dfrac{2}{7}$.

$$\begin{bmatrix} 1 & 1 & 1 & | & 2 \\ 0 & 1 & \frac{3}{2} & | & -\frac{1}{2} \\ 0 & 0 & 1 & | & -1 \end{bmatrix}$$

This corresponds to:

$$\begin{cases} x + y + z = 2 \\ y + \frac{3}{2} z = -\frac{1}{2} \\ z = -1 \end{cases}$$

$$y + \frac{3}{2}(-1) = -\frac{1}{2}$$
$$y - \frac{3}{2} = -\frac{1}{2}$$
$$y = 1$$
$$x + 1 + (-1) = 2$$
$$x = 2$$

The solution is $(2, 1, -1)$.

17. $\begin{cases} 5x - 2y = 27 \\ -3x + 5y = 18 \end{cases}$

$$\begin{bmatrix} 5 & -2 & | & 27 \\ -3 & 5 & | & 18 \end{bmatrix}$$

Divide row 1 by 5.

$$\begin{bmatrix} 1 & -\frac{2}{5} & | & \frac{27}{5} \\ -3 & 5 & | & 18 \end{bmatrix}$$

Multiply row 1 by 3 and add to row 2.

$$\begin{bmatrix} 1 & -\frac{2}{5} & | & \frac{27}{5} \\ 0 & \frac{19}{5} & | & \frac{171}{5} \end{bmatrix}$$

Multiply row 2 by $\dfrac{5}{19}$.

$$\begin{bmatrix} 1 & -\frac{2}{5} & | & \frac{27}{5} \\ 0 & 1 & | & 9 \end{bmatrix}$$

Multiply row 2 by $\dfrac{2}{5}$ and add to row 1.

$$\begin{bmatrix} 1 & 0 & | & 9 \\ 0 & 1 & | & 9 \end{bmatrix}$$

This corresponds to:

$$\begin{cases} x = 9 \\ y = 9 \end{cases}$$

The solution is $(9, 9)$.

19. $\begin{cases} 4x - 7y = 7 \\ 12x - 21y = 24 \end{cases}$

$\begin{bmatrix} 4 & -7 & | & 7 \\ 12 & -21 & | & 24 \end{bmatrix}$

Divide row 1 by 4.

$\begin{bmatrix} 1 & -\frac{7}{4} & | & \frac{7}{4} \\ 12 & -21 & | & 24 \end{bmatrix}$

Multiply row 1 by -12 and add to row 2.

$\begin{bmatrix} 1 & -\frac{7}{4} & | & \frac{7}{4} \\ 0 & 0 & | & 3 \end{bmatrix}$

This corresponds to:

$\begin{cases} x - \dfrac{7}{4}y = \dfrac{7}{4} \\ \qquad 0 = 3 \end{cases}$

This is an inconsistent system.
The solution set is $\varnothing$.

21. $\begin{cases} 4x - y + 2z = 5 \\ 2y + z = 4 \\ 4x + y + 3z = 10 \end{cases}$

$\begin{bmatrix} 4 & -1 & 2 & | & 5 \\ 0 & 2 & 1 & | & 4 \\ 4 & 1 & 3 & | & 10 \end{bmatrix}$

Divide row 1 by 4.

$\begin{bmatrix} 1 & -\frac{1}{4} & \frac{1}{2} & | & \frac{5}{4} \\ 0 & 2 & 1 & | & 4 \\ 4 & 1 & 3 & | & 10 \end{bmatrix}$

Multiply row 1 by -4 and add to row 3.

$\begin{bmatrix} 1 & -\frac{1}{4} & \frac{1}{2} & | & \frac{5}{4} \\ 0 & 2 & 1 & | & 4 \\ 0 & 2 & 1 & | & 5 \end{bmatrix}$

Divide row 2 by 2.

$\begin{bmatrix} 1 & -\frac{1}{4} & \frac{1}{2} & | & \frac{5}{4} \\ 0 & 1 & \frac{1}{2} & | & 2 \\ 0 & 2 & 1 & | & 5 \end{bmatrix}$

Multiply row 2 by -2 and add to row 3.

$\begin{bmatrix} 1 & -\frac{1}{4} & \frac{1}{2} & | & \frac{5}{4} \\ 0 & 1 & \frac{1}{2} & | & 2 \\ 0 & 0 & 0 & | & 1 \end{bmatrix}$

This corresponds to:

$\begin{cases} x - \dfrac{1}{4}y + \dfrac{1}{2}z = \dfrac{5}{4} \\ \qquad y + \dfrac{1}{2}z = 2 \\ \qquad\qquad 0 = 1 \end{cases}$

This is an inconsistent system.
The solution set is $\varnothing$.

23. $\begin{cases} 4x + y + z = 3 \\ -x + y - 2z = -11 \\ x + 2y + 2z = -1 \end{cases}$

$$\begin{bmatrix} 4 & 1 & 1 & | & 3 \\ -1 & 1 & -2 & | & -11 \\ 1 & 2 & 2 & | & -1 \end{bmatrix}$$

Interchange rows 1 and 3.

$$\begin{bmatrix} 1 & 2 & 2 & | & -1 \\ -1 & 1 & -2 & | & -11 \\ 4 & 1 & 1 & | & 3 \end{bmatrix}$$

Multiply row 1 by 1 and add to row 2.
Multiply row 1 by –4 and add to row 3.

$$\begin{bmatrix} 1 & 2 & 2 & | & -1 \\ 0 & 3 & 0 & | & -12 \\ 0 & -7 & -7 & | & 7 \end{bmatrix}$$

Divide row 2 by 3.

$$\begin{bmatrix} 1 & 2 & 2 & | & -1 \\ 0 & 1 & 0 & | & -4 \\ 0 & -7 & -7 & | & 7 \end{bmatrix}$$

Multiply row 2 by 7 and add to row 3.

$$\begin{bmatrix} 1 & 2 & 2 & | & -1 \\ 0 & 1 & 0 & | & -4 \\ 0 & 0 & -7 & | & -21 \end{bmatrix}$$

Divide row 3 by –7.

$$\begin{bmatrix} 1 & 2 & 2 & | & -1 \\ 0 & 1 & 0 & | & -4 \\ 0 & 0 & 1 & | & 3 \end{bmatrix}$$

This corresponds to:

$$\begin{cases} x + 2y + 2z = -1 \\ y = -4 \\ z = 3 \end{cases}$$

$$x + 2(-4) + 2(3) = -1$$
$$x - 8 + 6 = -1$$
$$x = 1$$

The solution is (1, –4, 3).

25. Answers may vary.

27. It is the graph of a function because it passes the vertical line test.

29. It is not the graph of a function because it fails the vertical line test.

31. $(-1)(-5) - (6)(3) = 5 - 18 = -13$

33. $(4)(-10) - (2)(-2) = -40 - (-4)$
$$= -40 + 4$$
$$= -36$$

35. $(-3)(-3) - (-1)(-9) = 9 - 9 = 0$

Exercise Set 4.5

1. $\begin{vmatrix} 3 & 5 \\ -1 & 7 \end{vmatrix} = 3(7) - 5(-1)$
$$= 21 + 5$$
$$= 26$$

3. $\begin{vmatrix} 9 & -2 \\ 4 & -3 \end{vmatrix} = 9(-3) - 4(-2)$
$$= -27 + 8$$
$$= -19$$

5. $\begin{vmatrix} -2 & 9 \\ 4 & -18 \end{vmatrix} = -2(-18) - 9(4)$
$$= 36 - 36$$
$$= 0$$

7. $\begin{cases} 2y - 4 = 0 \\ x + 2y = 5 \end{cases}$

or $\begin{cases} 2y = 4 \\ x + 2y = 5 \end{cases}$

$D = \begin{vmatrix} 0 & 2 \\ 1 & 2 \end{vmatrix} = 0(2) - 2(1) = 0 - 2 = -2$

$D_x = \begin{vmatrix} 4 & 2 \\ 5 & 2 \end{vmatrix} = 4(2) - 2(5) = 8 - 10 = -2$

$D_y = \begin{vmatrix} 0 & 4 \\ 1 & 5 \end{vmatrix} = 0(5) - 4(1) = 0 - 4 = -4$

$x = \dfrac{D_x}{D} = \dfrac{-2}{-2} = 1$

$y = \dfrac{D_y}{D} = \dfrac{-4}{-2} = 2$

The solution is $(1, 2)$.

9. $\begin{cases} 3x + y = 1 \\ \quad 2y = 2 - 6x \end{cases}$

or $\begin{cases} 3x + y = 1 \\ 6x + 2y = 2 \end{cases}$

$D = \begin{vmatrix} 3 & 1 \\ 6 & 2 \end{vmatrix} = 3(2) - 1(6) = 6 - 6 = 0$

Thus, this system cannot be solved using Cramer's rule. Since the second equation is 2 times the first equation, the system is dependent. The solution set is $\{(x, y) \mid 3x + y = 1\}$.

11. $\begin{cases} 5x - 2y = 27 \\ -3x + 5y = 18 \end{cases}$

$D = \begin{vmatrix} 5 & -2 \\ -3 & 5 \end{vmatrix}$

$= 5(5) - (-2)(-3)$

$= 25 - 6$

$= 19$

$D_x = \begin{vmatrix} 27 & -2 \\ 18 & 5 \end{vmatrix}$

$= 27(5) - (-2)18$

$= 135 + 36$

$= 171$

$D_y = \begin{vmatrix} 5 & 27 \\ -3 & 18 \end{vmatrix}$

$= 5(18) - 27(-3)$

$= 90 + 81$

$= 171$

$x = \dfrac{D_x}{D} = \dfrac{171}{19} = 9$

$y = \dfrac{D_y}{D} = \dfrac{171}{19} = 9$

The solution is $(9, 9)$.

13. Evaluate the determinant by expanding by the minors of the first row.

$\begin{vmatrix} 2 & 1 & 0 \\ 0 & 5 & -3 \\ 4 & 0 & 2 \end{vmatrix}$

$= 2\begin{vmatrix} 5 & -3 \\ 0 & 2 \end{vmatrix} - 1\begin{vmatrix} 0 & -3 \\ 4 & 2 \end{vmatrix} + 0\begin{vmatrix} 0 & 5 \\ 4 & 0 \end{vmatrix}$

$= 2[5(2) - (-3)(0)] - [0(2) - 4(-3)] + 0$

$= 2(10) - 12$

$= 8$

15. Evaluate the determinant by expanding by the minors of the third column.

$\begin{vmatrix} 4 & -6 & 0 \\ -2 & 3 & 0 \\ 4 & -6 & 1 \end{vmatrix}$

$= 0\begin{vmatrix} -2 & 3 \\ 4 & -6 \end{vmatrix} - 0\begin{vmatrix} 4 & -6 \\ 4 & -6 \end{vmatrix} + 1\begin{vmatrix} 4 & -6 \\ -2 & 3 \end{vmatrix}$

$= 0 - 0 + [4(3) - (-6)(-2)]$

$= 0$

17. Evaluate the determinant by expanding
by the minors of the first row.

$$\begin{vmatrix} 3 & 6 & -3 \\ -1 & -2 & 3 \\ 4 & -1 & 6 \end{vmatrix} = 3\begin{vmatrix} -2 & 3 \\ -1 & 6 \end{vmatrix} - 6\begin{vmatrix} -1 & 3 \\ 4 & 6 \end{vmatrix} + (-3)\begin{vmatrix} -1 & -2 \\ 4 & -1 \end{vmatrix}$$

$$= 3[-2(6) - 3(-1)] - 6[-1(6) - 3(4)] - 3[(-1)(-1) - (-2)4]$$
$$= 3(-9) - 6(-18) - 3(9)$$
$$= -27 + 108 - 27$$
$$= 54$$

19. $\begin{cases} 3x \quad\quad + z = -1 \\ -x - 3y + z = 7 \\ \quad\quad 3y + z = 5 \end{cases}$

$$D = \begin{vmatrix} 3 & 0 & 1 \\ -1 & -3 & 1 \\ 0 & 3 & 1 \end{vmatrix} = 3\begin{vmatrix} -3 & 1 \\ 3 & 1 \end{vmatrix} - 0\begin{vmatrix} -1 & 1 \\ 0 & 1 \end{vmatrix} + 1\begin{vmatrix} -1 & -3 \\ 0 & 3 \end{vmatrix} = 3(-3 - 3) - 0 + (-3) = -18 - 3 = -21$$

$$D_x = \begin{vmatrix} -1 & 0 & 1 \\ 7 & -3 & 1 \\ 5 & 3 & 1 \end{vmatrix} = -1\begin{vmatrix} -3 & 1 \\ 3 & 1 \end{vmatrix} - 0\begin{vmatrix} 7 & 1 \\ 5 & 1 \end{vmatrix} + 1\begin{vmatrix} 7 & -3 \\ 5 & 3 \end{vmatrix} = -(-3 - 3) - 0 + [21 - (-15)] = 6 + 36 = 42$$

$$D_y = \begin{vmatrix} 3 & -1 & 1 \\ -1 & 7 & 1 \\ 0 & 5 & 1 \end{vmatrix} = 3\begin{vmatrix} 7 & 1 \\ 5 & 1 \end{vmatrix} - (-1)\begin{vmatrix} -1 & 1 \\ 0 & 1 \end{vmatrix} + 1\begin{vmatrix} -1 & 7 \\ 0 & 5 \end{vmatrix} = 3(7 - 5) + (-1) + (-5) = 6 - 1 - 5 = 0$$

$$D_z = \begin{vmatrix} 3 & 0 & -1 \\ -1 & -3 & 7 \\ 0 & 3 & 5 \end{vmatrix} = 3\begin{vmatrix} -3 & 7 \\ 3 & 5 \end{vmatrix} - 0\begin{vmatrix} -1 & 7 \\ 0 & 5 \end{vmatrix} + (-1)\begin{vmatrix} -1 & -3 \\ 0 & 3 \end{vmatrix} = 3(-15 - 21) - 0 - (-3) = -108 + 3 = -105$$

$$x = \frac{D_x}{D} = \frac{42}{-21} = -2$$

$$y = \frac{D_y}{D} = \frac{0}{-21} = 0$$

$$z = \frac{D_z}{D} = \frac{-105}{-21} = 5$$

The solution is $(-2, 0, 5)$.

21. $\begin{cases} x + y + z = 8 \\ 2x - y - z = 10 \\ x - 2y + 3z = 22 \end{cases}$

$D = \begin{vmatrix} 1 & 1 & 1 \\ 2 & -1 & -1 \\ 1 & -2 & 3 \end{vmatrix}$

$= 1\begin{vmatrix} -1 & -1 \\ -2 & 3 \end{vmatrix} - 1\begin{vmatrix} 2 & -1 \\ 1 & 3 \end{vmatrix} + 1\begin{vmatrix} 2 & -1 \\ 1 & -2 \end{vmatrix}$

$= (-3 - 2) - [6 - (-1)] + [-4 - (-1)]$

$= -5 - 7 - 3$

$= -15$

$D_x = \begin{vmatrix} 8 & 1 & 1 \\ 10 & -1 & -1 \\ 22 & -2 & 3 \end{vmatrix}$

$= 8\begin{vmatrix} -1 & -1 \\ -2 & 3 \end{vmatrix} - 1\begin{vmatrix} 10 & -1 \\ 22 & 3 \end{vmatrix} + 1\begin{vmatrix} 10 & -1 \\ 22 & -2 \end{vmatrix}$

$= 8(-3 - 2) - [30 - (-22)] + [-20 - (-22)]$

$= 8(-5) - 52 + 2$

$= -40 - 52 + 2$

$= -90$

$D_y = \begin{vmatrix} 1 & 8 & 1 \\ 2 & 10 & -1 \\ 1 & 22 & 3 \end{vmatrix}$

$= 1\begin{vmatrix} 10 & -1 \\ 22 & 3 \end{vmatrix} - 8\begin{vmatrix} 2 & -1 \\ 1 & 3 \end{vmatrix} + 1\begin{vmatrix} 2 & 10 \\ 1 & 22 \end{vmatrix}$

$= [30 - (-22)] - 8[6 - (-1)] + [44 - 10]$

$= 52 - 8(7) + 34$

$= 52 - 56 + 34$

$= 30$

$D_z = \begin{vmatrix} 1 & 1 & 8 \\ 2 & -1 & 10 \\ 1 & -2 & 22 \end{vmatrix}$

$= 1\begin{vmatrix} -1 & 10 \\ -2 & 22 \end{vmatrix} - 1\begin{vmatrix} 2 & 10 \\ 1 & 22 \end{vmatrix} + 8\begin{vmatrix} 2 & -1 \\ 1 & -2 \end{vmatrix}$

$= [-22 - (-20)] - (44 - 10) + 8[-4 - (-1)]$

$= -2 - 34 + 8(-3)$

$= -2 - 34 - 24$

$= -60$

$x = \dfrac{D_x}{D} = \dfrac{-90}{-15} = 6$

$y = \dfrac{D_y}{D} = \dfrac{30}{-15} = -2$

$z = \dfrac{D_z}{D} = \dfrac{-60}{-15} = 4$

The solution is (6, –2, 4).

23. $\begin{vmatrix} 10 & -1 \\ -4 & 2 \end{vmatrix} = 10(2) - (-1)(-4)$

$= 20 - 4$

$= 16$

25. Evaluate the determinant by expanding by the minors of the first row.

$\begin{vmatrix} 1 & 0 & 4 \\ 1 & -1 & 2 \\ 3 & 2 & 1 \end{vmatrix} = 1\begin{vmatrix} -1 & 2 \\ 2 & 1 \end{vmatrix} - 0\begin{vmatrix} 1 & 2 \\ 3 & 1 \end{vmatrix} + 4\begin{vmatrix} 1 & -1 \\ 3 & 2 \end{vmatrix}$

$= [-1 - 4] - 0 + 4[2 - (-3)]$

$= -5 + 4(5)$

$= -5 + 20$

$= 15$

27. $\begin{vmatrix} \frac{3}{4} & \frac{5}{2} \\ -\frac{1}{6} & \frac{7}{3} \end{vmatrix} = \dfrac{3}{4}\left(\dfrac{7}{3}\right) - \dfrac{5}{2}\left(-\dfrac{1}{6}\right)$

$= \dfrac{21}{12} + \dfrac{5}{12}$

$= \dfrac{26}{12}$

$= \dfrac{13}{6}$

29. Evaluate the determinant by expanding by the minors of the first row.

$\begin{vmatrix} 4 & -2 & 2 \\ 6 & -1 & 3 \\ 2 & 1 & 1 \end{vmatrix}$

$= 4\begin{vmatrix} -1 & 3 \\ 1 & 1 \end{vmatrix} - (-2)\begin{vmatrix} 6 & 3 \\ 2 & 1 \end{vmatrix} + 2\begin{vmatrix} 6 & -1 \\ 2 & 1 \end{vmatrix}$

$= 4[-1 - 3] + 2[6 - 6] + 2[6 - (-2)]$

$= 4(-4) + 2 \cdot 0 + 2(8)$

$= -16 + 0 + 16$

$= 0$

31. Evaluate the determinant by expanding by the minors of the first row.

$$\begin{vmatrix} -2 & 5 & 4 \\ 5 & -1 & 3 \\ 4 & 1 & 2 \end{vmatrix} = -2\begin{vmatrix} -1 & 3 \\ 1 & 2 \end{vmatrix} - 5\begin{vmatrix} 5 & 3 \\ 4 & 2 \end{vmatrix} + 4\begin{vmatrix} 5 & -1 \\ 4 & 1 \end{vmatrix}$$

$$= -2[-2-3] - 5[10-12] + 4[5-(-4)]$$
$$= -2(-5) - 5(-2) + 4(9)$$
$$= 10 + 10 + 36$$
$$= 56$$

33. 0; Answers may vary.

35.
$$\begin{vmatrix} 1 & x \\ 2 & 7 \end{vmatrix} = -3$$
$$(1)(7) - 2x = -3$$
$$7 - 2x = -3$$
$$-2x = -10$$
$$x = 5$$

37. $\begin{cases} 2x - 5y = 4 \\ x + 2y = -7 \end{cases}$

$$D = \begin{vmatrix} 2 & -5 \\ 1 & 2 \end{vmatrix} = 2(2) - (-5)(1) = 4 + 5 = 9$$

$$D_x = \begin{vmatrix} 4 & -5 \\ -7 & 2 \end{vmatrix} = 4(2) - (-5)(-7) = 8 - 35 = -27$$

$$D_y = \begin{vmatrix} 2 & 4 \\ 1 & -7 \end{vmatrix} = 2(-7) - 4(1) = -14 - 4 = -18$$

$$x = \frac{D_x}{D} = \frac{-27}{9} = -3$$

$$y = \frac{D_y}{D} = \frac{-18}{9} = -2$$

The solution is $(-3, -2)$.

39. $\begin{cases} 4x + 2y = 5 \\ 2x + y = -1 \end{cases}$

$$D = \begin{vmatrix} 4 & 2 \\ 2 & 1 \end{vmatrix} = 4(1) - 2(2) = 4 - 4 = 0$$

Thus, Cramer's rule cannot be used to solve the system. Multiply the second equation by 2.
$$\begin{cases} 4x + 2y = 5 \\ 4x + 2y = -2 \end{cases}$$
The system is inconsistent. The solution set is $\varnothing$.

41. $\begin{cases} 2x + 2y + z = 1 \\ -x + y + 2z = 3 \\ x + 2y + 4z = 0 \end{cases}$

$D = \begin{vmatrix} 2 & 2 & 1 \\ -1 & 1 & 2 \\ 1 & 2 & 4 \end{vmatrix}$

$= 2\begin{vmatrix} 1 & 2 \\ 2 & 4 \end{vmatrix} - 2\begin{vmatrix} -1 & 2 \\ 1 & 4 \end{vmatrix} + 1\begin{vmatrix} -1 & 1 \\ 1 & 2 \end{vmatrix}$

$= 2(4-4) - 2(-4-2) + (-2-1)$

$= 2(0) - 2(-6) + (-3)$

$= 0 + 12 - 3$

$= 9$

$D_x = \begin{vmatrix} 1 & 2 & 1 \\ 3 & 1 & 2 \\ 0 & 2 & 4 \end{vmatrix}$

$= 1\begin{vmatrix} 1 & 2 \\ 2 & 4 \end{vmatrix} - 3\begin{vmatrix} 2 & 1 \\ 2 & 4 \end{vmatrix} + 0\begin{vmatrix} 2 & 1 \\ 1 & 2 \end{vmatrix}$

$= (4-4) - 3(8-2) + 0$

$= 0 - 3(6)$

$= -18$

$D_y = \begin{vmatrix} 2 & 1 & 1 \\ -1 & 3 & 2 \\ 1 & 0 & 4 \end{vmatrix}$

$= 1\begin{vmatrix} 1 & 1 \\ 3 & 2 \end{vmatrix} - 0\begin{vmatrix} 2 & 1 \\ -1 & 2 \end{vmatrix} + 4\begin{vmatrix} 2 & 1 \\ -1 & 3 \end{vmatrix}$

$= (2-3) - 0 + 4[6-(-1)]$

$= -1 + 4(7)$

$= -1 + 28$

$= 27$

$D_z = \begin{vmatrix} 2 & 2 & 1 \\ -1 & 1 & 3 \\ 1 & 2 & 0 \end{vmatrix}$

$= 1\begin{vmatrix} 2 & 1 \\ 1 & 3 \end{vmatrix} - 2\begin{vmatrix} 2 & 1 \\ -1 & 3 \end{vmatrix} + 0\begin{vmatrix} 2 & 2 \\ -1 & 1 \end{vmatrix}$

$= (6-1) - 2[6-(-1)] + 0$

$= 5 - 2(7)$

$= 5 - 14$

$= -9$

$x = \dfrac{D_x}{D} = -\dfrac{18}{9} = -2$

$y = \dfrac{D_y}{D} = \dfrac{27}{9} = 3$

$z = \dfrac{D_z}{D} = \dfrac{-9}{9} = -1$

The solution is $(-2, 3, -1)$.

43. $\begin{cases} \dfrac{2}{3}x - \dfrac{3}{4}y = -1 \\ -\dfrac{1}{6}x + \dfrac{3}{4}y = \dfrac{5}{2} \end{cases}$

$D = \begin{vmatrix} \dfrac{2}{3} & -\dfrac{3}{4} \\ -\dfrac{1}{6} & \dfrac{3}{4} \end{vmatrix}$

$= \dfrac{2}{3} \cdot \dfrac{3}{4} - \left(-\dfrac{3}{4}\right)\left(-\dfrac{1}{6}\right)$

$= \dfrac{1}{2} - \dfrac{1}{8}$

$= \dfrac{3}{8}$

$D_x = \begin{vmatrix} -1 & -\dfrac{3}{4} \\ \dfrac{5}{2} & \dfrac{3}{4} \end{vmatrix}$

$= (-1)\dfrac{3}{4} - \left(-\dfrac{3}{4}\right)\dfrac{5}{2}$

$= -\dfrac{3}{4} + \dfrac{15}{8}$

$= \dfrac{9}{8}$

$D_y = \begin{vmatrix} \dfrac{2}{3} & -1 \\ -\dfrac{1}{6} & \dfrac{5}{2} \end{vmatrix}$

$= \dfrac{2}{3} \cdot \dfrac{5}{2} - (-1)\left(-\dfrac{1}{6}\right)$

$= \dfrac{5}{3} - \dfrac{1}{6}$

$= \dfrac{3}{2}$

$x = \dfrac{D_x}{D} = \dfrac{\frac{9}{8}}{\frac{3}{8}} = 3$

$y = \dfrac{D_y}{D} = \dfrac{\frac{3}{2}}{\frac{3}{8}} = 4$

The solution is $(3, 4)$.

45. $\begin{cases} 0.7x - 0.2y = -1.6 \\ 0.2x - \quad y = -1.4 \end{cases}$

$D = \begin{vmatrix} 0.7 & -0.2 \\ 0.2 & -1 \end{vmatrix} = 0.7(-1) - (-0.2)(0.2) = -0.7 + 0.04 = -0.66$

$D_x = \begin{vmatrix} -1.6 & -0.2 \\ -1.4 & -1 \end{vmatrix} = (-1.6)(-1) - (-0.2)(-1.4) = 1.6 - 0.28 = 1.32$

$D_y = \begin{vmatrix} 0.7 & -1.6 \\ 0.2 & -1.4 \end{vmatrix} = (0.7)(-1.4) - (-1.6)(0.2) = -0.98 + 0.32 = -0.66$

$x = \dfrac{D_x}{D} = \dfrac{1.32}{-0.66} = -2$

$y = \dfrac{D_y}{D} = \dfrac{-0.66}{-0.66} = 1$

The solution is (–2, 1).

47. $\begin{cases} -2x + 4y - 2z = 6 \\ \quad x - 2y + z = -3 \\ 3x - 6y + 3z = -9 \end{cases}$

$D = \begin{vmatrix} -2 & 4 & -2 \\ 1 & -2 & 1 \\ 3 & -6 & 3 \end{vmatrix}$

$= -2\begin{vmatrix} -2 & 1 \\ -6 & 3 \end{vmatrix} - 4\begin{vmatrix} 1 & 1 \\ 3 & 3 \end{vmatrix} + (-2)\begin{vmatrix} 1 & -2 \\ 3 & -6 \end{vmatrix}$

$= -2[-6 - (-6)] - 4[3 - 3] - 2[-6 - (-6)]$

$= -2(0) - 4(0) - 2(0)$

$= 0$

Therefore, Cramer's rule will not provide the solution. Note that the first equation is –2 times the second equation and that the third equation is 3 times the second equation. Thus, the system is dependent. The solution set is $\{(x,\ y,\ z) \mid x - 2y + z = -3\}$.

49. $\begin{cases} x - 2y + z = -5 \\ 3y + 2z = 4 \\ 3x - y = -2 \end{cases}$

$D = \begin{vmatrix} 1 & -2 & 1 \\ 0 & 3 & 2 \\ 3 & -1 & 0 \end{vmatrix}$

$= 1\begin{vmatrix} 3 & 2 \\ -1 & 0 \end{vmatrix} - 0\begin{vmatrix} -2 & 1 \\ -1 & 0 \end{vmatrix} + 3\begin{vmatrix} -2 & 1 \\ 3 & 2 \end{vmatrix}$

$= [0 - (-2)] - 0 + 3[-4 - 3]$

$= 2 + 3(-7)$

$= 2 - 21$

$= -19$

$D_x = \begin{vmatrix} -5 & -2 & 1 \\ 4 & 3 & 2 \\ -2 & -1 & 0 \end{vmatrix}$

$= 1\begin{vmatrix} 4 & 3 \\ -2 & -1 \end{vmatrix} - 2\begin{vmatrix} -5 & -2 \\ -2 & -1 \end{vmatrix} + 0\begin{vmatrix} -5 & -2 \\ 4 & 3 \end{vmatrix}$

$= [-4 - (-6)] - 2[5 - 4] + 0$

$= 2 - 2(1)$

$= 0$

$D_y = \begin{vmatrix} 1 & -5 & 1 \\ 0 & 4 & 2 \\ 3 & -2 & 0 \end{vmatrix}$

$= 1\begin{vmatrix} 4 & 2 \\ -2 & 0 \end{vmatrix} - 0\begin{vmatrix} -5 & 1 \\ -2 & 0 \end{vmatrix} + 3\begin{vmatrix} -5 & 1 \\ 4 & 2 \end{vmatrix}$

$= [0 - (-4)] - 0 + 3[-10 - 4]$

$= 4 + 3(-14)$

$= 4 - 42$

$= -38$

$D_z = \begin{vmatrix} 1 & -2 & -5 \\ 0 & 3 & 4 \\ 3 & -1 & -2 \end{vmatrix}$

$= 1\begin{vmatrix} 3 & 4 \\ -1 & -2 \end{vmatrix} - 0\begin{vmatrix} -2 & -5 \\ -1 & -2 \end{vmatrix} + 3\begin{vmatrix} -2 & -5 \\ 3 & 4 \end{vmatrix}$

$= [-6 - (-4)] - 0 + 3[-8 - (-15)]$

$= -2 + 3(7)$

$= -2 + 21$

$= 19$

$x = \dfrac{D_x}{D} = \dfrac{0}{-19} = 0$

$y = \dfrac{D_y}{D} = \dfrac{-38}{-19} = 2$

$z = \dfrac{D_z}{D} = \dfrac{19}{-19} = -1$

The solution is $(0, 2, -1)$.

51. The array of signs for use with a 4×4 matrix is

$+ \quad - \quad + \quad -$

$- \quad + \quad - \quad +$

$+ \quad - \quad + \quad -$

$- \quad + \quad - \quad +$

53. $5x - 6 + x - 12 = 6x - 18$

55. $2(3x - 6) + 3(x - 1) = 6x - 12 + 3x - 3$
$\qquad\qquad\qquad\qquad\quad = 9x - 15$

57. $f(x) = 5x - 6$

59. $h(x) = 3$

61. Evaluate the determinant by expanding by the minors of the first row.

$$\begin{vmatrix} 5 & 0 & 0 & 0 \\ 0 & 4 & 2 & -1 \\ 1 & 3 & -2 & 0 \\ 0 & -3 & 1 & 2 \end{vmatrix} = 5\begin{vmatrix} 4 & 2 & -1 \\ 3 & -2 & 0 \\ -3 & 1 & 2 \end{vmatrix} - 0\begin{vmatrix} 0 & 2 & -1 \\ 1 & -2 & 0 \\ 0 & 1 & 2 \end{vmatrix} + 0\begin{vmatrix} 0 & 4 & -1 \\ 1 & 3 & 0 \\ 0 & -3 & 2 \end{vmatrix} - 0\begin{vmatrix} 0 & 4 & 2 \\ 1 & 3 & -2 \\ 0 & -3 & 1 \end{vmatrix}$$

$$= 5\begin{vmatrix} 4 & 2 & -1 \\ 3 & -2 & 0 \\ -3 & 1 & 2 \end{vmatrix} - 0 + 0 - 0$$

$$= 5\left((-1)\begin{vmatrix} 3 & -2 \\ -3 & 1 \end{vmatrix} - 0\begin{vmatrix} 4 & 2 \\ -3 & 1 \end{vmatrix} + 2\begin{vmatrix} 4 & 2 \\ 3 & -2 \end{vmatrix} \right)$$

$$= 5[-(3-6) - 0 + 2(-8-6)]$$
$$= 5[3 + 2(-14)]$$
$$= 5(3 - 28)$$
$$= 5(-25)$$
$$= -125$$

63. Evaluate D by expanding by the minors of the first column.

$$\begin{vmatrix} 4 & 0 & 2 & 5 \\ 0 & 3 & -1 & 1 \\ 0 & 0 & 2 & 0 \\ 0 & 0 & 0 & 1 \end{vmatrix} = 4\begin{vmatrix} 3 & -1 & 1 \\ 0 & 2 & 0 \\ 0 & 0 & 1 \end{vmatrix} - 0\begin{vmatrix} 0 & 2 & 5 \\ 0 & 2 & 0 \\ 0 & 0 & 1 \end{vmatrix} + 0\begin{vmatrix} 0 & 2 & 5 \\ 3 & -1 & 1 \\ 0 & 0 & 1 \end{vmatrix} - 0\begin{vmatrix} 0 & 2 & 5 \\ 3 & -1 & 1 \\ 0 & 2 & 0 \end{vmatrix}$$

$$= 4\begin{vmatrix} 3 & -1 & 1 \\ 0 & 2 & 0 \\ 0 & 0 & 1 \end{vmatrix} - 0 + 0 - 0$$

$$= 4\left(3\begin{vmatrix} 2 & 0 \\ 0 & 1 \end{vmatrix} - 0\begin{vmatrix} -1 & 1 \\ 0 & 1 \end{vmatrix} + 0\begin{vmatrix} -1 & 1 \\ 2 & 0 \end{vmatrix} \right)$$

$$= 4[3(2-0) - 0 + 0]$$
$$= 12(2)$$
$$= 24$$

Chapter 4 Review

1. $\begin{cases} 3x + 10y = 1 \\ x + 2y = -1 \end{cases}$

(1)

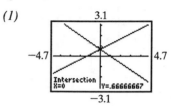

(2) $x = -2y - 1$
Replace x with $-2y - 1$ in the first equation.
$$3(-2y - 1) + 10y = 1$$
$$-6y - 3 + 10y = 1$$
$$4y - 3 = 1$$
$$4y = 4$$
$$y = 1$$
Replace y with 1 in the equation $x = -2y - 1$.
$$x = -2(1) - 1 = -2 - 1 = -3$$
The solution is $(-3, 1)$.

(3) Multiply the second equation by -3.
$$\begin{cases} 3x + 10y = 1 \\ -3x - 6y = 3 \end{cases}$$
Add the equations.
$$3x + 10y = 1$$
$$\underline{-3x - 6y = 3}$$
$$4y = 4$$
$$y = 1$$
Replace y with 1 in the second equation.
$$x + 2(1) = -1$$
$$x + 2 = -1$$
$$x = -3$$
The solution is $(-3, 1)$.

2. $\begin{cases} y = \dfrac{1}{2}x + \dfrac{2}{3} \\ 4x + 6y = 4 \end{cases}$

(1)

$\quad$ 3.1

$-4.7 \quad\quad\quad 4.7$

Intersection
X=0 $\quad$ Y=.66666667

$\quad -3.1$

(2) Replace y with $\dfrac{1}{2}x + \dfrac{2}{3}$ in the second equation.
$$4x + 6\left(\frac{1}{2}x + \frac{2}{3}\right) = 4$$
$$4x + 3x + 4 = 4$$
$$7x = 0$$
$$x = 0$$
Replace x with 0 in the first equation.
$$y = \frac{1}{2}(0) + \frac{2}{3}$$
$$y = \frac{2}{3}$$
The solution is $\left(0, \dfrac{2}{3}\right)$.

(3) $\begin{cases} -\dfrac{1}{2}x + y = \dfrac{2}{3} \\ 4x + 6y = 4 \end{cases}$
Multiply the first equation by -6.
$$\begin{cases} 3x - 6y = -4 \\ 4x + 6y = 4 \end{cases}$$
Add the equations.
$$3x - 6y = -4$$
$$\underline{4x + 6y = 4}$$
$$7x \quad\quad = 0$$
$$x = 0$$
Replace x with 0 in the first equation.
$$y = \frac{1}{2}(0) + \frac{2}{3}$$
$$y = \frac{2}{3}$$
The solution is $\left(0, \dfrac{2}{3}\right)$.

3. $\begin{cases} 2x - 4y = 22 \\ 5x - 10y = 16 \end{cases}$

(1)

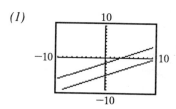

(2) Solve the first equation for x.

$2x = 4y + 22$

$x = 2y + 11$

Replace x with $2y + 11$ in the second equation.

$5(2y + 11) - 10y = 16$

$10y + 55 - 10y = 16$

$55 = 16$ False

This is an inconsistent system.

The solution set is $\varnothing$.

(3) Divide the first equation by 2.

$\begin{cases} x - 2y = 11 \\ 5x - 10y = 16 \end{cases}$

Multiply the first equation by -5.

$\begin{cases} -5x + 10y = -55 \\ 5x - 10y = 16 \end{cases}$

Add the equations.

$-5x + 10y = -55$

$\underline{5x - 10y = 16}$

$0 = -39$ False

This is an inconsistent system.

The solution set is $\varnothing$.

4. $\begin{cases} 3x - 6y = 12 \\ 2y = x - 4 \end{cases}$

(1)

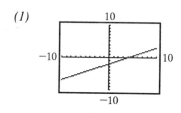

(2) Solve the second equation for x.

$x = 2y + 4$

Replace x with $2y + 4$ in the first equation.

$3(2y + 4) - 6y = 12$

$6y + 12 - 6y = 12$

$12 = 12$ True

This is a dependent system.

The solution set is

$\{(x,\ y) \mid 3x - 6y = 12\}$.

(3) $\begin{cases} 3x - 6y = 12 \\ -x + 2y = -4 \end{cases}$

Multiply the second equation by 3.

$\begin{cases} 3x - 6y = 12 \\ -3x + 6y = -12 \end{cases}$

Add the equations.

$3x - 6y = 12$

$\underline{-3x + 6y = -12}$

$0 = 0$ True

This is a dependent system.

The solution set is

$\{(x,\ y) \mid 3x - 6y = 12\}$.

5. $\begin{cases} \dfrac{1}{2}x - \dfrac{3}{4}y = -\dfrac{1}{2} \\ \dfrac{1}{8}x + \dfrac{3}{4}y = \dfrac{19}{8} \end{cases}$

(1)

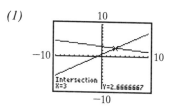

(2) Clear fractions by multiplying the first equation by 4 and the second equation by 8.
$$\begin{cases} 2x - 3y = -2 \\ x + 6y = 19 \end{cases}$$
Solve the new second equation for x.
$x = -6y + 19$
Replace x with $-6y + 19$ in the first equation.
$$2(-6y + 19) - 3y = -2$$
$$-12y + 38 - 3y = -2$$
$$-15y = -40$$
$$y = \frac{-40}{-15} = \frac{8}{3}$$
Replace y with $\frac{8}{3}$ in the equation
$x = -6y + 19$.
$$x = -6\left(\frac{8}{3}\right) + 19$$
$$x = -16 + 19$$
$$x = 3$$
The solution is $\left(3, \frac{8}{3}\right)$.

(3) Add the equations.
$$\frac{1}{2}x - \frac{3}{4}y = -\frac{1}{2}$$
$$\frac{1}{8}x + \frac{3}{4}y = \frac{19}{8}$$
$$\frac{5}{8}x = \frac{15}{8}$$
$$5x = 15$$
$$x = 3$$
Replace x with 3 in the first equation.
$$\frac{1}{2}(3) - \frac{3}{4}y = -\frac{1}{2}$$
$$-\frac{3}{4}y = -2$$
$$-3y = -8$$
$$y = \frac{8}{3}$$
The solution is $\left(3, \frac{8}{3}\right)$.

6. $\begin{cases} y = 32x \\ y = 15x + 25,500 \end{cases}$
Replace y with $32x$ in the second equation.
$$32x = 15x + 25,500$$
$$17x = 25,500$$
$$x = 1500$$
The number of backpacks that the company must sell is 1500.

7. $\begin{cases} x + z = 4 & (1) \\ 2x - y = 4 & (2) \\ x + y - z = 0 & (3) \end{cases}$
Add equations (2) and (3) to eliminate y.
$$2x - y = 4$$
$$x + y - z = 0$$
$$\overline{3x - z = 4} \quad (4)$$
Add equations (1) and (4).
$$x + z = 4$$
$$3x - z = 4$$
$$\overline{4x = 8}$$
$$x = 2$$

Replace x with 2 in equation (1).
$$2 + z = 4$$
$$z = 2$$
Replace x with 2 and z with 2 in equation (3).
$$2 + y - 2 = 0$$
$$y = 0$$
The solution is $(2, 0, 2)$.

8. $\begin{cases} 2x + 5y = 4 & (1) \\ x - 5y + z = -1 & (2) \\ 4x - z = 11 & (3) \end{cases}$

Add equations (2) and (3) to eliminate z.

$x - 5y + z = -1$

$\dfrac{4x - z = 11}{5x - 5y = 10} \quad (4)$

Add equations (1) and (4).

$2x + 5y = 4$

$\dfrac{5x - 5y = 10}{7x = 14}$

$x = 2$

Replace x with 2 in equation (1).

$2(2) + 5y = 4$

$4 + 5y = 4$

$5y = 0$

$y = 0$

Replace x with 2 in equation (3).

$4(2) - z = 11$

$8 - z = 11$

$-z = 3$

$z = -3$

The solution is $(2, 0, -3)$.

9. $\begin{cases} 4y + 2z = 5 & (1) \\ 2x + 8y = 5 & (2) \\ 6x + 4z = 1 & (3) \end{cases}$

Multiply equation (1) by -2 and add to equation (2).

$-8y - 4z = -10$

$\dfrac{2x + 8y = 5}{2x - 4z = -5} \quad (4)$

Add equations (3) and (4).

$6x + 4z = 1$

$\dfrac{2x - 4z = -5}{8x = -4}$

$x = -\dfrac{1}{2}$

Replace x with $-\dfrac{1}{2}$ in equation (2).

$2\left(-\dfrac{1}{2}\right) + 8y = 5$

$\phantom{2\left(-\dfrac{1}{2}\right)}-1 + 8y = 5$

$\phantom{2\left(-\dfrac{1}{2}\right) -1}8y = 6$

$\phantom{2\left(-\dfrac{1}{2}\right) -1 8}y = \dfrac{3}{4}$

Replace x with $-\dfrac{1}{2}$ in equation (3).

$6\left(-\dfrac{1}{2}\right) + 4z = 1$

$\phantom{6\left(-\dfrac{1}{2}\right)}-3 + 4z = 1$

$\phantom{6\left(-\dfrac{1}{2}\right) -3}4z = 4$

$\phantom{6\left(-\dfrac{1}{2}\right) -3 4}z = 1$

The solution is $\left(-\dfrac{1}{2}, \dfrac{3}{4}, 1\right)$.

10. $\begin{cases} 5x + 7y \quad = 9 & (1) \\ \quad 14y - z = 28 & (2) \\ 4x \quad + 2z = -4 & (3) \end{cases}$

Multiply equation (1) by –2 and add to equation (2).

$-10x - 14y \quad = -18$

$\underline{\qquad 14y - z = 28 \qquad}$

$-10x \qquad - z = 10 \quad (4)$

Multiply equation (4) by 2 and add to equation (3).

$-20x - 2z = 20$

$\underline{\quad 4x + 2z = -4 \quad}$

$-16x \qquad = 16$

$x = -1$

Replace x with –1 in equation (1).

$5(-1) + 7y = 9$

$7y = 14$

$y = 2$

Replace x with –1 in equation (3).

$4(-1) + 2z = -4$

$2z = 0$

$z = 0$

The solution is (–1, 2, 0).

11. $\begin{cases} 3x \quad - 2y + 2z = 5 & (1) \\ -x \quad + 6y \quad + z = 4 & (2) \\ 3x \quad + 14y + 7z = 20 & (3) \end{cases}$

Multiply equation (2) by 3 and add to equation (1).

$3x - 2y + 2z = 5$

$\underline{-3x + 18y + 3z = 12}$

$16y + 5z = 17 \quad (4)$

Multiply equation (3) by –1 and add to equation (1).

$3x - \quad 2y + 2z = 5$

$\underline{-3x - 14y - 7z = -20}$

$-16y - 5z = -15 \quad (5)$

Add equations (4) and (5).

$16y + 5z = 17$

$\underline{-16y - 5z = -15}$

$0 = 2 \qquad \text{False}$

The system is inconsistent.

The solution set is $\varnothing$.

12. $\begin{cases} x + 2y + 3z = 11 & (1) \\ \quad y + 2z = 3 & (2) \\ 2x \quad + 2z = 10 & (3) \end{cases}$

Multiply equation (2) by –2 and add to equation (1).

$x + 2y + 3z = 11$

$\underline{\quad -2y - 4z = -6 \quad}$

$x \qquad - z = 5 \quad (4)$

Multiply equation (4) by 2 and add to equation (3).

$2x + 2z = 10$

$\underline{2x - 2z = 10}$

$4x \qquad = 20$

$x = 5$

Replace x with 5 in equation (3).

$2(5) + 2z = 10$

$2z = 0$

$z = 0$

Replace z with 0 in equation (2).

$y + 2(0) = 3$

$y = 3$

The solution is (5, 3, 0).

13. $\begin{cases} 7x - 3y + 2z = 0 & (1) \\ 4x - 4y - z = 2 & (2) \\ 5x + 2y + 3z = 1 & (3) \end{cases}$

Multiply equation (2) by 2 and add to equation (1).

$7x - 3y + 2z = 0$

$\underline{8x - 8y - 2z = 4}$

$15x - 11y = 4 \quad (4)$

Multiply equation (2) by 3 and add to equation (3).

$12x - 12y - 3z = 6$

$\underline{5x + 2y + 3z = 1}$

$17x - 10y = 7 \quad (5)$

Multiply equation (4) by -10, multiply equation (5) by 11, and add.

$-150x + 110y = -40$

$\underline{187x - 110y = 77}$

$37x = 37$

$x = 1$

Replace x with 1 in equation (4).

$15(1) - 11y = 4$

$-11y = -11$

$y = 1$

Replace x with 1 and y with 1 in equation (1).

$7(1) - 3(1) + 2z = 0$

$4 + 2z = 0$

$2z = -4$

$z = -2$

The solution is $(1, 1, -2)$.

14. $\begin{cases} x - 3y - 5z = -5 & (1) \\ 4x - 2y + 3z = 13 & (2) \\ 5x + 3y + 4z = 22 & (3) \end{cases}$

Multiply equation (1) by -4 and add to equation (2).

$-4x + 12y + 20z = 20$

$\underline{4x - 2y + 3z = 13}$

$10y + 23z = 33 \quad (4)$

Multiply equation (1) by -5 and add to equation (3).

$-5x + 15y + 25z = 25$

$\underline{5x + 3y + 4z = 22}$

$18y + 29z = 47 \quad (5)$

Multiply equation (4) by 9, multiply equation (5) by -5 and add.

$90y + 207z = 297$

$\underline{-90y - 145z = -235}$

$62z = 62$

$z = 1$

Replace z with 1 in equation (4).

$10y + 23(1) = 33$

$10y = 10$

$y = 1$

Replace y with 1 and z with 1 in equation (1).

$x - 3(1) - 5(1) = -5$

$x - 8 = -5$

$x = 3$

The solution is $(3, 1, 1)$.

15. Let x = the first number
    y = the second number
    z = the third number
The conditions lead to:
$x + y + z = 98$

$x + y = z + 2$

$y = 4x$

$$\begin{cases} x + y + z = 98 & (1) \\ x + y - z = 2 & (2) \\ -4x + y = 0 & (3) \end{cases}$$

Add equations (1) and (2).
$$\begin{aligned} x + \; y + z &= 98 \\ \underline{x + \; y - z} &\underline{= 2} \\ 2x + 2y \quad\; &= 100 \quad (4) \end{aligned}$$

Multiply equation (3) by -2 and add to equation (4).
$$\begin{aligned} 8x - 2y &= 0 \\ \underline{2x + 2y} &\underline{= 100} \\ 10x \quad\;\; &= 100 \\ x &= 10 \end{aligned}$$

Replace x with 10 in equation (3).
$-4(10) + y = 0$
         $y = 40$

Replace x with 10 and y with 40 in equation (2).
$10 + 40 - z = 2$
      $50 - z = 2$
         $-z = -48$
           $z = 48$
The numbers are 10, 40, and 48.

16. Let x = the first number
    y = the second number

$$\begin{cases} x = 3y \\ 2(x + y) = 168 \end{cases}$$

Replace x with $3y$ in the second equation.
$2(3y + y) = 168$
      $8y = 168$
       $y = 21$

Replace y with 21 in the first equation.
$x = 3(21) = 63$
The numbers are 63 and 21.

17. Let x = the speed of the first car
    y = the speed of the second car
$$\begin{cases} 4x + 4y = 492 \\ y = x + 7 \end{cases}$$

Replace y with $x + 7$ in the first equation.
$4x + 4(x + 7) = 492$
     $8x + 28 = 492$
         $8x = 464$
          $x = 58$

Replace x with 58 in the second equation.
$y = 58 + 7 = 65$
The cars are going 58 and 65 miles
per hour.

18. Let w = the width of the foundation
    l = the length of the foundation
$$\begin{cases} l = 3w \\ 2w + 2l = 296 \end{cases}$$

Replace l with $3w$ in the second equation.
$2w + 2(3w) = 296$
    $2w + 6w = 296$
         $8w = 296$
          $w = 37$

Replace w with 37 in the first equation.
$l = 3(37) = 111$
The foundation is 37 feet wide and
111 feet long.

19. Let x = the number of liters of 10% solution
    y = the number of liters of 60% solution
$$\begin{cases} x + y = 50 \\ 0.10x + 0.60y = 0.40(50) \end{cases}$$

Solve the first equation for y.
$y = 50 - x$
Replace y with $50 - x$ in the second equation.
$0.10x + 0.60(50 - x) = 0.40(50)$
$10[0.10x + 0.60(50 - x)] = 10[0.40(50)]$
        $x + 6(50 - x) = 4(50)$
        $x + 300 - 6x = 200$
             $-5x = -100$
                 $x = 20$

Replace x with 20 in the equation
$y = 50 - x$.
$y = 50 - 20 = 30$
He should use 20 liters of 10% solution and 30 liters of 60% solution.

20. Let c = the number of pounds of
chocolates used
n = the number of pounds of nuts used
r = the number of pounds of raisins used

$$\begin{cases} r = 2n \\ c + n + r = 45 \\ 3.00c + 2.70n + 2.25r = 2.80(45) \end{cases}$$

$$\begin{cases} -2n + \quad r = 0 \quad (1) \\ c + n + \quad r = 45 \quad (2) \\ 3c + 2.7n + 2.25r = 126 \quad (3) \end{cases}$$

Multiply equation (2) by −3 and add to
equation (3).

$$-3c - 3n - 3r = -135$$
$$\underline{3c + 2.7n + 2.25r = 126}$$
$$-0.3n - 0.75r = -9 \quad (4)$$

Multiply equation (1) by 0.75 and add to
equation (4).

$$-1.5n + 0.75r = 0$$
$$\underline{-0.3n - 0.75r = -9}$$
$$-1.8n \quad\quad = -9$$
$$n = 5$$

Replace n with 5 in equation (1).
$$-2(5) + r = 0$$
$$r = 10$$

Replace n with 5 and r with 10 in
equation (2).
$$c + 5 + 10 = 45$$
$$c + 15 = 45$$
$$c = 30$$

She should use 30 pounds of creme-filled
chocolates, 5 pounds of chocolate-covered
nuts, and 10 pounds of chocolate-covered
raisins.

21. Let x = the number of pennies
y = the number of nickels
z = the number of dimes

$$\begin{cases} x + y + z = 53 \\ 0.01x + 0.05y + 0.10z = 2.77 \\ y = z + 4 \end{cases}$$

Multiply the second equation by 100, and
put the third equation in standard form.

$$\begin{cases} x + y + z = 53 \quad (1) \\ x + 5y + 10z = 277 \quad (2) \\ y - z = 4 \quad (3) \end{cases}$$

Multiply equation (1) by −1 and add to
equation (2).

$$-x - y - z = -53$$
$$\underline{x + 5y + 10z = 277}$$
$$4y + 9z = 224 \quad (4)$$

Multiply equation (3) by 9 and add to
equation (4).

$$9y - 9z = 36$$
$$\underline{4y + 9z = 224}$$
$$13y \quad\quad = 260$$
$$y = 20$$

Replace y with 20 in equation (3).
$$20 - z = 4$$
$$16 = z$$

Replace y with 20 and z with 16 in
equation (1).
$$x + 20 + 16 = 53$$
$$x + 36 = 53$$
$$x = 17$$

He has 17 pennies, 20 nickels, and
16 dimes in his jar.

22. Let l = the rate of interest on the larger investment, as a decimal
s = the rate of interest on the smaller investment, as a decimal

$$\begin{cases} 10,000l + 4000s = 1250 \\ l = s + 0.02 \end{cases}$$

Replace l with $s + 0.02$ in the first equation.

$$10,000(s + 0.02) + 4000s = 1250$$
$$10,000s + 200 + 4000s = 1250$$
$$14,000s = 1050$$
$$s = \frac{1050}{14,000} = 0.075$$

$l = 0.075 + 0.02 = 0.095$

The interest rate on the larger investment is 9.5% and the rate on the smaller investment is 7.5%.

23. Let x = the length of the equal sides
y = the length of the third side

$$\begin{cases} 2x + y = 73 \\ y = x + 7 \end{cases}$$

Replace y with $x + 7$ in the first equation.

$$2x + x + 7 = 73$$
$$3x = 66$$
$$x = 22$$

Replace x with 22 in the second equation.

$y = 22 + 7 = 29$

Two sides of the triangle have length 22 centimeters, and the third side has length 29 centimeters.

24. Let f = the first number
s = the second number
t = the third number

$$\begin{cases} f + s + t = 295 & (1) \\ f = s + 5 & (2) \\ f = 2t & (3) \end{cases}$$

Solve equations (2) and (3) for s and t, respectively, in terms of f.

$$s = f - 5$$
$$t = \frac{f}{2}$$

Replace s with $f - 5$ and t with $\frac{f}{2}$ in equation (1).

$$f + f - 5 + \frac{f}{2} = 295$$
$$\frac{5}{2}f - 5 = 295$$
$$\frac{5}{2}f = 300$$
$$f = 120$$

Replace f by 120 in the equation $s = f - 5$.

$s = 120 - 5 = 115$

Replace f by 120 in the equation $t = \frac{f}{2}$.

$$t = \frac{120}{2} = 60$$

The first number is 120, the second is 115, and the third is 60.

25. $\begin{cases} 3x+10y=1 \\ x+2y=-1 \end{cases}$

$\begin{bmatrix} 3 & 10 & | & 1 \\ 1 & 2 & | & -1 \end{bmatrix}$ *1* *2*

Interchange row 1 and row 2.

$\begin{bmatrix} 1 & 2 & | & -1 \\ 3 & 10 & | & 1 \end{bmatrix}$ *2* *1*

Multiply row 1 by -3 and add to row 2.

$-3\begin{bmatrix} 1 & 2 & | & -1 \\ 0 & 4 & | & 4 \end{bmatrix}$ $-3\ \ 2$ $-3\ -6\ |\ -3$ row1
                                          1 $3\ \ 10\ |\ \ 1$ row2

Divide row 2 by 4. $0\ \ 4\ |\ 4$ row 4

$\begin{bmatrix} 1 & 2 & | & -1 \\ 0 & 1 & | & 1 \end{bmatrix}$ *2* *1* $\dfrac{1\ \ 2\ |\ 1}{4\ \ 4\ |\ 4}$

This corresponds to:

$\begin{cases} x+2y=-1 \\ y=1 \end{cases}$

$x+2(1)=-1$

$x=-3$

The solution is $(-3, 1)$.

26. $\begin{cases} 3x-6y=12 \\ 2y=x-4 \end{cases}$

$\begin{cases} 3x-6y=12 \\ -x+2y=-4 \end{cases}$

$\begin{bmatrix} 3 & -6 & | & 12 \\ -1 & 2 & | & -4 \end{bmatrix}$

Divide row 1 by 3.

$\begin{bmatrix} 1 & -2 & | & 4 \\ -1 & 2 & | & -4 \end{bmatrix}$

Add row 1 to row 2.

$\begin{bmatrix} 1 & -2 & | & 4 \\ 0 & 0 & | & 0 \end{bmatrix}$

This corresponds to:

$\begin{cases} x-2y=4 \\ 0=0 \end{cases}$

This is a dependent system.

The solution set is $\{(x,\ y)|\ x-2y=4\}$.

27. $\begin{cases} 3x-2y=-8 \\ 6x+5y=11 \end{cases}$

$\begin{bmatrix} 3 & -2 & | & -8 \\ 6 & 5 & | & 11 \end{bmatrix}$

Divide row 1 by 3.

$\begin{bmatrix} 1 & -\frac{2}{3} & | & -\frac{8}{3} \\ 6 & 5 & | & 11 \end{bmatrix}$

Multiply row 1 by -6 and add to row 2.

$\begin{bmatrix} 1 & -\frac{2}{3} & | & -\frac{8}{3} \\ 0 & 9 & | & 27 \end{bmatrix}$

Divide row 2 by 9.

$\begin{bmatrix} 1 & -\frac{2}{3} & | & -\frac{8}{3} \\ 0 & 1 & | & 3 \end{bmatrix}$

This corresponds to:

$\begin{cases} x-\dfrac{2}{3}y=-\dfrac{8}{3} \\ y=3 \end{cases}$

$x-\dfrac{2}{3}(3)=-\dfrac{8}{3}$

$x-2=-\dfrac{8}{3}$

$x=-\dfrac{2}{3}$

The solution is $\left(-\dfrac{2}{3},\ 3\right)$.

28. $\begin{cases} 6x - 6y = -5 \\ 10x - 2y = 1 \end{cases}$

$\begin{bmatrix} 6 & -6 & \vdots & -5 \\ 10 & -2 & \vdots & 1 \end{bmatrix}$

Divide row 1 by 6.

$\begin{bmatrix} 1 & -1 & \vdots & -\frac{5}{6} \\ 10 & -2 & \vdots & 1 \end{bmatrix}$

Multiply row 1 by -10 and add to row 2.

$\begin{bmatrix} 1 & -1 & \vdots & -\frac{5}{6} \\ 0 & 8 & \vdots & \frac{28}{3} \end{bmatrix}$

Divide row 2 by 8.

$\begin{bmatrix} 1 & -1 & \vdots & -\frac{5}{6} \\ 0 & 1 & \vdots & \frac{7}{6} \end{bmatrix}$

This corresponds to:

$\begin{cases} x - y = -\dfrac{5}{6} \\ \quad y = \dfrac{7}{6} \end{cases}$

$x - \dfrac{7}{6} = -\dfrac{5}{6}$

$x = \dfrac{2}{6} = \dfrac{1}{3}$

The solution is $\left(\dfrac{1}{3}, \dfrac{7}{6} \right)$.

29. $\begin{cases} 3x - 6y = 0 \\ 2x + 4y = 5 \end{cases}$

$\begin{bmatrix} 3 & -6 & \vdots & 0 \\ 2 & 4 & \vdots & 5 \end{bmatrix}$

Divide row 1 by 3.

$\begin{bmatrix} 1 & -2 & \vdots & 0 \\ 2 & 4 & \vdots & 5 \end{bmatrix}$

Multiply row 1 by -2 and add to row 2.

$\begin{bmatrix} 1 & -2 & \vdots & 0 \\ 0 & 8 & \vdots & 5 \end{bmatrix}$

Divide row 2 by 8.

$\begin{bmatrix} 1 & -2 & \vdots & 0 \\ 0 & 1 & \vdots & \frac{5}{8} \end{bmatrix}$

This corresponds to:

$\begin{cases} x - 2y = 0 \\ \quad y = \dfrac{5}{8} \end{cases}$

$x - 2\left(\dfrac{5}{8}\right) = 0$

$x - \dfrac{5}{4} = 0$

$x = \dfrac{5}{4}$

The solution is $\left(\dfrac{5}{4}, \dfrac{5}{8} \right)$.

30. $\begin{cases} 5x - 3y = 10 \\ -2x + y = -1 \end{cases}$

$$\begin{bmatrix} 5 & -3 & \vdots & 10 \\ -2 & 1 & \vdots & -1 \end{bmatrix}$$

Divide row 1 by 5.

$$\begin{bmatrix} 1 & -\frac{3}{5} & \vdots & 2 \\ -2 & 1 & \vdots & -1 \end{bmatrix}$$

Multiply row 1 by 2 and add to row 2.

$$\begin{bmatrix} 1 & -\frac{3}{5} & \vdots & 2 \\ 0 & -\frac{1}{5} & \vdots & 3 \end{bmatrix}$$

Multiply row 2 by –5.

$$\begin{bmatrix} 1 & -\frac{3}{5} & \vdots & 2 \\ 0 & 1 & \vdots & -15 \end{bmatrix}$$

This corresponds to:

$$\begin{cases} x - \dfrac{3}{5}y = 2 \\ \qquad y = -15 \end{cases}$$

$$x - \frac{3}{5}(-15) = 2$$
$$x + 9 = 2$$
$$x = -7$$

The solution is $(-7, -15)$.

31. $\begin{cases} 0.2x - 0.3y = -0.7 \\ 0.5x + 0.3y = 1.4 \end{cases}$

$$\begin{bmatrix} 0.2 & -0.3 & \vdots & -0.7 \\ 0.5 & 0.3 & \vdots & 1.4 \end{bmatrix}$$

Multiply both rows by 10 to clear decimals.

$$\begin{bmatrix} 2 & -3 & \vdots & -7 \\ 5 & 3 & \vdots & 14 \end{bmatrix}$$

Divide row 1 by 2.

$$\begin{bmatrix} 1 & -\frac{3}{2} & \vdots & -\frac{7}{2} \\ 5 & 3 & \vdots & 14 \end{bmatrix}$$

Multiply row 1 by –5 and add to row 2.

$$\begin{bmatrix} 1 & -\frac{3}{2} & \vdots & -\frac{7}{2} \\ 0 & \frac{21}{2} & \vdots & \frac{63}{2} \end{bmatrix}$$

Multiply row 2 by $\dfrac{2}{21}$.

$$\begin{bmatrix} 1 & -\frac{3}{2} & \vdots & -\frac{7}{2} \\ 0 & 1 & \vdots & 3 \end{bmatrix}$$

This corresponds to:

$$\begin{cases} x - \dfrac{3}{2}y = -\dfrac{7}{2} \\ \qquad y = 3 \end{cases}$$

$$x - \frac{3}{2}(3) = -\frac{7}{2}$$
$$x - \frac{9}{2} = -\frac{7}{2}$$
$$x = 1$$

The solution is $(1, 3)$.

32. $\begin{cases} 3x + 2y = 8 \\ 3x - y = 5 \end{cases}$

$\begin{bmatrix} 3 & 2 & | & 8 \\ 3 & -1 & | & 5 \end{bmatrix}$

Divide row 1 by 3.

$\begin{bmatrix} 1 & \frac{2}{3} & | & \frac{8}{3} \\ 3 & -1 & | & 5 \end{bmatrix}$

Multiply row 1 by −3 and add to row 2.

$\begin{bmatrix} 1 & \frac{2}{3} & | & \frac{8}{3} \\ 0 & -3 & | & -3 \end{bmatrix}$

Divide row 2 by −3.

$\begin{bmatrix} 1 & \frac{2}{3} & | & \frac{8}{3} \\ 0 & 1 & | & 1 \end{bmatrix}$

This corresponds to:

$\begin{cases} x + \dfrac{2}{3}y = \dfrac{8}{3} \\ \qquad y = 1 \end{cases}$

$x + \dfrac{2}{3}(1) = \dfrac{8}{3}$

$\qquad x = 2$

The solution is (2, 1).

33. $\begin{cases} x \quad + z = 4 \\ 2x - y \quad = 0 \\ x + y - z = 0 \end{cases}$

$\begin{bmatrix} 1 & 0 & 1 & | & 4 \\ 2 & -1 & 0 & | & 0 \\ 1 & 1 & -1 & | & 0 \end{bmatrix}$

Multiply row 1 by −2 and add to row 2.
Multiply row 1 by −1 and add to row 3.

$\begin{bmatrix} 1 & 0 & 1 & | & 4 \\ 0 & -1 & -2 & | & -8 \\ 0 & 1 & -2 & | & -4 \end{bmatrix}$

Multiply row 2 by −1.

$\begin{bmatrix} 1 & 0 & 1 & | & 4 \\ 0 & 1 & 2 & | & 8 \\ 0 & 1 & -2 & | & -4 \end{bmatrix}$

Multiply row 2 by −1 and add to row 3.

$\begin{bmatrix} 1 & 0 & 1 & | & 4 \\ 0 & 1 & 2 & | & 8 \\ 0 & 0 & -4 & | & -12 \end{bmatrix}$

Divide row 3 by −4.

$\begin{bmatrix} 1 & 0 & 1 & | & 4 \\ 0 & 1 & 2 & | & 8 \\ 0 & 0 & 1 & | & 3 \end{bmatrix}$

This corresponds to:

$\begin{cases} x \quad + z = 4 \\ \quad y + 2z = 8 \\ \qquad z = 3 \end{cases}$

$y + 2(3) = 8$

$y + 6 = 8$

$\qquad y = 2$

$x + 3 = 4$

$\qquad x = 1$

The solution is (1, 2, 3).

34. $\begin{cases} 2x + 5y \quad = 4 \\ x - 5y + z = -1 \\ 4x \quad - z = 11 \end{cases}$

$$\begin{bmatrix} 2 & 5 & 0 & | & 4 \\ 1 & -5 & 1 & | & -1 \\ 4 & 0 & -1 & | & 11 \end{bmatrix}$$

Interchange row 1 and row 2.

$$\begin{bmatrix} 1 & -5 & 1 & | & -1 \\ 2 & 5 & 0 & | & 4 \\ 4 & 0 & -1 & | & 11 \end{bmatrix}$$

Multiply row 1 by –2 and add to row 2.
Multiply row 1 by –4 and add to row 3.

$$\begin{bmatrix} 1 & -5 & 1 & | & -1 \\ 0 & 15 & -2 & | & 6 \\ 0 & 20 & -5 & | & 15 \end{bmatrix}$$

Divide row 2 by 15.

$$\begin{bmatrix} 1 & -5 & 1 & | & -1 \\ 0 & 1 & -\frac{2}{15} & | & \frac{2}{5} \\ 0 & 20 & -5 & | & 15 \end{bmatrix}$$

Multiply row 2 by –20 and add to row 3.

$$\begin{bmatrix} 1 & -5 & 1 & | & -1 \\ 0 & 1 & -\frac{2}{15} & | & \frac{2}{5} \\ 0 & 0 & -\frac{7}{3} & | & 7 \end{bmatrix}$$

Multiply row 3 by $-\dfrac{3}{7}$.

$$\begin{bmatrix} 1 & -5 & 1 & | & -1 \\ 0 & 1 & -\frac{2}{15} & | & \frac{2}{5} \\ 0 & 0 & 1 & | & -3 \end{bmatrix}$$

This corresponds to:

$$\begin{cases} x - 5y \quad + z = -1 \\ y - \dfrac{2}{15}z = \dfrac{2}{5} \\ \quad z = -3 \end{cases}$$

$$y - \frac{2}{15}(-3) = \frac{2}{5}$$
$$y + \frac{2}{5} = \frac{2}{5}$$
$$y = 0$$

$$x - 5(0) + (-3) = -1$$
$$x - 3 = -1$$
$$x = 2$$

The solution is (2, 0, –3).

35. $\begin{cases} 3x - y \quad = 11 \\ x + 2z = 13 \\ y - z = -7 \end{cases}$

$$\begin{bmatrix} 3 & -1 & 0 & | & 11 \\ 1 & 0 & 2 & | & 13 \\ 0 & 1 & -1 & | & -7 \end{bmatrix}$$

Interchange row 1 and row 2.

$$\begin{bmatrix} 1 & 0 & 2 & | & 13 \\ 3 & -1 & 0 & | & 11 \\ 0 & 1 & -1 & | & -7 \end{bmatrix}$$

Interchange row 2 and row 3.

$$\begin{bmatrix} 1 & 0 & 2 & | & 13 \\ 0 & 1 & -1 & | & -7 \\ 3 & -1 & 0 & | & 11 \end{bmatrix}$$

Multiply row 1 by –3 and add to row 3.

$$\begin{bmatrix} 1 & 0 & 2 & | & 13 \\ 0 & 1 & -1 & | & -7 \\ 0 & -1 & -6 & | & -28 \end{bmatrix}$$

Add row 2 to row 3.

$$\begin{bmatrix} 1 & 0 & 2 & | & 13 \\ 0 & 1 & -1 & | & -7 \\ 0 & 0 & -7 & | & -35 \end{bmatrix}$$

Divide row 3 by –7.

$$\begin{bmatrix} 1 & 0 & 2 & | & 13 \\ 0 & 1 & -1 & | & -7 \\ 0 & 0 & 1 & | & 5 \end{bmatrix}$$

This corresponds to:

$$\begin{cases} x + 2z = 13 \\ y - z = -7 \\ z = 5 \end{cases}$$

$$y - 5 = -7$$
$$y = -2$$
$$x + 2(5) = 13$$
$$x + 10 = 13$$
$$x = 3$$

The solution is (3, –2, 5).

36. $\begin{cases} 5x + 7y + 3z = 9 \\ 14y - z = 28 \\ 4x + 2z = -4 \end{cases}$

$\begin{bmatrix} 5 & 7 & 3 & | & 9 \\ 0 & 14 & -1 & | & 28 \\ 4 & 0 & 2 & | & -4 \end{bmatrix}$

Divide row 1 by 5.

$\begin{bmatrix} 1 & \frac{7}{5} & \frac{3}{5} & | & \frac{9}{5} \\ 0 & 14 & -1 & | & 28 \\ 4 & 0 & 2 & | & -4 \end{bmatrix}$

Multiply row 1 by -4 and add to row 3.

$\begin{bmatrix} 1 & \frac{7}{5} & \frac{3}{5} & | & \frac{9}{5} \\ 0 & 14 & -1 & | & 28 \\ 0 & -\frac{28}{5} & -\frac{2}{5} & | & -\frac{56}{5} \end{bmatrix}$

Divide row 2 by 14.

$\begin{bmatrix} 1 & \frac{7}{5} & \frac{3}{5} & | & \frac{9}{5} \\ 0 & 1 & -\frac{1}{14} & | & 2 \\ 0 & -\frac{28}{5} & -\frac{2}{5} & | & -\frac{56}{5} \end{bmatrix}$

Multiply row 2 by $\dfrac{28}{5}$ and add to row 3.

$\begin{bmatrix} 1 & \frac{7}{5} & \frac{3}{5} & | & \frac{9}{5} \\ 0 & 1 & -\frac{1}{14} & | & 2 \\ 0 & 0 & -\frac{4}{5} & | & 0 \end{bmatrix}$

Multiply row 3 by $-\dfrac{5}{4}$.

$\begin{bmatrix} 1 & \frac{7}{5} & \frac{3}{5} & | & \frac{9}{5} \\ 0 & 1 & -\frac{1}{14} & | & 2 \\ 0 & 0 & 1 & | & 0 \end{bmatrix}$

This corresponds to:

$\begin{cases} x + \dfrac{7}{5}y + \dfrac{3}{5}z = \dfrac{9}{5} \\ y - \dfrac{1}{14}z = 2 \\ z = 0 \end{cases}$

$y - \dfrac{1}{14}(0) = 2$

$y = 2$

$x + \dfrac{7}{5}(2) + \dfrac{3}{5}(0) = \dfrac{9}{5}$

$x + \dfrac{14}{5} = \dfrac{9}{5}$

$x = -1$

The solution is $(-1, 2, 0)$.

37. $\begin{cases} 7x - 3y + 2z = 0 \\ 4x - 4y - z = 2 \\ 5x + 2y + 3z = 1 \end{cases}$

$$\begin{bmatrix} 7 & -3 & 2 & | & 0 \\ 4 & -4 & -1 & | & 2 \\ 5 & 2 & 3 & | & 1 \end{bmatrix}$$

Interchange row 1 and row 2.

$$\begin{bmatrix} 4 & -4 & -1 & | & 2 \\ 7 & -3 & 2 & | & 0 \\ 5 & 2 & 3 & | & 1 \end{bmatrix}$$

Divide row 1 by 4.

$$\begin{bmatrix} 1 & -1 & -\frac{1}{4} & | & \frac{1}{2} \\ 7 & -3 & 2 & | & 0 \\ 5 & 2 & 3 & | & 1 \end{bmatrix}$$

Multiply row 1 by -7 and add to row 2.
Multiply row 1 by -5 and add to row 3.

$$\begin{bmatrix} 1 & -1 & -\frac{1}{4} & | & \frac{1}{2} \\ 0 & 4 & \frac{15}{4} & | & -\frac{7}{2} \\ 0 & 7 & \frac{17}{4} & | & -\frac{3}{2} \end{bmatrix}$$

Divide row 2 by 4.

$$\begin{bmatrix} 1 & -1 & -\frac{1}{4} & | & \frac{1}{2} \\ 0 & 1 & \frac{15}{16} & | & -\frac{7}{8} \\ 0 & 7 & \frac{17}{4} & | & -\frac{3}{2} \end{bmatrix}$$

Multiply row 2 by -7 and add to row 3.

$$\begin{bmatrix} 1 & -1 & -\frac{1}{4} & | & \frac{1}{2} \\ 0 & 1 & \frac{15}{16} & | & -\frac{7}{8} \\ 0 & 0 & -\frac{37}{16} & | & \frac{37}{8} \end{bmatrix}$$

Multiply row 3 by $-\dfrac{16}{37}$.

$$\begin{bmatrix} 1 & -1 & -\frac{1}{4} & | & \frac{1}{2} \\ 0 & 1 & \frac{15}{16} & | & -\frac{7}{8} \\ 0 & 0 & 1 & | & -2 \end{bmatrix}$$

This corresponds to:

$$\begin{cases} x - y - \dfrac{1}{4}z = \dfrac{1}{2} \\ y + \dfrac{15}{16}z = -\dfrac{7}{8} \\ z = -2 \end{cases}$$

$$y + \frac{15}{16}(-2) = -\frac{7}{8}$$
$$y - \frac{15}{8} = -\frac{7}{8}$$
$$y = 1$$

$$x - 1 - \frac{1}{4}(-2) = \frac{1}{2}$$
$$x - 1 + \frac{1}{2} = \frac{1}{2}$$
$$x = 1$$

The solution is $(1, 1, -2)$.

38. $\begin{cases} x + 2y + 3z = 14 \\ y + 2z = 3 \\ 2x - 2z = 10 \end{cases}$

$$\begin{bmatrix} 1 & 2 & 3 & | & 14 \\ 0 & 1 & 2 & | & 3 \\ 2 & 0 & -2 & | & 10 \end{bmatrix}$$

Multiply row 1 by -2 and add to row 3.

$$\begin{bmatrix} 1 & 2 & 3 & | & 14 \\ 0 & 1 & 2 & | & 3 \\ 0 & -4 & -8 & | & -18 \end{bmatrix}$$

Multiply row 2 by 4 and add to row 3.

$$\begin{bmatrix} 1 & 2 & 3 & | & 14 \\ 0 & 1 & 2 & | & 3 \\ 0 & 0 & 0 & | & -6 \end{bmatrix}$$

This corresponds to:

$$\begin{cases} x + 2y + 3z = 14 \\ y + 2z = 3 \\ 0 = -6 \end{cases}$$

The system is inconsistent.
The solution set is $\varnothing$.

39. $\begin{vmatrix} -1 & 3 \\ 5 & 2 \end{vmatrix} = (-1)(2) - 3(5) = -2 - 15 = -17$

40. $\begin{vmatrix} 3 & -1 \\ 2 & 5 \end{vmatrix} = 3(5) - (-1)2 = 15 + 2 = 17$

41. $\begin{vmatrix} 2 & -1 & -3 \\ 1 & 2 & 0 \\ 3 & -2 & 2 \end{vmatrix}$

$= 2\begin{vmatrix} 2 & 0 \\ -2 & 2 \end{vmatrix} - (-1)\begin{vmatrix} 1 & 0 \\ 3 & 2 \end{vmatrix} + (-3)\begin{vmatrix} 1 & 2 \\ 3 & -2 \end{vmatrix}$

$= 2(4 - 0) + (2 - 0) - 3(-2 - 6)$

$= 2(4) + (2) - 3(-8)$

$= 8 + 2 + 24$

$= 34$

42. $\begin{vmatrix} -2 & 3 & 1 \\ 4 & 4 & 0 \\ 1 & -2 & 3 \end{vmatrix}$

$= 1\begin{vmatrix} 4 & 4 \\ 1 & -2 \end{vmatrix} - 0\begin{vmatrix} -2 & 3 \\ 1 & -2 \end{vmatrix} + 3\begin{vmatrix} -2 & 3 \\ 4 & 4 \end{vmatrix}$

$= (-8 - 4) - 0 + 3(-8 - 12)$

$= -12 + 3(-20)$

$= -12 - 60$

$= -72$

43. $\begin{cases} 3x - 2y = -8 \\ 6x + 5y = 11 \end{cases}$

$D = \begin{vmatrix} 3 & -2 \\ 6 & 5 \end{vmatrix} = 15 - (-12) = 27$

$D_x = \begin{vmatrix} -8 & -2 \\ 11 & 5 \end{vmatrix} = -40 - (-22) = -18$

$D_y = \begin{vmatrix} 3 & -8 \\ 6 & 11 \end{vmatrix} = 33 - (-48) = 81$

$x = \dfrac{D_x}{D} = \dfrac{-18}{27} = -\dfrac{2}{3}$

$y = \dfrac{D_y}{D} = \dfrac{81}{27} = 3$

The solution is $\left(-\dfrac{2}{3}, 3 \right)$.

44. $\begin{cases} 6x - 6y = -5 \\ 10x - 2y = 1 \end{cases}$

$D = \begin{vmatrix} 6 & -6 \\ 10 & -2 \end{vmatrix} = -12 - (-60) = 48$

$D_x = \begin{vmatrix} -5 & -6 \\ 1 & -2 \end{vmatrix} = 10 - (-6) = 16$

$D_y = \begin{vmatrix} 6 & -5 \\ 10 & 1 \end{vmatrix} = 6 - (-50) = 56$

$x = \dfrac{D_x}{D} = \dfrac{16}{48} = \dfrac{1}{3}$ and $y = \dfrac{D_y}{D} = \dfrac{56}{48} = \dfrac{7}{6}$

The solution is $\left(\dfrac{1}{3}, \dfrac{7}{6} \right)$.

45. $\begin{cases} 3x + 10y = 1 \\ x + 2y = -1 \end{cases}$

$D = \begin{vmatrix} 3 & 10 \\ 1 & 2 \end{vmatrix} = 6 - 10 = -4$

$D_x = \begin{vmatrix} 1 & 10 \\ -1 & 2 \end{vmatrix} = 2 - (-10) = 12$

$D_y = \begin{vmatrix} 3 & 1 \\ 1 & -1 \end{vmatrix} = -3 - 1 = -4$

$x = \dfrac{D_x}{D} = \dfrac{12}{-4} = -3$

$y = \dfrac{D_y}{D} = \dfrac{-4}{-4} = 1$

The solution is $(-3, 1)$.

46. $\begin{cases} y = \dfrac{1}{2}x + \dfrac{2}{3} \\ 4x + 6y = 4 \end{cases}$

or $\begin{cases} -\dfrac{1}{2}x + y = \dfrac{2}{3} \\ 4x + 6y = 4 \end{cases}$

$D = \begin{vmatrix} -\frac{1}{2} & 1 \\ 4 & 6 \end{vmatrix} = -3 - 4 = -7$

$D_x = \begin{vmatrix} \frac{2}{3} & 1 \\ 4 & 6 \end{vmatrix} = 4 - 4 = 0$

$D_y = \begin{vmatrix} -\frac{1}{2} & \frac{2}{3} \\ 4 & 4 \end{vmatrix} = -2 - \dfrac{8}{3} = -\dfrac{14}{3}$

$x = \dfrac{D_x}{D} = \dfrac{0}{-7} = 0$ and $y = \dfrac{D_y}{D} = \dfrac{-\frac{14}{3}}{-7} = \dfrac{2}{3}$

The solution is $\left(0, \dfrac{2}{3}\right)$.

47. $\begin{cases} 2x - 4y = 22 \\ 5x - 10y = 16 \end{cases}$

$D = \begin{vmatrix} 2 & -4 \\ 5 & -10 \end{vmatrix} = -20 - (-20) = 0$

This cannot be solved by Cramer's rule.
Multiply the first equation by –5, multiply
the second equation by 2, and add.
$-10x + 20y = -110$

$\underline{10x - 20y = 32}$

$\qquad 0 = -78$ False

The system is inconsistent.
The solution set is $\varnothing$.

48. $\begin{cases} 3x - 6y = 12 \\ 2y = x - 4 \end{cases}$

or $\begin{cases} 3x - 6y = 12 \\ -x + 2y = -4 \end{cases}$

$D = \begin{vmatrix} 3 & -6 \\ -1 & 2 \end{vmatrix} = 6 - 6 = 0$

Cramer's rule cannot be used to solve this
system. Since the first equation is –3 times
the second equation, the system is
dependent. The solution set is
$\{(x, \ y) \mid x - 2y = 4\}$.

49. $\begin{cases} x \quad\;\; + z = 4 \\ 2x - y \quad\;\; = 0 \\ x + y - z = 0 \end{cases}$

$D = \begin{vmatrix} 1 & 0 & 1 \\ 2 & -1 & 0 \\ 1 & 1 & -1 \end{vmatrix}$

$= 1\begin{vmatrix} -1 & 0 \\ 1 & -1 \end{vmatrix} - 0\begin{vmatrix} 2 & 0 \\ 1 & -1 \end{vmatrix} + 1\begin{vmatrix} 2 & -1 \\ 1 & 1 \end{vmatrix}$

$= (1 - 0) - 0 + [2 - (-1)]$

$= 1 + 3$

$= 4$

$D_x = \begin{vmatrix} 4 & 0 & 1 \\ 0 & -1 & 0 \\ 0 & 1 & -1 \end{vmatrix}$

$= 4\begin{vmatrix} -1 & 0 \\ 1 & -1 \end{vmatrix} - 0\begin{vmatrix} 0 & 1 \\ 1 & -1 \end{vmatrix} + 0\begin{vmatrix} 0 & 1 \\ -1 & 0 \end{vmatrix}$

$= 4(1 - 0) - 0 + 0$

$= 4$

$D_y = \begin{vmatrix} 1 & 4 & 1 \\ 2 & 0 & 0 \\ 1 & 0 & -1 \end{vmatrix}$

$= -4\begin{vmatrix} 2 & 0 \\ 1 & -1 \end{vmatrix} + 0\begin{vmatrix} 1 & 1 \\ 1 & -1 \end{vmatrix} - 0\begin{vmatrix} 1 & 1 \\ 2 & 0 \end{vmatrix}$

$= (-4)(-2 - 0) + 0 - 0$

$= 8$

$D_z = \begin{vmatrix} 1 & 0 & 4 \\ 2 & -1 & 0 \\ 1 & 1 & 0 \end{vmatrix}$

$= 4\begin{vmatrix} 2 & -1 \\ 1 & 1 \end{vmatrix} - 0\begin{vmatrix} 1 & 0 \\ 1 & 1 \end{vmatrix} + 0\begin{vmatrix} 1 & 0 \\ 2 & -1 \end{vmatrix}$

$= 4[2 - (-1)] - 0 + 0$

$= 4(3)$

$= 12$

$x = \dfrac{D_x}{D} = \dfrac{4}{4} = 1$

$y = \dfrac{D_y}{D} = \dfrac{8}{4} = 2$

$z = \dfrac{D_z}{D} = \dfrac{12}{4} = 3$

The solution is $(1, 2, 3)$.

50. $\begin{cases} 2x + 5y = 4 \\ x - 5y + z = -1 \\ 4x - z = 11 \end{cases}$

$D = \begin{vmatrix} 2 & 5 & 0 \\ 1 & -5 & 1 \\ 4 & 0 & -1 \end{vmatrix}$

$= 0\begin{vmatrix} 1 & -5 \\ 4 & 0 \end{vmatrix} - 1\begin{vmatrix} 2 & 5 \\ 4 & 0 \end{vmatrix} + (-1)\begin{vmatrix} 2 & 5 \\ 1 & -5 \end{vmatrix}$

$= 0 - (0 - 20) - (-10 - 5)$

$= 20 + 15$

$= 35$

$D_x = \begin{vmatrix} 4 & 5 & 0 \\ -1 & -5 & 1 \\ 11 & 0 & -1 \end{vmatrix}$

$= 0\begin{vmatrix} -1 & -5 \\ 11 & 0 \end{vmatrix} - 1\begin{vmatrix} 4 & 5 \\ 11 & 0 \end{vmatrix} + (-1)\begin{vmatrix} 4 & 5 \\ -1 & -5 \end{vmatrix}$

$= 0 - (0 - 55) - [-20 - (-5)]$

$= 55 + 15$

$= 70$

$D_y = \begin{vmatrix} 2 & 4 & 0 \\ 1 & -1 & 1 \\ 4 & 11 & -1 \end{vmatrix}$

$= 0\begin{vmatrix} 1 & -1 \\ 4 & 11 \end{vmatrix} - 1\begin{vmatrix} 2 & 4 \\ 4 & 11 \end{vmatrix} + (-1)\begin{vmatrix} 2 & 4 \\ 1 & -1 \end{vmatrix}$

$= 0 - (22 - 16) - (-2 - 4)$

$= -6 + 6$

$= 0$

$D_z = \begin{vmatrix} 2 & 5 & 4 \\ 1 & -5 & -1 \\ 4 & 0 & 11 \end{vmatrix}$

$= 4\begin{vmatrix} 1 & -5 \\ 4 & 0 \end{vmatrix} - (-1)\begin{vmatrix} 2 & 5 \\ 4 & 0 \end{vmatrix} + 11\begin{vmatrix} 2 & 5 \\ 1 & -5 \end{vmatrix}$

$= 4[0 - (-20)] + (0 - 20) + 11(-10 - 5)$

$= 4(20) - 20 + 11(-15) = 80 - 20 - 165$

$= -105$

$x = \dfrac{D_x}{D} = \dfrac{70}{35} = 2$

$y = \dfrac{D_y}{D} = \dfrac{0}{35} = 0$

$z = \dfrac{D_z}{D} = \dfrac{-105}{35} = -3$

The solution is $(2, 0, -3)$.

51. $\begin{cases} x + 3y - z = 5 \\ 2x - y - 2z = 3 \\ x + 2y + 3z = 4 \end{cases}$

$D = \begin{vmatrix} 1 & 3 & -1 \\ 2 & -1 & -2 \\ 1 & 2 & 3 \end{vmatrix}$

$= 1\begin{vmatrix} -1 & -2 \\ 2 & 3 \end{vmatrix} - 3\begin{vmatrix} 2 & -2 \\ 1 & 3 \end{vmatrix} + (-1)\begin{vmatrix} 2 & -1 \\ 1 & 2 \end{vmatrix}$

$= [-3 - (-4)] - 3[6 - (-2)] - [4 - (-1)]$

$= 1 - 3(8) - 5$

$= 1 - 24 - 5$

$= -28$

$D_x = \begin{vmatrix} 5 & 3 & -1 \\ 3 & -1 & -2 \\ 4 & 2 & 3 \end{vmatrix}$

$= 5\begin{vmatrix} -1 & -2 \\ 2 & 3 \end{vmatrix} - 3\begin{vmatrix} 3 & -2 \\ 4 & 3 \end{vmatrix} + (-1)\begin{vmatrix} 3 & -1 \\ 4 & 2 \end{vmatrix}$

$= 5[-3 - (-4)] - 3[9 - (-8)] - [6 - (-4)]$

$= 5(1) - 3(17) - 10$

$= 5 - 51 - 10$

$= -56$

$D_y = \begin{vmatrix} 1 & 5 & -1 \\ 2 & 3 & -2 \\ 1 & 4 & 3 \end{vmatrix}$

$= 1\begin{vmatrix} 3 & -2 \\ 4 & 3 \end{vmatrix} - 5\begin{vmatrix} 2 & -2 \\ 1 & 3 \end{vmatrix} + (-1)\begin{vmatrix} 2 & 3 \\ 1 & 4 \end{vmatrix}$

$= [9 - (-8)] - 5[6 - (-2)] - (8 - 3)$

$= 17 - 5(8) - 5$

$= 17 - 40 - 5$

$= -28$

$$D_z = \begin{vmatrix} 1 & 3 & 5 \\ 2 & -1 & 3 \\ 1 & 2 & 4 \end{vmatrix}$$

$$= 1\begin{vmatrix} -1 & 3 \\ 2 & 4 \end{vmatrix} - 3\begin{vmatrix} 2 & 3 \\ 1 & 4 \end{vmatrix} + 5\begin{vmatrix} 2 & -1 \\ 1 & 2 \end{vmatrix}$$

$$= (-4-6) - 3(8-3) + 5[4-(-1)]$$

$$= -10 - 3(5) + 5(5)$$

$$= -10 - 15 + 25$$

$$= 0$$

$$x = \frac{D_x}{D} = \frac{-56}{-28} = 2$$

$$y = \frac{D_y}{D} = \frac{-28}{-28} = 1$$

$$z = \frac{D_z}{D} = \frac{0}{-28} = 0$$

The solution is (2, 1, 0).

52. $\begin{cases} 2x & -z = 1 \\ 3x - y + 2z = 3 \\ x + y + 3z = -2 \end{cases}$

$$D = \begin{vmatrix} 2 & 0 & -1 \\ 3 & -1 & 2 \\ 1 & 1 & 3 \end{vmatrix}$$

$$= 2\begin{vmatrix} -1 & 2 \\ 1 & 3 \end{vmatrix} - 0\begin{vmatrix} 3 & 2 \\ 1 & 3 \end{vmatrix} + (-1)\begin{vmatrix} 3 & -1 \\ 1 & 1 \end{vmatrix}$$

$$= 2(-3-2) - 0 - [3-(-1)]$$

$$= 2(-5) - 4$$

$$= -10 - 4$$

$$= -14$$

$$D_x = \begin{vmatrix} 1 & 0 & -1 \\ 3 & -1 & 2 \\ -2 & 1 & 3 \end{vmatrix}$$

$$= 1\begin{vmatrix} -1 & 2 \\ 1 & 3 \end{vmatrix} - 0\begin{vmatrix} 3 & 2 \\ -2 & 3 \end{vmatrix} + (-1)\begin{vmatrix} 3 & -1 \\ -2 & 1 \end{vmatrix}$$

$$= (-3-2) - 0 - (3-2)$$

$$= -5 - 1$$

$$= -6$$

$$D_y = \begin{vmatrix} 2 & 1 & -1 \\ 3 & 3 & 2 \\ 1 & -2 & 3 \end{vmatrix}$$

$$= 2\begin{vmatrix} 3 & 2 \\ -2 & 3 \end{vmatrix} - 1\begin{vmatrix} 3 & 2 \\ 1 & 3 \end{vmatrix} + (-1)\begin{vmatrix} 3 & 3 \\ 1 & -2 \end{vmatrix}$$

$$= 2[9-(-4)] - (9-2) - (-6-3)$$

$$= 2(13) - 7 + 9$$

$$= 26 + 2$$

$$= 28$$

$$D_z = \begin{vmatrix} 2 & 0 & 1 \\ 3 & -1 & 3 \\ 1 & 1 & -2 \end{vmatrix}$$

$$= 2\begin{vmatrix} -1 & 3 \\ 1 & -2 \end{vmatrix} - 0\begin{vmatrix} 3 & 3 \\ 1 & -2 \end{vmatrix} + 1\begin{vmatrix} 3 & -1 \\ 1 & 1 \end{vmatrix}$$

$$= 2(2-3) - 0 + [3-(-1)]$$

$$= 2(-1) + 4$$

$$= -2 + 4$$

$$= 2$$

$$x = \frac{D_x}{D} = \frac{-6}{-14} = \frac{3}{7}$$

$$y = \frac{D_y}{D} = \frac{28}{-14} = -2$$

$$z = \frac{D_z}{D} = \frac{2}{-14} = -\frac{1}{7}$$

The solution is $\left(\frac{3}{7}, -2, -\frac{1}{7} \right)$.

53. $\begin{cases} x + 2y + 3z = 14 & (1) \\ y + 2z = 3 & (2) \\ 2x \quad\quad - 2z = 10 & (3) \end{cases}$

$D = \begin{vmatrix} 1 & 2 & 3 \\ 0 & 1 & 2 \\ 2 & 0 & -2 \end{vmatrix}$

$= 1\begin{vmatrix} 1 & 2 \\ 0 & -2 \end{vmatrix} - 0\begin{vmatrix} 2 & 3 \\ 0 & -2 \end{vmatrix} + 2\begin{vmatrix} 2 & 3 \\ 1 & 2 \end{vmatrix}$

$= (-2 - 0) - 0 + 2(4 - 3)$

$= -2 + 2$

$= 0$

This cannot be solved by Cramer's rule.
Multiply equation (2) by –2 and add to
equation (1).

$x + 2y + 3z = 14$

$\underline{\quad -2y - 4z = -6 \quad}$

$x \quad\quad - z = 8 \quad (4)$

Multiply equation (4) by –2 and add to
equation (3).

$2x - 2z = 10$

$\underline{-2x + 2z = -16}$

$\quad\quad\quad 0 = -6 \quad$ False

The system is inconsistent.
The solution set is $\varnothing$.

54. $\begin{cases} 5x + 7y \quad\quad = 9 \\ 14y - z = 28 \\ 4x \quad\quad + 2z = -4 \end{cases}$

$D = \begin{vmatrix} 5 & 7 & 0 \\ 0 & 14 & -1 \\ 4 & 0 & 2 \end{vmatrix}$

$= 5\begin{vmatrix} 14 & -1 \\ 0 & 2 \end{vmatrix} - 7\begin{vmatrix} 0 & -1 \\ 4 & 2 \end{vmatrix} + 0\begin{vmatrix} 0 & 14 \\ 4 & 0 \end{vmatrix}$

$= 5(28 - 0) - 7[0 - (-4)] + 0$

$= 140 - 28$

$= 112$

$D_x = \begin{vmatrix} 9 & 7 & 0 \\ 28 & 14 & -1 \\ -4 & 0 & 2 \end{vmatrix}$

$= 9\begin{vmatrix} 14 & -1 \\ 0 & 2 \end{vmatrix} - 7\begin{vmatrix} 28 & -1 \\ -4 & 2 \end{vmatrix} + 0\begin{vmatrix} 28 & 14 \\ -4 & 0 \end{vmatrix}$

$= 9(28 - 0) - 7(56 - 4) + 0$

$= 252 - 7(52)$

$= 252 - 364$

$= -112$

$D_y = \begin{vmatrix} 5 & 9 & 0 \\ 0 & 28 & -1 \\ 4 & -4 & 2 \end{vmatrix}$

$= 5\begin{vmatrix} 28 & -1 \\ -4 & 2 \end{vmatrix} - 9\begin{vmatrix} 0 & -1 \\ 4 & 2 \end{vmatrix} + 0\begin{vmatrix} 0 & 28 \\ 4 & -4 \end{vmatrix}$

$= 5(56 - 4) - 9[0 - (-4)] + 0$

$= 5(52) - 9(4)$

$= 260 - 36$

$= 224$

$D_z = \begin{vmatrix} 5 & 7 & 9 \\ 0 & 14 & 28 \\ 4 & 0 & -4 \end{vmatrix}$

$= 5\begin{vmatrix} 14 & 28 \\ 0 & -4 \end{vmatrix} - 0\begin{vmatrix} 7 & 9 \\ 0 & -4 \end{vmatrix} + 4\begin{vmatrix} 7 & 9 \\ 14 & 28 \end{vmatrix}$

$= 5(-56 - 0) - 0 + 4(196 - 126)$

$= -280 + 4(70)$

$= -280 + 280$

$= 0$

$x = \dfrac{D_x}{D} = \dfrac{-112}{112} = -1$

$y = \dfrac{D_y}{D} = \dfrac{224}{112} = 2$

$z = \dfrac{D_z}{D} = \dfrac{0}{112} = 0$

The solution is $(-1, 2, 0)$.

Chapter 4 Test

1. $\begin{vmatrix} 4 & -7 \\ 2 & 5 \end{vmatrix} = 4(5) - (-7)(2) = 20 + 14 = 34$

2. Evaluate the determinant by expanding by the minors of the first column.

$\begin{vmatrix} 4 & 0 & 2 \\ 1 & -3 & 5 \\ 0 & -1 & 2 \end{vmatrix}$

$= 4 \begin{vmatrix} -3 & 5 \\ -1 & 2 \end{vmatrix} - 1 \begin{vmatrix} 0 & 2 \\ -1 & 2 \end{vmatrix} + 0 \begin{vmatrix} 0 & 2 \\ -3 & 5 \end{vmatrix}$

$= 4[-6 - (-5)] - [0 - (-2)] + 0$

$= 4(-1) - 2$

$= -4 - 2$

$= -6$

3. $\begin{cases} 2x - y = -1 \\ 5x + 4y = 17 \end{cases}$

Graphically:

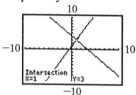

By elimination:
Multiply the first equation by 4 and add to the second equation.

$8x - 4y = -4$

$\underline{5x + 4y = 17}$

$13x \qquad = 13$

$x = 1$

Replace x with 1 in the second equation.

$5(1) + 4y = 17$

$4y = 12$

$y = 3$

The solution is (1, 3).

4. $\begin{cases} 7x - 14y = 5 \\ \quad\quad x = 2y \end{cases}$

Graphically:

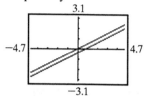

By substitution:
Replace x with $2y$ in the first equation.

$7(2y) - 14y = 5$

$14y - 14y = 5$

$0 = 5$ False

The system is inconsistent.
The solution set is $\varnothing$.

5. $\begin{cases} 4x - 7y = 29 \\ 2x + 5y = -11 \end{cases}$

Multiply the second equation by –2 and add to the first equation.

$4x - 7y = 29$

$\underline{-4x - 10y = 22}$

$-17y = 51$

$y = -3$

Replace y with –3 in the first equation.

$4x - 7(-3) = 29$

$4x + 21 = 29$

$4x = 8$

$x = 2$

The solution is (2, –3).

6. $\begin{cases} 15x + 6y = 15 \\ 10x + 4y = 10 \end{cases}$

Divide the first equation by 3 and the second equation by 2.

$\begin{cases} 5x + 2y = 5 \\ 5x + 2y = 5 \end{cases}$

The system is dependent.
The solution set is $\{(x, y) \mid 10x + 4y = 10\}$.

7. $\begin{cases} 2x - 3y \quad\;\; = 4 \quad (1) \\ \quad\; 3y + 2z = 2 \quad (2) \\ \; x \quad\;\; - z = -5 \quad (3) \end{cases}$

Add equations (1) and (2).

$2x - 3y \qquad = 4$

$\underline{\quad\; 3y + 2z = 2 \quad\;\;}$

$2x \qquad + 2z = 6 \quad (4)$

Multiply equation (3) by 2 and add to equation (4).

$2x - 2z = -10$

$\underline{2x + 2z = 6 \qquad}$

$4x \qquad = -4$

$\qquad x = -1$

Replace x with -1 in equation (3).

$-1 - z = -5$

$\quad -z = -4$

$\quad\;\; z = 4$

Replace x with -1 in equation (1).

$2(-1) - 3y = 4$

$-2 - 3y = 4$

$-3y = 6$

$y = -2$

The solution is $(-1, -2, 4)$.

8. $\begin{cases} 3x - 2y - z = -1 \quad (1) \\ 2x - 2y \quad\;\; = 4 \quad (2) \\ 2x \quad\;\; - 2z = -12 \quad (3) \end{cases}$

Multiply equation (2) by -1 and add to equation (1).

$3x - 2y - z = -1$

$\underline{-2x + 2y \qquad = -4}$

$x \qquad - z = -5 \quad (4)$

Multiply equation (4) by -2 and add to equation (3).

$2x - 2z = -12$

$\underline{-2x + 2z = 10 \qquad}$

$0 = -2 \quad$ False

The system is inconsistent.

The solution set is $\varnothing$.

9. $\begin{cases} \dfrac{x}{2} + \dfrac{y}{4} = -\dfrac{3}{4} \\ x + \dfrac{3}{4}y = -4 \end{cases}$

Clear fractions by multiplying both equations by 4.

$\begin{cases} 2x + \; y = -3 \quad (1) \\ 4x + 3y = -16 \quad (2) \end{cases}$

Multiply equation (1) by -2 and add to equation (2).

$-4x - 2y = 6$

$\underline{4x + 3y = -16}$

$y = -10$

Replace y with -10 in equation (1).

$2x + (-10) = -3$

$2x = 7$

$x = \dfrac{7}{2}$

The solution is $\left(\dfrac{7}{2}, -10 \right)$.

10. $\begin{cases} 3x - y = 7 \\ 2x + 5y = -1 \end{cases}$

$D = \begin{vmatrix} 3 & -1 \\ 2 & 5 \end{vmatrix}$

$\quad = 3(5) - (-1)(2)$

$\quad = 15 + 2$

$\quad = 17$

$D_x = \begin{vmatrix} 7 & -1 \\ -1 & 5 \end{vmatrix}$

$\quad = 7(5) - (-1)(-1)$

$\quad = 35 - 1$

$\quad = 34$

$D_y = \begin{vmatrix} 3 & 7 \\ 2 & -1 \end{vmatrix}$

$\quad = 3(-1) - 7(2)$

$\quad = -3 - 14$

$\quad = -17$

$x = \dfrac{D_x}{D} = \dfrac{34}{17} = 2$ and $y = \dfrac{D_y}{D} = \dfrac{-17}{17} = -1$

The solution is $(2, -1)$.

11. $\begin{cases} 4x - 3y = -6 \\ -2x + y = 0 \end{cases}$

$D = \begin{vmatrix} 4 & -3 \\ -2 & 1 \end{vmatrix}$

$\quad = 4(1) - (-3)(-2)$

$\quad = 4 - 6$

$\quad = -2$

$D_x = \begin{vmatrix} -6 & -3 \\ 0 & 1 \end{vmatrix}$

$\quad = (-6)(1) - (-3)(0)$

$\quad = -6 - 0$

$\quad = -6$

$D_y = \begin{vmatrix} 4 & -6 \\ -2 & 0 \end{vmatrix}$

$\quad = 4(0) - (-6)(-2)$

$\quad = 0 - 12$

$\quad = -12$

$x = \dfrac{D_x}{D} = \dfrac{-6}{-2} = 3$ and $y = \dfrac{D_y}{D} = \dfrac{-12}{-2} = 6$

The solution is $(3, 6)$.

12. $\begin{cases} x + y + z = 4 \\ 2x + 5y = 1 \\ x - y - 2z = 0 \end{cases}$

$D = \begin{vmatrix} 1 & 1 & 1 \\ 2 & 5 & 0 \\ 1 & -1 & -2 \end{vmatrix}$

$\quad = 1\begin{vmatrix} 2 & 5 \\ 1 & -1 \end{vmatrix} - 0\begin{vmatrix} 1 & 1 \\ 1 & -1 \end{vmatrix} + (-2)\begin{vmatrix} 1 & 1 \\ 2 & 5 \end{vmatrix}$

$\quad = (-2 - 5) - 0 - 2(5 - 2)$

$\quad = -7 - 2(3)$

$\quad = -7 - 6$

$\quad = -13$

$D_x = \begin{vmatrix} 4 & 1 & 1 \\ 1 & 5 & 0 \\ 0 & -1 & -2 \end{vmatrix}$

$\quad = 1\begin{vmatrix} 1 & 5 \\ 0 & -1 \end{vmatrix} - 0\begin{vmatrix} 4 & 1 \\ 0 & -1 \end{vmatrix} + (-2)\begin{vmatrix} 4 & 1 \\ 1 & 5 \end{vmatrix}$

$\quad = (-1 - 0) - 0 - 2(20 - 1)$

$\quad = -1 - 2(19)$

$\quad = -1 - 38$

$\quad = -39$

$D_y = \begin{vmatrix} 1 & 4 & 1 \\ 2 & 1 & 0 \\ 1 & 0 & -2 \end{vmatrix}$

$\quad = 1\begin{vmatrix} 2 & 1 \\ 1 & 0 \end{vmatrix} - 0\begin{vmatrix} 1 & 4 \\ 1 & 0 \end{vmatrix} + (-2)\begin{vmatrix} 1 & 4 \\ 2 & 1 \end{vmatrix}$

$\quad = (0 - 1) - 0 - 2(1 - 8)$

$\quad = -1 - 2(-7)$

$\quad = -1 + 14$

$\quad = 13$

$D_z = \begin{vmatrix} 1 & 1 & 4 \\ 2 & 5 & 1 \\ 1 & -1 & 0 \end{vmatrix}$

$\quad = 1\begin{vmatrix} 1 & 4 \\ 5 & 1 \end{vmatrix} - (-1)\begin{vmatrix} 1 & 4 \\ 2 & 1 \end{vmatrix} + 0\begin{vmatrix} 1 & 1 \\ 2 & 5 \end{vmatrix}$

$\quad = (1 - 20) + (1 - 8) + 0$

$\quad = -19 - 7$

$\quad = -26$

$x = \dfrac{D_x}{D} = \dfrac{-39}{-13} = 3$

$y = \dfrac{D_y}{D} = \dfrac{13}{-13} = -1$

$z = \dfrac{D_z}{D} = \dfrac{-26}{-13} = 2$

The solution is $(3, -1, 2)$.

13. $\begin{cases} 3x + 2y + 3z = 3 \\ x \qquad - z = 9 \\ \quad 4y + \ z = -4 \end{cases}$

$D = \begin{vmatrix} 3 & 2 & 3 \\ 1 & 0 & -1 \\ 0 & 4 & 1 \end{vmatrix}$

$= -1\begin{vmatrix} 2 & 3 \\ 4 & 1 \end{vmatrix} + 0\begin{vmatrix} 3 & 3 \\ 0 & 1 \end{vmatrix} - (-1)\begin{vmatrix} 3 & 2 \\ 0 & 4 \end{vmatrix}$

$= -(2 - 12) + 0 + (12 - 0)$

$= -(-10) + 12$

$= 10 + 12$

$= 22$

$D_x = \begin{vmatrix} 3 & 2 & 3 \\ 9 & 0 & -1 \\ -4 & 4 & 1 \end{vmatrix}$

$= -9\begin{vmatrix} 2 & 3 \\ 4 & 1 \end{vmatrix} + 0\begin{vmatrix} 3 & 3 \\ -4 & 1 \end{vmatrix} - (-1)\begin{vmatrix} 3 & 2 \\ -4 & 4 \end{vmatrix}$

$= -9(2 - 12) + 0 + [12 - (-8)]$

$= -9(-10) + 20$

$= 90 + 20$

$= 110$

$D_y = \begin{vmatrix} 3 & 3 & 3 \\ 1 & 9 & -1 \\ 0 & -4 & 1 \end{vmatrix}$

$= 3\begin{vmatrix} 9 & -1 \\ -4 & 1 \end{vmatrix} - 1\begin{vmatrix} 3 & 3 \\ -4 & 1 \end{vmatrix} + 0\begin{vmatrix} 3 & 3 \\ 9 & -1 \end{vmatrix}$

$= 3(9 - 4) - [3 - (-12)] + 0$

$= 3(5) - (15)$

$= 15 - 15$

$= 0$

$D_z = \begin{vmatrix} 3 & 2 & 3 \\ 1 & 0 & 9 \\ 0 & 4 & -4 \end{vmatrix}$

$= 3\begin{vmatrix} 0 & 9 \\ 4 & -4 \end{vmatrix} - 1\begin{vmatrix} 2 & 3 \\ 4 & -4 \end{vmatrix} + 0\begin{vmatrix} 2 & 3 \\ 0 & 9 \end{vmatrix}$

$= 3(0 - 36) - (-8 - 12) + 0$

$= 3(-36) - (-20)$

$= -108 + 20$

$= -88$

$x = \dfrac{D_x}{D} = \dfrac{110}{22} = 5$

$y = \dfrac{D_y}{D} = \dfrac{0}{22} = 0$

$z = \dfrac{D_z}{D} = \dfrac{-88}{22} = -4$

The solution is $(5, 0, -4)$.

14. $\begin{cases} x - \ y = -2 \\ 3x - 3y = -6 \end{cases}$

$\begin{bmatrix} 1 & -1 & \vdots & -2 \\ 3 & -3 & \vdots & -6 \end{bmatrix}$

Multiply row 1 by -3 and add to row 2.

$\begin{bmatrix} 1 & -1 & \vdots & -2 \\ 0 & 0 & \vdots & 0 \end{bmatrix}$

This corresponds to:

$\begin{cases} x - y = -2 \\ \quad 0 = 0 \end{cases}$

This is a dependent system.

The solution set is $\{(x, y) \mid x - y = -2\}$.

15. $\begin{cases} x + 2y = -1 \\ 2x + 5y = -5 \end{cases}$

$\begin{bmatrix} 1 & 2 & \vdots & -1 \\ 2 & 5 & \vdots & -5 \end{bmatrix}$

Multiply row 1 by -2 and add to row 2.

$\begin{bmatrix} 1 & 2 & \vdots & -1 \\ 0 & 1 & \vdots & -3 \end{bmatrix}$

This corresponds to:

$\begin{cases} x + 2y = -1 \\ \quad y = -3 \end{cases}$

$x + 2(-3) = -1$

$x - 6 = -1$

$x = 5$

The solution is $(5, -3)$.

16. $\begin{cases} x - y - z = 0 \\ 3x - y - 5z = -2 \\ 2x + 3y \quad = -5 \end{cases}$

$$\begin{bmatrix} 1 & -1 & -1 & | & 0 \\ 3 & -1 & -5 & | & -2 \\ 2 & 3 & 0 & | & -5 \end{bmatrix}$$

Multiply row 1 by –3 and add to row 2.
Multiply row 1 by –2 and add to row 3.

$$\begin{bmatrix} 1 & -1 & -1 & | & 0 \\ 0 & 2 & -2 & | & -2 \\ 0 & 5 & 2 & | & -5 \end{bmatrix}$$

Divide row 2 by 2.

$$\begin{bmatrix} 1 & -1 & -1 & | & 0 \\ 0 & 1 & -1 & | & -1 \\ 0 & 5 & 2 & | & -5 \end{bmatrix}$$

Multiply row 2 by –5 and add to row 3.

$$\begin{bmatrix} 1 & -1 & -1 & | & 0 \\ 0 & 1 & -1 & | & -1 \\ 0 & 0 & 7 & | & 0 \end{bmatrix}$$

Divide row 3 by 7.

$$\begin{bmatrix} 1 & -1 & -1 & | & 0 \\ 0 & 1 & -1 & | & -1 \\ 0 & 0 & 1 & | & 0 \end{bmatrix}$$

This corresponds to:

$$\begin{cases} x - y - z = 0 \\ y - z = -1 \\ z = 0 \end{cases}$$

$y - 0 = -1$
$y = -1$
$x - (-1) - 0 = 0$
$x + 1 = 0$
$x = -1$

The solution is $(-1, -1, 0)$.

17. $\begin{cases} 2x - y + 3z = 4 \\ 3x \quad - 3z = -2 \\ -5x + y \quad = 0 \end{cases}$

$$\begin{bmatrix} 2 & -1 & 3 & | & 4 \\ 3 & 0 & -3 & | & -2 \\ -5 & 1 & 0 & | & 0 \end{bmatrix}$$

Divide row 1 by 2.

$$\begin{bmatrix} 1 & -\frac{1}{2} & \frac{3}{2} & | & 2 \\ 3 & 0 & -3 & | & -2 \\ -5 & 1 & 0 & | & 0 \end{bmatrix}$$

Multiply row 1 by –3 and add to row 2.
Multiply row 1 by 5 and add to row 3.

$$\begin{bmatrix} 1 & -\frac{1}{2} & \frac{3}{2} & | & 2 \\ 0 & \frac{3}{2} & -\frac{15}{2} & | & -8 \\ 0 & -\frac{3}{2} & \frac{15}{2} & | & 10 \end{bmatrix}$$

Multiply row 2 by $\dfrac{2}{3}$.

$$\begin{bmatrix} 1 & -\frac{1}{2} & \frac{3}{2} & | & 2 \\ 0 & 1 & -5 & | & -\frac{16}{3} \\ 0 & -\frac{3}{2} & \frac{15}{2} & | & 10 \end{bmatrix}$$

Multiply row 2 by $\dfrac{3}{2}$ and add to row 3.

$$\begin{bmatrix} 1 & -\frac{1}{2} & \frac{3}{2} & | & 2 \\ 0 & 1 & -5 & | & -\frac{16}{3} \\ 0 & 0 & 0 & | & 2 \end{bmatrix}$$

This corresponds to:

$$\begin{cases} x - \dfrac{1}{2}y + \dfrac{3}{2}z = 2 \\ y - 5z = -\dfrac{16}{3} \\ 0 = 2 \quad \text{False} \end{cases}$$

This is an inconsistent system.
The solution set is $\varnothing$.

18. $R(x) = 38x$
$C(x) = 18x + 5500$
$38x = 18x + 5500$
$20x = 5500$
$x = 275$
275 frames must be sold.

19. Let x = the number of double occupancy rooms
y = the number of single occupancy rooms
$$\begin{cases} x + y = 80 \\ 90x + 80y = 6930 \end{cases}$$
Multiply the first equation by -80 and add to the second equation.
$-80x - 80y = -6400$
$\underline{90x + 80y = 6930}$
$10x \quad\quad = 530$
$x = 53$
Replace x with 53 in the first equation.
$53 + y = 80$
$y = 27$
53 double-occupancy and 27 single-occupancy rooms are occupied.

20. Let x = the number of gallons of 10% solution
y = the number of gallons of 20% solution
$$\begin{cases} x + y = 20 \\ 0.10x + 0.20y = 0.175(20) \end{cases}$$
Multiply the first equation by -0.10 and add to the second equation.
$-0.10x - 0.10y = -2.0$
$\underline{0.10x + 0.20y = 3.5}$
$\quad\quad\quad 0.10y = 1.5$
$\quad\quad\quad\quad y = 15$
Replace y with 15 in the first equation.
$x + 15 = 20$
$x = 5$
They should use 5 gallons of 10% fructose solution and 15 gallons of the 20% solution.

21. $R(x) = 4x$
$C(x) = 1.5x + 2000$
$4x = 1.5x + 2000$
$2.5x = 2000$
$x = 800$
The company must sell 800 packages to break even.

Chapter 5

Section 5.1

Mental Math

1. $5x^{-1}y^{-2} = \dfrac{5}{xy^2}$

2. $7xy^{-4} = \dfrac{7x}{y^4}$

3. $a^2b^{-1}c^{-5} = \dfrac{a^2}{bc^5}$

4. $a^{-4}b^2c^{-6} = \dfrac{b^2}{a^4c^6}$

5. $\dfrac{y^{-2}}{x^{-4}} = \dfrac{x^4}{y^2}$

6. $\dfrac{x^{-7}}{z^{-3}} = \dfrac{z^3}{x^7}$

Exercise Set 5.1

1. $4^2 \cdot 4^3 = 4^{2+3} = 4^5$

3. $x^5 \cdot x^3 = x^{5+3} = x^8$

5. $-7x^3 \cdot 20x^9 = -140x^{3+9} = -140x^{12}$

7. $(4xy)(-5x) = -20x^{1+1}y = -20x^2y$

9. $(-4x^3p^2)(4y^3x^3) = -16x^{3+3}y^3p^2$
$= -16x^6y^3p^2$

11. $-8^0 = -(8^0) = -1$

13. $(4x+5)^0 = 1$

15. $(5x)^0 + 5x^0 = 1 + 5(1) = 1 + 5 = 6$

17. Answers may vary.

19. $\dfrac{a^5}{a^2} = a^{5-2} = a^3$

21. $\dfrac{x^9y^6}{x^8y^6} = x^{9-8}y^{6-6} = x^1y^0 = x$

23. $-\dfrac{26z^{11}}{2z^7} = -13z^{11-7} = -13z^4$

25. $\dfrac{-36a^5b^7c^{10}}{6ab^3c^4} = -6a^{5-1}b^{7-3}c^{10-4}$
$= -6a^4b^4c^6$

27. $4^{-2} = \dfrac{1}{4^2} = \dfrac{1}{16}$

29. $\dfrac{x^7}{x^{15}} = x^{7-15} = x^{-8} = \dfrac{1}{x^8}$

31. $5a^{-4} = 5\dfrac{1}{a^4} = \dfrac{5}{a^4}$

33. $\dfrac{x^{-2}}{x^5} = x^{-2-5} = x^{-7} = \dfrac{1}{x^7}$

35. $\dfrac{8r^4}{2r^{-4}} = 4r^{4-(-4)} = 4r^8$

37. $\dfrac{x^{-9}x^4}{x^{-5}} = \dfrac{x^{-9+4}}{x^{-5}}$
$= \dfrac{x^{-5}}{x^{-5}}$
$= x^{-5-(-5)}$
$= x^0$
$= 1$

39. $4^{-1} + 3^{-2} = \dfrac{1}{4^1} + \dfrac{1}{3^2}$

$\qquad = \dfrac{1}{4} + \dfrac{1}{9}$

$\qquad = \dfrac{9}{36} + \dfrac{4}{36}$

$\qquad = \dfrac{13}{36}$

41. $4x^0 + 5 = 4(1) + 5 = 4 + 5 = 9$

43. $x^7 \cdot x^8 = x^{7+8} = x^{15}$

45. $2x^3 \cdot 5x^7 = 2 \cdot 5 x^{3+7} = 10x^{10}$

47. $\dfrac{z^{12}}{z^{15}} = z^{12-15} = z^{-3} = \dfrac{1}{z^3}$

49. $\dfrac{y^{-3}}{y^{-7}} = y^{-3-(-7)} = y^4$

51. $3x^{-1} = 3 \cdot \dfrac{1}{x} = \dfrac{3}{x}$

53. $3^0 - 3t^0 = 1 - 3 \cdot 1 = 1 - 3 = -2$

55. $\dfrac{r^4}{r^{-4}} = r^{4-(-4)} = r^8$

57. $\dfrac{x^{-7}y^{-2}}{x^2y^2} = x^{-7-2}y^{-2-2} = x^{-9}y^{-4} = \dfrac{1}{x^9 y^4}$

59. $\dfrac{2a^{-6}b^2}{18ab^{-5}} = \dfrac{a^{-6-1}b^{2-(-5)}}{9} = \dfrac{a^{-7}b^7}{9} = \dfrac{b^7}{9a^7}$

61. $\dfrac{(24x^8)x}{20x^{-7}} = \dfrac{6x^{8+1}}{5x^{-7}} = \dfrac{6x^{9-(-7)}}{5} = \dfrac{6x^{16}}{5}$

63. $31,250,000 = 3.125 \times 10^7$

65. $0.016 = 1.6 \times 10^{-2}$

67. $67,413 = 6.7413 \times 10^4$

69. $0.0125 = 1.25 \times 10^{-2}$

71. $0.000053 = 5.3 \times 10^{-5}$

73. $3.6 \times 10^{-9} = 0.0000000036$

75. $9.3 \times 10^7 = 93,000,000$

77. $1.278 \times 10^6 = 1,278,000$

79. $7.35 \times 10^{12} = 7,350,000,000,000$

81. $4.03 \times 10^{-7} = 0.000000403$

83. Answers may vary.

85. **a.** $3.5 \times 10^{-5} = 0.000035 < 1$

$\quad$ **b.** $3.5 \times 10^5 = 350,000 > 1$

$\quad$ **c.** $-3.5 \times 10^5 = -350,000 < 0 < 1$

$\quad$ **d.** $-3.5 \times 10^{-5} = -0.000035 < 0 < 1$

$\quad$ a, c, and d have values that are less than one.

87. $778,300,000 = 7.783 \times 10^8$

89. $43,141,000 = 4.3141 \times 10^7$

91. $1,130,000,000 = 1.13 \times 10^9$

93. $0.001 = 1 \times 10^{-3}$

95. $x^5 \cdot x^{7a} = x^{5+7a} = x^{7a+5}$

97. $\dfrac{x^{3t-1}}{x^t} = x^{(3t-1)-t} = x^{2t-1}$

99. $x^{4a} \cdot x^7 = x^{4a+7}$

101. $\dfrac{z^{6x}}{z^7} = z^{6x-7}$

103. $\dfrac{x^{3t} \cdot x^{4t-1}}{x^t} = \dfrac{x^{3t+(4t-1)}}{x^t} = x^{(7t-1)-t} = x^{6t-1}$

105. $x^{9+b} \cdot x^{3a-b} = x^{(9+b)+(3a-b)} = x^{9+3a}$

107. 7^{13} is larger.

109. 7^{-11} is larger.

111. $(5 \cdot 2)^2 = (10)^2 = 100$

113. $\left(\dfrac{3}{4}\right)^3 = \dfrac{3}{4} \cdot \dfrac{3}{4} \cdot \dfrac{3}{4} = \dfrac{27}{64}$

115. $(2^3)^2 = 8^2 = 64$

117. $(2^{-1})^4 = \left(\dfrac{1}{2}\right)^4 = \dfrac{1^4}{2^4} = \dfrac{1}{16}$

Section 5.2

Mental Math

1. $(x^4)^5 = x^{4(5)} = x^{20}$

2. $(5^6)^2 = 5^{6(2)} = 5^{12}$

3. $x^4 \cdot x^5 = x^{4+5} = x^9$

4. $x^7 \cdot x^8 = x^{7+8} = x^{15}$

5. $(y^6)^7 = y^{6(7)} = y^{42}$

6. $(x^3)^4 = x^{3(4)} = x^{12}$

7. $(z^4)^5 = z^{4(5)} = z^{20}$

8. $(z^3)^7 = z^{3(7)} = z^{21}$

9. $(z^{-6})^{-3} = z^{-6(-3)} = z^{18}$

10. $(y^{-4})^{-2} = y^{-4(-2)} = y^8$

Exercise Set 5.2

1. $(3^{-1})^2 = 3^{-1(2)} = 3^{-2} = \dfrac{1}{3^2} = \dfrac{1}{9}$

3. $(x^4)^{-9} = x^{4(-9)} = x^{-36} = \dfrac{1}{x^{36}}$

5. $(y)^{-5} = y^{-5} = \dfrac{1}{y^5}$

7. $(3x^2 y^3)^2 = 3^2 (x^2)^2 (y^3)^2$
$$= 9x^{2(2)} y^{3(2)}$$
$$= 9x^4 y^6$$

9. $\left(\dfrac{2x^5}{y^{-3}}\right)^4 = \dfrac{2^4 (x^5)^4}{(y^{-3})^4}$
$$= \dfrac{16x^{5(4)}}{y^{-3(4)}}$$
$$= \dfrac{16x^{20}}{y^{-12}}$$
$$= 16x^{20} y^{12}$$

11. $(a^2 bc^{-3})^{-6} = (a^2)^{-6} b^{-6} (c^{-3})^{-6}$
$$= a^{2(-6)} b^{-6} c^{-3(-6)}$$
$$= a^{-12} b^{-6} c^{18}$$
$$= \dfrac{c^{18}}{a^{12} b^6}$$

13. $\left(\dfrac{x^7 y^{-3}}{z^{-4}}\right)^{-5} = \dfrac{(x^7)^{-5} (y^{-3})^{-5}}{(z^{-4})^{-5}}$
$$= \dfrac{x^{-35} y^{15}}{z^{20}}$$
$$= \dfrac{y^{15}}{x^{35} z^{20}}$$

15. $\left(\dfrac{a^{-4}}{a^{-5}}\right)^{-2} = \dfrac{(a^{-4})^{-2}}{(a^{-5})^{-2}}$
$$= \dfrac{a^8}{a^{10}}$$
$$= a^{8-10}$$
$$= a^{-2}$$
$$= \dfrac{1}{a^2}$$

17. $\left(\dfrac{2a^{-2}b^5}{4a^2b^7}\right)^{-2} = \left(\dfrac{1}{2a^4b^2}\right)^{-2}$

$\qquad = \dfrac{(1)^{-2}}{(2)^{-2}(a^4)^{-2}(b^2)^{-2}}$

$\qquad = \dfrac{2^2}{1^2 a^{-8}b^{-4}}$

$\qquad = 4a^8b^4$

19. $\dfrac{4^{-1}x^2yz}{x^{-2}yz^3} = \dfrac{x^{2-(-2)}y^0}{4z^{3-1}} = \dfrac{x^4}{4z^2}$

21. Yes; $a = \pm 1$.

23. $(5^{-1})^3 = 5^{-3} = \dfrac{1}{5^3} = \dfrac{1}{125}$

25. $(x^7)^{-9} = x^{-63} = \dfrac{1}{x^{63}}$

27. $\left(\dfrac{7}{8}\right)^3 = \dfrac{7^3}{8^3} = \dfrac{343}{512}$

29. $(4x^2)^2 = 4^2(x^2)^2 = 16x^4$

31. $(-2^{-2}y)^3 = (-2^{-2})^3 y^3$

$\qquad = -2^{-6}y^3$

$\qquad = -\dfrac{y^3}{2^6}$

$\qquad = -\dfrac{y^3}{64}$

33. $\left(\dfrac{4^{-4}}{y^3 x}\right)^{-2} = \dfrac{(4^{-4})^{-2}}{(y^3)^{-2}x^{-2}}$

$\qquad = \dfrac{4^8}{y^{-6}x^{-2}}$

$\qquad = 4^8 x^2 y^6$

35. $\left(\dfrac{6p^6}{p^{12}}\right)^2 = \dfrac{6^2 p^{12}}{p^{24}}$

$\qquad = 6^2 p^{12-24}$

$\qquad = 36p^{-12}$

$\qquad = \dfrac{36}{p^{12}}$

37. $(-8y^3xa^{-2})^{-3} = (-8)^{-3}(y^3)^{-3}x^{-3}(a^{-2})^{-3}$

$\qquad = -\dfrac{y^{-9}a^6}{8^3 x^3}$

$\qquad = -\dfrac{a^6}{512x^3y^9}$

39. $\left(\dfrac{x^{-2}y^{-2}}{a^{-3}}\right)^{-7} = \dfrac{(x^{-2})^{-7}(y^{-2})^{-7}}{(a^{-3})^{-7}}$

$\qquad = \dfrac{x^{14}y^{14}}{a^{21}}$

41. $\left(\dfrac{3x^5}{6x^4}\right)^4 = \left(\dfrac{x^{5-4}}{2}\right)^4 = \dfrac{(x^1)^4}{2^4} = \dfrac{x^4}{16}$

43. $\left(\dfrac{1}{4}\right)^{-3} = (4^{-1})^{-3} = 4^3 = 64$

45. $\dfrac{(y^3)^{-4}}{y^3} = y^{-12-3} = y^{-15} = \dfrac{1}{y^{15}}$

47. $\dfrac{8p^7}{4p^9} = 2p^{7-9} = 2p^{-2} = \dfrac{2}{p^2}$

49. $(4x^6y^5)^{-2}(6x^4y^3)$

$\qquad = 4^{-2}(x^6)^{-2}(y^5)^{-2}(6x^4y^3)$

$\qquad = \dfrac{1}{4^2}x^{-12}y^{-10} \cdot 6x^4y^3$

$\qquad = \dfrac{6}{16}x^{-12+4}y^{-10+3}$

$\qquad = \dfrac{3x^{-8}y^{-7}}{8}$

$\qquad = \dfrac{3}{8x^8y^7}$

51. $x^6(x^6bc)^{-6} = x^6(x^6)^{-6}b^{-6}c^{-6}$

$$= \frac{x^6 \cdot x^{-36}}{b^6c^6}$$

$$= \frac{x^{-30}}{b^6c^6}$$

$$= \frac{1}{x^{30}b^6c^6}$$

53. $\dfrac{2^{-3}x^2y^{-5}}{5^{-2}x^7y^{-1}} = \dfrac{5^2x^{2-7}y^{-5-(-1)}}{2^3}$

$$= \frac{25x^{-5}y^{-4}}{8}$$

$$= \frac{25}{8x^5y^4}$$

55. $\left(\dfrac{2x^2}{y^4}\right)^3 \cdot \left(\dfrac{2x^5}{y}\right)^{-2} = \dfrac{2^3(x^2)^3 2^{-2}(x^5)^{-2}}{(y^4)^3 y^{-2}}$

$$= \frac{8x^6x^{-10}}{2^2y^{12}y^{-2}}$$

$$= \frac{8x^{6-10}}{4y^{12-2}}$$

$$= \frac{2x^{-4}}{y^{10}}$$

$$= \frac{2}{x^4y^{10}}$$

57. $(5\times10^{11})(2.9\times10^{-3}) = 5\times2.9\times10^{11-3}$

$$= 14.5\times10^8$$

$$= 1.45\times10^1\times10^8$$

$$= 1.45\times10^9$$

59. $(2\times10^5)^3 = 2^3\times10^{5\cdot3} = 8\times10^{15}$

61. $\dfrac{3.6\times10^{-4}}{9\times10^2} = 0.4\times10^{-4-2}$

$$= 0.4\times10^{-6}$$

$$= 4\times10^{-1}\times10^{-6}$$

$$= 4\times10^{-7}$$

63. $\dfrac{0.0069}{0.023} = \dfrac{6.9\times10^{-3}}{2.3\times10^{-2}}$

$$= 3\times10^{-3-(-2)}$$

$$= 3\times10^{-1}$$

65. $\dfrac{18,200\times100}{91,000} = \dfrac{(1.82\times10^4)(1\times10^2)}{9.1\times10^4}$

$$= \frac{1.82\times10^6}{9.1\times10^4}$$

$$= 0.2\times10^{6-4}$$

$$= 2\times10^{-1}\times10^2$$

$$= 2\times10^{2-1}$$

$$= 2\times10^1$$

67.
$$\frac{6000 \times 0.006}{0.009 \times 400} = \frac{(6 \times 10^3)(6 \times 10^{-3})}{(9 \times 10^{-3})(4 \times 10^2)}$$
$$= \frac{36 \times 10^{3-3}}{36 \times 10^{-3+2}}$$
$$= \frac{10^0}{10^{-1}}$$
$$= 10^{0-(-1)}$$
$$= 10^1$$
$$= 1 \times 10^1$$

```
(6000*.006)/(.00
9*400)
            1E1
```

69.
$$\frac{0.00064 \times 2000}{16,000} = \frac{(6.4 \times 10^{-4})(2 \times 10^3)}{1.6 \times 10^4}$$
$$= \frac{12.8 \times 10^{-4+3}}{1.6 \times 10^4}$$
$$= 8 \times 10^{-1-4}$$
$$= 8 \times 10^{-5}$$

```
(.00064*2000)/16
000
            8E-5
```

71.
$$\frac{66,000 \times 0.001}{0.002 \times 0.003} = \frac{(6.6 \times 10^4)(1 \times 10^{-3})}{(2 \times 10^{-3})(3 \times 10^{-3})}$$
$$= \frac{6.6 \times 10^{4-3}}{6 \times 10^{-3-3}}$$
$$= \frac{1.1 \times 10^1}{10^{-6}}$$
$$= 1.1 \times 10^{1-(-6)}$$
$$= 1.1 \times 10^7$$

```
(66000*.001)/(.0
02*.003)
         1.1E7
```

73.
$$\frac{8.25 \times 10^{15}}{(2.5 \times 10^{-2})(2.2 \times 10^{-5})}$$
$$= \frac{8.25 \times 10^{15}}{5.5 \times 10^{-7}}$$
$$= 1.5 \times 10^{15-(-7)}$$
$$= 1.5 \times 10^{22}$$

```
(8.25E15)/(2.5E-
2*2.2E-5)
          1.5E22
```

75. $200,000 \times 10^{-8} = 2 \times 10^5 \times 10^{-8}$
$$= 2 \times 10^{-3}$$
$$= 0.002$$
It would take 0.002 second.

77. $D = \dfrac{M}{V}$ or $M = DV$
$$M = (3.12 \times 10^{-2})(4.269 \times 10^{14})$$
$$= 13.31928 \times 10^{12}$$
$$= 1.331928 \times 10^{13}$$
The mass is 1.331928×10^{13} tons.

79. Recall $V = s^3$
$$= \left(\frac{2x^{-2}}{y}\right)^3$$
$$= \frac{2^3(x^{-2})^3}{y^3}$$
$$= \frac{8x^{-6}}{y^3}$$
$$= \frac{8}{x^6 y^3}$$
The volume is $\dfrac{8}{x^6 y^3}$ cubic meters.

81. $(4 \times 10^{-2})(6.452 \times 10^{-4})$
$$= 25.808 \times 10^{-6}$$
$$= 2.5808 \times 10^{-5}$$
The area is 2.5808×10^{-5} square meters.

83. Answers may vary.

85. $\dfrac{2.93 \times 10^6}{4.25 \times 10^5} \approx 0.7 \times 10^{6-5}$

$\qquad = 7.0 \times 10^{-1} \times 10^1$

$\qquad = 7 \times 10^0$

$\qquad = 7$

China's fighting force numbers approximately 7 times Taiwan's.

87. $(x^{2b+7})^2 = x^{(2b+7)2} = x^{4b+14}$

89. $\dfrac{x^{-5y+2} x^{2y}}{x} = x^{(-5y+2)+2y-1} = x^{-3y+1}$

91. $(c^{2a+3})^3 = c^{(2a+3)3} = c^{6a+9}$

93. $\dfrac{(y^{4a})^7}{y^{2a-1}} = \dfrac{y^{28a}}{y^{2a-1}}$

$\qquad = y^{28a-(2a-1)}$

$\qquad = y^{28a-2a+1}$

$\qquad = y^{26a+1}$

95. $\left(\dfrac{3y^{5a}}{y^{-a+1}}\right)^2 = (3y^{5a-(-a+1)})^2$

$\qquad = (3y^{6a-1})^2$

$\qquad = 3^2 y^{(6a-1)2}$

$\qquad = 9y^{12a-2}$

97. $\dfrac{(y^{3-a})^b}{(y^{1-b})^a} = \dfrac{y^{3b-ab}}{y^{a-ab}}$

$\qquad = y^{(3b-ab)-(a-ab)}$

$\qquad = y^{3b-ab-a+ab}$

$\qquad = y^{3b-a}$

99. $\dfrac{x^{-5-3a} y^{-2a-b}}{x^{-5+3b} y^{-2b-a}}$

$\qquad = x^{(-5-3a)-(-5+3b)} y^{(-2a-b)-(-2b-a)}$

$\qquad = x^{-5-3a+5-3b} y^{-2a-b+2b+a}$

$\qquad = x^{-3a-3b} y^{b-a}$

101. $12m - 14 - 15m - 1 = -3m - 15$

103. $-9y - (5 - 6y) = -9y - 5 + 6y = -3y - 5$

105. $5(x - 3) - 4(2x - 5) = 5x - 15 - 8x + 20$

$\qquad\qquad\qquad\qquad = -3x + 5$

Exercise Set 5.3

1. The degree of 4, which can be written as $4x^0$, is 0.

3. The exponent on x is 2, so the degree of $5x^2$ is 2.

5. The degree of $-3xy^2$ is the sum of the exponents on the variables, or $1 + 2 = 3$.

7. $6x + 3$ has degree 1 because the largest degree of any term is 1. It is a binomial because it has two terms.

9. $3x^2 - 2x + 5$ has degree 2 because the largest degree of any term is 2. It is a trinomial because it has three terms.

11. $-xyz$ has degree 3 because the sum of the exponents on the variables is $1 + 1 + 1 = 3$. It is a monomial because it is one term.

13. $x^2y - 4xy^2 + 5x + y$ has degree 3 because the largest degree of any term is 3. It is none of these because it has four terms.

15. Answers may vary.

17. $P(x) = x^2 + x + 1$

$P(7) = 7^2 + 7 + 1$

$\qquad = 49 + 7 + 1$

$\qquad = 57$

19. $Q(x) = 5x^2 - 1$

$Q(-10) = 5(-10)^2 - 1$

$\qquad\quad = 5(100) - 1$

$\qquad\quad = 500 - 1$

$\qquad\quad = 499$

21.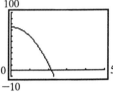

a. Between the x-values of 2 and 3, the y-values change from positive to negative.

b.

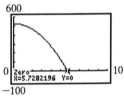

The graph crosses the x-axis between 2 and 3.

23. $P(x) = -16x^2 + 525$

Graph $y_1 = -16x^2 + 525$.

The object hits the ground when $y = 0$.

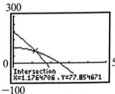

To the nearest tenth, the object hits the ground at 5.7 seconds.

25. Graph $y_1 = -16x^2 + 100$ and

$y_2 = -16x^2 - 85x + 200$.

300

![graph with intersection X=1.1764706, Y=77.854671]

a. The second egg hits the ground first because its graph crosses the x-axis at a smaller x-value than the graph of the first egg.

b. The intersection point is approximately (1, 78). The eggs are the same distance above the ground at approximately 1 second.

27. $5y + y = (5 + 1)y = 6y$

29. $4x + 7x - 3 = (4 + 7)x - 3 = 11x - 3$

31. $4xy + 2x - 3xy - 1 = (4 - 3)xy + 2x - 1 = xy + 2x - 1$

33. $(9y^2 - 8) + (9y^2 - 9) = 9y^2 - 8 + 9y^2 - 9$
$$= 9y^2 + 9y^2 - 8 - 9$$
$$= 18y^2 - 17$$

35. $\begin{array}{l} x^2 + \ xy - \ y^2 \\ +(2x^2 - 4xy + 7y^2) \\ \hline 3x^2 - 3xy + 6y^2 \end{array}$

37. $\begin{array}{l} x^2 - 6x + 3 \\ + \quad (2x + 5) \\ \hline x^2 - 4x + 8 \end{array}$

39. $(9y^2 - 7y + 5) - (8y^2 - 7y + 2)$
$$= 9y^2 - 7y + 5 - 8y^2 + 7y - 2$$
$$= 9y^2 - 8y^2 - 7y + 7y + 5 - 2$$
$$= y^2 + 3$$

41. $\begin{array}{l} 4x^2 + 2x \\ - \ (6x^2 - 3x) \\ \hline -2x^2 + 5x \end{array}$ or $\begin{array}{l} 4x^2 + 2x \\ +(-6x^2 + 3x) \\ \hline -2x^2 + 5x \end{array}$

43. $\begin{array}{l} 3x^2 - 4x + 8 \\ - \ (5x^2 \qquad - 7) \\ \hline -2x^2 - 4x + 15 \end{array}$ or $\begin{array}{l} 3x^2 - 4x + 8 \\ +(-5x^2 \qquad + 7) \\ \hline -2x^2 - 4x + 15 \end{array}$

45. $(5x - 11) + (-x - 2)$
$= 5x - 11 - x - 2$
$= 5x - x - 11 - 2$
$= 4x - 13$

47. $(7x^2 + x + 1) - (6x^2 + x - 1)$
$= 7x^2 + x + 1 - 6x^2 - x + 1$
$= 7x^2 - 6x^2 + x - x + 1 + 1$
$= x^2 + 2$

49. $(7x^3 - 4x + 8) + (5x^3 + 4x + 8x)$
$= 7x^3 - 4x + 8 + 5x^3 + 12x$
$= 12x^3 + 8x + 8$

51. $(9x^3 - 2x^2 + 4x - 7) - (2x^3 - 6x^2 - 4x + 3)$
$= 9x^3 - 2x^2 + 4x - 7 - 2x^3 + 6x^2 + 4x - 3$
$= 7x^3 + 4x^2 + 8x - 10$

53.
$$\begin{array}{r} y^2 + 4yx + 7 \\ + (-19y^2 + 7yx + 7) \\ \hline -18y^2 + 11yx + 14 \end{array}$$

55. $(3x^3 - b + 2a - 6) + (-4x^3 + b + 6a - 6)$
$= 3x^3 - 4x^3 - b + b + 2a + 6a - 6 - 6$
$= -x^3 + 8a - 12$

57. $(4x^2 - 6x + 2) - (-x^2 + 3x + 5)$
$= 4x^2 - 6x + 2 + x^2 - 3x - 5$
$= 5x^2 - 9x - 3$

59. $(-3x + 8) + (-3x^2 + 3x - 5)$
$= -3x + 8 - 3x^2 + 3x - 5$
$= -3x^2 + 3$

61. $(-3 + 4x^2 + 7xy^2) + (2x^3 - x^2 + xy^2)$
$= -3 + 4x^2 - x^2 + 7xy^2 + xy^2 + 2x^3$
$= 8xy^2 + 2x^3 + 3x^2 - 3$

63.
$$\begin{array}{r} 6y^2 - 6y + 4 \\ -(-y^2 - 6y + 7) \\ \hline 7y^2 \qquad -3 \end{array} \text{ or } \begin{array}{r} 6y^2 - 6y + 4 \\ + (y^2 + 6y - 7) \\ \hline 7y^2 \qquad -3 \end{array}$$

65.
$$\begin{array}{r} 3x^2 + 15x + 8 \\ + (2x^2 + 7x + 8) \\ \hline 5x^2 + 22x + 16 \end{array}$$

67.
$$\begin{array}{r} 5q^4 - 2q^2 - 3q \\ + (-6q^4 + 3q^2 \qquad + 5) \\ \hline -q^4 + q^2 - 3q + 5 \end{array}$$

69.
$$\begin{array}{r} 7x^2 + 4x + 9 \\ + (8x^2 + 7x - 8) \\ \hline 15x^2 + 11x + 1 \end{array}$$
$$\begin{array}{r} - \qquad (3x + 7) \\ \hline 15x^2 + 8x - 6 \end{array}$$

71.
$$\begin{array}{r} 4x^4 - 7x^2 + 3 \\ + (-3x^4 \qquad + 2) \\ \hline x^4 - 7x^2 + 5 \end{array}$$

73. $(8x^{2y} - 7x^y + 3) + (-4x^{2y} + 9x^y - 14)$
$= 8x^{2y} - 7x^y + 3 - 4x^{2y} + 9x^y - 14$
$= 4x^{2y} + 2x^y - 11$

75. $P(t) = -32t + 500$
$P(3) = -32(3) + 500$
$\qquad = -96 + 500$
$\qquad = 404$
The accrued velocity after 3 seconds is 404 feet per second.

77. $f(x) = 0.43x^2 + 164.6x + 949.3$

   a. $f(5) = 0.43(5)^2 + 164.6(5) + 949.3 = 1783.05$
       In 1985, \$1783.05 was spent.

   b. $f(15) = 0.43(15)^2 + 164.6(15) + 949.3 = 3515.05$
       In 1995, \$3515.05 was spent.

   c. $f(30) = 0.43(30)^2 + 164.6(30) + 949.3 = 6274.30$
       In 2010, \$6274.30 will be spent.

   d. No, the amount of money spent is not rising at a steady rate. The first 15 years after 1980 it increased \$3515.05 − \$949.3 = \$2565.75. The next 15 years it increased \$6274.30 − \$3515.05 = \$2759.25, which is more than the increase for the first 15 years.

79. $C(x) = 0.8x + 10,000$
     $C(20,000) = 0.8(20,000) + 10,000 = 26,000$
     The cost in producing 20,000 tapes per week is \$26,000.

81. **a.** At $x = 1$, $y_1 = 284$.
       The height at 1 second is 284 feet.

   b. At $x = 2$, $y_1 = 536$.
       The height at 2 seconds is 536 feet.

   c. At $x = 3$, $y_1 = 756$.
       The height at 3 seconds is 756 feet.

   d. At $x = 4$, $y_1 = 944$.
       The height at 4 seconds is 944 feet.

83.

X	Y1	
17	476	
18	216	
19	-76	
20	-400	

$Y_1 \blacksquare -16X^2 + 300X$

The object hits the ground at approximately 19 seconds.

85. $f(x) = -0.85x^3 + 14.28x^2 - 49.38x + 574.16$

    a. $f(5) = -0.85(5)^3 + 14.28(5)^2 - 49.38(5) + 574.16 \approx 578$
        There were approximately 578 HMOs in 1995.

    b. $f(8) = -0.85(8)^3 + 14.28(8)^2 - 49.38(8) + 574.16 \approx 658$
        There were approximately 658 HMOs in 1998.

    c. $f(13) = -0.85(13)^3 + 14.28(13)^2 - 49.38(13) + 574.16 \approx 478$
        There will be approximately 478 HMOs in 2003.

87. $P(x) + Q(x) = (3x + 3) + (4x^2 - 6x + 3) = 4x^2 - 3x + 6$

89. $Q(x) - R(x) = (4x^2 - 6x + 3) - (5x^2 - 7) = 4x^2 - 6x + 3 - 5x^2 + 7 = -x^2 - 6x + 10$

91. $2[Q(x)] - R(x) = 2(4x^2 - 6x + 3) - (5x^2 - 7) = 8x^2 - 12x + 6 - 5x^2 + 7 = 3x^2 - 12x + 13$

93. $3[R(x)] + 4[P(x)] = 3(5x^2 - 7) + 4(3x + 3) = 15x^2 - 21 + 12x + 12 = 15x^2 + 12x - 9$

95. **a.** $P(x) = R(x) - C(x) = 5.5x - (3x + 3000) = 2.5x - 3000$

    b. $P(2000) = 2.5(2000) - 3000 = 5000 - 3000 = 2000$
        The profit is $2000.

97. $h(x) = 5x^3 - 6x + 2$ has degree 3, so the shape of its graph is like option
     B or D. Since the coefficient of x^3 is positive, its graph is that of option B.

99. $g(x) = -2x^2 - 6x + 2$ has degree 2, so the shape of its graph is a parabola.
     Since the coefficient of x^2 is negative, its parabola opens downward. The answer is C.

101. $P = (z + 2) + (2z^2 + z) + (z^3 - 4z + 1) = z^3 + 2z^2 - 2z + 3$
      The perimeter is $(z^3 + 2z^2 - 2z + 3)$ units.

103. $-7(2z - 6y) = -7(2z) - 7(-6y) = -14z + 42y$

105. $5(-3y^2 - 2y + 7) = 5(-3y^2) + 5(-2y) + 5(7) = -15y^2 - 10y + 35$

107. $P(x) = 8x + 3$

    a. $P(a) = 8a + 3$

    b. $P(-x) = 8(-x) + 3 = -8x + 3$

    c. $P(x + h) = 8(x + h) + 3 = 8x + 8h + 3$

109. $P(x) = -4x$

    a. $P(a) = -4a$

    b. $P(-x) = -4(-x) = 4x$

    c. $P(x + h) = -4(x + h) = -4x - 4h$

111. $P(x) = 3x - 2$

    a. $P(a) = 3a - 2$

    b. $P(-x) = 3(-x) - 2 = -3x - 2$

    c. $P(x + h) = 3(x + h) - 2 = 3x + 3h - 2$

Exercise Set 5.4

  1. $(-4x^3)(3x^2) = (-4)(3)(x^3 x^2) = -12x^5$

  3. $3x(4x + 7) = 3x(4x) + 3x(7) = 12x^2 + 21x$

  5. $-6xy(4x + y) = -6xy(4x) - 6xy(y) = -24x^2 y - 6xy^2$

  7. $-4ab(xa^2 + ya^2 - 3) = -4ab(xa^2) - 4ab(ya^2) - 4ab(-3) = -4a^3 bx - 4a^3 by + 12ab$

  9. $\begin{aligned}(x - 3)(2x + 4) &= x(2x + 4) - 3(2x + 4) \\ &= x(2x) + x(4) - 3(2x) - 3(4) \\ &= 2x^2 + 4x - 6x - 12 \\ &= 2x^2 - 2x - 12\end{aligned}$

11. $\begin{aligned}(2x + 3)(x^3 - x + 2) &= 2x(x^3 - x + 2) + 3(x^3 - x + 2) \\ &= 2x(x^3) + 2x(-x) + 2x(2) + 3(x^3) + 3(-x) + 3(2) \\ &= 2x^4 - 2x^2 + 4x + 3x^3 - 3x + 6 \\ &= 2x^4 + 3x^3 - 2x^2 + x + 6\end{aligned}$

13.

$$\begin{array}{r} 3x - 2 \\ \underline{\times \quad\quad 5x + 1} \\ 3x - 2 \\ \underline{15x^2 - 10x \quad\quad} \\ 15x^2 - \ 7x - 2 \end{array}$$

15.

$$3m^2 + 2m - 1$$
$$\underline{\times \qquad\qquad 5m + 2}$$
$$6m^2 + 4m - 2$$
$$\underline{15m^3 + 10m^2 - 5m}$$
$$15m^3 + 16m^2 - \;\;m - 2$$

17. Answers may vary.

19. $(x-3)(x+4) = x \cdot x + x \cdot 4 - 3 \cdot x - 3 \cdot 4$
$$= x^2 + 4x - 3x - 12$$
$$= x^2 + x - 12$$

21. $(5x+8y)(2x-y)$
$$= 5x(2x) + 5x(-y) + 8y(2x) + 8y(-y)$$
$$= 10x^2 - 5xy + 16xy - 8y^2$$
$$= 10x^2 + 11xy - 8y^2$$

23. $(3x-1)(x+3) = 3x(x) + 3x(3) - 1(x) - 1(3)$
$$= 3x^2 + 9x - x - 3$$
$$= 3x^2 + 8x - 3$$

25. $\left(3x + \dfrac{1}{2}\right)\left(3x - \dfrac{1}{2}\right)$
$$= 3x(3x) + 3x\left(-\dfrac{1}{2}\right) + \dfrac{1}{2}(3x) + \dfrac{1}{2}\left(-\dfrac{1}{2}\right)$$
$$= 9x^2 - \dfrac{3}{2}x + \dfrac{3}{2}x - \dfrac{1}{4}$$
$$= 9x^2 - \dfrac{1}{4}$$

27. $(x+4)^2 = x^2 + 2(x)4 + 4^2 = x^2 + 8x + 16$

29. $(6y-1)(6y+1) = (6y)^2 - 1^2 = 36y^2 - 1$

31. $(3x-y)^2 = (3x)^2 - 2(3x)y + y^2$
$$= 9x^2 - 6xy + y^2$$

33. $(3b-6y)(3b+6y) = (3b)^2 - (6y)^2$
$$= 9b^2 - 36y^2$$

35. $[3 + (4b+1)]^2$
$$= 3^2 + 2(3)(4b+1) + (4b+1)^2$$
$$= 9 + 6(4b+1) + (4b)^2 + 2(4b)1 + 1^2$$
$$= 9 + 24b + 6 + 16b^2 + 8b + 1$$
$$= 16b^2 + 32b + 16$$

37. $[(2s-3)-1][(2s-3)+1]$
$$= (2s-3)^2 - 1^2$$
$$= (2s)^2 - 2(2s)3 + 3^2 - 1$$
$$= 4s^2 - 12s + 9 - 1$$
$$= 4s^2 - 12s + 8$$

39. $[(xy+4)-6]^2$
$$= (xy+4)^2 - 2(xy+4)6 + 6^2$$
$$= (xy)^2 + 2(xy)4 + 4^2 - 12(xy+4) + 36$$
$$= x^2y^2 + 8xy + 16 - 12xy - 48 + 36$$
$$= x^2y^2 - 4xy + 4$$

41. Answers may vary.

43. $(3x+1)(3x+5)$
$$= 3x(3x) + 3x(5) + 1(3x) + 1(5)$$
$$= 9x^2 + 15x + 3x + 5$$
$$= 9x^2 + 18x + 5$$

45. $(2x^3 + 5)(5x^2 + 4x + 1) = 2x^3(5x^2) + 2x^3(4x) + 2x^3(1) + 5(5x^2) + 5(4x) + 5(1)$
$$= 10x^5 + 8x^4 + 2x^3 + 25x^2 + 20x + 5$$

47. $(7x + 3)(7x - 3) = (7x)^2 - 3^2 = 49x^2 - 9$

49.
$$\begin{array}{r} 3x^2 + 4x - 4 \\ \times \quad\quad 3x + 6 \\ \hline 18x^2 + 24x - 24 \\ 9x^3 + 12x^2 - 12x \quad\quad \\ \hline 9x^3 + 30x^2 + 12x - 24 \end{array}$$

51. $\left(4x + \dfrac{1}{3}\right)\left(4x - \dfrac{1}{2}\right) = 4x(4x) + 4x\left(-\dfrac{1}{2}\right) + \dfrac{1}{3}(4x) + \dfrac{1}{3}\left(-\dfrac{1}{2}\right) = 16x^2 - 2x + \dfrac{4}{3}x - \dfrac{1}{6} = 16x^2 - \dfrac{2}{3}x - \dfrac{1}{6}$

53. $(6x + 1)^2 = (6x)^2 + 2(6x)1 + 1^2 = 36x^2 + 12x + 1$

55. $(x^2 + 2y)(x^2 - 2y) = (x^2)^2 - (2y)^2 = x^4 - 4y^2$

57. $-6a^2b^2[5a^2b^2 - 6a - 6b] = -6a^2b^2(5a^2b^2) - 6a^2b^2(-6a) - 6a^2b^2(-6b)$
$$= -30a^4b^4 + 36a^3b^2 + 36a^2b^3$$

59. $(a - 4)(2a - 4) = a(2a) + a(-4) - 4(2a) - 4(-4) = 2a^2 - 4a - 8a + 16 = 2a^2 - 12a + 16$

61. $(7ab + 3c)(7ab - 3c) = (7ab)^2 - (3c)^2 = 49a^2b^2 - 9c^2$

63. $(m - 4)^2 = m^2 - 2(m)4 + 4^2 = m^2 - 8m + 16$

65. $(3x + 1)^2 = (3x)^2 + 2(3x)1 + 1^2 = 9x^2 + 6x + 1$

67. $(y - 4)(y - 3) = y(y) + y(-3) - 4(y) - 4(-3) = y^2 - 3y - 4y + 12 = y^2 - 7y + 12$

69. $(x + y)(2x - 1)(x + 1) = (2x^2 - x + 2xy - y)(x + 1)$
$$= 2x^3 + 2x^2 - x^2 - x + 2x^2y + 2xy - xy - y$$
$$= 2x^3 + 2x^2y + x^2 + xy - x - y$$

71. $(3x^2 + 2x - 1)^2 = [(3x^2 + 2x) - 1]^2$
$$= (3x^2 + 2x)^2 + 2(3x^2 + 2x)(-1) + (-1)^2$$
$$= (3x^2)^2 + 2(3x^2)(2x) + (2x)^2 - 6x^2 - 4x + 1$$
$$= 9x^4 + 12x^3 + 4x^2 - 6x^2 - 4x + 1$$
$$= 9x^4 + 12x^3 - 2x^2 - 4x + 1$$

73. $(3x+1)(4x^2-2x+5) = 3x(4x^2)+3x(-2x)+3x(5)+1(4x^2)+1(-2x)+1(5)$
$$= 12x^3 - 6x^2 + 15x + 4x^2 - 2x + 5$$
$$= 12x^3 - 2x^2 + 13x + 5$$

75. $P(x) \cdot R(x) = 5x(x+5) = 5x(x)+5x(5) = 5x^2 + 25x$

77. $[Q(x)]^2 = (x^2-2)^2 = (x^2)^2 - 2(x^2)(2)+(2)^2 = x^4 - 4x^2 + 4$

79. $R(x) \cdot Q(x) = (x+5)(x^2-2)$
$$= x(x^2)+x(-2)+5(x^2)+5(-2)$$
$$= x^3 - 2x + 5x^2 - 10$$
$$= x^3 + 5x^2 - 2x - 10$$

81. a. $(3x+5)+(3x+7) = 3x+3x+5+7 = 6x+12$

b. $(3x+5)(3x+7) = 3x(3x)+3x(7)+5(3x)+5(7) = 9x^2 + 21x + 15x + 35 = 9x^2 + 36x + 35$

Part a involves adding polynomials while part b involves multiplying.

83. $A = \pi(5x-2)^2 = \pi[(5x)^2 - 2(5x)2 + 2^2] = \pi(25x^2 - 20x + 4)$
The area is $\pi(25x^2 - 20x + 4)$ square kilometers.

85. $f(x) = x^2 - 3x$
$f(a) = a^2 - 3a$

87. $f(x) = x^2 - 3x$
$f(a+h) = (a+h)^2 - 3(a+h) = a^2 + 2ah + h^2 - 3a - 3h$

89. $f(x) = x^2 - 3x$
$f(b-2) = (b-2)^2 - 3(b-2) = b^2 - 2 \cdot b \cdot 2 + 2^2 - 3b + 6 = b^2 - 4b + 4 - 3b + 6 = b^2 - 7b + 10$

91. $F(x) = x^2 + 3x + 2$

a. $F(a+h) = (a+h)^2 + 3(a+h)+2 = a^2 + 2ah + h^2 + 3a + 3h + 2$

b. $F(a) = a^2 + 3a + 2$

c. $F(a+h) - F(a) = (a^2 + 2ah + h^2 + 3a + 3h + 2) - (a^2 + 3a + 2) = 2ah + h^2 + 3h$

93. $5x^2 y^n (6y^{n+1} - 2) = 30x^2 y^{2n+1} - 10x^2 y^n$

95. $(x^a + 5)(x^{2a} - 3)$

$= x^a(x^{2a}) + x^a(-3) + (5)x^{2a} - 15$

$= x^{3a} - 3x^a + 5x^{2a} - 15$

$= x^{3a} + 5x^{2a} - 3x^2 - 15$

97. $y = -2x + 7$

$m = -2$

99. $3x - 5y = 14$

$-5y = -3x + 14$

$y = \dfrac{3}{5}x - \dfrac{14}{5}$

$m = \dfrac{3}{5}$

101. Since any vertical line only crosses the graph once it is a function.

Section 5.5

Mental Math

1. $6 = 2 \cdot 3$

$12 = 2 \cdot 2 \cdot 3$

$\text{GCF} = 2 \cdot 3 = 6$

2. $9 = 3 \cdot 3$

$27 = 3 \cdot 3 \cdot 3$

$\text{GCF} = 3 \cdot 3 = 9$

3. $15x = 3 \cdot 5 \cdot x$

$10 = 2 \cdot 5$

$\text{GCF} = 5$

4. $9x = 3 \cdot 3 \cdot x$

$12 = 2 \cdot 2 \cdot 3$

$\text{GCF} = 3$

5. $13x = 13 \cdot x$

$2x = 2 \cdot x$

$\text{GCF} = x$

6. $4y = 2 \cdot 2 \cdot y$

$5y = 5 \cdot y$

$\text{GCF} = y$

7. $7x = 7 \cdot x$

$14x = 2 \cdot 7 \cdot x$

$\text{GCF} = 7x$

8. $8z = 2 \cdot 2 \cdot 2 \cdot z$

$4z = 2 \cdot 2 \cdot z$

$\text{GCF} = 2 \cdot 2 \cdot z = 4z$

Exercise Set 5.5

1. a^8, a^5, and a^3 have a GCF a^3.

3. $x^2y^3z^3$, y^2z^3, and xy^2z^2 have GCF y^2z^2.

5. $6x^3y$, $9x^2y^2$, and $12x^2y$ have. GCF $3x^2y$

7. $10x^3yz^3$, $20x^2z^5$, and $45xz^3$ have GCF $5xz^3$.

9. $18x - 12 = 6 \cdot 3x - 6 \cdot 2 = 6(3x - 2)$

11. $4y^2 - 16xy^3 = 4y^2(1) + 4y^2(-4xy)$

$= 4y^2(1 - 4xy)$

13. $6x^5 - 8x^4 + 2x^3$

$= 2x^3(3x^2) + 2x^3(-4x) + 2x^3(1)$

$= 2x^3(3x^2 - 4x + 1)$

15. $8a^3b^3 - 4a^2b^2 + 4ab + 16ab^2 = 4ab(2a^2b^2) + 4ab(-ab) + 4ab(1) + 4ab(4b)$
$$= 4ab(2a^2b^2 - ab + 1 + 4b)$$

17. $6(x+3) + 5a(x+3) = (x+3)(6+5a)$

19. $2x(z+7) + (z+7) = 2x(z+7) + 1(z+7)$
$$= (z+7)(2x+1)$$

21. $3x(x^2+5) - 2(x^2+5) = (x^2+5)(3x-2)$

23. Answers may vary.

25. $ab + 3a + 2b + 6 = a(b+3) + 2(b+3)$
$$= (a+2)(b+3)$$

27. $ac + 4a - 2c - 8 = a(c+4) - 2(c+4)$
$$= (a-2)(c+4)$$

29. $2xy - 3x - 4y + 6 = x(2y-3) - 2(2y-3)$
$$= (x-2)(2y-3)$$

31. $12xy - 8x - 3y + 2 = 4x(3y-2) - 1(3y-2)$
$$= (4x-1)(3y-2)$$

33. $6x^3 + 9 = 3(2x^3) + 3(3) = 3(2x^3+3)$

35. $x^3 + 3x^2 = x^2 \cdot x + x^2 \cdot 3$
$$= x^2(x+3)$$

37. $8a^3 - 4a = 4a(2a^2) + 4a(-1)$
$$= 4a(2a^2 - 1)$$

39. $-20x^2y + 16xy^3 = 4xy(-5x) + 4xy(4y^2)$
$$= 4xy(-5x + 4y^2)$$
$$\text{or } -4xy(5x - 4y^2)$$

41. $10a^2b^3 + 5ab^2 - 15ab^3$
$$= 5ab^2(2ab) + 5ab^2(1) + 5ab^2(-3b)$$
$$= 5ab^2(2ab + 1 - 3b)$$

43. $9abc^2 + 6a^2bc - 6ab + 3bc$
$$= 3b(3ac^2) + 3b(2a^2c) + 3b(-2a) + 3b(c)$$
$$= 3b(3ac^2 + 2a^2c - 2a + c)$$

45. $4x(y-2) - 3(y-2) = (y-2)(4x-3)$

47. $6xy + 10x + 9y + 15$
$$= 2x(3y+5) + 3(3y+5)$$
$$= (2x+3)(3y+5)$$

49. $xy + 3y - 5x - 15 = y(x+3) - 5(x+3)$
$$= (x+3)(y-5)$$

51. $6ab - 2a - 9b + 3 = 2a(3b-1) - 3(3b-1)$
$$= (2a-3)(3b-1)$$

53. $12xy + 18x + 2y + 3$
$$= 6x(2y+3) + 1(2y+3)$$
$$= (6x+1)(2y+3)$$

55. $2m(n-8) - (n-8) = 2m(n-8) - 1(n-8)$
$$= (n-8)(2m-1)$$

57. $15x^3y^2 - 18x^2y^2$
$$= 3x^2y^2(5x) + 3x^2y^2(-6)$$
$$= 3x^2y^2(5x-6)$$

59. $2x^2 + 3xy + 4x + 6y$
$$= x(2x+3y) + 2(2x+3y)$$
$$= (2x+3y)(x+2)$$

61. $5x^2 + 5xy - 3x - 3y = 5x(x+y) - 3(x+y)$
$$= (5x-3)(x+y)$$

63. $x^3 + 3x^2 + 4x + 12 = x^2(x+3) + 4(x+3)$
$$= (x^2+4)(x+3)$$

65. $x^3 - x^2 - 2x + 2 = x^2(x-1) - 2(x-1)$
$$= (x^2-2)(x-1)$$

67. $2\pi r^2 + 2\pi rh = 2\pi r \cdot r + 2\pi r \cdot h$
$$= 2\pi r(r+h)$$

69. $A = P + PRT$
$A = P \cdot 1 + P \cdot RT$
$A = P(1 + RT)$

71. $h(t) = -16t^2 + 64t$

   a. $h(t) = -16t(t) - 16t(-4) = -16t(t-4)$

   b. $h(1) = -16(1)^2 + 64(1) = -16 + 64 = 48$
       $h(1) = -16(1)(1-4) = -16(-3) = 48$
       The height of the object after
       1 second is 48 feet.

   c. Answers may vary.

73. None

   a. $(2-x)(3-y) = 6 - 2y - 3x + xy$
              $= xy - 3x - 2y + 6$

   b. $(-2+x)(-3+y) = 6 - 2y - 3x + xy$
                 $= xy - 3x - 2y + 6$

   c. $(x-2)(y-3) = xy - 3x - 2y + 6$

   d. $(-x+2)(-y+3) = xy - 3x - 2y + 6$

75. a is correct;
    $3(4x^2 + 3x + 1) = 12x^2 + 9x + 3$

77. $(7y)(-2y^3) = 7 \cdot -2 \cdot y \cdot y^3 = -14y^{1+3} = -14y^4$

79. $(-2y^3)^4 = (-2)^4(y^3)^4 = 16y^{3 \cdot 4} = 16y^{12}$

81. $(x-7)(x-1) = x^2 - x - 7x + 7$
                   $= x^2 - 8x + 7$

83. $(x-4)(x+2) = x^2 + 2x - 4x - 8$
                   $= x^2 - 2x - 8$

85. $(s+8)(s+10) = s^2 + 10s + 8s + 80$
                    $= s^2 + 18s + 80$

87. $3y^n + 3y^{2n} + 5y^{8n}$
    $= y^n(3) + y^n(3y^n) + y^n(5y^{7n})$
    $= y^n(3 + 3y^n + 5y^{7n})$

89. $3x^{5a} - 6x^{3a} + 9x^{2a}$
    $= 3x^{2a}(x^{3a}) + 3x^{2a}(-2x^a) + 3x^{2a}(3)$
    $= 3x^{2a}(x^{3a} - 2x^a + 3)$

Section 5.6

Mental Math

1. $10 = 5 \cdot 2$
    $7 = 5 + 2$
    5 and 2

2. $12 = 2 \cdot 6$
    $8 = 2 + 6$
    2 and 6

3. $24 = 8 \cdot 3$
    $11 = 8 + 3$
    8 and 3

4. $30 = 10 \cdot 3$
    $13 = 10 + 3$
    10 and 3

Exercise Set 5.6

1. $x^2 + 9x + 18 = (x+3)(x+6)$

3. $x^2 - 12x + 32 = (x-8)(x-4)$

5. $x^2 + 10x - 24 = (x+12)(x-2)$

7. $x^2 - 2x - 24 = (x-6)(x+4)$

9. Note that the GCF is 3, so that
    $3x^2 - 18x + 24 = 3(x^2 - 6x + 8)$
                        $= 3(x-2)(x-4)$.

11. Note that the GCF is 4z, so that
    $4x^2z + 28xz + 40z = 4z(x^2 + 7x + 10)$
                        $= 4z(x+2)(x+5)$.

13. Note that the GCF is 2, so that
    $2x^2 + 30x - 108 = 2(x^2 + 15x - 54)$
                        $= 2(x+18)(x-3)$.

15. $x^2 + bx + 6$
$6 = 2 \cdot 3$ or $6 = (-2)(-3)$
$6 = 1 \cdot 6$ or $6 = (-1)(-6)$
$(x + 2)(x + 3) = x^2 + 5x + 6$
$(x - 2)(x - 3) = x^2 - 5x + 6$
$(x + 1)(x + 6) = x^2 + 7x + 6$
$(x - 1)(x - 6) = x^2 - 7x + 6$
$b = \pm 5$ or ± 7

17. The x-intercepts of the graph are -5 and 3. Thus, $(x + 5)$ and $(x - 3)$ are factors of the related polynomial.

19. The x-intercepts of the graph are 2 and 6. Thus, $(x - 2)$ and $(x - 6)$ are factors of the related polynomial.

21. $2x^2 + 25x - 20$ is prime.

23. $4x^2 - 12x + 9 = (2x - 3)(2x - 3)$
$\qquad = (2x - 3)^2$

25. Note that the GCF is 2, so that
$12x^2 + 10x - 50 = 2(6x^2 + 5x - 25)$
$\qquad = 2(3x - 5)(2x + 5).$

27. Note that the GCF is y^2, so that
$3y^4 - y^3 - 10y^2 = y^2(3y^2 - y - 10)$
$\qquad = y^2(3y + 5)(y - 2).$

29. Note that the GCF is $2x$, so that
$6x^3 + 8x^2 + 24x = 2x(3x^2 + 4x + 12).$

31. $x^2 + 8xz + 7z^2 = (x + 7z)(x + z)$

33. $2x^2 - 5xy - 3y^2 = (2x + y)(x - 3y)$

35. $x^2 - x - 12 = (x - 4)(x + 3)$

37. Note that the GCF is 2, so that
$28y^2 + 22y + 4 = 2(14y^2 + 11y + 2)$
$\qquad = 2(7y + 2)(2y + 1).$

39. $2x^2 + 15x - 27 = (2x - 3)(x + 9)$

41. $3x^2 + bx + 5$
$3 = 1 \cdot 3$ or $3 = (-1)(-3)$
$5 = 1 \cdot 5$ or $5 = (-1)(-5)$
$(3x + 1)(x + 5) = 3x^2 + 16x + 5$
$(3x - 1)(x - 5) = 3x^2 - 16x + 5$
$(-3x + 1)(-x + 5) = 3x^2 - 16x + 5$
$(-3x - 1)(-x - 5) = 3x^2 + 16x + 5$
$(3x + 5)(x + 1) = 3x^2 + 8x + 5$
$(3x - 5)(x - 1) = 3x^2 - 8x + 5$
$(-3x + 5)(-x + 1) = 3x^2 - 8x + 5$
$(-3x - 5)(-x - 1) = 3x^2 + 8x + 5$
$b = \pm 8$ or ± 16

43. First, let $y = x^2$. Then we have
$x^4 + x^2 - 6 = y^2 + y - 6 = (y + 3)(y - 2).$
This yields $(x^2 + 3)(x^2 - 2).$

45. First, let $y = 5x + 1$. Then we have
$(5x + 1)^2 + 8(5x + 1) + 7 = y^2 + 8y + 7$
$\qquad\qquad\qquad = (y + 7)(y + 1).$
This yields
$[(5x + 1) + 7][(5x + 1) + 1] = (5x + 8)(5x + 2).$

47. First, let $y = x^3$. Then we have
$x^6 - 7x^3 + 12 = y^2 - 7y + 12 = (y - 4)(y - 3).$
This yields $(x^3 - 4)(x^3 - 3).$

49. First, let $y = a + 5$. Then we have
$(a + 5)^2 - 5(a + 5) - 24 = y^2 - 5y - 24$
$\qquad\qquad\qquad = (y - 8)(y + 3).$
This yields
$[(a + 5) - 8][(a + 5) + 3] = (a - 3)(a + 8).$

51. Note that the GCF is x, so that
$V(x) = 3x^3 - 2x^2 - 8x = x(3x^2 - 2x - 8)$
$\qquad = x(3x + 4)(x - 2).$

53. $x^2 - 24x - 81 = (x - 27)(x + 3)$

55. $x^2 - 15x - 54 = (x - 18)(x + 3)$

57. $3x^2 - 6x + 3 = 3(x^2 - 2x + 1)$
$$= 3(x - 1)(x - 1)$$
$$= 3(x - 1)^2$$

59. $3x^2 - 5x - 2 = (3x + 1)(x - 2)$

61. $8x^2 - 26x + 15 = (4x - 3)(2x - 5)$

63. $18x^4 + 21x^3 + 6x^2 = 3x^2(6x^2 + 7x + 2)$
$$= 3x^2(2x + 1)(3x + 2)$$

65. $3a^2 + 12ab + 12b^2 = 3(a^2 + 4ab + 4b^2)$
$$= 3(a + 2b)(a + 2b)$$
$$= 3(a + 2b)^2$$

67. $x^2 + 4x + 5$ is prime.

69. First, let $y = x + 4$. Then we have
$$2(x + 4)^2 + 3(x + 4) - 5 = 2y^2 + 3y - 5$$
$$= (2y + 5)(y - 1).$$
This yields
$$[2(x + 4) + 5][(x + 4) - 1] = (2x + 13)(x + 3).$$

71. $6x^2 - 49x + 30 = (3x - 2)(2x - 15)$

73. $x^4 - 5x^2 - 6 = (x^2 - 6)(x^2 + 1)$

75. $6x^3 - x^2 - x = x(6x^2 - x - 1)$
$$= x(3x + 1)(2x - 1)$$

77. $12a^2 - 29ab + 15b^2 = (4a - 3b)(3a - 5b)$

79. $9x^2 + 30x + 25 = (3x + 5)(3x + 5)$
$$= (3x + 5)^2$$

81. $3x^2y - 11xy + 8y = y(3x^2 - 11x + 8)$
$$= y(3x - 8)(x - 1)$$

83. $2x^2 + 2x - 12 = 2(x^2 + x - 6)$
$$= 2(x + 3)(x - 2)$$

85. First let $y = x - 4$. Then we have
$$(x - 4)^2 + 3(x - 4) - 18 = y^2 + 3y - 18$$
$$= (y + 6)(y - 3).$$
This yields
$$[(x - 4) + 6][(x - 4) - 3] = (x + 2)(x - 7).$$

87. First, let $y = x^3$. Then we have
$$2x^6 + 3x^3 - 9 = 2y^2 + 3y - 9$$
$$= (2y - 3)(y + 3).$$
This yields $(2x^3 - 3)(x^3 + 3)$.

89. $72xy^4 - 24xy^2z + 2xz^2$
$$= 2x(36y^4 - 12y^2z + z^2)$$
$$= 2x(6y^2 - z)(6y^2 - z)$$
$$= 2x(6y^2 - z)^2$$

91. $x^4 + 6x^3 + 5x^2 = x^2(x^2 + 6x + 5)$
$$= x^2(x + 5)(x + 1)$$
$$y_1 = x^4 + 6x^3 + 5x^2,$$
$$y_2 = x^2(x + 5)(x + 1)$$

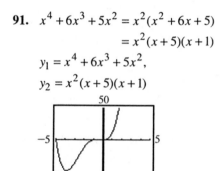

93. $30x^3 + 9x^2 - 3x = 3x(10x^2 + 3x - 1)$
$$= 3x(5x - 1)(2x + 1)$$
$$y_1 = 30x^3 + 9x^2 - 3x,$$
$$y_2 = 3x(5x - 1)(2x + 1)$$

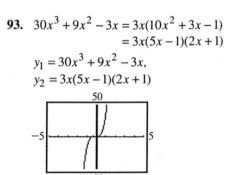

95. $(x-2)(x^2+2x+4) = x \cdot x^2 + x \cdot 2x + x \cdot 4 - 2 \cdot x^2 - 2 \cdot 2x - 2 \cdot 4$
$$= x^3 + 2x^2 + 4x - 2x^2 - 4x - 8$$
$$= x^3 - 8$$

97. $P(x) = 3x^2 + 2x - 9$
$P(0) = 3(0)^2 + 2(0) - 9 = 0 + 0 - 9 = -9$

99. $P(x) = 3x^2 + 2x - 9$
$P(-1) = 3(-1)^2 + 2(-1) - 9 = 3 - 2 - 9 = -8$

101. $x^{2n} + 10x^n + 16 = (x^n + 8)(x^n + 2)$

103. $x^{2n} - 3x^n - 18 = (x^n - 6)(x^n + 3)$

105. $2x^{2n} + 11x^n + 5 = (2x^n + 1)(x^n + 5)$

107. $4x^{2n} - 12x^n + 9 = (2x^n - 3)(2x^n - 3)$
$$= (2x^n - 3)^2$$

Exercise Set 5.7

1. $x^2 + 6x + 9 = (x+3)^2$

3. $4x^2 - 12x + 9 = (2x-3)^2$

5. $3x^2 - 24x + 48 = 3(x^2 - 8x + 16)$
$$= 3(x-4)^2$$

7. $9y^2x^2 + 12yx^2 + 4x^2 = x^2(9y^2 + 12y + 4)$
$$= x^2(3y+2)^2$$

9. $x^2 - 25 = x^2 - 5^2 = (x+5)(x-5)$

11. $9 - 4z^2 = 3^2 - (2z)^2 = (3+2z)(3-2z)$

13. $(y+2)^2 - 49 = (y+2)^2 - 7^2$
$$= [(y+2)+7][(y+2)-7]$$
$$= (y+9)(y-5)$$

15. $64x^2 - 100 = 4(16x^2 - 25)$
$$= 4[(4x)^2 - 5^2]$$
$$= 4(4x+5)(4x-5)$$

17. $x^3 + 27 = x^3 + 3^3 = (x+3)(x^2 - 3x + 9)$

19. $z^3 - 1 = z^3 - 1^3 = (z-1)(z^2 + z + 1)$

21. $m^3 + n^3 = (m+n)(m^2 - mn + n^2)$

23. $x^3y^2 - 27y^2 = y^2(x^3 - 27)$
$$= y^2(x^3 - 3^3)$$
$$= y^2(x-3)(x^2 + 3x + 9)$$

25. $a^3b + 8b^4 = b(a^3 + 8b^3)$
$$= b(a^3 + (2b)^3)$$
$$= b(a+2b)(a^2 - 2ab + 4b^2)$$

27. $125y^3 - 8x^3$
$$= (5y)^3 - (2x)^3$$
$$= (5y - 2x)(25y^2 + 10yx + 4x^2)$$

29. $x^2 + 6x + 9 - y^2 = (x+3)^2 - y^2$
$$= (x+3+y)(x+3-y)$$

31. $x^2 - 10x + 25 - y^2$
$$= (x-5)^2 - y^2$$
$$= (x-5+y)(x-5-y)$$

33. $4x^2 + 4x + 1 - z^2$
$$= (2x+1)^2 - z^2$$
$$= (2x+1+z)(2x+1-z)$$

35. $9x^2 - 49 = (3x)^2 - 7^2 = (3x+7)(3x-7)$

37. $x^2 - 12x + 36 = (x-6)^2$

39. $x^4 - 81 = (x^2)^2 - 9^2 = (x^2 + 9)(x^2 - 9)$
$$= (x^2 + 9)(x+3)(x-3)$$

41. $x^2 + 8x + 16 - 4y^2$

$= (x+4)^2 - (2y)^2$

$= (x+4+2y)(x+4-2y)$

43. $(x+2y)^2 - 9 = (x+2y+3)(x+2y-3)$

45. $x^3 - 216 = x^3 - 6^3 = (x-6)(x^2+6x+36)$

47. $x^3 + 125 = x^3 + 5^3 = (x+5)(x^2-5x+25)$

49. $4x^2 + 25$ is prime.

51. $4a^2 + 12a + 9 = (2a+3)^2$

53. $18x^2y - 2y = 2y(9x^2 - 1)$

$= 2y[(3x)^2 - 1^2]$

$= 2y(3x+1)(3x-1)$

55. $8x^3 + y^3 = (2x)^3 + y^3$

$= (2x+y)(4x^2 - 2xy + y^2)$

57. $x^6 - y^3 = (x^2)^3 - y^3$

$= (x^2 - y)(x^4 + x^2y + y^2)$

59. $x^2 + 16x + 64 - x^4$

$= (x+8)^2 - (x^2)^2$

$= (x+8+x^2)(x+8-x^2)$

61. $3x^6y^2 + 81y^2 = 3y^2(x^6 + 27)$

$= 3y^2[(x^2)^3 + 3^3]$

$= 3y^2(x^2+3)(x^4 - 3x^2 + 9)$

63. $(x+y)^3 + 125$

$= (x+y)^3 + 5^3$

$= [(x+y)+5][(x+y)^2 - 5(x+y) + 25]$

$= (x+y+5)(x^2 + 2xy + y^2 - 5x - 5y + 25)$

65. $(2x+3)^3 - 64$

$= (2x+3)^3 - 4^3$

$= [(2x+3)-4][(2x+3)^2 + 4(2x+3) + 16]$

$= (2x-1)(4x^2 + 12x + 9 + 8x + 12 + 16)$

$= (2x-1)(4x^2 + 20x + 37)$

67. $A = \pi R^2 - \pi r^2$

$= \pi(R^2 - r^2)$

$= \pi(R+r)(R-r)$

69. The coating of the candy is the difference between the volume of the outer sphere $\frac{4}{3}\pi R^3$ and the inner sphere $\frac{4}{3}\pi(6)^3$.

Thus,

$V = \frac{4}{3}\pi R^3 - \frac{4}{3}\pi(6)^3$

$V = \frac{4}{3}\pi(R^3 - 6^3)$

$V = \frac{4}{3}\pi(R-6)(R^2 + 6R + 36)$

71. $1 - y^3 = 1^3 - y^3 = (1-y)(1+y+y^2)$

73. $9x^2 + 6x + 1 = (3x+1)^2$

75. $x^2 - 8x + 16 - y^2 = (x-4)^2 - y^2$

$= (x-4+y)(x-4-y)$

77. $x^4 - x = x(x^3 - 1)$

$= x(x-1)(x^2 + x + 1)$

79. $14x^2y - 2xy = 2xy(7x-1)$

81. $4x^2 - 16 = 4(x^2 - 4) = 4(x+2)(x-2)$

83. $128a^3 - 2b^3$

$= 2(64a^3 - b^3)$

$= 2[(4a)^3 - b^3]$

$= 2(4a-b)(16a^2 + 4ab + b^2)$

85. $3x^2 - 8x - 11 = (3x-11)(x+1)$

87. $4x^2 + 8x - 12 = 4(x^2 + 2x - 3)$
$$= 4(x + 3)(x - 1)$$

89. $4x^2 + 36x + 81 = (2x + 9)^2$

91. $8x^3 + 27y^3$
$$= (2x)^3 + (3y)^3$$
$$= (2x + 3y)(4x^2 - 6xy + 9y^2)$$

93. $64x^2y^3 - 8x^2$
$$= 8x^2(8y^3 - 1)$$
$$= 8x^2[(2y)^3 - (1)^3]$$
$$= 8x^2(2y - 1)(4y^2 + 2y + 1)$$

95. $(x + 5)^3 + y^3$
$$= [(x + 5) + y][(x + 5)^2 - (x + 5)y + y^2]$$
$$= (x + 5 + y)(x^2 + 10x + 25 - xy - 5y + y^2)$$

97. First let $y = 5a - 3$. Then we have
$$(5a - 3)^2 - 6(5a - 3) + 9 = y^2 - 6y + 9$$
$$= (y - 3)^2.$$
This yields $[(5a - 3) - 3]^2 = (5a - 6)^2$.

99. $\left(\dfrac{6}{2}\right)^2 = 3^2 = 9$
$$x^2 + 6x + 9 = (x + 3)^2$$
$$c = 9$$

101. $\left(\dfrac{14}{2}\right)^2 = 7^2 = 49$
$$m^2 - 14m + 49 = (m - 7)^2$$
$$c = 49$$

103. $\pm 2\sqrt{16} = \pm 2 \cdot 4 = \pm 8$
$$x^2 + 8x + 16 = (x + 4)^2$$
$$x^2 - 8x + 16 = (x - 4)^2$$
$$c = \pm 8$$

105. a. $x^6 - 1$
$$= (x^3)^2 - 1^2$$
$$= (x^3 + 1)(x^3 - 1)$$
$$= (x + 1)(x^2 - x + 1)(x - 1)(x^2 + x + 1)$$

b. $x^6 - 1 = (x^2)^3 - 1^3$
$$= (x^2 - 1)((x^2)^2 + 1(x^2) + 1^2)$$
$$= (x + 1)(x - 1)(x^4 + x^2 + 1)$$

c. Answers may vary.

107. $x + 7 = 0$
$$x = -7$$
The solution is -7.

109. $5x - 15 = 0$
$$5x = 15$$
$$x = 3$$
The solution is 3.

111. $3x = 0$
$$x = 0$$
The solution is 0.

113. $-4x - 16 = 0$
$$-4x = 16$$
$$x = -4$$
The solution is -4.

115. $x^{2n} - 36 = (x^n)^2 - 6^2 = (x^n + 6)(x^n - 6)$

117. $25x^{2n} - 81 = (5x^n)^2 - 9^2$
$$= (5x^n + 9)(5x^n - 9)$$

119. $x^{4n} - 625 = (x^{2n})^2 - (25)^2$
$$= (x^{2n} + 25)(x^{2n} - 25)$$
$$= (x^{2n} + 25)[(x^n)^2 - 5^2]$$
$$= (x^{2n} + 25)(x^n + 5)(x^n - 5)$$

Section 5.8

Mental Math

1. $(x - 3)(x + 5) = 0$
$x - 3 = 0$ or $x + 5 = 0$
$x = 3$ or $x = -5$
The solutions are 3, –5.

2. $(y + 5)(y + 3) = 0$
$y + 5 = 0$ or $y + 3 = 0$
$y = -5$ or $y = -3$
The solutions are –5, –3.

3. $(z - 3)(z + 7) = 0$
$z - 3 = 0$ or $z + 7 = 0$
$z = 3$ or $z = -7$
The solutions are 3, –7.

4. $(c - 2)(c - 4) = 0$
$c - 2 = 0$ or $c - 4 = 0$
$c = 2$ or $c = 4$
The solutions are 2, 4.

5. $x(x - 9) = 0$
$x = 0$ or $x - 9 = 0$
$x = 9$
The solutions are 0, 9.

6. $w(w + 7) = 0$
$w = 0$ or $w + 7 = 0$
$w = -7$
The solutions are 0, –7.

Exercise Set 5.8

1. $(x + 3)(3x - 4) = 0$
$x + 3 = 0$ or $3x - 4 = 0$
$x = -3$ or $3x = 4$
$x = \dfrac{4}{3}$
The solutions are $-3, \dfrac{4}{3}$.

3. $3(2x - 5)(4x + 3) = 0$
$2x - 5 = 0$ or $4x + 3 = 0$
$2x = 5$ or $4x = -3$
$x = \dfrac{5}{2}$ or $x = -\dfrac{3}{4}$
The solutions are $\dfrac{5}{2}, -\dfrac{3}{4}$.

5. $x^2 + 11x + 24 = 0$
$(x + 3)(x + 8) = 0$
$x + 3 = 0$ or $x + 8 = 0$
$x = -3$ or $x = -8$
The solutions are –3, –8.

7. $12x^2 + 5x - 2 = 0$
$(4x - 1)(3x + 2) = 0$
$4x - 1 = 0$ or $3x + 2 = 0$
$4x = 1$ or $3x = -2$
$x = \dfrac{1}{4}$ or $x = -\dfrac{2}{3}$
The solutions are $\dfrac{1}{4}, -\dfrac{2}{3}$.

9. $z^2 + 9 = 10z$
$z^2 - 10z + 9 = 0$
$(z - 1)(z - 9) = 0$
$z - 1 = 0$ or $z - 9 = 0$
$z = 1$ or $z = 9$
The solutions are 1, 9.

11. $x(5x + 2) = 3$
$5x^2 + 2x - 3 = 0$
$(5x - 3)(x + 1) = 0$

$5x - 3 = 0$ or $x + 1 = 0$
$5x = 3$ or $x = -1$
$x = \dfrac{3}{5}$
The solutions are $\dfrac{3}{5}, -1$.

13. $x^2 - 6x = x(8 + x)$
$x^2 - 6x = 8x + x^2$
$0 = 14x$
$x = 0$
The solution is 0.

15. $\dfrac{z^2}{6} - \dfrac{z}{2} - 3 = 0$

$z^2 - 3z - 18 = 0$

$(z - 6)(z + 3) = 0$

$z - 6 = 0$ or $z + 3 = 0$

$z = 6$ or $z = -3$

The solutions are $6, -3$.

17. $\dfrac{x^2}{2} + \dfrac{x}{20} = \dfrac{1}{10}$

$10x^2 + x = 2$

$10x^2 + x - 2 = 0$

$(5x - 2)(2x + 1) = 0$

$5x - 2 = 0$ or $2x + 1 = 0$

$5x = 2$ or $2x = -1$

$x = \dfrac{2}{5}$ or $x = -\dfrac{1}{2}$

The solutions are $\dfrac{2}{5}, -\dfrac{1}{2}$.

19. $\dfrac{4t^2}{5} = \dfrac{t}{5} + \dfrac{3}{10}$

$8t^2 = 2t + 3$

$8t^2 - 2t - 3 = 0$

$(4t - 3)(2t + 1) = 0$

$4t - 3 = 0$ or $2t + 1 = 0$

$4t = 3$ or $2t = -1$

$t = \dfrac{3}{4}$ or $t = -\dfrac{1}{2}$

The solutions are $\dfrac{3}{4}, -\dfrac{1}{2}$.

21. $(x + 2)(x - 7)(3x - 8) = 0$

$x + 2 = 0$ or $x - 7 = 0$ or $3x - 8 = 0$

$x = -2$ or $x = 7$ or $3x = 8$

$x = \dfrac{8}{3}$

The solutions are $-2, 7, \dfrac{8}{3}$.

23. $y^3 = 9y$

$y^3 - 9y = 0$

$y(y^2 - 9) = 0$

$y(y + 3)(y - 3) = 0$

$y = 0$ or $y + 3 = 0$ or $y - 3 = 0$

$y = -3$ or $y = 3$

The solutions are $0, -3, 3$.

25. $x^3 - x = 2x^2 - 2$

$x^3 - 2x^2 - x + 2 = 0$

$x^2(x - 2) - (x - 2) = 0$

$(x^2 - 1)(x - 2) = 0$

$(x + 1)(x - 1)(x - 2) = 0$

$x + 1 = 0$ or $x - 1 = 0$ or $x - 2 = 0$

$x = -1$ or $x = 1$ or $x = 2$

The solutions are $-1, 1, 2$.

27. Answers may vary.

29. $(2x + 7)(x - 10) = 0$

$2x + 7 = 0$ or $x - 10 = 0$

$2x = -7$ or $x = 10$

$x = -\dfrac{7}{2}$

The solutions are $-\dfrac{7}{2}, 10$.

31. $3x(x - 5) = 0$

$3x = 0$ or $x - 5 = 0$

$x = 0$ or $x = 5$

The solutions are $0, 5$.

33. $x^2 - 2x - 15 = 0$

$(x - 5)(x + 3) = 0$

$x - 5 = 0$ or $x + 3 = 0$

$x = 5$ or $x = -3$

The solutions are $5, -3$.

35. $12x^2 + 2x - 2 = 0$

$6x^2 + x - 1 = 0$

$(3x - 1)(2x + 1) = 0$

$3x - 1 = 0$ or $2x + 1 = 0$

$3x = 1$ or $2x = -1$

$x = \dfrac{1}{3}$ or $x = -\dfrac{1}{2}$

The solutions are $\dfrac{1}{3}, -\dfrac{1}{2}$.

37. $w^2 - 5w = 36$

$w^2 - 5w - 36 = 0$

$(w - 9)(w + 4) = 0$

$w - 9 = 0$ or $w + 4 = 0$

$w = 9$ or $w = -4$

The solutions are 9, –4.

39. $25x^2 - 40x + 16 = 0$

$(5x - 4)^2 = 0$

$5x - 4 = 0$

$5x = 4$

$x = \dfrac{4}{5}$

The solution is $\dfrac{4}{5}$.

41. $2r^3 + 6r^2 = 20r$

$r^3 + 3r^2 = 10r$

$r^3 + 3r^2 - 10r = 0$

$r(r^2 + 3r - 10) = 0$

$r(r + 5)(r - 2) = 0$

$r = 0$ or $r + 5 = 0$ or $r - 2 = 0$

$r = -5$ or $r = 2$

The solutions are 0, –5, 2.

43. $z(5z - 4)(z + 3) = 0$

$z = 0$ or $5z - 4 = 0$ or $z + 3 = 0$

$5z = 4$ or $z = -3$

$z = \dfrac{4}{5}$

The solutions are 0, $\dfrac{4}{5}$, –3.

45. $2z(z + 6) = 2z^2 + 12z - 8$

$2z^2 + 12z = 2z^2 + 12z - 8$

$0 = -8$

No solution exists. The solution is $\varnothing$.

47. $(x - 1)(x + 4) = 24$

$x^2 + 3x - 4 = 24$

$x^2 + 3x - 28 = 0$

$(x + 7)(x - 4) = 0$

$x + 7 = 0$ or $x - 4 = 0$

$x = -7$ or $x = 4$

The solutions are –7, 4.

49. $\dfrac{x^2}{4} - \dfrac{5}{2}x + 6 = 0$

$x^2 - 10x + 24 = 0$

$(x - 4)(x - 6) = 0$

$x - 4 = 0$ or $x - 6 = 0$

$x = 4$ or $x = 6$

The solutions are 4, 6.

51. $y^2 + \dfrac{1}{4} = -y$

$4y^2 + 1 = -4y$

$4y^2 + 4y + 1 = 0$

$(2y + 1)^2 = 0$

$2y + 1 = 0$

$2y = -1$

$y = -\dfrac{1}{2}$

The solution is $-\dfrac{1}{2}$.

53. $y^3 + 4y^2 = 9y + 36$

$y^3 + 4y^2 - 9y - 36 = 0$

$y^2(y + 4) - 9(y + 4) = 0$

$(y^2 - 9)(y + 4) = 0$

$(y + 3)(y - 3)(y + 4) = 0$

$y + 3 = 0$ or $y - 3 = 0$ or $y + 4 = 0$

$y = -3$ or $y = 3$ or $y = -4$

The solutions are –3, 3, –4.

55.
$$2x^3 = 50x$$
$$x^3 = 25x$$
$$x^3 - 25x = 0$$
$$x(x^2 - 25) = 0$$
$$x(x+5)(x-5) = 0$$
$$x = 0 \quad\text{or}\quad x+5 = 0 \quad\text{or}\quad x-5 = 0$$
$$x = -5 \quad\text{or}\quad x = 5$$
The solutions are 0, –5, 5.

57.
$$x^2 + (x+1)^2 = 61$$
$$x^2 + x^2 + 2x + 1 = 61$$
$$2x^2 + 2x - 60 = 0$$
$$x^2 + x - 30 = 0$$
$$(x+6)(x-5) = 0$$
$$x+6 = 0 \quad\text{or}\quad x-5 = 0$$
$$x = -6 \quad\text{or}\quad x = 5$$
The solutions are –6, 5.

59.
$$m^2(3m-2) = m$$
$$3m^3 - 2m^2 = m$$
$$3m^3 - 2m^2 - m = 0$$
$$m(3m^2 - 2m - 1) = 0$$
$$m(3m+1)(m-1) = 0$$
$$m = 0 \quad\text{or}\quad 3m+1 = 0 \quad\text{or}\quad m-1 = 0$$
$$3m = -1 \quad\text{or}\quad m = 1$$
$$m = -\frac{1}{3}$$
The solutions are $0, -\frac{1}{3}, 1$.

61.
$$3x^2 = -x$$
$$3x^2 + x = 0$$
$$x(3x+1) = 0$$
$$x = 0 \quad\text{or}\quad 3x+1 = 0$$
$$3x = -1$$
$$x = -\frac{1}{3}$$
The solutions are $0, -\frac{1}{3}$.

63.
$$x(x-3) = x^2 + 5x + 7$$
$$x^2 - 3x = x^2 + 5x + 7$$
$$-7 = 8x$$
$$x = -\frac{7}{8}$$
The solution is $-\frac{7}{8}$.

65.
$$3(t-8) + 2t = 7 + t$$
$$3t - 24 + 2t = 7 + t$$
$$5t - 24 = 7 + t$$
$$4t = 31$$
$$t = \frac{31}{4}$$
The solution is $\frac{31}{4}$.

67.
$$-3(x-4) + x = 5(3-x)$$
$$-3x + 12 + x = 15 - 5x$$
$$-2x + 12 = 15 - 5x$$
$$3x = 3$$
$$x = 1$$
The solution is 1.

69. a and d are incorrect because the right side of the equation is not zero.

71. Let n = the one number and $n + 5$ = the other number.
Then: $n(n + 5) = 66$
$$n^2 + 5n - 66 = 0$$
$$(n+11)(n-6) = 0$$
$$n+11 = 0 \quad\text{or}\quad n-6 = 0$$
$$n = -11 \quad\text{or}\quad n = 6$$
There are two solutions: –11 and –6 and 6 and 11.

73. Let d = the amount of cable needed. Then from the Pythagorean theorem,
$$d^2 = 45^2 + 60^2 = 5625.$$
So, $d = \sqrt{5625} = 75$
He should make the cable 75 feet.

75. $C(x) = x^2 - 15x + 50$

$9500 = x^2 - 15x + 50$

$0 = x^2 - 15x - 9450$

$0 = (x - 105)(x + 90)$

$x - 105 = 0$ or $x + 90 = 0$

$x = 105$ or $x = -90$

Reject the negative. 105 units were manufactured at a cost of $9500.

77. Let x = one leg of a right triangle and $x - 3$ = the other leg of the right triangle. Then from the Pythagorean theorem,

$15^2 = x^2 + (x - 3)^2$

$225 = x^2 + x^2 - 6x + 9$

$0 = 2x^2 - 6x - 216$

$0 = x^2 - 3x - 108$

$0 = (x - 12)(x + 9)$

$x - 12 = 0$ or $x + 9 = 0$

$x = 12$ or $x = -9$.

Rejecting the extraneous solution of –9, we find that one leg of the right triangle is 12 centimeters and the other leg is 9 centimeters.

79. Note that the outer rectangle has lengths of $2x + 12$ and $2x + 16$. Thus, the area of the border is $(2x + 12)(2x + 16) - 12 \cdot 16$. We set this equal to 128 and solve for x.

$(2x + 12)(2x + 16) - 12 \cdot 16 = 128$

$4x^2 + 56x = 128$

$4x^2 + 56x - 128 = 0$

$x^2 + 14x - 32 = 0$

$(x + 16)(x - 2) = 0$

$x + 16 = 0$ or $x - 2 = 0$

$x = -16$ or $x = 2$

Since x must be positive, we see that the width is 2 inches.

81. The sunglasses will hit the ground when the height $h(t)$ equals 0.

$-16t^2 + 1600 = 0$

$t^2 - 100 = 0$

$(t - 10)(t + 10) = 0$

$t - 10 = 0$ or $t + 10 = 0$

$t = 10$ or $t = -10$

The sunglasses hit the ground 10 seconds after being dropped.

83. Let the width of the floor = w. Then the length is $w + 6$. So the area is

$91 = w(w + 6)$

$91 = w^2 + 6w$

$w^2 + 6w - 91 = 0$

$(w + 13)(w - 7) = 0$

$w + 13 = 0$ or $w - 7 = 0$

$w = -13$ or $w = 7$.

Since the width cannot be negative, the width is 7 feet and the length is 13 feet.

85. $0.5x^2 = 50$

$0.5x^2 - 50 = 0$

$x^2 - 100 = 0$

$(x + 10)(x - 10) = 0$

$x + 10 = 0$ or $x - 10 = 0$

$x = -10$ or $x = 10$

Reject the negative solution. A 10-inch square tier is needed, provided each person has one serving.

87. E; Two x-intercepts $(-5, 0), (2, 0)$

89. F; Three x-intercepts, $(-3, 0), (0, 0), (3, 0)$

91. B; $2x^2 + 9x + 4 = (2x + 1)(x + 4)$;

Two x-intercepts, $(-4, 0), \left(-\dfrac{1}{2}, 0\right)$

93. Answers may vary.
Example: $f(x) = (x-5)(x-3)$
$= x^2 - 8x + 15$

95. Answers may vary.
Example: $f(x) = (x+1)(x-2)$
$= x^2 - x - 2$

97. $(-3, 0)$, $(0, 2)$; It is a function since it passes the vertical line test.

99. $(-4, 0)$, $(4, 0)$, $(0, -2)$, $(0, 2)$; It is not a function since it fails the vertical line test.

101. Answers may vary.

Chapter 5 Review

1. $(-2)^2 = (-2)(-2) = 4$

2. $(-3)^4 = (-3)(-3)(-3)(-3) = 81$

3. $-2^2 = -(2 \cdot 2) = -4$

4. $-3^4 = -(3 \cdot 3 \cdot 3 \cdot 3) = -81$

5. $8^0 = 1$

6. $-9^0 = -1$

7. $-4^{-2} = -\dfrac{1}{4^2} = -\dfrac{1}{16}$

8. $(-4)^{-2} = \dfrac{1}{(-4)^2} = \dfrac{1}{16}$

9. $-xy^2 \cdot y^3 \cdot xy^2 z = -x^{1+1} y^{2+3+2} z = -x^2 y^7 z$

10. $(-4xy)(-3xy^2 b) = (-4)(-3)x^{1+1} y^{1+2} b$
$= 12x^2 y^3 b$

11. $a^{-14} a^5 = a^{-14+5} = a^{-9} = \dfrac{1}{a^9}$

12. $\dfrac{a^{16}}{a^{17}} = a^{16-17} = a^{-1} = \dfrac{1}{a}$

13. $\dfrac{x^{-7}}{x^4} = x^{-7-4} = x^{-11} = \dfrac{1}{x^{11}}$

14. $\dfrac{9a(a^{-3})}{18a^{15}} = \dfrac{a^{1-3-15}}{2} = \dfrac{a^{-17}}{2} = \dfrac{1}{2a^{17}}$

15. $\dfrac{y^{6p-3}}{y^{6p+2}} = y^{(6p-3)-(6p+2)}$
$= y^{6p-3-6p-2}$
$= y^{-5}$
$= \dfrac{1}{y^5}$

16. $36,890,000 = 3.689 \times 10^7$

17. $-0.000362 = -3.62 \times 10^{-4}$

18. $1.678 \times 10^{-6} = 0.000001678$

19. $4.1 \times 10^5 = 410,000$

20. $(8^5)^3 = 8^{5 \cdot 3} = 8^{15}$

21. $\left(\dfrac{a}{4}\right)^2 = \dfrac{a^2}{4^2} = \dfrac{a^2}{16}$

22. $(3x^3) = 3^3 x^3 = 27x^3$

23. $(-4x)^{-2} = \dfrac{1}{(-4x)^2} = \dfrac{1}{(-4)^2 x^2} = \dfrac{1}{16x^2}$

24. $\left(\dfrac{6x}{5}\right)^2 = \dfrac{(6x)^2}{(5)^2} = \dfrac{36x^2}{25}$

25. $(8^6)^{-3} = 8^{-18} = \dfrac{1}{8^{18}}$

26. $\left(\dfrac{4}{3}\right)^{-2} = \dfrac{4^{-2}}{3^{-2}} = \dfrac{3^2}{4^2} = \dfrac{9}{16}$

27. $(-2x^3)^{-3} = \dfrac{1}{(-2x^3)^3}$

$\qquad = \dfrac{1}{(-2)^3(x^3)^3}$

$\qquad = \dfrac{1}{-8x^9}$

$\qquad = -\dfrac{1}{8x^9}$

28. $\left(\dfrac{8p^6}{4p^4}\right)^{-2} = (2p^2)^{-2} = 2^{-2}p^{-4} = \dfrac{1}{4p^4}$

29. $(-3x^{-2}y^2)^3 = (-3)^3(x^{-2})^3(y^2)^3$

$\qquad = -27x^{-6}y^6$

$\qquad = \dfrac{-27y^6}{x^6}$

30. $\left(\dfrac{x^{-5}y^{-3}}{z^3}\right)^{-5} = \dfrac{x^{25}y^{15}}{z^{-15}} = x^{25}y^{15}z^{15}$

31. $\dfrac{4^{-1}x^3yz}{x^{-2}yx^4} = \dfrac{x^{3-(-2)-4}z}{4} = \dfrac{x^{5-4}z}{4} = \dfrac{xz}{4}$

32. $(5xyz)^{-4}(x^{-2})^{-3} = \dfrac{1}{(5xyz)^4}x^6$

$\qquad = \dfrac{x^6}{625x^4y^4z^4}$

$\qquad = \dfrac{x^2}{625y^4z^4}$

33. $\dfrac{2(3yz)^{-3}}{y^{-3}} = \dfrac{2(3)^{-3}y^{-3}z^{-3}}{y^{-3}}$

$\qquad = \dfrac{2}{3^3z^3}$

$\qquad = \dfrac{2}{27z^3}$

34. $x^{4a}(3x^{5a})^3 = x^{4a}(3^3x^{15a})$

$\qquad = 27x^{4a+15a}$

$\qquad = 27x^{19a}$

35. $\dfrac{4y^{3x-3}}{2y^{2x+4}} = 2y^{(3x-3)-(2x+4)}$

$\qquad = 2y^{3x-3-2x-4}$

$\qquad = 2y^{x-7}$

36. $\dfrac{(0.00012)(144,000)}{0.0003}$

$\qquad = \dfrac{(1.2\times10^{-4})(1.44\times10^5)}{3\times10^{-4}}$

$\qquad = 0.576\times10^5$

$\qquad = 5.76\times10^4$

37. $\dfrac{(-0.00017)(0.00039)}{3000}$

$\qquad = \dfrac{(-1.7\times10^{-4})(3.9\times10^{-4})}{3\times10^3}$

$\qquad = -2.21\times10^{-4-4-3}$

$\qquad = -2.21\times10^{-11}$

38. $\dfrac{27x^{-5}y^5}{18x^{-6}y^2}\cdot\dfrac{x^4y^{-2}}{x^{-2}y^3} = \dfrac{3x^{-5+4}y^{5-2}}{2x^{-6-2}y^{2+3}}$

$\qquad = \dfrac{3x^{-1}y^3}{2x^{-8}y^5}$

$\qquad = \dfrac{3}{2}x^{-1+8}y^{3-5}$

$\qquad = \dfrac{3x^7}{2y^2}$

39. $\dfrac{3x^5}{y^{-4}}\cdot\dfrac{(3xy^{-3})^{-2}}{(z^{-3})^{-4}} = \dfrac{3x^5\cdot3^{-2}x^{-2}(y^{-3})^{-2}}{y^{-4}z^{12}}$

$\qquad = \dfrac{3^{1-2}x^{5-2}y^{6-(-4)}}{z^{12}}$

$\qquad = \dfrac{3^{-1}x^3y^{10}}{z^{12}}$

$\qquad = \dfrac{x^3y^{10}}{3z^{12}}$

40. $\dfrac{(x^w)^2}{(x^{w-4})^{-2}} = \dfrac{x^{2w}}{x^{-2w+8}} = x^{2w-(-2w+8)} = x^{4w-8}$

41. The degree of the polynomial $x^2y - 3xy^3z + 5x + 7y$ is the degree of the term $-3xy^3z$ which is 5.

42. $3x + 2$ has degree 1. x^2

43. $4x + 8x - 6x^2 - 6x^2y = (4+8)x - 6x^2 - 6x^2y = 12x - 6x^2 - 6x^2y$

44. $-8xy^3 + 4xy^3 - 3x^3y = (-8+4)xy^3 - 3x^3y = -4xy^3 - 3x^3y$

45. $(3x + 7y) + (4x^2 - 3x + 7) + (y - 1) = 4x^2 + (3-3)x + (7+1)y + (7-1) = 4x^2 + 8y + 6$

46. $(4x^2 - 6xy + 9y^2) - (8x^2 - 6xy - y^2) = 4x^2 - 6xy + 9y^2 - 8x^2 + 6xy + y^2$
$$= (4-8)x^2 + (-6+6)xy + (9+1)y^2$$
$$= -4x^2 + 10y^2$$

47. $(3x^2 - 4b + 28) + (9x^2 - 30) - (4x^2 - 6b + 20) = 3x^2 - 4b + 28 + 9x^2 - 30 - 4x^2 + 6b - 20$
$$= (3+9-4)x^2 + (-4+6)b + (28-30-20)$$
$$= 8x^2 + 2b - 22$$

48. $(9xy + 4x^2 + 18) + (7xy - 4x^3 - 9x) = -4x^3 + 4x^2 + (9+7)xy - 9x + 18$
$$= -4x^3 + 4x^2 + 16xy - 9x + 18$$

49. $(3x^2y - 7xy - 4) + (9x^2y + x) - (x - 7) = 3x^2y - 7xy - 4 + 9x^2y + x - x + 7$
$$= (3+9)x^2y - 7xy + (1-1)x + (-4+7)$$
$$= 12x^2y - 7xy + 3$$

50.
$$\begin{array}{ccc} x^2 - 5x + 7 & \text{or} & x^2 - 5x + 7 \\ -\quad (x+4) & & +\quad (-x-4) \\ \hline x^2 - 6x + 3 & & x^2 - 6x + 3 \end{array}$$

51.

$$x^3 \quad + 2xy^2 - y$$
$$+ \quad \underline{x - 4xy^2 \quad - 7}$$
$$x^3 + x - 2xy^2 - y - 7$$

52. $P(6) = 9(6)^2 - 7(6) + 8 = 290$

53. $P(-2) = 9(-2)^2 - 7(-2) + 8 = 58$

54. $P(-3) = 9(-3)^2 - 7(-3) + 8 = 110$

55. $P(x) + Q(x) = (2x - 1) + (x^2 + 2x - 5) = x^2 + 4x - 6$

56. $2[P(x)] - Q(x) = 2(2x - 1) - (x^2 + 2x - 5) = 4x - 2 - x^2 - 2x + 5 = -x^2 + 2x + 3$

57. $P = 2(x^2 y + 5) + 2(2x^2 y - 6x + 1) = 2x^2 y + 10 + 4x^2 y - 12x + 2 = 6x^2 y - 12x + 12$
The perimeter is $(6x^2 y - 12x + 12)$ centimeters.

58. $-6x(4x^2 - 6x + 1) = -6x(4x^2) - 6x(-6x) - 6x(1) = -24x^3 + 36x^2 - 6x$

59. $-4ab^2(3ab^3 + 7ab + 1) = (-4ab^2)(3ab^3) - 4ab^2(7ab) - 4ab^2(1) = -12a^2 b^5 - 28a^2 b^3 - 4ab^2$

60. $(x - 4)(2x + 9) = 2x^2 + 9x - 8x - 36 = 2x^2 + x - 36$

61. $(-3xa + 4b)^2 = (-3xa)^2 + 2(-3xa)(4b) + (4b)^2 = 9x^2 a^2 - 24xab + 16b^2$

62.

$$9x^2 + \quad 4x + 1$$
$$\times \quad \underline{\qquad\qquad 4x - 3}$$
$$-27x^2 - 12x - 3$$
$$\underline{36x^3 + 16x^2 + \ 4x}$$
$$36x^3 - 11x^2 - \ 8x - 3$$

63. $(5x - 9y)(3x + 9y) = 15x^2 + 45xy - 27xy - 81y^2 = 15x^2 + 18xy - 81y^2$

64. $\left(x - \dfrac{1}{3}\right)\left(x + \dfrac{2}{3}\right) = (x)(x) + (x)\left(\dfrac{2}{3}\right) - \left(\dfrac{1}{3}\right)(x) - \left(\dfrac{1}{3}\right)\left(\dfrac{2}{3}\right) = x^2 + \dfrac{2}{3}x - \dfrac{1}{3}x - \dfrac{2}{9} = x^2 + \dfrac{1}{3}x - \dfrac{2}{9}$

65. $(x^2 + 9x + 1)^2 = [(x^2 + 9x) + 1]^2$
$$= (x^2 + 9x)^2 + 2(x^2 + 9x)(1) + 1^2$$
$$= (x^2)^2 + 2(x^2)(9x) + (9x)^2 + 2x^2 + 18x + 1$$
$$= x^4 + 18x^3 + 81x^2 + 2x^2 + 18x + 1$$

66. $(3x - y)^2 = (3x)^2 - 2(3x)y + y^2$
$= 9x^2 - 6xy + y^2$

67. $(4x + 9)^2 = (4x)^2 + 2(4x)(9) + (9)^2$
$= 16x^2 + 72x + 81$

68. $(x + 3y)(x - 3y) = x^2 - (3y)^2 = x^2 - 9y^2$

69. $[4 + (3a - b)][4 - (3a - b)]$
$= 4^2 - (3a - b)^2$
$= 16 - (9a^2 - 6ab + b^2)$
$= 16 - 9a^2 + 6ab - b^2$

70.
$$\begin{array}{r} x^2 + 2x - 5 \\ \times \qquad 2x - 1 \\ \hline -x^2 - 2x + 5 \\ 2x^3 + 4x^2 - 10x \\ \hline 2x^3 + 3x^2 - 12x + 5 \end{array}$$

71. $A = (3y - 7z)(3y + 7z)$
$= (3y)^2 - (7z)^2$
$= 9y^2 - 49z^2$

This area of the rectangle is $(9y^2 - 49z^2)$ square units.

72. $4a^b(3a^{b+2} - 7) = 4a^b(3a^{b+2}) + 4a^b(-7)$
$= 12a^{b+b+2} - 28a^b$
$= 12a^{2b+2} - 28a^b$

73. $(4xy^z - b)^2 = (4xy^z)^2 - 2(4xy^z)b + b^2$
$= 4^2 x^2 (y^z)^2 - 8xy^z b + b^2$
$= 16x^2 y^{2z} - 8xy^z b + b^2$

74. $(3x^a - 4)(3x^a + 4) = (3x^a)^2 - 4^2$
$= 9x^{2a} - 16$

75. $16x^3 - 24x^2 = 8x^2(2x - 3)$

76. $36y - 24y^2 = 12y(3 - 2y)$

77. $6ab^2 + 8ab - 4a^2b^2 = 2ab(3b + 4 - 2ab)$

78. $14a^2b^2 - 21ab^2 + 7ab$
$= 7ab(2ab - 3b + 1)$

79. $6a(a + 3b) - 5(a + 3b) = (6a - 5)(a + 3b)$

80. $4x(x - 2y) - 5(x - 2y) = (x - 2y)(4x - 5)$

81. $xy - 6y + 3x - 18 = y(x - 6) + 3(x - 6)$
$= (x - 6)(y + 3)$

82. $ab - 8b + 4a - 32 = b(a - 8) + 4(a - 8)$
$= (a - 8)(b + 4)$

83. $pq - 3p - 5q + 15 = p(q - 3) - 5(q - 3)$
$= (p - 5)(q - 3)$

84. $x^3 - x^2 - 2x + 2 = x^2(x - 1) - 2(x - 1)$
$= (x^2 - 2)(x - 1)$

85. $2xy - x^2 = x(2y - x)$

86. $x^2 - 14x - 72 = (x - 18)(x + 4)$

87. $x^2 + 16x - 80 = (x - 4)(x + 20)$

88. $2x^2 - 18x + 28 = 2(x^2 - 9x + 14)$
$= 2(x - 2)(x - 7)$

89. $3x^2 + 33x + 54 = 3(x^2 + 11x + 18)$
$= 3(x + 2)(x + 9)$

90. $2x^3 - 7x^2 - 9x = x(2x^2 - 7x - 9)$
$= x(2x - 9)(x + 1)$

91. $3x^2 + 2x - 16 = (3x + 8)(x - 2)$

92. $6x^2 + 17x + 10 = (6x + 5)(x + 2)$

93. $15x^2 - 91x + 6 = (15x - 1)(x - 6)$

94. $4x^2 + 2x - 12 = 2(2x^2 + x - 6)$
$= 2(2x - 3)(x + 2)$

95. $9x^2 - 12x - 12 = 3(3x^2 - 4x - 4)$
$= 3(x - 2)(3x + 2)$

96. $y^2(x+6)^2 - 2y(x+6)^2 - 3(x+6)^2$
$= (y^2 - 2y - 3)(x+6)^2$
$= (y-3)(y+1)(x+6)^2$

97. Using the substitution $y = x + 5$, we have
$(x+5)^2 + 6(x+5) + 8$
$= y^2 + 6y + 8$
$= (y+2)(y+4)$
$= [(x+5)+2][(x+5)+4]$
$= (x+7)(x+9).$

98. $x^4 - 6x^2 - 16 = (x^2 - 8)(x^2 + 2)$

99. Using the substitution $y = x^2$, we have
$x^4 + 8x^2 - 20 = y^2 + 8y - 20$
$= (y-2)(y+10)$
$= (x^2 - 2)(x^2 + 10).$

100. $x^2 - 100 = (x+10)(x-10)$

101. $x^2 - 81 = x^2 - 9^2 = (x-9)(x+9)$

102. $2x^2 - 32 = 2(x^2 - 16) = 2(x+4)(x-4)$

103. $6x^2 - 54 = 6(x^2 - 9) = 6(x^2 - 3^2)$
$= 6(x-3)(x+3)$

104. $81 - x^4 = (9 + x^2)(9 - x^2)$
$= (9 + x^2)(3 + x)(3 - x)$

105. $16 - y^4 = 4^2 - (y^2)^2$
$= (4 + y^2)(4 - y^2)$
$= (4 + y^2)(2^2 - y^2)$
$= (4 + y^2)(2 - y)(2 + y)$

106. $(y+2)^2 - 25 = [(y+2)+5][(y+2)-5]$
$= (y+7)(y-3)$

107. $(x-3)^2 - 16 = (x-3)^2 - 4^2$
$= [(x-3)-4][(x-3)+4]$
$= (x-7)(x+1)$

108. $x^3 + 216 = x^3 + 6^3$
$= (x+6)(x^2 - 6x + 36)$

109. $y^3 + 512 = y^3 + 8^3$
$= (y+8)(y^2 - 8y + 64)$

110. $8 - 27y^3 = 2^3 - (3y)^3$
$= (2 - 3y)(4 + 2 \cdot 3y + 9y^2)$
$= (2 - 3y)(4 + 6y + 9y^2)$

111. $1 - 64y^3 = 1^3 - (4y)^3$
$= (1 - 4y)[1 + 4y + (4y)^2]$
$= (1 - 4y)(1 + 4y + 16y^2)$

112. $6x^4 y + 48xy = 6xy(x^3 + 8)$
$= 6xy(x^3 + 2^3)$
$= 6xy(x+2)(x^2 - 2x + 4)$

113. $2x^5 + 16x^2 y^3$
$= 2x^2(x^3 + 8y^3)$
$= 2x^2(x^3 + (2y)^3)$
$= 2x^2(x+2y)[x^2 - x \cdot 2y + (2y)^2]$
$= 2x^2(x+2y)(x^2 - 2xy + 4y^2)$

114. $x^2 - 2x + 1 - y^2 = (x-1)^2 - y^2$
$= (x - 1 - y)(x - 1 + y)$

115. $x^2 - 6x + 9 - 4y^2 = (x-3)^2 - 4y^2$
$= (x-3)^2 - (2y)^2$
$= (x - 3 - 2y)(x - 3 + 2y)$

116. $4x^2 + 12x + 9 = (2x+3)^2$

117. $16a^2 - 40ab + 25b^2 = (4a - 5b)^2$

118. $\pi R^2 h - \pi r^2 h$
$= \pi h(R^2 - r^2)$
$= \pi h(R - r)(R + r)$ cubic units

119. $(3x-1)(x+7)=0$
$3x-1=0 \quad \text{or} \quad x+7=0$
$3x=1 \quad \text{or} \quad x=-7$
$x=\dfrac{1}{3}$

The solutions are $\dfrac{1}{3}$, -7.

120. $3(x+5)(8x-3)=0$
$x+5=0 \quad \text{or} \quad 8x-3=0$
$x=-5 \quad \text{or} \quad x=\dfrac{3}{8}$

The solutions are -5, $\dfrac{3}{8}$.

121. $5x(x-4)(2x-9)=0$
$5x=0 \quad \text{or} \quad x-4=0 \quad \text{or} \quad 2x-9=0$
$x=0 \quad \text{or} \quad x=4 \quad \text{or} \quad 2x=9$
$x=\dfrac{9}{2}$

The solutions are 0, 4, $\dfrac{9}{2}$.

122. $6(x+3)(x-4)(5x+1)=0$
$x+3=0 \quad \text{or} \quad x-4=0 \quad \text{or} \quad 5x+1=0$
$x=-3 \quad \text{or} \quad x=4 \quad \text{or} \quad x=-\dfrac{1}{5}$

The solutions are -3, 4, $-\dfrac{1}{5}$.

123. $2x^2=12x$
$2x^2-12x=0$
$2x(x-6)=0$
$2x=0 \quad \text{or} \quad x-6=0$
$x=0 \quad \text{or} \quad x=6$
The solutions are 0, 6.

124. $4x^3-36x=0$
$4x(x^2-9)=0$
$4x(x-3)(x+3)=0$
$4x=0 \quad \text{or} \quad x-3=0 \quad \text{or} \quad x+3=0$
$x=0 \quad \text{or} \quad x=3 \quad \text{or} \quad x=-3$
The solutions are 0, 3, -3.

125. $(1-x)(3x+2)=-4x$
$3x+2-3x^2-2x=-4x$
$-3x^2+x+2=-4x$
$0=3x^2-5x-2$
$0=(3x+1)(x-2)$
$3x+1=0 \quad \text{or} \quad x-2=0$
$3x=-1 \quad \text{or} \quad x=2$
$x=-\dfrac{1}{3}$

The solutions are $-\dfrac{1}{3}$, 2.

126. $2x(x-12)=-40$
$2x^2-24x+40=0$
$2(x^2-12x+20)=0$
$2(x-2)(x-10)=0$
$x-2=0 \quad \text{or} \quad x-10=0$
$x=2 \quad \text{or} \quad x=10$
The solutions are 2, 10.

127. $3x^2+2x=12-7x$
$3x^2+9x-12=0$
$x^2+3x-4=0$
$(x+4)(x-1)=0$
$x+4=0 \quad \text{or} \quad x-1=0$
$x=-4 \quad \text{or} \quad x=1$
The solutions are -4, 1.

128. $2x^2+3x=35$
$2x^2+3x-35=0$
$(2x-7)(x+5)=0$
$2x-7=0 \quad \text{or} \quad x+5=0$
$x=\dfrac{7}{2} \quad \text{or} \quad x=-5$

The solutions are $\dfrac{7}{2}$, -5.

129. $x^3-18x=3x^2$
$x^3-3x^2-18x=0$
$x(x^2-3x-18)=0$
$x(x-6)(x+3)=0$
$x=0 \quad \text{or} \quad x-6=0 \quad \text{or} \quad x+3=0$
$x=6 \quad \text{or} \quad x=-3$
The solutions are 0, 6, -3.

130.
$$19x^2 - 42x = -x^3$$
$$x^3 + 19x^2 - 42x = 0$$
$$x(x^2 + 19x - 42) = 0$$
$$x(x - 2)(x + 21) = 0$$
$x = 0$ or $x - 2 = 0$ or $x + 21 = 0$
$x = 0$ or $x = 2$ or $x = -21$
The solutions are 0, 2, –21.

131.
$$12x = 6x^3 + 6x^2$$
$$6x^3 + 6x^2 - 12x = 0$$
$$6x(x^2 + x - 2) = 0$$
$$6x(x + 2)(x - 1) = 0$$
$6x = 0$ or $x + 2 = 0$ or $x - 1 = 0$
$x = 0$ or $x = -2$ or $x = 1$
The solutions are 0, –2, 1.

132.
$$8x^3 + 10x^2 = 3x$$
$$8x^3 + 10x^2 - 3x = 0$$
$$x(8x^2 + 10x - 3) = 0$$
$$x(4x - 1)(2x + 3) = 0$$
$x = 0$ or $4x - 1 = 0$ or $2x + 3 = 0$
$$x = \frac{1}{4} \text{ or } \qquad x = -\frac{3}{2}$$
The solutions are $0, \frac{1}{4}, -\frac{3}{2}$.

133. Let x be the number.
$$x + 2x^2 = 105$$
$$2x^2 + x - 105 = 0$$
$$(2x + 15)(x - 7) = 0$$
$2x + 15 = 0$ or $x - 7 = 0$
$$x = -\frac{15}{2} \text{ or } \qquad x = 7$$
The number is $-\frac{15}{2}$ or 7.

134. Let x be the width. Then $5x - 2$ is the length.
$$x(5x - 2) = 16$$
$$5x^2 - 2x - 16 = 0$$
$$(5x + 8)(x - 2) = 0$$
$5x + 8 = 0$ or $x - 2 = 0$
$$x = -\frac{8}{5} \text{ or } \qquad x = 2$$
Reject the negative.
Width = 2 meters
Length = 5(2) – 2 = 8 meters

135.
$$h(t) = -16t^2 + 400$$
$$0 = -16t^2 + 400$$
$$t^2 - 25 = 0$$
$$t^2 = 25$$
$$t = \pm 5$$
Reject the negative. The dummy will hit the ground after 5 seconds.

Chapter 5 Test

1. $(-9x)^{-2} = \dfrac{1}{(-9x)^2} = \dfrac{1}{81x^2}$

2. $-3xy^{-2}(4xy^2)z = (-3)(4)x^{1+1}y^{-2+2}z$
$$= -12x^2 z$$

3. $\dfrac{6^{-1}a^2b^{-3}}{3^{-2}a^{-5}b^2} = \dfrac{3^2 a^{2+5}}{6^1 b^{2+3}} = \dfrac{9a^7}{6b^5} = \dfrac{3a^7}{2b^5}$

4. $\left(\dfrac{-xy^{-5}z}{xy^3}\right)^{-5} = \dfrac{-x^{-5}y^{25}z^{-5}}{x^{-5}y^{-15}}$
$$= \dfrac{-x^{-5+5}y^{25+15}}{z^5}$$
$$= -\dfrac{y^{40}}{z^5}$$

5. $630,000,000 = 6.3 \times 10^8$

6. $0.01200 = 1.2 \times 10^{-2}$

7. $5.0 \times 10^{-6} = 0.000005$

8. $\dfrac{(2.4\times 10^{-3})(1.2\times 10^{-4})}{(3.2\times 10^{-4})}$

$= \dfrac{(2.4)(1.2)}{(3.2)}\times 10^{-3-4+4}$

$= 0.9\times 10^{-3}$

$= 9\times 10^{-4}$

$= 0.0009$

9. $(4x^3 - 3x - 4) - (9x^3 + 8x + 5)$

$= 4x^3 - 3x - 4 - 9x^3 - 8x - 5$

$= (4 - 9)x^3 + (-3 - 8)x + (-4 - 5)$

$= -5x^3 - 11x - 9$

10. $-3xy(4x + y) = -12x^2y - 3xy^2$

11. $(3x + 4)(4x - 7)$

$= 3x(4x) + 3x(-7) + 4(4x) + 4(-7)$

$= 12x^2 - 21x + 16x - 28$

$= 12x^2 - 5x - 28$

12. $(5a - 2b)(5a + 2b) = (5a)^2 - (2b)^2$

$= 25a^2 - 4b^2$

13. $(6m + n)^2 = (6m)^2 + 2(6m)(n) + (n)^2$

$= 36m^2 + 12mn + n^2$

14.

$$
\begin{array}{r}
x^2 - 6x + 4 \\
\times \qquad 2x - 1 \\
\hline
-x^2 + 6x - 4 \\
2x^3 - 12x^2 + 8x \\
\hline
2x^3 - 13x^2 + 14x - 4
\end{array}
$$

15. $16x^3y - 12x^2y^4 = 4x^2y(4x - 3y^3)$

16. $x^2 - 13x - 30 = (x - 15)(x + 2)$

17. $4y^2 + 20y + 25 = (2y + 5)^2$

18. $6x^2 - 15x - 9 = 3(2x^2 - 5x - 3)$

$= 3(2x + 1)(x - 3)$

19. $4x^2 - 25 = (2x + 5)(2x - 5)$

20. $x^3 + 64 = x^3 + 4^3 = (x + 4)(x^2 - 4x + 16)$

21. $3x^2y - 27y^3 = 3y(x^2 - 9y^2)$

$= 3y(x + 3y)(x - 3y)$

22. $6x^2 + 24 = 6(x^2 + 4)$

23. $16y^3 - 2 = 2(8y^3 - 1)$

$= 2[(2y)^3 - (1)^3]$

$= 2(2y - 1)(4y^2 + 2y + 1)$

24. $x^2y - 9y - 3x^2 + 27 = y(x^2 - 9) - 3(x^2 - 9)$

$= (x^2 - 9)(y - 3)$

$= (x + 3)(x - 3)(y - 3)$

25. $3(n - 4)(7n + 8) = 0$

$3(n - 4) = 0 \quad \text{or} \quad 7n + 8 = 0$

$n - 4 = 0 \quad \text{or} \qquad 7n = -8$

$n = 4 \quad \text{or} \qquad n = -\dfrac{8}{7}$

The solutions are $4, -\dfrac{8}{7}$.

26. $(x + 2)(x - 2) = 5(x + 4)$

$x^2 - 4 = 5x + 20$

$x^2 - 5x - 24 = 0$

$(x - 8)(x + 3) = 0$

$x - 8 = 0 \quad \text{or} \quad x + 3 = 0$

$x = 8 \quad \text{or} \qquad x = -3$

The solutions are $8, -3$.

27. $2x^3 + 5x^2 - 8x - 20 = 0$

$x^2(2x + 5) - 4(2x + 5) = 0$

$(2x + 5)(x^2 - 4) = 0$

$(2x + 5)(x + 2)(x - 2) = 0$

$2x + 5 = 0 \quad \text{or} \quad x + 2 = 0 \quad \text{or} \quad x - 2 = 0$

$2x = -5 \quad \text{or} \qquad x = -2 \quad \text{or} \qquad x = 2$

$x = -\dfrac{5}{2}$

The solutions are $-\dfrac{5}{2}, -2, 2$.

28. $x^2 - (2y)^2 = (x - 2y)(x + 2y)$

29. $h(t) = -16t^2 + 96t + 880$

    a. $h(1) = -16(1)^2 + 96(1) + 880$

            $= -16 + 96 + 880$

            $= 960$

        After 1 second, the pebble will be at a height of 960 feet.

    b. $h(5.1) = -16(5.1)^2 + 96(5.1) + 880$

            $= -416.16 + 489.6 + 880$

            $= 953.44$

        After 5.1 second, the pebble will be at a height of 953.44 feet.

    c. $0 = -16t^2 + 96t + 880$

        $0 = t^2 - 6t - 55$

        $0 = (t + 5)(t - 11)$

        $t + 5 = 0$ or $t - 11 = 0$

          $t = -5$ or $t = 11$

        Reject the negative. The pebble will hit the ground at 11 seconds.

Chapter 6

Exercise Set 6.1

1. $f(x) = \dfrac{x+8}{2x-1}$

$f(2) = \dfrac{2+8}{2(2)-1} = \dfrac{10}{4-1} = \dfrac{10}{3}$

$f(0) = \dfrac{0+8}{2(0)-1} = \dfrac{8}{-1} = -8$

$f(-1) = \dfrac{-1+8}{2(-1)-1} = \dfrac{7}{-3} = -\dfrac{7}{3}$

3. $g(x) = \dfrac{x^2+8}{x^3-25x}$

$g(3) = \dfrac{3^2+8}{3^3-25(3)}$

$= \dfrac{9+8}{27-75}$

$= \dfrac{17}{-48}$

$= -\dfrac{17}{48}$

$g(-2) = \dfrac{(-2)^2+8}{(-2)^3-25(-2)}$

$= \dfrac{4+8}{-8+50}$

$= \dfrac{12}{42}$

$= \dfrac{2}{7}$

$g(1) = \dfrac{1^2+8}{1^3-25(1)} = \dfrac{1+8}{1-25} = \dfrac{9}{-24} = -\dfrac{3}{8}$

5. $f(x) = \dfrac{5x-7}{4}$

Domain: $\{x \mid x \text{ is a real number}\}$

7. $s(t) = \dfrac{t^2+1}{2t}$

Undefined when $2t = 0$.

$2t = 0$

$t = \dfrac{0}{2} = 0$

Domain: $\{t \mid t \text{ is a real number and } t \neq 0\}$

9. $f(x) = \dfrac{3x}{7-x}$

Undefined when $7 - x = 0$.

$7 - x = 0$

$7 = x$

Domain: $\{x \mid x \text{ is a real number and } x \neq 7\}$

11. $R(x) = \dfrac{3+2x}{x^3+x^2-2x}$

Undefined when $x^3 + x^2 - 2x = 0$.

$x^3 + x^2 - 2x = 0$

$x(x^2 + x - 2) = 0$

$x(x+2)(x-1) = 0$

$x = 0$ or $x+2 = 0$ or $x-1 = 0$

$x = 0$ or $x = -2$ or $x = 1$

Domain:

$\{x \mid x \text{ is a real number } x \neq 0,\ x \neq -2,\ x \neq 1\}$

13. $C(x) = \dfrac{x+3}{x^2-4}$

Undefined when $x^2 - 4 = 0$.

$x^2 - 4 = 0$

$(x+2)(x-2) = 0$

$x+2 = 0$ or $x-2 = 0$

$x = -2$ or $x = 2$

Domain:

$\{x \mid x \text{ is a real number and } x \neq 2,\ x \neq -2\}$

15. Answers may vary.

17. $\dfrac{4x-8}{3x-6} = \dfrac{4(x-2)}{3(x-2)} = \dfrac{4}{3}$

19. $\dfrac{2x-14}{7-x} = \dfrac{2(x-7)}{-(x-7)} = -2$

21. $\dfrac{x^2-2x-3}{x^2-6x+9} = \dfrac{(x-3)(x+1)}{(x-3)^2} = \dfrac{x+1}{x-3}$

23. $\dfrac{2x^2+12x+18}{x^2-9} = \dfrac{2(x^2+6x+9)}{x^2-3^2}$
$\qquad = \dfrac{2(x+3)^2}{(x+3)(x-3)}$
$\qquad = \dfrac{2(x+3)}{x-3}$

25. $\dfrac{3x+6}{x^2+2x} = \dfrac{3(x+2)}{x(x+2)} = \dfrac{3}{x}$

27. $\dfrac{2x^2-x-3}{2x^3-3x^2+2x-3}$
$\qquad = \dfrac{(2x-3)(x+1)}{x^2(2x-3)+1(2x-3)}$
$\qquad = \dfrac{(2x-3)(x+1)}{(2x-3)(x^2+1)}$
$\qquad = \dfrac{x+1}{x^2+1}$

29. $\dfrac{8q^2}{16q^3-16q^2} = \dfrac{8q^2}{16q^2(q-1)} = \dfrac{1}{2(q-1)}$

31. $\dfrac{x^2+6x-40}{10+x} = \dfrac{(x+10)(x-4)}{x+10} = x-4$

33. $\dfrac{x^3-125}{5-x} = \dfrac{x^3-5^3}{-(x-5)}$
$\qquad = \dfrac{(x-5)(x^2+5x+25)}{-(x-5)}$
$\qquad = -(x^2+5x+25)$
$\qquad = -x^2-5x-25$

35. $\dfrac{8x^3-27}{4x-6} = \dfrac{(2x)^3-3^3}{2(2x-3)}$
$\qquad = \dfrac{(2x-3)(4x^2+6x+9)}{2(2x-3)}$
$\qquad = \dfrac{4x^2+6x+9}{2}$

37. $\dfrac{3xy^3}{4x^3y^2} \cdot \dfrac{-8x^3y^4}{9x^4y^7} = \dfrac{-24x^4y^7}{36x^7y^9} = -\dfrac{2}{3x^3y^2}$

39. $\dfrac{8a}{3a^4b^2} \div \dfrac{4b^5}{6a^2b} = \dfrac{8a}{3a^4b^2} \cdot \dfrac{6a^2b}{4b^5}$
$\qquad = \dfrac{48a^3b}{12a^4b^7}$
$\qquad = \dfrac{4}{ab^6}$

41. $\dfrac{a^2b}{a^2-b^2} \cdot \dfrac{a+b}{4a^3b} = \dfrac{a^2b(a+b)}{4a^3b(a^2-b^2)}$
$\qquad = \dfrac{a^2b(a+b)}{4a^3b(a+b)(a-b)}$
$\qquad = \dfrac{1}{4a(a-b)}$

43. $\dfrac{x^2-9}{4} \div \dfrac{x^2-6x+9}{x^2-x-6}$
$\qquad = \dfrac{(x+3)(x-3)}{4} \cdot \dfrac{(x-3)(x+2)}{(x-3)^2}$
$\qquad = \dfrac{(x+3)(x+2)}{4}$

45. $\dfrac{9x+9}{4x+8} \cdot \dfrac{2x+4}{3x^2-3} = \dfrac{9(x+1)\cdot 2(x+2)}{4(x+2)\cdot 3(x^2-1)}$
$\qquad = \dfrac{18(x+1)(x+2)}{12(x+2)(x+1)(x-1)}$
$\qquad = \dfrac{3}{2(x-1)}$

47. $\dfrac{a+b}{ab} \div \dfrac{a^2-b^2}{4a^3b} = \dfrac{a+b}{ab} \cdot \dfrac{4a^3b}{(a+b)(a-b)}$

$\qquad = \dfrac{4a^3b(a+b)}{ab(a+b)(a-b)}$

$\qquad = \dfrac{4a^2}{a-b}$

49. $\dfrac{2x^2-4x-30}{5x^2-40x-75} \div \dfrac{x^2-8x+15}{x^2-6x+9}$

$\qquad = \dfrac{2(x^2-2x-15)}{5(x^2-8x-15)} \cdot \dfrac{(x-3)^2}{(x-5)(x-3)}$

$\qquad = \dfrac{2(x-5)(x+3)(x-3)}{5(x^2-8x-15)(x-5)}$

$\qquad = \dfrac{2(x+3)(x-3)}{5(x^2-8x-15)}$

51. $\dfrac{2x^3-16}{6x^2+6x-36} \cdot \dfrac{9x+18}{3x^2+6x+12}$

$\qquad = \dfrac{2(x^3-8)\cdot 9(x+2)}{6(x^2+x-6)\cdot 3(x^2+2x+4)}$

$\qquad = \dfrac{18(x-2)(x^2+2x+4)(x+2)}{18(x+3)(x-2)(x^2+2x+4)}$

$\qquad = \dfrac{x+2}{x+3}$

53. $\dfrac{15b-3a}{b^2-a^2} \div \dfrac{a-5b}{ab+b^2}$

$\qquad = \dfrac{3(5b-a)}{(b+a)(b-a)} \cdot \dfrac{b(a+b)}{-(5b-a)}$

$\qquad = \dfrac{3}{b-a} \cdot \dfrac{b}{-1}$

$\qquad = \dfrac{3b}{a-b}$

55. $\dfrac{a^3+a^2b+a+b}{a^3+a} \cdot \dfrac{6a^2}{2a^2-2b^2}$

$\qquad = \dfrac{a^2(a+b)+1(a+b)}{a(a^2+1)} \cdot \dfrac{6a^2}{2(a^2-b^2)}$

$\qquad = \dfrac{(a^2+1)(a+b)}{a(a^2+1)} \cdot \dfrac{3a^2}{(a+b)(a-b)}$

$\qquad = \dfrac{3a}{a-b}$

57. $\dfrac{5a}{12} \cdot \dfrac{2}{25a^2} \cdot \dfrac{15a}{2} = \dfrac{150a^2}{600a^2} = \dfrac{1}{4}$

59. $\dfrac{3x-x^2}{x^3-27} \div \dfrac{x}{x^2+3x+9}$

$\qquad = \dfrac{x(3-x)}{(x-3)(x^2+3x+9)} \cdot \dfrac{x^2+3x+9}{x}$

$\qquad = \dfrac{3-x}{x-3}$

$\qquad = \dfrac{-(x-3)}{x-3}$

$\qquad = -1$

61. $\dfrac{4a}{7} \div \left(\dfrac{a^2}{14} \cdot \dfrac{3}{a} \right) = \dfrac{4a}{7} \div \dfrac{3a}{14}$

$\qquad = \dfrac{4a}{7} \cdot \dfrac{14}{3a}$

$\qquad = \dfrac{4}{1} \cdot \dfrac{2}{3}$

$\qquad = \dfrac{8}{3}$

63. $\dfrac{8b+24}{3a+6} \div \dfrac{ab-2b+3a-6}{a^2-4a+4}$

$\qquad = \dfrac{8(b+3)}{3(a+2)} \cdot \dfrac{(a-2)^2}{b(a-2)+3(a-2)}$

$\qquad = \dfrac{8(b+3)}{3(a+2)} \cdot \dfrac{(a-2)^2}{(a-2)(b+3)}$

$\qquad = \dfrac{8(a-2)}{3(a+2)}$

65. $\dfrac{4}{x} \div \dfrac{3xy}{x^2} \cdot \dfrac{6x^2}{x^4} = \dfrac{4}{x} \cdot \dfrac{x^2}{3xy} \cdot \dfrac{6}{x^2} = \dfrac{24x^2}{3x^4 y} = \dfrac{8}{x^2 y}$

67. $\dfrac{3x^2 - 5x - 2}{y^2 + y - 2} \cdot \dfrac{y^2 + 4y - 5}{12x^2 + 7x + 1} \div \dfrac{5x^2 - 9x - 2}{8x^2 - 2x - 1} = \dfrac{(3x+1)(x-2)}{(y+2)(y-1)} \cdot \dfrac{(y+5)(y-1)}{(4x+1)(3x+1)} \cdot \dfrac{(4x+1)(2x-1)}{(5x+1)(x-2)}$

$$= \dfrac{(y+5)(2x-1)}{(y+2)(5x+1)}$$

69. $\dfrac{5a^2 - 20}{3a^2 - 12a} \div \dfrac{a^3 + 2a^2}{2a^2 - 8a} \cdot \dfrac{9a^3 + 6a^2}{2a^2 - 4a} = \dfrac{5(a^2 - 4)}{3a(a-4)} \cdot \dfrac{2a(a-4)}{a^2(a+2)} \cdot \dfrac{3a^2(3a+2)}{2a(a-2)}$

$$= \dfrac{5(a+2)(a-2)}{a} \cdot \dfrac{1}{a+2} \cdot \dfrac{3a+2}{a-2}$$

$$= \dfrac{5(3a+2)}{a}$$

71. $\dfrac{5x^4 + 3x^2 - 2}{x - 1} \cdot \dfrac{x+1}{x^4 - 1} = \dfrac{(5x^2 - 2)(x^2 + 1)(x+1)}{(x-1)(x^2 + 1)(x^2 - 1)} = \dfrac{(5x^2 - 2)(x+1)}{(x-1)(x+1)(x-1)} = \dfrac{5x^2 - 2}{(x-1)^2}$

73. Recall that $A = l \cdot w$. So, $A = \dfrac{x+2}{x} \cdot \dfrac{5x}{x^2 - 4} = \dfrac{x+2}{x} \cdot \dfrac{5x}{(x+2)(x-2)} = \dfrac{5}{x-2}$.

The area is $\dfrac{5}{x-2}$ square meters.

75. Answers may vary.

77. Since $A = b \cdot h$, $b = \dfrac{A}{h}$ or $b = A \div h$.

$b = \dfrac{x^2 + x - 2}{x^3} \div \dfrac{x^2}{x-1} = \dfrac{(x+2)(x-1)}{x^3} \cdot \dfrac{(x-1)}{x^2} = \dfrac{(x+2)(x-1)^2}{x^5}$

The length of its base is $\dfrac{(x+2)(x-1)^2}{x^5}$ feet.

79. $f(x) = \dfrac{20x}{100 - x}$

x	0	10	30	50	70	90	95	99
y	0	$\frac{20}{9}$	$\frac{60}{7}$	20	$\frac{140}{3}$	180	380	1980

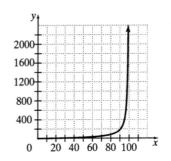

81. $f(x) = \dfrac{100,000x}{100-x}$

    a. $\{x \mid 0 \le x < 100\}$

    b. $f(30) = \dfrac{100,000(30)}{100-30} = 42,857.14$

    The cost of removing 30% of the pollutants is $42,857.14$.

    c. $f(60) = \dfrac{100,000(60)}{100-60} = 150,000$

    The cost of removing 60% of the pollutants is $150,000$.

    $f(80) = \dfrac{100,000(80)}{100-80} = 400,000$

    The cost of removing 80% of the pollutants is $400,000$.

    d. $f(90) = \dfrac{100,000(90)}{100-90} = 900,000$

    The cost of removing 90% of the pollutants is $900,000$.

    $f(95) = \dfrac{100,000(95)}{100-95} = 1,900,000$

    The cost of removing 95% of the pollutants is $1,900,000$.

    $f(99) = \dfrac{100,000(99)}{100-99} = 9,900,000$

    The cost of removing 99% of the pollutants is $9,900,000$.

    Answers may vary.

83. $\dfrac{4}{5} + \dfrac{3}{5} = \dfrac{7}{5}$

85. $\dfrac{5}{28} - \dfrac{2}{21}$

The LCD is 84.

$\dfrac{5}{28} \cdot \dfrac{3}{3} - \dfrac{2}{21} \cdot \dfrac{4}{4} = \dfrac{15}{84} - \dfrac{8}{84} = \dfrac{7}{84} = \dfrac{1}{12}$

87. $\dfrac{3}{8} + \dfrac{1}{2} - \dfrac{3}{16}$

The LCD is 16.

$\dfrac{3}{8} \cdot \dfrac{2}{2} + \dfrac{1}{2} \cdot \dfrac{8}{8} - \dfrac{3}{16} = \dfrac{6}{16} + \dfrac{8}{16} - \dfrac{3}{16} = \dfrac{11}{16}$

89. $\dfrac{x^{2n}-4}{7x} \cdot \dfrac{14x^3}{x^n-2} = \dfrac{(x^n+2)(x^n-2) \cdot 14x^3}{7x \cdot (x^n-2)}$

$= 2x^2(x^n+2)$

91. $\dfrac{y^{2n}+9}{10y} \cdot \dfrac{y^n-3}{y^{4n}-81}$

$= \dfrac{(y^{2n}+9)(y^n-3)}{10y \cdot (y^{2n}+9)(y^{2n}-9)}$

$= \dfrac{y^n-3}{10y(y^n+3)(y^n-3)}$

$= \dfrac{1}{10y(y^n+3)}$

93. $\dfrac{y^{2n}-y^n-2}{2y^n-4} \div \dfrac{y^{2n}-1}{1+y^n}$

$= \dfrac{(y^n-2)(y^n+1)}{2(y^n-2)} \cdot \dfrac{y^n+1}{(y^n+1)(y^n-1)}$

$= \dfrac{y^n+1}{2(y^n-1)}$

Exercise Set 6.2

1. $\dfrac{2}{x} - \dfrac{5}{x} = \dfrac{2-5}{x} = -\dfrac{3}{x}$

3. $\dfrac{2}{x-2} + \dfrac{x}{x-2} = \dfrac{2+x}{x-2} = \dfrac{x+2}{x-2}$

5. $\dfrac{x^2}{x+2} - \dfrac{4}{x+2} = \dfrac{x^2-4}{x+2}$

$= \dfrac{(x+2)(x-2)}{x+2}$

$= x-2$

7. $\dfrac{2x-6}{x^2+x-6}+\dfrac{3-3x}{x^2+x-6}=\dfrac{2x-6+3-3x}{x^2+x-6}$

$\phantom{\dfrac{2x-6}{x^2+x-6}}=\dfrac{-x-3}{x^2+x-6}$

$\phantom{\dfrac{2x-6}{x^2+x-6}}=\dfrac{-(x+3)}{(x+3)(x-2)}$

$\phantom{\dfrac{2x-6}{x^2+x-6}}=\dfrac{-1}{x-2}$

$\phantom{\dfrac{2x-6}{x^2+x-6}}=\dfrac{1}{2-x}$

9. Recall that $P=4\cdot s$.

$P=4\left(\dfrac{x}{x+5}\right)=\dfrac{4x}{x+5}$

The perimeter is $\dfrac{4x}{x+5}$ feet.

Recall that $A=s^2$.

$A=\left(\dfrac{x}{x+5}\right)^2=\dfrac{x^2}{(x+5)^2}=\dfrac{x^2}{x^2+10x+25}$

The area is $\dfrac{x^2}{x^2+10x+25}$ square feet.

11. $7=7$

$5x=5\cdot x$

$\text{LCD}=7\cdot5\cdot x=35x$

13. $x=x$

$x+1=x+1$

$\text{LCD}=x(x+1)$

15. $x+7=x+7$

$x-7=x-7$

$\text{LCD}=(x+7)(x-7)$

17. $3x+6=3(x+2)$

$2x-4=2(x-2)$

$\text{LCD}=2\cdot3\cdot(x+2)(x-2)$

$\phantom{\text{LCD}}=6(x+2)(x-2)$

19. $(3x-1)(x+2)=(3x-1)(x+2)$

$3x-1=3x-1$

$\text{LCD}=(3x-1)(x+2)$

21. $a^2-b^2=(a+b)(a-b)$

$a^2-2ab+b^2=(a-b)(a-b)$

$\text{LCD}=(a+b)(a-b)(a-b)$

$\phantom{\text{LCD}}=(a+b)(a-b)^2$

23. $x^2-9=(x+3)(x-3)$

$x=x$

$12-4x=4(3-x)=-4(x-3)$

$\text{LCD}=-4x(x+3)(x-3)$

25. Answers may vary.

27. $\dfrac{4}{3x}+\dfrac{3}{2x}=\dfrac{(4)2}{(3x)2}+\dfrac{(3)3}{(2x)3}$

$\phantom{\dfrac{4}{3x}+\dfrac{3}{2x}}=\dfrac{8}{6x}+\dfrac{9}{6x}$

$\phantom{\dfrac{4}{3x}+\dfrac{3}{2x}}=\dfrac{8+9}{6x}$

$\phantom{\dfrac{4}{3x}+\dfrac{3}{2x}}=\dfrac{17}{6x}$

29. $\dfrac{5}{2y^2}-\dfrac{2}{7y}=\dfrac{(5)7}{(2y^2)7}-\dfrac{2(2y)}{(7y)2y}$

$\phantom{\dfrac{5}{2y^2}-\dfrac{2}{7y}}=\dfrac{35}{14y^2}-\dfrac{4y}{14y^2}$

$\phantom{\dfrac{5}{2y^2}-\dfrac{2}{7y}}=\dfrac{35-4y}{14y^2}$

31. $\dfrac{x-3}{x+4}-\dfrac{x+2}{x-4}$

$=\dfrac{(x-3)(x-4)}{(x+4)(x-4)}-\dfrac{(x+2)(x+4)}{(x-4)(x+4)}$

$=\dfrac{(x^2-7x+12)-(x^2+6x+8)}{(x+4)(x-4)}$

$=\dfrac{-13x+4}{(x+4)(x-4)}$

33. $\dfrac{1}{x-5} + \dfrac{x}{x^2-x-20}$

$= \dfrac{1}{x-5} + \dfrac{x}{(x-5)(x+4)}$

$= \dfrac{1(x+4)}{(x-5)(x+4)} + \dfrac{x}{(x-5)(x+4)}$

$= \dfrac{x+4+x}{(x-5)(x+4)}$

$= \dfrac{2x+4}{(x-5)(x+4)}$

35. $\dfrac{1}{a-b} + \dfrac{1}{b-a} = \dfrac{1}{a-b} + \dfrac{1}{-(a-b)}$

$= \dfrac{1}{a-b} - \dfrac{1}{a-b}$

$= 0$

37. $\dfrac{x+1}{1-x} + \dfrac{1}{x-1} = \dfrac{x+1}{-(x-1)} + \dfrac{1}{x-1}$

$= \dfrac{-(x+1)}{x-1} + \dfrac{1}{x-1}$

$= \dfrac{-x-1+1}{x-1}$

$= -\dfrac{x}{x-1}$

39. $\dfrac{5}{x-2} + \dfrac{x+4}{2-x} = \dfrac{5}{x-2} + \dfrac{x+4}{-(x-2)}$

$= \dfrac{5}{x-2} - \dfrac{x+4}{x-2}$

$= \dfrac{5-(x+4)}{x-2}$

$= \dfrac{5-x-4}{x-2}$

$= \dfrac{-x+1}{x-2}$

41. $\dfrac{y+1}{y^2-6y+8} - \dfrac{3}{y^2-16}$

$= \dfrac{y+1}{(y-2)(y-4)} - \dfrac{3}{(y+4)(y-4)}$

$= \dfrac{(y+1)(y+4) - 3(y-2)}{(y-2)(y-4)(y+4)}$

$= \dfrac{y^2+5y+4-3y+6}{(y-2)(y-4)(y+4)}$

$= \dfrac{y^2+2y+10}{(y-2)(y-4)(y+4)}$

43. $\dfrac{x+4}{3x^2+11x+6} + \dfrac{x}{2x^2+x-15}$

$= \dfrac{x+4}{(3x+2)(x+3)} + \dfrac{x}{(2x-5)(x+3)}$

$= \dfrac{(x+4)(2x-5) + x(3x+2)}{(3x+2)(x+3)(2x-5)}$

$= \dfrac{2x^2+3x-20+3x^2+2x}{(3x+2)(x+3)(2x-5)}$

$= \dfrac{5x^2+5x-20}{(3x+2)(x+3)(2x-5)}$

$= \dfrac{5(x^2+x-4)}{(3x+2)(x+3)(2x-5)}$

45. $\dfrac{7}{x^2-x-2}+\dfrac{x}{x^2+4x+3}=\dfrac{7}{(x-2)(x+1)}+\dfrac{x}{(x+1)(x+3)}$

$$=\dfrac{7(x+3)+x(x-2)}{(x-2)(x+1)(x+3)}$$

$$=\dfrac{7x+21+x^2-2x}{(x-2)(x+1)(x+3)}$$

$$=\dfrac{x^2+5x+21}{(x-2)(x+1)(x+3)}$$

47. $\dfrac{2}{x+1}-\dfrac{3x}{3x+3}+\dfrac{1}{2x+2}=\dfrac{2}{x+1}-\dfrac{3x}{3(x+1)}+\dfrac{1}{2(x+1)}$

$$=\dfrac{2}{x+1}-\dfrac{x}{x+1}+\dfrac{1}{2(x+1)}$$

$$=\dfrac{2(2)-2x+1}{2(x+1)}$$

$$=\dfrac{4-2x+1}{2(x+1)}$$

$$=\dfrac{-2x+5}{2(x+1)}$$

49. $\dfrac{3}{x+3}+\dfrac{5}{x^2+6x+9}-\dfrac{x}{x^2-9}=\dfrac{3}{x+3}+\dfrac{5}{(x+3)^2}-\dfrac{x}{(x+3)(x-3)}$

$$=\dfrac{3(x+3)(x-3)}{(x+3)^2(x-3)}+\dfrac{5(x-3)}{(x+3)^2(x-3)}-\dfrac{x(x+3)}{(x+3)^2(x-3)}$$

$$=\dfrac{3x^2-27+5x-15-x^2-3x}{(x+3)^2(x-3)}$$

$$=\dfrac{2x^2+2x-42}{(x+3)^2(x-3)}$$

$$=\dfrac{2(x^2+x-21)}{(x+3)^2(x-3)}$$

51. $\dfrac{4}{3x^2y^3}+\dfrac{5}{3x^2y^3}=\dfrac{4+5}{3x^2y^3}=\dfrac{9}{3x^2y^3}=\dfrac{3}{x^2y^3}$

53. $\dfrac{x-5}{2x}-\dfrac{x+5}{2x}=\dfrac{x-5-x-5}{2x}=\dfrac{-10}{2x}=-\dfrac{5}{x}$

55. $\dfrac{3}{2x+10}+\dfrac{8}{3x+15}=\dfrac{3}{2(x+5)}+\dfrac{8}{3(x+5)}=\dfrac{9}{6(x+5)}+\dfrac{16}{6(x+5)}=\dfrac{9+16}{6(x+5)}=\dfrac{25}{6(x+5)}$

57. $\dfrac{-2}{x^2-3x}-\dfrac{1}{x^3-3x^2}=\dfrac{-2}{x(x-3)}-\dfrac{1}{x^2(x-3)}=\dfrac{-2x}{x^2(x-3)}-\dfrac{1}{x^2(x-3)}=\dfrac{-2x-1}{x^2(x-3)}$

59. $\dfrac{ab}{a^2-b^2}+\dfrac{b}{a+b}$

$$=\dfrac{ab}{(a+b)(a-b)}+\dfrac{b}{a+b}$$

$$=\dfrac{ab}{(a+b)(a-b)}+\dfrac{b(a-b)}{(a+b)(a-b)}$$

$$=\dfrac{ab+ab-b^2}{(a+b)(a-b)}$$

$$=\dfrac{2ab-b^2}{(a+b)(a-b)}$$

61. $\dfrac{5}{x^2-4}-\dfrac{3}{x^2+4x+4}$

$$=\dfrac{5}{(x+2)(x-2)}-\dfrac{3}{(x+2)^2}$$

$$=\dfrac{5(x+2)}{(x+2)^2(x-2)}-\dfrac{3(x-2)}{(x+2)^2(x-2)}$$

$$=\dfrac{5x+10-3x+6}{(x+2)^2(x-2)}$$

$$=\dfrac{2x+16}{(x+2)^2(x-2)}$$

63. $\dfrac{2}{a^2+2a+1}+\dfrac{3}{a^2-1}$

$$=\dfrac{2}{(a+1)^2}+\dfrac{3}{(a+1)(a-1)}$$

$$=\dfrac{2(a-1)}{(a+1)^2(a-1)}+\dfrac{3(a+1)}{(a+1)^2(a-1)}$$

$$=\dfrac{2a-2+3a+3}{(a+1)^2(a-1)}$$

$$=\dfrac{5a+1}{(a+1)^2(a-1)}$$

65. Answers may vary.

67. Answers may vary.

69. $\left(\dfrac{2}{3}-\dfrac{1}{x}\right)\left(\dfrac{3}{x}+\dfrac{1}{2}\right)=\left(\dfrac{2x}{3x}-\dfrac{3}{3x}\right)\left(\dfrac{6}{2x}+\dfrac{x}{2x}\right)$

$$=\dfrac{2x-3}{3x}\cdot\dfrac{x+6}{2x}$$

$$=\dfrac{2x^2+9x-18}{6x^2}$$

71. $\left(\dfrac{1}{x}+\dfrac{2}{3}\right)-\left(\dfrac{1}{x}-\dfrac{2}{3}\right)=\dfrac{1}{x}+\dfrac{2}{3}-\dfrac{1}{x}+\dfrac{2}{3}=\dfrac{4}{3}$

73. $\left(\dfrac{2a}{3}\right)^2\div\left(\dfrac{a^2}{a+1}-\dfrac{1}{a+1}\right)$

$$=\dfrac{4a^2}{9}\div\dfrac{a^2-1}{a+1}$$

$$=\dfrac{4a^2}{9}\cdot\dfrac{a+1}{(a+1)(a-1)}$$

$$=\dfrac{4a^2}{9}\cdot\dfrac{1}{a-1}$$

$$=\dfrac{4a^2}{9(a-1)}$$

75. $\left(\dfrac{2x}{3}\right)^2\div\left(\dfrac{x}{3}\right)^2=\dfrac{4x^2}{9}\cdot\dfrac{9}{x^2}=4$

77. $\dfrac{x}{x^2-9}+\dfrac{3}{x^2-6x+9}-\dfrac{1}{x+3}$

$$=\dfrac{x}{(x+3)(x-3)}+\dfrac{3}{(x-3)^2}-\dfrac{1}{(x+3)}$$

$$=\dfrac{x(x-3)+3(x+3)-(x-3)^2}{(x+3)(x-3)^2}$$

$$=\dfrac{x^2-3x+3x+9-x^2+6x-9}{(x-3)(x-3)^2}$$

$$=\dfrac{6x}{(x+3)(x-3)^2}$$

79. $\left(\dfrac{x}{x+1}-\dfrac{x}{x-1}\right)\div\dfrac{x}{2x+2}$

$$=\dfrac{x(x-1)-x(x+1)}{(x-1)(x+1)}\cdot\dfrac{2(x+1)}{x}$$

$$=\dfrac{x^2-x-x^2-x}{(x+1)(x-1)}\cdot\dfrac{2(x+1)}{x}$$

$$=\dfrac{-2x}{(x+1)(x-1)}\cdot\dfrac{2(x+1)}{x}$$

$$=\dfrac{-4}{x-1}$$

81. $\dfrac{4}{x} \cdot \left(\dfrac{2}{x+2} - \dfrac{2}{x-2} \right)$

$= \dfrac{4 \cdot 2}{x} \left(\dfrac{1}{x+2} - \dfrac{1}{x-2} \right)$

$= \dfrac{8}{x} \left[\dfrac{(x-2)-(x+2)}{(x+2)(x-2)} \right]$

$= \dfrac{8}{x} \cdot \dfrac{x-2-x-2}{(x+2)(x-2)}$

$= \dfrac{8}{x} \cdot \dfrac{-4}{(x+2)(x-2)}$

$= -\dfrac{32}{x(x+2)(x-2)}$

83.

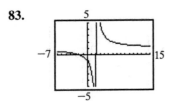

85.

87. $12\left(\dfrac{2}{3} + \dfrac{1}{6} \right) = \dfrac{24}{3} + \dfrac{12}{6} = 8 + 2 = 10$

89. $x^2 \left(\dfrac{4}{x^2} + 1 \right) = x^2 \cdot \dfrac{4}{x^2} + x^2 \cdot 1 = 4 + x^2$

91. $\sqrt{100} = 10$ since $10^2 = 100$.

93. $\sqrt[3]{8} = 2$ since $2^3 = 8$.

95. $\sqrt[4]{81} = 3$ since $3^4 = 81$.

97. $a^2 + b^2 = c^2$

$3^2 + 4^2 = c^2$

$9 + 16 = c^2$

$25 = c^2$

$c = 5$

The unknown length is 5 meters.

99. $x^{-1} + (2x)^{-1} = \dfrac{1}{x} + \dfrac{1}{2x}$

$= \dfrac{2}{2x} + \dfrac{1}{2x}$

$= \dfrac{3}{2x}$

101. $4x^{-2} - 3x^{-1} = \dfrac{4}{x^2} - \dfrac{3}{x}$

$= \dfrac{4}{x^2} - \dfrac{3x}{x^2}$

$= \dfrac{4 - 3x}{x^2}$

103. $x^{-3}(2x+1) - 5x^{-2} = \dfrac{2x+1}{x^3} - \dfrac{5}{x^2}$

$= \dfrac{2x+1}{x^3} - \dfrac{5x}{x^3}$

$= \dfrac{2x+1-5x}{x^3}$

$= \dfrac{1-3x}{x^3}$

Exercise Set 6.3

1. $\dfrac{\frac{1}{3}}{\frac{2}{5}} = \dfrac{1}{3} \cdot \dfrac{5}{2} = \dfrac{5}{6}$

3. $\dfrac{\frac{4}{x}}{\frac{5}{2x}} = \dfrac{4}{x} \cdot \dfrac{2x}{5} = \dfrac{4}{1} \cdot \dfrac{2}{5} = \dfrac{8}{5}$

5. $\dfrac{\frac{10}{3x}}{\frac{5}{6x}} = \dfrac{10}{3x} \cdot \dfrac{6x}{5} = \dfrac{2}{1} \cdot \dfrac{2}{1} = 4$

7. $\dfrac{1+\frac{2}{5}}{2+\frac{3}{5}} = \dfrac{\left(1+\frac{2}{5}\right)5}{\left(2+\frac{3}{5}\right)5} = \dfrac{5+2}{10+3} = \dfrac{7}{13}$

9. $\dfrac{\frac{4}{x-1}}{\frac{x}{x-1}} = \dfrac{4}{x-1} \cdot \dfrac{x-1}{x} = \dfrac{4}{x}$

11. $\dfrac{1-\frac{2}{x}}{x-\frac{4}{9x}} = \dfrac{\left(1-\frac{2}{x}\right)9x}{\left(x-\frac{4}{9x}\right)9x} = \dfrac{9x-18}{9x^2-4}$

13. $\dfrac{\frac{1}{x+1}-1}{\frac{1}{x-1}+1} = \dfrac{\frac{1-(x+1)}{x+1}}{\frac{1+(x-1)}{x-1}}$

$= \dfrac{\frac{-x}{x+1}}{\frac{x}{x-1}}$

$= \dfrac{-x}{x+1} \cdot \dfrac{x-1}{x}$

$= \dfrac{-x(x-1)}{x(x+1)}$

$= \dfrac{-(x-1)}{x+1}$

$= \dfrac{1-x}{x+1}$

15. $\dfrac{x^{-1}}{x^{-2}+y^{-2}} = \dfrac{\frac{1}{x}}{\frac{1}{x^2}+\frac{1}{y^2}}$

$= \dfrac{\frac{1}{x}(x^2 y^2)}{\left(\frac{1}{x^2}+\frac{1}{y^2}\right)(x^2 y^2)}$

$= \dfrac{xy^2}{y^2+x^2}$

17. $\dfrac{2a^{-1}+3b^{-2}}{a^{-1}-b^{-1}} = \dfrac{\frac{2}{a}+\frac{3}{b^2}}{\frac{1}{a}-\frac{1}{b}}$

$= \dfrac{\left(\frac{2}{a}+\frac{3}{b^2}\right)ab^2}{\left(\frac{1}{a}-\frac{1}{b}\right)ab^2}$

$= \dfrac{2b^2+3a}{b^2-ab}$

19. $\dfrac{1}{x-x^{-1}} = \dfrac{1}{x-\frac{1}{x}}$

$= \dfrac{1 \cdot x}{\left(x-\frac{1}{x}\right) \cdot x}$

$= \dfrac{x}{x^2-1}$

21. $\dfrac{\frac{x+1}{7}}{\frac{x+2}{7}} = \dfrac{x+1}{7} \cdot \dfrac{7}{x+2} = \dfrac{x+1}{x+2}$

23. $\dfrac{\frac{1}{2}-\frac{1}{3}}{\frac{3}{4}+\frac{2}{5}} = \dfrac{\left(\frac{1}{2}-\frac{1}{3}\right)60}{\left(\frac{3}{4}+\frac{2}{5}\right)60} = \dfrac{30-20}{45+24} = \dfrac{10}{69}$

25. $\dfrac{\frac{x+1}{3}}{\frac{2x-1}{6}} = \dfrac{x+1}{3} \cdot \dfrac{6}{2x-1}$

$= \dfrac{x+1}{1} \cdot \dfrac{2}{2x-1}$

$= \dfrac{2(x+1)}{2x-1}$

27. $\dfrac{\frac{x}{3}}{\frac{2}{x+1}} = \dfrac{x}{3} \cdot \dfrac{x+1}{2} = \dfrac{x(x+1)}{6}$

29. $\dfrac{\frac{2}{x}+3}{\frac{4}{x^2}-9} = \dfrac{\left(\frac{2}{x}+3\right)x^2}{\left(\frac{4}{x^2}-9\right)x^2}$

$= \dfrac{2x+3x^2}{4-9x^2}$

$= \dfrac{x(2+3x)}{(2-3x)(2+3x)}$

$= \dfrac{x}{2-3x}$

31.
$$\frac{1 - \frac{x}{y}}{\frac{x^2}{y^2} - 1} = \frac{\left(1 - \frac{x}{y}\right)y^2}{\left(\frac{x^2}{y^2} - 1\right)y^2}$$
$$= \frac{y^2 - xy}{x^2 - y^2}$$
$$= \frac{y(y - x)}{(x + y)(x - y)}$$
$$= \frac{y(-1)}{x + y}$$
$$= -\frac{y}{x + y}$$

33.
$$\frac{\frac{-2x}{x - y}}{\frac{y}{x^2}} = \frac{-2x}{x - y} \cdot \frac{x^2}{y} = -\frac{2x^3}{y(x - y)}$$

35.
$$\frac{\frac{2}{x} + \frac{1}{x^2}}{\frac{y}{x^2}} = \frac{\left(\frac{2}{x} + \frac{1}{x^2}\right)x^2}{\left(\frac{y}{x^2}\right)x^2} = \frac{2x + 1}{y}$$

37.
$$\frac{\frac{x}{9} - \frac{1}{x}}{1 + \frac{3}{x}} = \frac{\left(\frac{x}{9} - \frac{1}{x}\right)9x}{\left(1 + \frac{3}{x}\right)9x}$$
$$= \frac{x^2 - 9}{9x + 27}$$
$$= \frac{(x + 3)(x - 3)}{9(x + 3)}$$
$$= \frac{x - 3}{9}$$

39.
$$\frac{\frac{x - 1}{x^2 - 4}}{1 + \frac{1}{x - 2}} = \frac{\frac{x - 1}{(x + 2)(x - 2)}(x + 2)(x - 2)}{\left(1 + \frac{1}{x - 2}\right)(x - 2)(x + 2)}$$
$$= \frac{x - 1}{(x - 2 + 1)(x + 2)}$$
$$= \frac{x - 1}{(x - 1)(x + 2)}$$
$$= \frac{1}{x + 2}$$

41.
$$\frac{\frac{4}{5 - x} + \frac{5}{x - 5}}{\frac{2}{x} + \frac{3}{x - 5}} = \frac{\left(\frac{4}{5 - x} + \frac{5}{x - 5}\right)[x(x - 5)]}{\left(\frac{2}{x} + \frac{3}{x - 5}\right)[x(x - 5)]}$$
$$= \frac{-4x + 5x}{2x - 10 + 3x}$$
$$= \frac{x}{5x - 10}$$

43.
$$\frac{\frac{x + 2}{x} - \frac{2}{x - 1}}{\frac{x + 1}{x} + \frac{x + 1}{x - 1}} = \frac{\left(\frac{x + 2}{x} - \frac{2}{x - 1}\right)[x(x - 1)]}{\left(\frac{x + 1}{x} + \frac{x + 1}{x - 1}\right)[x(x - 1)]}$$
$$= \frac{(x + 2)(x - 1) - 2x}{(x + 1)(x - 1) + x(x + 1)}$$
$$= \frac{(x^2 + x - 2) - 2x}{(x^2 - 1) + (x^2 + x)}$$
$$= \frac{x^2 - x - 2}{2x^2 + x - 1}$$
$$= \frac{(x - 2)(x + 1)}{(2x - 1)(x + 1)}$$
$$= \frac{x - 2}{2x - 1}$$

45.
$$\frac{\frac{x - 2}{x + 2} + \frac{x + 2}{x - 2}}{\frac{x - 2}{x + 2} - \frac{x + 2}{x - 2}} = \frac{\left(\frac{x - 2}{x + 2} + \frac{x + 2}{x - 2}\right)(x + 2)(x - 2)}{\left(\frac{x - 2}{x + 2} - \frac{x + 2}{x - 2}\right)(x + 2)(x - 2)}$$
$$= \frac{(x - 2)^2 + (x + 2)^2}{(x - 2)^2 - (x + 2)^2}$$
$$= \frac{(x^2 - 4x + 4) + (x^2 + 4x + 4)}{(x^2 - 4x + 4) - (x^2 + 4x + 4)}$$
$$= \frac{2x^2 + 8}{-8x}$$
$$= -\frac{x^2 + 4}{4x}$$

47.
$$\frac{\frac{2}{y^2}-\frac{5}{xy}-\frac{3}{x^2}}{\frac{2}{y^2}+\frac{7}{xy}+\frac{3}{x^2}}=\frac{\left(\frac{2}{y^2}-\frac{5}{xy}-\frac{3}{x^2}\right)x^2y^2}{\left(\frac{2}{y^2}+\frac{7}{xy}+\frac{3}{x^2}\right)x^2y^2}$$
$$=\frac{2x^2-5xy-3y^2}{2x^2+7xy+3y^2}$$
$$=\frac{(2x+y)(x-3y)}{(2x+y)(x+3y)}$$
$$=\frac{x-3y}{x+3y}$$

49.
$$\frac{a^{-1}+1}{a^{-1}-1}=\frac{\left(\frac{1}{a}+1\right)a}{\left(\frac{1}{a}-1\right)a}=\frac{1+a}{1-a}$$

51.
$$\frac{3x^{-1}+(2y)^{-1}}{x^{-2}}=\frac{\left(\frac{3}{x}+\frac{1}{2y}\right)\cdot 2x^2y}{\left(\frac{1}{x^2}\right)\cdot 2x^2y}$$
$$=\frac{6xy+x^2}{2y}$$

53.
$$\frac{2a^{-1}+(2a)^{-1}}{a^{-1}+2a^{-2}}=\frac{\left(\frac{2}{a}+\frac{1}{2a}\right)\cdot 2a^2}{\left(\frac{1}{a}+\frac{2}{a^2}\right)\cdot 2a^2}$$
$$=\frac{4a+a}{2a+4}$$
$$=\frac{5a}{2a+4}$$

55.
$$\frac{5x^{-1}+2y^{-1}}{x^{-2}y^{-2}}=\frac{\left(\frac{5}{x}+\frac{2}{y}\right)\cdot x^2y^2}{\left(\frac{1}{x^2}\cdot\frac{1}{y^2}\right)\cdot x^2y^2}$$
$$=\frac{5xy^2+2x^2y}{1}$$
$$=5xy^2+2x^2y$$

57.
$$\frac{5x^{-1}-2y^{-1}}{25x^{-2}-4y^{-2}}=\frac{\left(\frac{5}{x}-\frac{2}{y}\right)\cdot x^2y^2}{\left(\frac{25}{x^2}-\frac{4}{y^2}\right)\cdot x^2y^2}$$
$$=\frac{5xy^2-2x^2y}{25y^2-4x^2}$$
$$=\frac{xy(5y-2x)}{(5y-2x)(5y+2x)}$$
$$=\frac{xy}{5y+2x}$$

59.
$$(x^{-1}+y^{-1})^{-1}=\left(\frac{1}{x}+\frac{1}{y}\right)^{-1}$$
$$=\frac{1}{\frac{1}{x}+\frac{1}{y}}$$
$$=\frac{1}{\frac{1}{x}+\frac{1}{y}}\cdot\frac{xy}{xy}$$
$$=\frac{xy}{y+x}$$

61.
$$\frac{x}{1-\frac{1}{1+\frac{1}{x}}}=\frac{x}{1-\frac{1\cdot x}{\left(1+\frac{1}{x}\right)\cdot x}}$$
$$=\frac{x}{1-\frac{x}{x+1}}$$
$$=\frac{x(x+1)}{\left(1-\frac{x}{x+1}\right)(x+1)}$$
$$=\frac{x(x+1)}{(x+1)-x}$$
$$=\frac{x^2+x}{1}$$
$$=x^2+x$$

63.
$$\frac{a}{1-\frac{s}{770}}=\frac{a\cdot 770}{\left(1-\frac{s}{770}\right)\cdot 770}=\frac{770a}{770-s}$$

65. $f(x) = \dfrac{1}{x}$

  a. $f(a+h) = \dfrac{1}{a+h}$

  b. $f(a) = \dfrac{1}{a}$

  c. $\dfrac{f(a+h) - f(a)}{h} = \dfrac{\frac{1}{a+h} - \frac{1}{a}}{h}$

  d. $\dfrac{f(a+h) - f(a)}{h} = \dfrac{\frac{1}{a+h} - \frac{1}{a}}{h}$

$$= \dfrac{\left(\frac{1}{a+h} - \frac{1}{a}\right)[a(a+h)]}{h[a(a+h)]}$$

$$= \dfrac{a - (a+h)}{ah(a+h)}$$

$$= \dfrac{-h}{ah(a+h)}$$

$$= \dfrac{-1}{a(a+h)}$$

67. $f(x) = \dfrac{3}{x+1}$

  a. $f(a+h) = \dfrac{3}{a+h+1}$

  b. $f(a) = \dfrac{3}{a+1}$

  c. $\dfrac{f(a+h) - f(a)}{h} = \dfrac{\frac{3}{a+h+1} - \frac{3}{a+1}}{h}$

  d. $\dfrac{f(a+h) - f(a)}{h}$

$$= \dfrac{\frac{3}{a+h+1} - \frac{3}{a+1}}{h}$$

$$= \dfrac{\left(\frac{3}{a+h+1} - \frac{3}{a+1}\right)[(a+h+1)(a+1)]}{h[(a+h+1)(a+1)]}$$

$$= \dfrac{3(a+1) - 3(a+h+1)}{h(a+h+1)(a+1)}$$

$$= \dfrac{3a+3 - 3a - 3h - 3}{h(a+h+1)(a+1)}$$

$$= \dfrac{-3h}{h(a+h+1)(a+1)}$$

$$= \dfrac{-3}{(a+h+1)(a+1)}$$

69. $\dfrac{3x^3 y^2}{12x} = \dfrac{x^{3-1} y^2}{4} = \dfrac{x^2 y^2}{4}$

71. $\dfrac{144 x^5 y^5}{-16 x^2 y} = -9 x^{5-2} y^{5-1} = -9 x^3 y^4$

73. $|x - 5| = 9$

  $x - 5 = 9$ or $x - 5 = -9$

  $x = 14$ or $x = -4$

  The solutions are -4 and 14.

75. $|x - 5| < 9$

  $x - 5 < 9$ and $x - 5 > -9$

  $x < 14$ and $x > -4$

  The solution is $(-4, 14)$.

77. $\dfrac{1}{1 - (1-x)^{-1}} = \dfrac{1 \cdot (1-x)}{\left(1 - \frac{1}{1-x}\right) \cdot (1-x)}$

$$= \dfrac{1-x}{(1-x) - 1}$$

$$= \dfrac{1-x}{-x}$$

$$= \dfrac{x-1}{x}$$

79. $\dfrac{(x+2)^{-1} + (x-2)^{-1}}{(x^2 - 4)^{-1}}$

$$= \dfrac{\left(\frac{1}{x+2} + \frac{1}{x-2}\right) \cdot (x+2)(x-2)}{\left(\frac{1}{x^2-4}\right) \cdot (x+2)(x-2)}$$

$$= \dfrac{(x-2) + (x+2)}{1}$$

$$= 2x$$

81. $\dfrac{3(a+1)^{-1} + 4a^{-2}}{(a^3 + a^2)^{-1}}$

$$= \dfrac{\left(\frac{3}{a+1} + \frac{4}{a^2}\right) \cdot a^2(a+1)}{\left(\frac{1}{a^3+a^2}\right) \cdot a^2(a+1)}$$

$$= \dfrac{3a^2 + 4(a+1)}{1}$$

$$= 3a^2 + 4a + 4$$

Exercise Set 6.4

1. $\dfrac{4a^2+8a}{2a} = \dfrac{4a^2}{2a} + \dfrac{8a}{2a} = 2a+4$

3. $\dfrac{12a^5b^2+16a^4b}{4a^4b} = \dfrac{12a^5b^2}{4a^4b} + \dfrac{16a^4b}{4a^4b}$
$\qquad\qquad = 3ab+4$

5. $\dfrac{4x^2y^2+6xy^2-4y^2}{2x^2y}$

$= \dfrac{4x^2y^2}{2x^2y} + \dfrac{6xy^2}{2x^2y} - \dfrac{4y^2}{2x^2y}$

$= 2y + \dfrac{3y}{x} - \dfrac{2y}{x^2}$

7. $\dfrac{4x^2+8x+4}{4} = \dfrac{4x^2}{4} + \dfrac{8x}{4} + \dfrac{4}{4}$
$\qquad\qquad = x^2+2x+1$

9. $\dfrac{3x^4+6x-18}{3} = \dfrac{3x^4}{3} + \dfrac{6x}{3} - \dfrac{18}{3}$
$\qquad\qquad = x^4+2x-6$

The length of each piece is (x^4+2x-6) meters.

11.
$$
\begin{array}{r}
x+1 \\
x+2 \overline{)\,x^2+3x+2} \\
\underline{x^2+2x} \\
x+2 \\
\underline{x+2} \\
0
\end{array}
$$
Answer: $x+1$

13.
$$
\begin{array}{r}
2x-8 \\
x+1 \overline{)\,2x^2-6x-8} \\
\underline{2x^2+2x} \\
-8x-8 \\
\underline{-8x-8} \\
0
\end{array}
$$
Answer: $2x-8$

15.
$$
\begin{array}{r}
x-\tfrac{1}{2} \\
2x+4 \overline{)\,2x^2+3x-2} \\
\underline{2x^2+4x} \\
-x-2 \\
\underline{-\;x-2} \\
0
\end{array}
$$
Answer: $x-\dfrac{1}{2}$

17.
$$
\begin{array}{r}
2x^2-\tfrac{1}{2}x+5 \\
2x+4 \overline{)\,4x^3+7x^2+8x+20} \\
\underline{4x^3+8x^2} \\
-x^2+8x \\
\underline{-\;x^2-2x} \\
10x+20 \\
\underline{10x+20} \\
0
\end{array}
$$
Answer: $2x^2-\dfrac{1}{2}x+5$

19. Recall that $A=\ell\cdot w$, so

$w = \dfrac{A}{\ell} = \dfrac{15x^2-29x-14}{5x+2}$.

$$
\begin{array}{r}
3x-7 \\
5x+2 \overline{)\,15x^2-29x-14} \\
\underline{15x^2+6x} \\
-35x-14 \\
\underline{-35x-14} \\
0
\end{array}
$$
The width is $(3x-7)$ inches.

21. $\dfrac{25a^2b^{12}}{10a^5b^7} = \dfrac{5b^{12-7}}{2a^{5-2}} = \dfrac{5b^5}{2a^3}$

23. $\dfrac{x^6y^6-x^3y^3}{x^3y^3} = \dfrac{x^6y^6}{x^3y^3} - \dfrac{x^3y^3}{x^3y^3} = x^3y^3-1$

25.

$$
\begin{array}{r}
a+3 \\
a+1\overline{\smash{)}a^2+4a+3} \\
\underline{a^2+\ a} \\
3a+3 \\
\underline{3a+3} \\
0
\end{array}
$$

Answer: $a+3$

27.

$$
\begin{array}{r}
2x+\ 5 \\
x-2\overline{\smash{)}2x^2+\ x-10} \\
\underline{2x^2-4x} \\
5x-10 \\
\underline{5x-10} \\
0
\end{array}
$$

Answer: $2x+5$

29. $\dfrac{-16y^3+24y^4}{-4y^2}=\dfrac{-16y^3}{-4y^2}+\dfrac{24y^4}{-4y^2}$

$$= 4y-6y^2$$

31.

$$
\begin{array}{r}
2x+23 \\
x-5\overline{\smash{)}2x^2+13x+\ 15} \\
\underline{2x^2-10x} \\
23x+\ 15 \\
\underline{23x-115} \\
130
\end{array}
$$

Answer: $2x+23+\dfrac{130}{x-5}$

33. $\dfrac{20x^2y^3+6xy^4-12x^3y^5}{2xy^3}$

$$=\dfrac{20x^2y^3}{2xy^3}+\dfrac{6xy^4}{2xy^3}-\dfrac{12x^3y^5}{2xy^3}$$

$$=10x+3y-6x^2y^2$$

35.

$$
\begin{array}{r}
2x+4 \\
3x+2\overline{\smash{)}6x^2+16x+8} \\
\underline{6x^2+\ 4x} \\
12x+8 \\
\underline{12x+8} \\
0
\end{array}
$$

Answer: $2x+4$

37.

$$
\begin{array}{r}
y+\ 5 \\
2y-3\overline{\smash{)}2y^2+7y-15} \\
\underline{2y^2-3y} \\
10y-15 \\
\underline{10y-15} \\
0
\end{array}
$$

Answer: $y+5$

39.

$$
\begin{array}{r}
2x+3 \\
2x-3\overline{\smash{)}4x^2+0x-9} \\
\underline{4x^2-6x} \\
6x-9 \\
\underline{6x-9} \\
0
\end{array}
$$

Answer: $2x+3$

41.

$$
\begin{array}{r}
2x^2-8x\ +38 \\
x+4\overline{\smash{)}2x^3+0x^2+6x\ \ \ -4} \\
\underline{2x^3+8x^2} \\
-8x^2+\ 6x \\
\underline{-8x^2-32x} \\
38x-\ \ 4 \\
\underline{38x\ +152} \\
-156
\end{array}
$$

Answer: $2x^2-8x+38-\dfrac{156}{x+4}$

43.

$$\begin{array}{r} 3x+3 \\ x-1\overline{)3x^2+0x-4} \\ \underline{3x^2-3x} \\ 3x-4 \\ \underline{3x-3} \\ -1 \end{array}$$

Answer: $3x+3-\dfrac{1}{x-1}$

45.

$$\begin{array}{r} -2x^3+3x^2-x+4 \\ -x+5\overline{)2x^4-13x^3+16x^2-9x+20} \\ \underline{2x^4-10x^3} \\ -3x^3+16x^2 \\ \underline{-3x^3+15x^2} \\ x^2-9x \\ \underline{x^2-5x} \\ -4x+20 \\ \underline{-4x+20} \\ 0 \end{array}$$

Answer: $-2x^3+3x^2-x+4$

47.

$$\begin{array}{r} 3x^3+5x+4 \\ x^2-2\overline{)3x^5+0x^4-x^3+4x^2-12x-8} \\ \underline{3x^5+0x^4-6x^3} \\ 5x^3+4x^2-12x \\ \underline{5x^3+0x^2-10x} \\ 4x^2-2x-8 \\ \underline{4x^2+0x-8} \\ -2x \end{array}$$

Answer: $3x^3+5x+4-\dfrac{2x}{x^2-2}$

49. $\dfrac{3x^3-5}{3x^2} = \dfrac{3x^3}{3x^2} - \dfrac{5}{3x^2} = x - \dfrac{5}{3x^2}$

51. $P(x) = 3x^3 + 2x^2 - 4x + 3$

$\begin{aligned} P(1) &= 3(1)^3 + 2(1)^2 - 4(1) + 3 \\ &= 3 + 2 - 4 + 3 \\ &= 4 \end{aligned}$

$$\begin{array}{r} 3x^2+5x+1 \\ x-1\overline{)3x^3+2x^2-4x+3} \\ \underline{3x^3-3x^2} \\ 5x^2-4x \\ \underline{5x^2-5x} \\ x+3 \\ \underline{x-1} \\ 4 \end{array}$$

Remainder $= 4$

53. $P(x) = 5x^4 - 2x^2 + 3x - 6$

$\begin{aligned} P(-3) &= 5(-3)^4 - 2(-3)^2 + 3(-3) - 6 \\ &= 5(81) - 2(9) - 9 - 6 \\ &= 405 - 18 - 9 - 6 \\ &= 372 \end{aligned}$

$$\begin{array}{r} 5x^3-15x^2+43x-126 \\ x+3\overline{)5x^4+0x^3-2x^2+3x-6} \\ \underline{5x^4+15x^3} \\ -15x^3-2x^2 \\ \underline{-15x^3-45x^2} \\ 43x^2+3x \\ \underline{43x^2+129x} \\ -126x-6 \\ \underline{-126x-378} \\ 372 \end{array}$$

Remainder $= 372$

55. Answers may vary.

57.

$$x - 2 \overline{\smash{\big)}\ \begin{array}{l} 3x^2 + 10x + 8 \\ \hline 4x^2 - 12x - 12 + 3x^3 \end{array}}$$

$$\underline{-6x^2 \qquad\qquad + 3x^3}$$
$$10x^2 - 12x$$
$$\underline{10x^2 - 20x}$$
$$8x - 12$$
$$\underline{8x - 16}$$
$$4$$

$$x - 2 \overline{\smash{\big)}\ \begin{array}{l} 3x^2 + 10x + \ 8 \\ \hline 3x^3 + \ 4x^2 - 12x - 12 \end{array}}$$

$$\underline{3x^3 - \ 6x^2}$$
$$10x^2 - 12x$$
$$\underline{10x^2 - 20x}$$
$$8x - 12$$
$$\underline{8x - 16}$$
$$4$$

Answer: $3x^2 + 10x + 8 + \dfrac{4}{x - 2}$

59. $3^2 = (-3)^2$ since $3^2 = 3 \cdot 3 = 9$ and
$(-3)^2 = (-3)(-3) = 9$.

61. $-2^3 = (-2)^3$ since
$-2^3 = -(2 \cdot 2 \cdot 2) = -8$ and
$(-2)^3 = (-2)(-2)(-2) = -8$.

63. $|x + 5| < 4$
$x + 5 < 4 \quad$ and $\quad x + 5 > -4$
$\quad x < -1 \quad$ and $\qquad x > -9$
The solution is $(-9, -1)$.

65. $|2x + 7| \geq 9$
$2x + 7 \geq 9 \quad$ or $\quad 2x + 7 \leq -9$
$\quad 2x \geq 2 \quad$ or $\qquad 2x \leq -16$
$\quad x \geq 1 \quad$ or $\qquad x \leq -8$
The solution is $(-\infty, -8] \cup [1, \infty)$.

67.

$$x - 1 \overline{\smash{\big)}\ \begin{array}{l} x^3 + \frac{5}{3}x^2 + \frac{5}{3}x + \frac{8}{3} \\ \hline x^4 + \frac{2}{3}x^3 + 0x^2 + \ x + 0 \end{array}}$$

$$\underline{x^4 - \ x^3}$$
$$\frac{5}{3}x^3 + 0x^2$$
$$\underline{\frac{5}{3}x^3 - \frac{5}{3}x^2}$$
$$\frac{5}{3}x^2 + \ x$$
$$\underline{\frac{5}{3}x^2 - \frac{5}{3}x}$$
$$\frac{8}{3}x + 0$$
$$\underline{\frac{8}{3}x - \frac{8}{3}}$$
$$\frac{8}{3}$$

Answer: $x^3 + \dfrac{5}{3}x^2 + \dfrac{5}{3}x + \dfrac{8}{3} + \dfrac{8}{3(x-1)}$

69.

$$2x - 1 \overline{\smash{\big)}\ \begin{array}{l} \frac{3}{2}x^3 + \frac{1}{4}x^2 + \frac{1}{8}x - \frac{7}{16} \\ \hline 3x^4 - \ x^3 + 0x^2 - \ x + \frac{1}{2} \end{array}}$$

$$\underline{3x^4 - \frac{3}{2}x^3}$$
$$\frac{1}{2}x^3 + 0x^2$$
$$\underline{\frac{1}{2}x^3 - \frac{1}{4}x^2}$$
$$\frac{1}{4}x^2 - \ x$$
$$\underline{\frac{1}{4}x^2 - \frac{1}{8}x}$$
$$-\frac{7}{8}x + \frac{1}{2}$$
$$\underline{-\frac{7}{8}x + \frac{7}{16}}$$
$$\frac{1}{16}$$

Answer:
$$\frac{3}{2}x^3 + \frac{1}{4}x^2 + \frac{1}{8}x - \frac{7}{16} + \frac{1}{16(2x-1)}$$

71.
$$\require{enclose}\begin{array}{r} x^3 \qquad\quad -\frac{2}{5}x \\[2pt] 5x+10 \enclose{longdiv}{5x^4+10x^3-2x^2-4x+0} \end{array}$$

$$\underline{5x^4+10x^3}$$
$$-2x^2-4x$$
$$\underline{-2x^2-4x}$$
$$0$$

Answer: $x^3 - \dfrac{2}{5}x$

Exercise Set 6.5

1.
$$\begin{array}{r|rrr} 5 & 1 & 3 & -40 \\ & & 5 & 40 \\ \hline & 1 & 8 & 0 \end{array}$$
$(x^2+3x-40)\div(x-5)=x+8$

3.
$$\begin{array}{r|rrr} -6 & 1 & 5 & -6 \\ & & -6 & 6 \\ \hline & 1 & -1 & 0 \end{array}$$
$(x^2+5x-6)\div(x+6)=x-1$

5.
$$\begin{array}{r|rrrr} 2 & 1 & -7 & -13 & 5 \\ & & 2 & -10 & -46 \\ \hline & 1 & -5 & -23 & -41 \end{array}$$
$(x^3-7x^2-13x+5)\div(x-2)$
$$=x^2-5x-23-\frac{41}{x-2}$$

7.
$$\begin{array}{r|rrr} 2 & 4 & 0 & -9 \\ & & 8 & 16 \\ \hline & 4 & 8 & 7 \end{array}$$
$(4x^2-9)\div(x-2)=4x+8+\dfrac{7}{x-2}$

9. a. $P(2)=3(2)^2-4(2)-1$
   $\qquad = 12-8-1=3$

b.
$$\begin{array}{r|rrr} 2 & 3 & -4 & -1 \\ & & 6 & 4 \\ \hline & 3 & 2 & 3 \end{array}$$
$P(2)=3$

11. a. $P(-2)=4(-2)^4+7(-2)^2+9(-2)-1$
   $\qquad = 64+28-18-1$
   $\qquad = 73$

b.
$$\begin{array}{r|rrrrr} -2 & 4 & 0 & 7 & 9 & -1 \\ & & -8 & 16 & -46 & 74 \\ \hline & 4 & -8 & 23 & -37 & 73 \end{array}$$
$P(-2)=73$

13. a. $P(-1)=(-1)^5+3(-1)^4+3(-1)-7$
   $\qquad = -1+3-3-7$
   $\qquad = -8$

b.
$$\begin{array}{r|rrrrrr} -1 & 1 & 3 & 0 & 0 & 3 & -7 \\ & & -1 & -2 & 2 & -2 & -1 \\ \hline & 1 & 2 & -2 & 2 & 1 & -8 \end{array}$$
$P(-1)=-8$

15.
$$\begin{array}{r|rrrr} 3 & 1 & -3 & 0 & 2 \\ & & 3 & 0 & 0 \\ \hline & 1 & 0 & 0 & 2 \end{array}$$
$(x^3-3x^2+2)\div(x-3)=x^2+\dfrac{2}{x-3}$

17.
$$\begin{array}{r|rrr} -1 & 6 & 13 & 8 \\ & & -6 & -7 \\ \hline & 6 & 7 & 1 \end{array}$$
$(6x^2+13x+8)\div(x+1)=6x+7+\dfrac{1}{x+1}$

19.

$$\begin{array}{r|rrrrr} 5 & 2 & -13 & 16 & -9 & 20 \\ & & 10 & -15 & 5 & -20 \\ \hline & 2 & -3 & 1 & -4 & 0 \end{array}$$

$(2x^4 - 13x^3 + 16x^2 - 9x + 20) \div (x - 5)$
$= 2x^3 - 3x^2 + x - 4$

21.

$$\begin{array}{r|rrr} -3 & 3 & 0 & -15 \\ & & -9 & 27 \\ \hline & 3 & -9 & 12 \end{array}$$

$(3x^2 - 15) \div (x + 3) = 3x - 9 + \dfrac{12}{x+3}$

23.

$$\begin{array}{r|rrrr} \frac{1}{2} & 3 & -6 & 4 & 5 \\ & & \frac{3}{2} & -\frac{9}{4} & \frac{7}{8} \\ \hline & 3 & -\frac{9}{2} & \frac{7}{4} & \frac{47}{8} \end{array}$$

$(3x^3 - 6x^2 + 4x + 5) \div \left(x - \dfrac{1}{2}\right)$

$= 3x^2 - \dfrac{9}{2}x + \dfrac{7}{4} + \dfrac{47}{8\left(x - \frac{1}{2}\right)}$

25.

$$\begin{array}{r|rrrr} \frac{1}{3} & 3 & 2 & -4 & 1 \\ & & 1 & 1 & -1 \\ \hline & 3 & 3 & -3 & 0 \end{array}$$

$(3x^3 + 2x^2 - 4x + 1) \div \left(x - \dfrac{1}{3}\right)$

$= 3x^2 + 3x - 3$

27.

$$\begin{array}{r|rrrr} -1 & 3 & 7 & -4 & 12 \\ & & -3 & -4 & 8 \\ \hline & 3 & 4 & -8 & 20 \end{array}$$

$(7x^2 - 4x + 12 + 3x^3) \div (x + 1)$

$= 3x^2 + 4x - 8 + \dfrac{20}{x+1}$

29.

$$\begin{array}{r|rrrr} 1 & 1 & 0 & 0 & -1 \\ & & 1 & 1 & 1 \\ \hline & 1 & 1 & 1 & 0 \end{array}$$

$(x^3 - 1) \div (x - 1) = x^2 + x + 1$

31.

$$\begin{array}{r|rrr} -6 & 1 & 0 & -36 \\ & & -6 & 36 \\ \hline & 1 & -6 & 0 \end{array}$$

$(x^2 - 36) \div (x + 6) = x - 6$

33.

$$\begin{array}{r|rrrr} 1 & 1 & 3 & -7 & 4 \\ & & 1 & 4 & -3 \\ \hline & 1 & 4 & -3 & 1 \end{array}$$

Thus, $P(1) = 1$.

35.

$$\begin{array}{r|rrrr} -3 & 3 & -7 & -2 & 5 \\ & & -9 & 48 & -138 \\ \hline & 3 & -16 & 46 & -133 \end{array}$$

Thus, $P(-3) = -133$.

37.

$$\begin{array}{r|rrrrr} -1 & 4 & 0 & 1 & 0 & -2 \\ & & -4 & 4 & -5 & 5 \\ \hline & 4 & -4 & 5 & -5 & 3 \end{array}$$

Thus, $P(-1) = 3$.

39.

$$\begin{array}{r|rrrrr} \frac{1}{3} & 2 & 0 & -3 & 0 & -2 \\ & & \frac{2}{3} & \frac{2}{9} & -\frac{25}{27} & -\frac{25}{81} \\ \hline & 2 & \frac{2}{3} & -\frac{25}{9} & -\frac{25}{27} & -\frac{187}{81} \end{array}$$

Thus, $P\left(\dfrac{1}{3}\right) = -\dfrac{187}{81}$.

41.

$$\begin{array}{r|rrrrrr} \tfrac{1}{2} & 1 & 1 & -1 & 0 & 0 & 3 \\ & & \tfrac{1}{2} & \tfrac{3}{4} & -\tfrac{1}{8} & -\tfrac{1}{16} & -\tfrac{1}{32} \\ \hline & 1 & \tfrac{3}{2} & -\tfrac{1}{4} & -\tfrac{1}{8} & -\tfrac{1}{16} & \tfrac{95}{32} \end{array}$$

Thus, $P\left(\dfrac{1}{2}\right) = \dfrac{95}{32}$.

43. Answers may vary.

45.

$$\begin{array}{r|rrrr} -3 & 1 & 3 & 4 & 12 \\ & & -3 & 0 & -12 \\ \hline & 1 & 0 & 4 & 0 \end{array}$$

Remainder = 0. Therefore, $x + 3$ is a factor of $x^3 + 3x^2 + 4x + 12$.

47. $P(x)$ is equal to the remainder when $P(x)$ is divided by $x - c$. Therefore, $P(c) = 0$.

49. Multiply $(x^2 - x + 10)$ by $(x + 3)$ and add the remainder, -2.
$(x^2 - x + 10)(x + 3) - 2$
$= x^3 - x^2 + 10x + 3x^2 - 3x + 30 - 2$
$= x^3 + 2x^2 + 7x + 28$

51. $V = lwh$
$w = \dfrac{V}{lh}$
$ = \dfrac{x^4 + 6x^3 - 7x^2}{x^2(x+7)}$
$ = \dfrac{x^4 + 6x^3 - 7x^2}{x^3 + 7x^2}$

$$\begin{array}{r} x - 1 \\ x^3 + 7x^2 \overline{\smash{\big)}\, x^4 + 6x^3 - 7x^2 + 0x + 0} \\ \underline{x^4 + 7x^3 } \\ -1x^3 - 7x^2 \\ \underline{-1x^3 - 7x^2 } \\ 0 \end{array}$$

The width is $(x - 1)$ meters.

53. $4 - 2x = 17 - 5x$
$5x - 2x = 17 - 4$
$3x = 13$
$x = \dfrac{13}{3}$
The solution is $\dfrac{13}{3}$.

55. $5x^2 + 10x = 15$
$5x^2 + 10x - 15 = 0$
$x^2 + 2x - 3 = 0$
$(x + 3)(x - 1) = 0$
$x + 3 = 0$ or $x - 1 = 0$
$x = -3$ or $ x = 1$
The solutions are -3 and 1.

57. $\dfrac{2x}{9} + 1 = \dfrac{7}{9}$
$2x + 9 = 7$
$2x = -2$
$x = -1$
The solution is -1.

59. $8y^3 + 1 = (2y)^3 + 1^3$
$ = (2y + 1)(4y^2 - 2y + 1)$

61. $a^3 - 27 = a^3 - 3^3 = (a - 3)(a^2 + 3a + 9)$

63. $x^2 - x + xy - y = x(x - 1) + y(x - 1)$
$ = (x + y)(x - 1)$

65. $2x^3 - 32x = 2x(x^2 - 16)$
$ = 2x(x + 4)(x - 4)$

Exercise Set 6.6

1. $\dfrac{x}{2} - \dfrac{x}{3} = 12$

$6\left(\dfrac{x}{2} - \dfrac{x}{3}\right) = 6 \cdot 12$

$3x - 2x = 72$

$x = 72$

The solution is 72.

3. $\dfrac{x}{3} = \dfrac{1}{6} + \dfrac{x}{4}$

$12\left(\dfrac{x}{3}\right) = 12\left(\dfrac{1}{6} + \dfrac{x}{4}\right)$

$4x = 2 + 3x$

$x = 2$

The solution is 2.

5. $\dfrac{2}{x} + \dfrac{1}{2} = \dfrac{5}{x}$

$2x\left(\dfrac{2}{x} + \dfrac{1}{2}\right) = 2x\left(\dfrac{5}{x}\right)$

$4 + x = 10$

$x = 6$

The solution is 6.

7. $\dfrac{x+3}{x} = \dfrac{5}{x}$

$x\left(\dfrac{x+3}{x}\right) = x\left(\dfrac{5}{x}\right)$

$x + 3 = 5$

$x = 2$

The solution is 2.

9. $\dfrac{x+5}{x+3} = \dfrac{8}{x+3}$

$(x+3) \cdot \dfrac{x+5}{x+3} = (x+3) \cdot \dfrac{8}{x+3}$

$x + 5 = 8$

$x = 3$

The solution is 3.

11.

$$\frac{1}{x-1}+\frac{1}{x+1}=\frac{2}{x^2-1}$$

$$(x+1)(x-1)\left(\frac{1}{x-1}+\frac{1}{x+1}\right)=(x+1)(x-1)\left[\frac{2}{(x+1)(x-1)}\right]$$

$$(x+1)+(x-1)=2$$

$$2x=2$$

$$x=1$$

Note: 1 is an extraneous solution. Therefore, the solution is $\varnothing$.

13.

$$\frac{6}{x+3}=\frac{4}{x-3}$$

$$(x+3)(x-3)\left(\frac{6}{x+3}\right)=(x+3)(x-3)\left(\frac{4}{x-3}\right)$$

$$6(x-3)=4(x+3)$$

$$6x-18=4x+12$$

$$2x=30$$

$$x=15$$

The solution is 15.

15.

$$\frac{3}{2x+3}-\frac{1}{2x-3}=\frac{4}{4x^2-9}$$

$$(2x+3)(2x-3)\left(\frac{3}{2x+3}-\frac{1}{2x-3}\right)=(2x+3)(2x-3)\left[\frac{4}{(2x+3)(2x-3)}\right]$$

$$3(2x-3)-(2x+3)=4$$

$$6x-9-2x-3=4$$

$$4x-12=4$$

$$4x=16$$

$$x=4$$

The solution is 4.

17.

$$\frac{2}{x^2-4}=\frac{1}{2x-4}$$

$$2(x+2)(x-2)\left[\frac{2}{(x+2)(x-2)}\right]=2(x+2)(x-2)\left[\frac{1}{2(x-2)}\right]$$

$$4=x+2$$

$$2=x$$

Note: 2 is an extraneous solution. Therefore, the solution is $\varnothing$.

19.

$$\frac{12}{3x^2 + 12x} = 1 - \frac{1}{x+4}$$

$$3x(x+4)\left[\frac{12}{3x(x+4)}\right] = 3x(x+4)\left(1 - \frac{1}{x+4}\right)$$

$$12 = 3x(x+4) - 3x$$

$$12 = 3x^2 + 12x - 3x$$

$$12 = 3x^2 + 9x$$

$$4 = x^2 + 3x$$

$$x^2 + 3x - 4 = 0$$

$$(x+4)(x-1) = 0$$

$$x + 4 = 0 \qquad \text{or} \qquad x - 1 = 0$$

$$x = -4 \qquad \text{or} \qquad x = 1$$

Note: -4 is an extraneous solution. Therefore, the solution is 1.

21.

$$\frac{2}{x} = \frac{10}{5}$$

$$5x\left(\frac{2}{x}\right) = 5x\left(\frac{10}{5}\right)$$

$$10 = 10x$$

$$1 = x$$

The solution is 1.

23.

$$7 + \frac{6}{a} = 5$$

$$a\left(7 + \frac{6}{a}\right) = a \cdot 5$$

$$7a + 6 = 5a$$

$$2a = -6$$

$$a = -3$$

The solution is -3.

25.

$$\frac{2}{x-5} + \frac{1}{2x} = \frac{5}{3x^2 - 15x}$$

$$6x(x-5)\left(\frac{2}{x-5} + \frac{1}{2x}\right) = 6x(x-5)\left[\frac{5}{3x(x-5)}\right]$$

$$2(6x) + 3(x-5) = 2(5)$$

$$12x + 3x - 15 = 10$$

$$15x = 25$$

$$x = \frac{25}{15} = \frac{5}{3}$$

The solution is $\frac{5}{3}$.

27.
$$\frac{x}{4} + \frac{5}{x} = 3$$
$$4x\left(\frac{x}{4} + \frac{5}{x}\right) = 4x(3)$$
$$x^2 + 20 = 12x$$
$$x^2 - 12x + 20 = 0$$
$$(x - 10)(x - 2) = 0$$
$$x - 10 = 0 \quad \text{or} \quad x - 2 = 0$$
$$x = 10 \quad \text{or} \quad x = 2$$
The solutions are 10 and 2.

29.
$$1 - \frac{5}{y + 7} = \frac{4}{y + 7}$$
$$(y + 7)\left(1 - \frac{5}{y + 7}\right) = (y + 7)\left(\frac{4}{y + 7}\right)$$
$$(y + 7) - 5 = 4$$
$$y + 2 = 4$$
$$y = 2$$
The solution is 2.

31.
$$\frac{6x + 7}{2x + 9} = \frac{5}{3}$$
$$3(2x + 9)\left(\frac{6x + 7}{2x + 9}\right) = 3(2x + 9)\left(\frac{5}{3}\right)$$
$$3(6x + 7) = 5(2x + 9)$$
$$18x + 21 = 10x + 45$$
$$8x = 24$$
$$x = 3$$
The solution is 3.

33.
$$\frac{2x + 1}{4 - x} = \frac{9}{4 - x}$$
$$(4 - x)\left(\frac{2x + 1}{4 - x}\right) = (4 - x)\left(\frac{9}{4 - x}\right)$$
$$2x + 1 = 9$$
$$2x = 8$$
$$x = 4$$
Note: 4 is an extraneous solution.
Therefore, the solution is $\varnothing$.

35.

$$\frac{12}{9-a^2}+\frac{3}{3+a}=\frac{2}{3-a}$$

$$(3+a)(3-a)\left[\frac{12}{(3+a)(3-a)}+\frac{3}{3+a}\right]=(3+a)(3-a)\left(\frac{2}{3-a}\right)$$

$$12+3(3-a)=2(3+a)$$

$$12+9-3a=6+2a$$

$$21-3a=6+2a$$

$$15=5a$$

$$3=a$$

Note: 3 is an extraneous solution. Therefore, the solution is $\varnothing$.

37.

$$2+\frac{3}{x}=\frac{2x}{x+3}$$

$$x(x+3)\left(2+\frac{3}{x}\right)=x(x+3)\left(\frac{2x}{x+3}\right)$$

$$2x(x+3)+3(x+3)=2x(x)$$

$$2x^2+6x+3x+9=2x^2$$

$$2x^2+9x+9=2x^2$$

$$9x+9=0$$

$$9x=-9$$

$$x=-1$$

The solution is -1.

39.

$$\frac{36}{x^2-9}+1=\frac{2x}{x+3}$$

$$(x+3)(x-3)\left[\frac{36}{(x+3)(x-3)}+1\right]=(x+3)(x-3)\left(\frac{2x}{x+3}\right)$$

$$36+(x+3)(x-3)=2x(x-3)$$

$$36+(x^2-9)=2x^2-6x$$

$$x^2+27=2x^2-6x$$

$$x^2-6x-27=0$$

$$(x-9)(x+3)=0$$

$$x-9=0 \quad \text{or} \quad x+3=0$$

$$x=9 \quad \text{or} \quad x=-3$$

Note: -3 is an extraneous solution. Therefore, the solution is 9.

41.

$$\frac{x^2-20}{x^2-7x+12}=\frac{3}{x-3}+\frac{5}{x-4}$$

$$(x-3)(x-4)\left[\frac{x^2-20}{(x-3)(x-4)}\right]=(x-3)(x-4)\left(\frac{3}{x-3}+\frac{5}{x-4}\right)$$

$$x^2-20=3(x-4)+5(x-3)$$

$$x^2-20=3x-12+5x-15$$

$$x^2-20=8x-27$$

$$x^2-8x+7=0$$

$$(x-1)(x-7)=0$$

$$x-1=0 \quad \text{or} \quad x-7=0$$

$$x=1 \quad \text{or} \quad x=7$$

The solutions are 1 and 7.

43.

$$\frac{3}{2x-5}+\frac{2}{2x+3}=0$$

$$(2x-5)(2x+3)\left(\frac{3}{2x-5}+\frac{2}{2x+3}\right)=(2x-5)(2x+3)\cdot 0$$

$$3(2x+3)+2(2x-5)=0$$

$$6x+9+4x-10=0$$

$$10x-1=0$$

$$10x=1$$

$$x=\frac{1}{10}$$

The solution is $\frac{1}{10}$.

45. $f(x)=20+\frac{4000}{x}$

$$25=20+\frac{4000}{x}$$

$$25-20=\frac{4000}{x}$$

$$5=\frac{4000}{x}$$

$$x(5)=x\left(\frac{4000}{x}\right)$$

$$5x=4000$$

$$x=\frac{4000}{5}$$

$$x=800$$

800 pencil sharpeners must be produced.

47. $x^{-2} - 5x^{-1} - 36 = 0$

$$\frac{1}{x^2} - \frac{5}{x} - 36 = 0$$

$$x^2\left(\frac{1}{x^2} - \frac{5}{x} - 36\right) = x^2 \cdot 0$$

$$1 - 5x - 36x^2 = 0$$

$$36x^2 + 5x - 1 = 0$$

$$(9x - 1)(4x + 1) = 0$$

$9x - 1 = 0$ or $4x + 1 = 0$

$9x = 1$ or $4x = -1$

$x = \dfrac{1}{9}$ or $x = -\dfrac{1}{4}$

The solutions are $\dfrac{1}{9}$ and $-\dfrac{1}{4}$.

49. $6p^{-2} - 5p^{-1} + 1 = 0$

$$\frac{6}{p^2} - \frac{5}{p} + 1 = 0$$

$$p^2\left(\frac{6}{p^2} - \frac{5}{p} + 1\right) = p^2 \cdot 0$$

$$6 - 5p + p^2 = 0$$

$$p^2 - 5p + 6 = 0$$

$$(p - 3)(p - 2) = 0$$

$p - 3 = 0$ or $p - 2 = 0$

$p = 3$ or $p = 2$

The solutions are 3 and 2.

51.
$$\frac{-8.5}{x + 1.9} = \frac{5.7}{x - 3.6}$$

$$(x + 1.9)(x - 3.6)\left(\frac{-8.5}{x + 1.9}\right) = (x + 1.9)(x - 3.6)\left(\frac{5.7}{x - 3.6}\right)$$

$$-8.5(x - 3.6) = 5.7(x + 1.9)$$

$$-8.5x + 30.6 = 5.7x + 10.83$$

$$30.6 - 10.83 = 5.7x + 8.5x$$

$$19.77 = 14.2x$$

$$1.39 \approx x$$

The solution is approximately 1.39.

53. $\dfrac{12.2}{x} + 17.3 = \dfrac{9.6}{x} - 14.7$

$x\left(\dfrac{12.2}{x} + 17.3\right) = x\left(\dfrac{9.6}{x} - 14.7\right)$

$12.2 + 17.3x = 9.6 - 14.7x$

$32x = -2.6$

$x \approx -0.08$

The solution is approximately −0.08.

55. $y_1 = \dfrac{2}{x}$, $y_2 = 2$

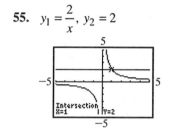

$x = 1$

The solution is 1.

57. $y_1 = \dfrac{6x+7}{2x+9}$, $y_2 = \dfrac{5}{3}$

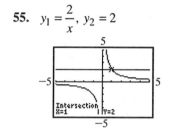

$x = 3$

The solution is 3.

59. Let n = the first integer.

Then $n + 1$ = the second integer.

$n + (n+1) = 147$

$2n + 1 = 147$

$2n = 146$

$n = 73$

The integers are 73 and 74.

61. Let x = the number.

Then $\dfrac{1}{x}$ = the reciprocal of the number.

$x + \dfrac{1}{x} = \dfrac{5}{2}$

$2x\left(x + \dfrac{1}{x}\right) = 2x\left(\dfrac{5}{2}\right)$

$2x^2 + 2 = 5x$

$2x^2 - 5x + 2 = 0$

$(2x - 1)(x - 2) = 0$

$2x - 1 = 0$ or $x - 2 = 0$

$2x = 1$ or $x = 2$

$x = \dfrac{1}{2}$

The reciprocal pair is 2 and $\dfrac{1}{2}$.

63. 3% (reading from the graph);
3% of state and federal prison inmates are 55 years old or older.

65. Adding the percents for the three bars,
16% + 19% + 19% = 54%;
54% of state and federal prison inmates are 20 to 34 years old.

67. $(x - 1)^2 + 3(x - 1) + 2 = 0$

Let $u = x - 1$.

$u^2 + 3u + 2 = 0$

$(u + 1)(u + 2) = 0$

$u + 1 = 0$ or $u + 2 = 0$

$u = -1$ or $u = -2$

Substituting $x - 1$ for u,

$x - 1 = -1$ or $x - 1 = -2$

$x = 0$ or $x = -1$

The solutions are 0 and −1.

69. $\left(\dfrac{3}{x-1}\right)^2 + 2\left(\dfrac{3}{x-1}\right) + 1 = 0$

Let $u = \dfrac{3}{x-1}$.

$u^2 + 2u + 1 = 0$

$(u+1)^2 = 0$

$u + 1 = 0$

$u = -1$

Substituting $\dfrac{3}{x-1}$ for u,

$\dfrac{3}{x-1} = -1$

$(x-1)\left(\dfrac{3}{x-1}\right) = -1(x-1)$

$3 = -x + 1$

$x = -2$

The solution is -2.

Supplementary Exercises on Expressions and Equations

1. $\dfrac{x}{2} = \dfrac{1}{8} + \dfrac{x}{4}$

$8\left(\dfrac{x}{2}\right) = 8\left(\dfrac{1}{8} + \dfrac{x}{4}\right)$

$4x = 1 + 2x$

$2x = 1$

$x = \dfrac{1}{2}$

The solution is $\dfrac{1}{2}$.

3. $\dfrac{1}{8} + \dfrac{x}{4} = \dfrac{1}{8} + \dfrac{2 \cdot x}{2 \cdot 4} = \dfrac{1}{8} + \dfrac{2x}{8} = \dfrac{1+2x}{8}$

5. $\dfrac{4}{x+2} - \dfrac{2}{x-1}$

$= \dfrac{4(x-1)}{(x+2)(x-1)} - \dfrac{2(x+2)}{(x-1)(x+2)}$

$= \dfrac{4(x-1) - 2(x+2)}{(x+2)(x-1)}$

$= \dfrac{4x - 4 - 2x - 4}{(x+2)(x-1)}$

$= \dfrac{2x - 8}{(x+2)(x-1)}$

$= \dfrac{2(x-4)}{(x+2)(x-1)}$

7. $\dfrac{4}{x+2} = \dfrac{2}{x-1}$

$(x-1)(x+2)\left(\dfrac{4}{x+2}\right) = (x-1)(x+2)\left(\dfrac{2}{x-1}\right)$

$4(x-1) = 2(x+2)$

$4x - 4 = 2x + 4$

$2x = 8$

$x = 4$

The solution is 4.

9.
$$\frac{2}{x^2-4} = \frac{1}{x+2} - \frac{3}{x-2}$$

$$(x-2)(x+2)\left[\frac{2}{(x-2)(x+2)}\right] = (x-2)(x+2)\left(\frac{1}{x+2} - \frac{3}{x-2}\right)$$

$$2 = (x-2) - 3(x+2)$$
$$2 = x-2-3x-6$$
$$10 = -2x$$
$$-5 = x$$

The solution is –5.

11. $\dfrac{5}{x^2-3x} + \dfrac{4}{2x-6} = \dfrac{5}{x(x-3)} + \dfrac{4}{2(x-3)} = \dfrac{5\cdot 2}{2x(x-3)} + \dfrac{4\cdot x}{2x(x-3)} = \dfrac{10+4x}{2x(x-3)} = \dfrac{2(5+2x)}{2x(x-3)} = \dfrac{2x+5}{x(x-3)}$

13.
$$\frac{x-1}{x+1} + \frac{x+7}{x-1} = \frac{4}{x^2-1}$$

$$(x+1)(x-1)\left(\frac{x-1}{x+1} + \frac{x+7}{x-1}\right) = (x+1)(x-1)\left[\frac{4}{(x+1)(x-1)}\right]$$

$$(x-1)^2 + (x+7)(x+1) = 4$$
$$(x^2-2x+1) + (x^2+8x+7) = 4$$
$$2x^2 + 6x + 4 = 0$$
$$x^2 + 3x + 2 = 0$$
$$(x+2)(x+1) = 0$$
$$x+2 = 0 \quad \text{or} \quad x+1 = 0$$
$$x = -2 \quad \text{or} \quad x = -1$$

Note: –1 is an extraneous solution. Therefore, the solution is –2.

15. $\dfrac{a^2-9}{a-6} \cdot \dfrac{a^2-5a-6}{a^2-a-6} = \dfrac{(a+3)(a-3)}{a-6} \cdot \dfrac{(a-6)(a+1)}{(a-3)(a+2)} = \dfrac{(a+3)(a+1)}{a+2}$

17.
$$\frac{2x+3}{3x-2} = \frac{4x+1}{6x+1}$$

$$(3x-2)(6x+1)\left(\frac{2x+3}{3x-2}\right) = (3x-2)(6x+1)\left(\frac{4x+1}{6x+1}\right)$$

$$(6x+1)(2x+3) = (3x-2)(4x+1)$$
$$12x^2 + 20x + 3 = 12x^2 - 5x - 2$$
$$25x = -5$$
$$x = -\frac{1}{5}$$

The solution is $-\dfrac{1}{5}$.

19. $\dfrac{a}{9a^2-1}+\dfrac{2}{6a-2}$

$=\dfrac{a}{(3a-1)(3a+1)}+\dfrac{2}{2(3a-1)}$

$=\dfrac{2a+2(3a+1)}{2(3a-1)(3a+1)}$

$=\dfrac{2a+6a+2}{2(3a-1)(3a+1)}$

$=\dfrac{8a+2}{2(3a-1)(3a+1)}$

$=\dfrac{2(4a+1)}{2(3a-1)(3a+1)}$

$=\dfrac{4a+1}{(3a-1)(3a+1)}$

21. $-\dfrac{3}{x^2}-\dfrac{1}{x}+2=0$

$x^2\left(-\dfrac{3}{x^2}-\dfrac{1}{x}+2\right)=x^2\cdot 0$

$-3-x+2x^2=0$

$2x^2-x-3=0$

$(2x-3)(x+1)=0$

$2x-3=0\quad\text{or}\quad x+1=0$

$x=\dfrac{3}{2}\quad\text{or}\quad x=-1$

The solutions are $\dfrac{3}{2}$ and -1.

23. $\dfrac{x-8}{x^2-x-2}+\dfrac{2}{x-2}$

$=\dfrac{x-8}{(x-2)(x+1)}+\dfrac{2}{x-2}$

$=\dfrac{(x-8)+2(x+1)}{(x-2)(x+1)}$

$=\dfrac{x-8+2x+2}{(x-2)(x+1)}$

$=\dfrac{3x-6}{(x-2)(x+1)}$

$=\dfrac{3(x-2)}{(x-2)(x+1)}$

$=\dfrac{3}{x+1}$

25. $\dfrac{3}{a}-5=\dfrac{7}{a}-1$

$a\left(\dfrac{3}{a}-5\right)=a\left(\dfrac{7}{a}-1\right)$

$3-5a=7-a$

$-4a=4$

$a=-1$

The solution is -1.

Exercise Set 6.7

1. $F=\dfrac{9}{5}C+32$

$F-32=\dfrac{9}{5}C$

$C=\dfrac{5}{9}(F-32)$

3. $Q=\dfrac{A-I}{L}$

$QL=L\left(\dfrac{A-I}{L}\right)$

$QL=A-I$

$I=A-QL$

5. $\dfrac{1}{R}=\dfrac{1}{R_1}+\dfrac{1}{R_2}$

$RR_1R_2\left(\dfrac{1}{R}\right)=RR_1R_2\left(\dfrac{1}{R_1}+\dfrac{1}{R_2}\right)$

$R_1R_2=RR_2+RR_1$

$R_1R_2=R(R_2+R_1)$

$R=\dfrac{R_1R_2}{R_1+R_2}$

7. $S=\dfrac{n(a+L)}{2}$

$2S=n(a+L)$

$n=\dfrac{2S}{a+L}$

9.
$$A = \frac{h(a+b)}{2}$$
$$2A = ah + bh$$
$$2A - ah = bh$$
$$\frac{2A - ah}{h} = b$$

11.
$$\frac{P_1 V_1}{T_1} = \frac{P_2 V_2}{T_2}$$
$$(T_1 T_2)\left(\frac{P_1 V_1}{T_1}\right) = \left(\frac{P_2 V_2}{T_2}\right)(T_1 T_2)$$
$$P_1 V_1 T_2 = P_2 V_2 T_1$$
$$T_2 = \frac{P_2 V_2 T_1}{P_1 V_1}$$

13.
$$f = \frac{f_1 f_2}{f_1 + f_2}$$
$$(f_1 + f_2)f = \left(\frac{f_1 f_2}{f_1 + f_2}\right)(f_1 + f_2)$$
$$f_1 f + f_2 f = f_1 f_2$$
$$f_1 f = f_1 f_2 - f_2 f$$
$$f_1 f = f_2 (f_1 - f)$$
$$\frac{f_1 f}{f_1 - f} = f_2$$

15.
$$\lambda = \frac{2L}{n}$$
$$n\lambda = \left(\frac{2L}{n}\right)(n)$$
$$n\lambda = 2L$$
$$\frac{n\lambda}{2} = L$$

17.
$$\frac{\theta}{\omega} = \frac{2L}{c}$$
$$c\omega\left(\frac{\theta}{\omega}\right) = \frac{2L}{c}(c\omega)$$
$$c\theta = 2L\omega$$
$$c = \frac{2L\omega}{\theta}$$

19. Let n = the number. Then
$$\frac{1}{n} = \text{the reciprocal of the number.}$$
$$n + 5\left(\frac{1}{n}\right) = 6$$
$$n + \frac{5}{n} = 6$$
$$n\left(n + \frac{5}{n}\right) = n \cdot 6$$
$$n^2 + 5 = 6n$$
$$n^2 - 6n + 5 = 0$$
$$(n-1)(n-5) = 0$$
$$n - 1 = 0 \quad \text{or} \quad n - 5 = 0$$
$$n = 1 \quad \text{or} \quad n = 5$$
The number is either 1 or 5.

21. Let x = the number.
$$\frac{12 + x}{41 + 2x} = \frac{1}{3}$$
$$3(41 + 2x)\left(\frac{12 + x}{41 + 2x}\right) = 3(41 + x)\left(\frac{1}{3}\right)$$
$$3(12 + x) = (41 + 2x) \cdot 1$$
$$36 + 3x = 41 + 2x$$
$$x = 5$$
The number is 5.

23. $R = 2$ ohms and $R_1 = 3$ ohms.
$$\frac{1}{R} = \frac{1}{R_1} + \frac{1}{R_2}$$
$$\frac{1}{2} = \frac{1}{3} + \frac{1}{R_2}$$
$$6R_2\left(\frac{1}{2}\right) = 6R_2\left(\frac{1}{3} + \frac{1}{R_2}\right)$$
$$3R_2 = 2R_2 + 6$$
$$R_2 = 6$$
The other resistance is 6 ohms.

25. For three resistances, R_1, R_2, and R_3, wired in a parallel circuit, the combined resistance R is given by

$$\frac{1}{R} = \frac{1}{R_1} + \frac{1}{R_2} + \frac{1}{R_3}$$

$$\frac{1}{R} = \frac{1}{5} + \frac{1}{6} + \frac{1}{2}$$

$$30R\left(\frac{1}{R}\right) = 30R\left(\frac{1}{5} + \frac{1}{6} + \frac{1}{2}\right)$$

$$30 = 6R + 5R + 15R$$

$$30 = 26R$$

$$\frac{30}{26} = R$$

$$\frac{15}{13} = R$$

The combined resistance is $\frac{15}{13}$ ohms.

27. Let x = the number of hours for Steve to complete the work alone.

$$\frac{1}{x} + \frac{1}{6} = \frac{1}{4}$$

$$12x\left(\frac{1}{x} + \frac{1}{6}\right) = 12x\left(\frac{1}{4}\right)$$

$$12 + 2x = 3x$$

$$12 = x$$

Steve would take 12 hours to word process the paper alone.

29. Let x = the number of hours to complete the job if both printers are operating.

$$\frac{1}{4} + \frac{2}{4} = \frac{1}{x}$$

$$4x\left(\frac{1}{4} + \frac{2}{4}\right) = 4x\left(\frac{1}{x}\right)$$

$$x + 2x = 4$$

$$3x = 4$$

$$x = \frac{4}{3} = 1\frac{1}{3}$$

It will take $1\frac{1}{3}$ hours.

31. Let r = the speed of the truck. Then $3r$ = the speed of the plane.

$$d = rt \text{ or } t = \frac{d}{r}$$

$$\frac{450}{r} - 6 = \frac{450}{3r}$$

$$3r\left(\frac{450}{r} - 6\right) = 3r\left(\frac{450}{3r}\right)$$

$$3(450) - 3r(6) = 450$$

$$1350 - 18r = 450$$

$$-18r = -900$$

$$r = 50$$

The truck's speed is 50 miles per hour.

33. Let $r =$ the speed of the current. Then $24 + r =$ the speed of the boat downstream, and $24 - r =$ the speed of the boat upstream.

$$d = rt \text{ or } t = \frac{d}{r}$$

$$\frac{54}{24 - r} = \frac{90}{24 + r}$$

$$(24 - r)(24 + r)\left(\frac{54}{24 - r}\right) = (24 - r)(24 + r)\left(\frac{90}{24 + r}\right)$$

$$54(24 + r) = 90(24 - r)$$

$$1296 + 54r = 2160 - 90r$$

$$144r = 864$$

$$r = 6$$

The speed of the current is 6 miles per hour.

35. Let $n =$ the first odd integer. Then $n + 2 =$ the next odd integer.

$$\frac{1}{n} + \frac{1}{n + 2} = \frac{20}{99}$$

$$99n(n + 2)\left(\frac{1}{n} + \frac{1}{n + 2}\right) = 99n(n + 2)\left(\frac{20}{99}\right)$$

$$99(n + 2) + 99n = 20n(n + 2)$$

$$99n + 198 + 99n = 20n^2 + 40n$$

$$20n^2 - 158n - 198 = 0$$

$$10n^2 - 79n - 99 = 0$$

$$(10n + 11)(n - 9) = 0$$

$$10n + 11 = 0 \quad \text{or} \quad n - 9 = 0$$

$$n = -\frac{11}{10} \quad \text{or} \quad n = 9$$

Disregard the noninteger solution.

$n = 9$

$n + 2 = 11$

The integers are 9 and 11.

37. Let $x =$ the number of hours for Dick to do the job alone.

$$\frac{1}{5} + \frac{1}{x} = \frac{1}{2}$$

$$10x\left(\frac{1}{5} + \frac{1}{x}\right) = 10x\left(\frac{1}{2}\right)$$

$$2x + 10 = 5x$$

$$10 = 3x$$

$$\frac{10}{3} = x$$

$$3\frac{1}{3} = x$$

It takes Dick $3\frac{1}{3}$ hours to do the job alone.

39. Let $r =$ the speed of the bicyclist. Then $r - 10 =$ the speed of the walker.

Recall $d = rt$ or $t = \frac{d}{r}$.

$$\frac{26}{r} = \frac{6}{r - 10}$$

$$r(r - 10)\left(\frac{26}{r}\right) = r(r - 10)\left(\frac{6}{r - 10}\right)$$

$$26(r - 10) = 6r$$

$$26r - 260 = 6r$$

$$20r = 260$$

$$r = 13$$

The bicyclist travels at 13 miles per hour.

41. Let d = the denominator. Then
$d - 4$ = the numerator.

$$\frac{(d-4)+2}{d+2} = \frac{2}{3}$$

$$\frac{d-2}{d+2} = \frac{2}{3}$$

$$3(d+2)\left(\frac{d-2}{d+2}\right) = 3(d+2)\left(\frac{2}{3}\right)$$

$$3(d-2) = 2(d+2)$$

$$3d - 6 = 2d + 4$$

$$d = 10$$

The denominator is 10. The numerator is
$d - 4 = 10 - 4 = 6$. Therefore, the fraction
is $\frac{6}{10}$.

43. Let r = the cyclist's rate during the first
portion.

	Distance = Rate $\cdot$ Time $\left(\frac{d}{r}\right)$		
First portion	20	r	$\frac{20}{r}$
Cool-down portion	16	$r - 2$	$\frac{16}{r-2}$

$$\frac{20}{r} = \frac{16}{r-2}$$

$$r(r-2)\left(\frac{20}{r}\right) = r(r-2)\left(\frac{16}{r-2}\right)$$

$$20(r-2) = 16r$$

$$20r - 40 = 16r$$

$$4r = 40$$

$$r = \frac{40}{4} = 10$$

The first portion is at 10 miles per hour,
and the cool-down portion is at 8 miles
per hour.

45. Let t = the number of hours required to
empty the tank.

$$-\frac{1}{1.5} + \frac{1}{1} = \frac{1}{t}$$

$$1.5t\left(-\frac{1}{1.5} + 1\right) = 1.5t\left(\frac{1}{t}\right)$$

$$-t + 1.5t = 1.5$$

$$0.5t = 1.5$$

$$t = 3$$

It will take 3 hours to empty the tank.

47. Let d = the distance at which the rockets
are an equal distance from Earth.

Recall $d = rt$ or $t = \frac{d}{r}$.

$$\frac{d}{9000} - \frac{1}{4} = \frac{d}{10,000}$$

$$90,000\left(\frac{d}{9000} - \frac{1}{4}\right) = 90,000\left(\frac{d}{10,000}\right)$$

$$10d - 22,500 = 9d$$

$$d = 22,500$$

The distance is 22,500 miles.

49. Let x = the time in hours it takes both surveyors to complete the job together.

	Hours to Complete the Job	Part of Job Completed in 1 Hour
Experienced	4	$\frac{1}{4}$
Apprentice	5	$\frac{1}{5}$
Together	x	$\frac{1}{x}$

$$\frac{1}{4}+\frac{1}{5}=\frac{1}{x}$$
$$20x\left(\frac{1}{4}+\frac{1}{5}\right)=20x\left(\frac{1}{x}\right)$$
$$20x\left(\frac{1}{4}\right)+20x\left(\frac{1}{5}\right)=20$$
$$5x+4x=20$$
$$9x=20$$
$$x=\frac{20}{9}=2\frac{2}{9}$$

It will take $2\frac{2}{9}$ hours to complete the job together.

51. Let x = the number of hours to complete the job together.
$$\frac{1}{3}+\frac{1}{6}=\frac{1}{x}$$
$$6x\left(\frac{1}{3}+\frac{1}{6}\right)=6x\left(\frac{1}{x}\right)$$
$$2x+x=6$$
$$3x=6$$
$$x=2$$
It will take 2 hours.

53. Let t = the number of hours to complete the job together.
$$\frac{1}{6}+\frac{1}{4}=\frac{1}{t}$$
$$12t\left(\frac{1}{6}+\frac{1}{4}\right)=12t\left(\frac{1}{t}\right)$$
$$2t+3t=12$$
$$5t=12$$
$$t=\frac{12}{5}$$
$$45\cdot\frac{12}{5}=108$$
The labor estimate should be \$108.

55. Let t = the number of days to complete the job together.
$$\frac{1}{4}+\frac{1}{5}=\frac{1}{t}$$
$$20t\left(\frac{1}{4}+\frac{1}{5}\right)=20t\left(\frac{1}{t}\right)$$
$$5t+4t=20$$
$$9t=20$$
$$t=\frac{20}{9}=2\frac{2}{9}$$

It will take $2\frac{2}{9}$ days.

57. Let r = the rowing rate. Then
$r + 6$ = the rate of the boat downstream,
and $r - 6$ = the rate of the boat upstream.

Recall $d = rt$ or $t = \dfrac{d}{r}$.

$$\frac{9}{r+6} = \frac{3}{r-6}$$

$$(r-6)(r+6)\left(\frac{9}{r+6}\right) = (r-6)(r+6)\left(\frac{3}{r-6}\right)$$

$$9(r-6) = 3(r+6)$$
$$9r - 54 = 3r + 18$$
$$6r = 72$$
$$r = 12$$

$$t = \frac{\text{distance downstream}}{\text{rate downstream}} + \frac{\text{distance upstream}}{\text{rate upstream}}$$

$$= \frac{9}{12+6} + \frac{3}{12-6}$$

$$= \frac{9}{18} + \frac{3}{6}$$

$$= 1$$

It will take 1 hour.

59.
$$B = \frac{705w}{h^2}$$

$$47 = \frac{705(240)}{h^2}$$

$$h^2 = \frac{705(240)}{47}$$

$$h^2 = 3600$$

$h = 60$ or -60 (disregard)
The patient is 60 inches, or 5 feet, tall.

61.
$$\frac{x}{5} = \frac{x+2}{3}$$

$$15\left(\frac{x}{5}\right) = 15\left(\frac{x+2}{3}\right)$$

$$3x = 5(x+2)$$
$$3x = 5x + 10$$
$$-10 = 2x$$
$$-5 = x$$

The solution is -5.

63.
$$\frac{x-3}{2} = \frac{x-5}{6}$$

$$6\left(\frac{x-3}{2}\right) = 6\left(\frac{x-5}{6}\right)$$

$$3(x-3) = x-5$$
$$3x - 9 = x - 5$$
$$2x = 4$$
$$x = 2$$

The solution is 2.

Exercise Set 6.8

1.　$y = kx$
$$4 = k(20)$$
$$k = \frac{4}{20}$$
$$k = \frac{1}{5}$$
$$y = \frac{1}{5}x$$

3.　$y = kx$
$$6 = k(4)$$
$$k = \frac{6}{4}$$
$$k = \frac{3}{2}$$
$$y = \frac{3}{2}x$$

5.　$y = kx$
$$7 = k\left(\frac{1}{2}\right)$$
$$k = 14$$
$$y = 14x$$

7. $y = kx$
$0.2 = k(0.8)$
$k = 0.25$
$y = 0.25x$

9. $W = kr^3$
$1.2 = k \cdot 2^3$
$k = \dfrac{1.2}{8} = 0.15$
$W = 0.15r^3 = 0.15(3)^3 = 0.15(27) = 4.05$
The weight is 4.05 pounds.

11. $P = kN$
$260,000 = k(450,000)$
$k = \dfrac{260,000}{450,000}$
$k = \dfrac{26}{45}$
$P = \dfrac{26}{45}N = \dfrac{26}{45}(980,000) \approx 566,222$
There are approximately 566,222 tons of pollution.

13. $y = \dfrac{k}{x}$
$6 = \dfrac{k}{5}$
$k = 30$
$y = \dfrac{30}{x}$

15. $y = \dfrac{k}{x}$
$100 = \dfrac{k}{7}$
$k = 700$
$y = \dfrac{700}{x}$

17. $y = \dfrac{k}{x}$
$\dfrac{1}{8} = \dfrac{k}{16}$
$k = 2$
$y = \dfrac{2}{x}$

19. $y = \dfrac{k}{x}$
$0.2 = \dfrac{k}{0.7}$
$k = 0.14$
$y = \dfrac{0.14}{x}$

21. $R = \dfrac{k}{T}$
$45 = \dfrac{k}{6}$
$k = 270$
$R = \dfrac{270}{T} = \dfrac{270}{5} = 54$
The car's speed is 54 miles per hour.

23. $I = \dfrac{k}{R}$
$40 = \dfrac{k}{270}$
$k = 10,800$
$I = \dfrac{10,800}{R} = \dfrac{10,800}{150} = 72$
The current is 72 amperes.

25. $I_1 = \dfrac{k}{d^2}$
Replace d by $2d$.
$I_2 = \dfrac{k}{(2d)^2} = \dfrac{k}{4d^2} = \dfrac{1}{4}I_1$
Thus, the intensity is divided by 4.

27. $x = kyz$

29. $r = kst^3$

31. $W = \dfrac{kwh^2}{l}$
$12 = \dfrac{k\left(\frac{1}{2}\right)\left(\frac{1}{3}\right)^2}{10}$
$120 = \dfrac{1}{18}k$
$k = 2160$
$W = \dfrac{2160wh^2}{l} = \dfrac{2160\left(\frac{2}{3}\right)\left(\frac{1}{2}\right)^2}{16} = \dfrac{45}{2}$ or 22.5

It can support 22.5 tons.

33. $V = kr^2h$

$32\pi = k(4)^2(6)$

$32\pi = 96k$

$k = \dfrac{32\pi}{96}$

$k = \dfrac{\pi}{3}$

$V = \dfrac{\pi}{3}r^2h = \dfrac{\pi}{3}(3)^2(5) = 15\pi$

The volume is 15π cubic inches.

35. $H = ksd^3$

$40 = k(120)(2)^3$

$40 = 960k$

$k = \dfrac{40}{960}$

$k = \dfrac{1}{24}$

$H = \dfrac{1}{24}sd^3 = \dfrac{1}{24}(80)(3)^3 = 90$

90 horsepower can be transmitted.

37. $y = \dfrac{k}{x}$

$400 = \dfrac{k}{8}$

$k = 3200$

$y = \dfrac{3200}{x} = \dfrac{3200}{4} = 800$

The atmospheric pressure is 800 millibars.

39. $V_1 = khr^2$

$V_2 = k\left(\dfrac{1}{2}h\right)(2r)^2 = 2khr^2 = 2V_1$

It is multiplied by 2.

41. $y_1 = kx^2$

$y_2 = k(2x)^2 = 4kx^2 = 4y_1$

It is multiplied by 4.

43.

x	$\frac{1}{4}$	$\frac{1}{2}$	1	2	4
$y = \frac{3}{x}$	12	6	3	$\frac{3}{2}$	$\frac{3}{4}$

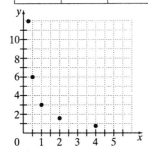

45.

x	$\frac{1}{4}$	$\frac{1}{2}$	1	2	4
$y = \frac{1}{2x}$	2	1	$\frac{1}{2}$	$\frac{1}{4}$	$\frac{1}{8}$

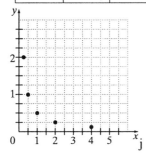

47. $r = 6$
$C = 2\pi r = 2\pi(6) = 12\pi$
$A = \pi r^2 = \pi(6^2) = 36\pi$
The circumference is 12π centimeters and the area is 36π square centimeters.

49. $r = 7$
$C = 2\pi r = 2\pi(7) = 14\pi$
$A = \pi r^2 = \pi(7^2) = 49\pi$
The circumference is 14π meters and the area is 49π square meters.

51. $\sqrt{36} = 6$ since $6^2 = 36$.

53. $\sqrt{4} = 2$ since $2^2 = 4$.

55. $\sqrt{\frac{1}{25}} = \frac{1}{5}$ since $\left(\frac{1}{5}\right)^2 = \frac{1}{25}$.

57. $\sqrt{\frac{25}{121}} = \frac{5}{11}$ since $\left(\frac{5}{11}\right)^2 = \frac{25}{121}$.

Chapter 6 Review

1. $f(x) = \dfrac{3 - 5x}{7}$
Domain: $\{x \mid x \text{ is a real number}\}$

2. $g(x) = \dfrac{2x + 4}{11}$
Domain: $\{x \mid x \text{ is a real number}\}$

3. $F(x) = \dfrac{-3x^2}{x-5}$

Undefined when $x - 5 = 0$.

$x - 5 = 0$

$x = 5$

Domain:

$\{x | x \text{ is a real number and } x \neq 5\}$

4. $h(x) = \dfrac{4x}{3x - 12}$

Undefined when $3x - 12 = 0$.

$3x - 12 = 0$

$3x = 12$

$x = 4$

Domain:

$\{x | x \text{ is a real number and } x \neq 4\}$

5. $f(x) = \dfrac{x^3 + 2}{x^2 + 8x}$

Undefined when $x^2 + 8x = 0$.

$x^2 + 8x = 0$

$x(x + 8) = 0$

$x = 0 \text{ or } x + 8 = 0$

$x = -8$

Domain:

$\{x | x \text{ is a real number and } x \neq 0, \ x \neq -8\}$

6. $G(x) = \dfrac{20}{3x^2 - 48}$

Undefined when $3x^2 - 48 = 0$.

$3x^2 - 48 = 0$

$x^2 - 16 = 0$

$(x + 4)(x - 4) = 0$

$x + 4 = 0 \quad \text{or} \quad x - 4 = 0$

$x = -4 \quad \text{or} \quad x = 4$

Domain:

$\{x | x \text{ is a real number and } x \neq -4, \ x \neq 4\}$

7. $\dfrac{15x^4}{45x^2} = \dfrac{15x^{4-2}}{15 \cdot 3} = \dfrac{x^2}{3}$

8. $\dfrac{x+2}{2+x} = \dfrac{x+2}{x+2} = 1$

9. $\dfrac{18m^6 p^2}{10m^4 p} = \dfrac{2 \cdot 9 m^{6-4} p^{2-1}}{2 \cdot 5} = \dfrac{9m^2 p}{5}$

10. $\dfrac{x-12}{12-x} = \dfrac{-1(12-x)}{12-x} = -1$

11. $\dfrac{5x-15}{25x-75} = \dfrac{5(x-3)}{5 \cdot 5(x-3)} = \dfrac{1}{5}$

12. $\dfrac{22x+8}{11x+4} = \dfrac{2(11x+4)}{11x+4} = 2$

13. $\dfrac{2x}{2x^2 - 2x} = \dfrac{2x}{2x(x-1)} = \dfrac{1}{x-1}$

14. $\dfrac{x+7}{x^2 - 49} = \dfrac{x+7}{(x+7)(x-7)} = \dfrac{1}{x-7}$

15. $\dfrac{2x^2 + 4x - 30}{x^2 + x - 20} = \dfrac{2(x^2 + 2x - 15)}{x^2 + x - 20}$

$= \dfrac{2(x+5)(x-3)}{(x+5)(x-4)}$

$= \dfrac{2(x-3)}{x-4}$

16. $\dfrac{xy - 3x + 2y - 6}{x^2 + 4x + 4} = \dfrac{x(y-3) + 2(y-3)}{(x+2)^2}$

$= \dfrac{(x+2)(y-3)}{(x+2)(x+2)}$

$= \dfrac{y-3}{x+2}$

17. $C(x) = \dfrac{35x + 4200}{x}$

a. $C(50) = \dfrac{35(50) + 4200}{50} = 119$

The cost is \$119 per bookcase.

b. $C(100) = \dfrac{35(100) + 4200}{100} = 77$

The cost is \$77 per bookcase.

c. The average cost per bookcase decreases as the number of bookcases increases.

18. $\dfrac{5}{x^3} \cdot \dfrac{x^2}{15} = \dfrac{1}{3x}$

19. $\dfrac{3x^4yz^3}{15x^2y^2} \cdot \dfrac{10xy}{z^6} = \dfrac{30x^{4+1}y^{1+1}z^3}{15x^2y^2z^6}$

$$= \dfrac{30x^5y^2z^3}{15x^2y^2z^6}$$

$$= \dfrac{2x^{5-2}}{z^{6-3}}$$

$$= \dfrac{2x^3}{z^3}$$

20. $\dfrac{4-x}{5} \cdot \dfrac{15}{2x-8} = \dfrac{-1(x-4)}{5} \cdot \dfrac{15}{2(x-4)}$

$$= -\dfrac{3}{2}$$

21. $\dfrac{x^2-6x+9}{2x^2-18} \cdot \dfrac{4x+12}{5x-15}$

$$= \dfrac{(x-3)^2}{2(x^2-9)} \cdot \dfrac{4(x+3)}{5(x-3)}$$

$$= \dfrac{x-3}{(x+3)(x-3)} \cdot \dfrac{2(x+3)}{5}$$

$$= \dfrac{2}{5}$$

22. $\dfrac{a-4b}{a^2+ab} \cdot \dfrac{b^2-a^2}{8b-2a}$

$$= \dfrac{a-4b}{a(a+b)} \cdot \dfrac{(b-a)(b+a)}{2(4b-a)}$$

$$= \dfrac{-(b-a)}{2a}$$

$$= \dfrac{a-b}{2a}$$

23. $\dfrac{x^2-x-12}{2x^2-32} \cdot \dfrac{x^2+8x+16}{3x^2+21x+36}$

$$= \dfrac{(x-4)(x+3)}{2(x^2-16)} \cdot \dfrac{(x+4)^2}{3(x^2+7x+12)}$$

$$= \dfrac{(x-4)(x+3)(x+4)^2}{6(x+4)(x-4)(x+3)(x+4)}$$

$$= \dfrac{1}{6}$$

24. $\dfrac{2x^3+54}{5x^2+5x-30} \cdot \dfrac{6x+12}{3x^2-9x+27}$

$$= \dfrac{2(x^3+27)}{5(x^2+x-6)} \cdot \dfrac{6(x+2)}{3(x^2-3x+9)}$$

$$= \dfrac{2(x+3)(x^2-3x+9)}{5(x-2)(x+3)} \cdot \dfrac{6(x+2)}{3(x^2-3x+9)}$$

$$= \dfrac{4(x+2)}{5(x-2)}$$

25. $\dfrac{3}{4x} \div \dfrac{8}{2x^2} = \dfrac{3}{4x} \cdot \dfrac{2x^2}{8} = \dfrac{6x^2}{32x} = \dfrac{3x}{16}$

26. $\dfrac{4x+8y}{3} \div \dfrac{5x+10y}{9} = \dfrac{4(x+2y)}{3} \cdot \dfrac{9}{5(x+2y)}$

$$= \dfrac{12}{5}$$

27. $\dfrac{5ab}{14c^3} \div \dfrac{10a^4b^2}{6ac^5} = \dfrac{5ab}{14c^3} \cdot \dfrac{6ac^5}{10a^4b^2}$

$$= \dfrac{30a^2bc^5}{140a^4b^2c^3}$$

$$= \dfrac{3c^{5-3}}{14a^{4-2}b^{2-1}}$$

$$= \dfrac{3c^2}{14a^2b}$$

28. $\dfrac{2}{5x} \div \dfrac{4-18x}{6-27x} = \dfrac{2}{5x} \cdot \dfrac{3(2-9x)}{2(2-9x)}$

$$= \dfrac{3}{5x}$$

29. $\dfrac{x^2-25}{3} \div \dfrac{x^2-10x+25}{x^2-x-20}$

$$= \dfrac{(x+5)(x-5)}{3} \cdot \dfrac{(x-5)(x+4)}{(x-5)^2}$$

$$= \dfrac{(x+5)(x+4)}{3}$$

30. $\dfrac{a-4b}{a^2+ab} \div \dfrac{20b-5a}{b^2-a^2}$

$$= \dfrac{a-4b}{a(a+b)} \cdot \dfrac{(b-a)(b+a)}{5(4b-a)}$$

$$= \dfrac{-(b-a)}{5a}$$

$$= \dfrac{a-b}{5a}$$

31. $\dfrac{7x+28}{2x+4} \div \dfrac{x^2+2x-8}{x^2-2x-8}$

$$= \dfrac{7(x+4)}{2(x+2)} \cdot \dfrac{(x-4)(x+2)}{(x+4)(x-2)}$$

$$= \dfrac{7(x-4)}{2(x-2)}$$

32. $\dfrac{3x+3}{x-1} \div \dfrac{x^2-6x-7}{x^2-1}$

$$= \dfrac{3(x+1)}{x-1} \cdot \dfrac{(x-1)(x+1)}{(x+1)(x-7)}$$

$$= \dfrac{3(x+1)}{x-7}$$

33. $\dfrac{2x-x^2}{x^3-8} \div \dfrac{x^2}{x^2+2x+4}$

$$= \dfrac{x(2-x)}{(x-2)(x^2+2x+4)} \cdot \dfrac{x^2+2x+4}{x^2}$$

$$= \dfrac{-(x-2)}{x-2} \cdot \dfrac{1}{x}$$

$$= -\dfrac{1}{x}$$

34. $\dfrac{5a^2-20}{a^3+2a^2+a+2} \div \dfrac{7a}{a^3+a}$

$$= \dfrac{5(a^2-4)}{a^2(a+2)+(a+2)} \cdot \dfrac{a(a^2+1)}{7a}$$

$$= \dfrac{5(a+2)(a-2)}{(a^2+1)(a+2)} \cdot \dfrac{a(a^2+1)}{7a}$$

$$= \dfrac{5(a-2)}{7}$$

35. $\dfrac{2a}{21} \div \dfrac{3a^2}{7} \cdot \dfrac{4}{a} = \dfrac{2a}{21} \cdot \dfrac{7}{3a^2} \cdot \dfrac{4}{a}$

$$= \dfrac{56a}{63a^3}$$

$$= \dfrac{8}{9a^2}$$

36. $\dfrac{5x-15}{3-x} \cdot \dfrac{x+2}{10x+20} \cdot \dfrac{x^2-9}{x^2-x-6}$

$$= \dfrac{5(x-3)}{3-x} \cdot \dfrac{x+2}{10(x+2)} \cdot \dfrac{(x+3)(x-3)}{(x+2)(x-3)}$$

$$= -\dfrac{x+3}{2(x+2)}$$

37. $\dfrac{4a+8}{5a^2-20} \cdot \dfrac{3a^2-6a}{a+3} \div \dfrac{2a^2}{5a+15}$

$$= \dfrac{4(a+2)}{5(a^2-4)} \cdot \dfrac{3a(a-2)}{a+3} \cdot \dfrac{5(a+3)}{2a^2}$$

$$= \dfrac{2(a+2)}{(a+2)(a-2)} \cdot \dfrac{3(a-2)}{1} \cdot \dfrac{1}{a}$$

$$= \dfrac{6}{a}$$

38. $\dfrac{4}{9}, \dfrac{5}{2}$

The LCD is 18.

39. $\dfrac{5}{4x^2y^5}, \dfrac{3}{10x^2y^4}, \dfrac{x}{6y^4}$

The LCD is $60x^2y^5$.

40. $\dfrac{5}{2x}, \dfrac{7}{x-2}$

The LCD is $2x(x-2)$.

41. $\dfrac{3}{5x}, \dfrac{2}{x-5}$

The LCD is $5x(x-5)$.

42. $\dfrac{1}{5x^3}, \dfrac{4}{(x+7)(x-4)}, \dfrac{11}{10x(x-3)}$

The LCD is $10x^3(x-4)(x+7)(x-3)$.

43. $\dfrac{2}{15} + \dfrac{4}{15} = \dfrac{2+4}{15} = \dfrac{6}{15} = \dfrac{2}{5}$

44. $\dfrac{4}{x-4} + \dfrac{x}{x-4} = \dfrac{4+x}{x-4}$

45. $\dfrac{4}{3x^2} + \dfrac{2}{3x^2} = \dfrac{4+2}{3x^2} = \dfrac{6}{3x^2} = \dfrac{2}{x^2}$

46. $\dfrac{1}{x-2} - \dfrac{1}{4-2x} = \dfrac{1}{x-2} - \dfrac{1}{2(2-x)}$

$= \dfrac{1(2)}{(x-2)(2)} + \dfrac{1}{2(x-2)}$

$= \dfrac{2+1}{2(x-2)}$

$= \dfrac{3}{2(x-2)}$

47. $\dfrac{2x+1}{x^2+x-6} + \dfrac{2-x}{x^2+x-6} = \dfrac{(2x+1)+(2-x)}{x^2+x-6}$

$= \dfrac{x+3}{(x+3)(x-2)}$

$= \dfrac{1}{x-2}$

48. $\dfrac{7}{2x} + \dfrac{5}{6x} = \dfrac{7\cdot3}{2x\cdot3} + \dfrac{5}{6x}$

$= \dfrac{21}{6x} + \dfrac{5}{6x}$

$= \dfrac{26}{6x}$

$= \dfrac{13}{3x}$

49. $\dfrac{1}{3x^2y^3} - \dfrac{1}{5x^4y}$

$= \dfrac{1\cdot5x^2}{3x^2y^3\cdot5x^2} - \dfrac{1\cdot3y^2}{5x^4y\cdot3y^2}$

$= \dfrac{5x^2}{15x^4y^3} - \dfrac{3y^2}{15x^4y^3}$

$= \dfrac{5x^2-3y^2}{15x^4y^3}$

50. $\dfrac{1}{10-x} + \dfrac{x-1}{x-10} = \dfrac{-1}{x-10} + \dfrac{x-1}{x-10}$

$= \dfrac{x-2}{x-10}$

51. $\dfrac{x-2}{x+1} - \dfrac{x-3}{x-1}$

$= \dfrac{(x-2)(x-1)-(x-3)(x+1)}{(x+1)(x-1)}$

$= \dfrac{(x^2-3x+2)-(x^2-2x-3)}{(x+1)(x-1)}$

$= \dfrac{x^2-3x+2-x^2+2x+3}{(x+1)(x-1)}$

$= \dfrac{-x+5}{(x+1)(x-1)}$

52. $\dfrac{x}{9-x^2} - \dfrac{2}{5x-15}$

$= \dfrac{x}{(3-x)(3+x)} - \dfrac{2}{5(x-3)}$

$= \dfrac{-5x}{5(x-3)(x+3)} - \dfrac{2(x+3)}{5(x-3)(x+3)}$

$= \dfrac{-5x-2x-6}{5(x-3)(x+3)}$

$= \dfrac{-7x-6}{5(x-3)(x+3)}$

53. $2x+1 - \dfrac{1}{x-3} = \dfrac{(2x+1)(x-3)-1}{x-3}$

$= \dfrac{2x^2-5x-3-1}{x-3}$

$= \dfrac{2x^2-5x-4}{x-3}$

54. $\dfrac{2}{a^2-2a+1} + \dfrac{3}{a^2-1}$

$= \dfrac{2}{(a-1)^2} + \dfrac{3}{(a+1)(a-1)}$

$= \dfrac{2(a+1)}{(a-1)^2(a+1)} + \dfrac{3(a-1)}{(a-1)^2(a+1)}$

$= \dfrac{2a+2+3a-3}{(a-1)^2(a+1)}$

$= \dfrac{5a-1}{(a-1)^2(a+1)}$

55. $\dfrac{x}{9x^2+12x+16} - \dfrac{3x+4}{27x^3-64}$

$= \dfrac{x}{9x^2+12x+16} - \dfrac{3x+4}{(3x-4)(9x^2+12x+16)}$

$= \dfrac{(3x-4)x-(3x+4)}{(3x-4)(9x^2+12x+16)}$

$= \dfrac{3x^2-4x-3x-4}{(3x-4)(9x^2+12x+16)}$

$= \dfrac{3x^2-7x-4}{(3x-4)(9x^2+12x+16)}$

56. $\dfrac{2}{x-1} - \dfrac{3x}{3x-3} + \dfrac{1}{2x-2}$

$= \dfrac{2}{x-1} - \dfrac{3x}{3(x-1)} + \dfrac{1}{2(x-1)}$

$= \dfrac{2}{x-1} - \dfrac{x}{x-1} + \dfrac{1}{2(x-1)}$

$= \dfrac{4}{2(x-1)} - \dfrac{2x}{2(x-1)} + \dfrac{1}{2(x-1)}$

$= \dfrac{5-2x}{2(x-1)}$

57. $\dfrac{3}{2x}\left(\dfrac{2}{x+1} - \dfrac{2}{x-3}\right)$

$= \dfrac{3}{2x}\left[2\left(\dfrac{1}{x+1} - \dfrac{1}{x-3}\right)\right]$

$= \dfrac{3}{x}\left(\dfrac{1}{x+1} - \dfrac{1}{x-3}\right)$

$= \dfrac{3}{x}\left[\dfrac{(x-3)-(x+1)}{(x+1)(x-3)}\right]$

$= \dfrac{3}{x}\cdot\dfrac{-4}{(x+1)(x-3)}$

$= -\dfrac{12}{x(x+1)(x-3)}$

58. $\left(\dfrac{2}{x} - \dfrac{1}{5}\right)\cdot\left(\dfrac{2}{x} + \dfrac{1}{3}\right) = \left(\dfrac{10-x}{5x}\right)\cdot\left(\dfrac{6+x}{3x}\right)$

$= \dfrac{(10-x)(6+x)}{15x^2}$

59. $\dfrac{2}{x^2-16}-\dfrac{3x}{x^2+8x+16}+\dfrac{3}{x+4}=\dfrac{2}{(x+4)(x-4)}-\dfrac{3x}{(x+4)^2}+\dfrac{3}{x+4}$

$$=\dfrac{2(x+4)}{(x+4)^2(x-4)}-\dfrac{3x(x-4)}{(x+4)^2(x-4)}+\dfrac{3(x+4)(x-4)}{(x+4)^2(x-4)}$$

$$=\dfrac{2x+8-3x^2+12x+3x^2-48}{(x+4)^2(x-4)}$$

$$=\dfrac{14x-40}{(x+4)^2(x-4)}$$

60. $P=\dfrac{1}{x}+\dfrac{1}{x}+\dfrac{1}{x}+\dfrac{2}{x}+\dfrac{5}{2x}+\dfrac{2}{x}+\dfrac{3}{2x}$

$$=\dfrac{1+1+1+2+2}{x}+\dfrac{5+3}{2x}$$

$$=\dfrac{7}{x}+\dfrac{8}{2x}$$

$$=\dfrac{7}{x}+\dfrac{4}{x}$$

$$=\dfrac{11}{x}$$

The perimeter is $\dfrac{11}{x}$.

61. $\dfrac{\frac{2}{5}}{\frac{3}{5}}=\dfrac{2}{5}\cdot\dfrac{5}{3}=\dfrac{2}{1}\cdot\dfrac{1}{3}=\dfrac{2}{3}$

62. $\dfrac{1-\frac{3}{4}}{2+\frac{1}{4}}=\dfrac{\left(1-\frac{3}{4}\right)(4)}{\left(2+\frac{1}{4}\right)(4)}=\dfrac{4-3}{8+1}=\dfrac{1}{9}$

63. $\dfrac{\frac{1}{x}-\frac{2}{3x}}{\frac{5}{2x}-\frac{1}{3}}=\dfrac{6x\left(\frac{1}{x}-\frac{2}{3x}\right)}{6x\left(\frac{5}{2x}-\frac{1}{3}\right)}$

$$=\dfrac{6-4}{15-2x}$$

$$=\dfrac{2}{15-2x}$$

64. $\dfrac{\frac{x^2}{15}}{\frac{x+1}{5x}}=\dfrac{x^2}{15}\cdot\dfrac{5x}{x+1}=\dfrac{x^3}{3(x+1)}$

65. $\dfrac{\frac{3}{y^2}}{\frac{6}{y^3}}=\dfrac{3}{y^2}\cdot\dfrac{y^3}{6}=\dfrac{y}{2}$

66. $\dfrac{\frac{x+2}{3}}{\frac{5}{x-2}}=\dfrac{x+2}{3}\cdot\dfrac{x-2}{5}=\dfrac{(x+2)(x-2)}{15}$

67. $\dfrac{2-\frac{3}{2x}}{x-\frac{2}{5x}}=\dfrac{10x\left(2-\frac{3}{2x}\right)}{10x\left(x-\frac{2}{5x}\right)}=\dfrac{20x-15}{10x^2-4}$

68. $\dfrac{1+\frac{x}{y}}{\frac{x^2}{y^2}-1}=\dfrac{\left(1+\frac{x}{y}\right)\cdot y^2}{\left(\frac{x^2}{y^2}-1\right)\cdot y^2}$

$$=\dfrac{y^2+xy}{x^2-y^2}$$

$$=\dfrac{y(y+x)}{(x+y)(x-y)}$$

$$=\dfrac{y}{x-y}$$

69. $\dfrac{\frac{5}{x}+\frac{1}{xy}}{\frac{3}{x^2}}=\dfrac{\left(\frac{5}{x}+\frac{1}{xy}\right)x^2y}{\left(\frac{3}{x^2}\right)x^2y}=\dfrac{5xy+x}{3y}$

70. $\dfrac{\frac{x}{3}-\frac{3}{x}}{1+\frac{3}{x}}=\dfrac{\left(\frac{x}{3}-\frac{3}{x}\right)3x}{\left(1+\frac{3}{x}\right)3x}$

$$=\dfrac{x^2-9}{3x+9}$$

$$=\dfrac{(x+3)(x-3)}{3(x+3)}$$

$$=\dfrac{x-3}{3}$$

71.
$$\frac{\frac{1}{x-1}+1}{\frac{1}{x+1}-1} = \frac{\frac{1+(x-1)}{x-1}}{\frac{1-(x+1)}{x+1}}$$

$$= \frac{\frac{x}{x-1}}{\frac{-x}{x+1}}$$

$$= \frac{x}{x-1} \cdot \frac{x+1}{-x}$$

$$= \frac{x(x+1)}{-x(x-1)}$$

$$= \frac{x+1}{-(x-1)}$$

$$= \frac{1+x}{1-x}$$

72. $\dfrac{2}{1-\dfrac{2}{x}} = \dfrac{(2)x}{\left(1-\frac{2}{x}\right)x} = \dfrac{2x}{x-2}$

73.
$$\frac{1}{1+\frac{2}{1-\frac{1}{x}}} = \frac{1}{1+\frac{2x}{\left(1-\frac{1}{x}\right)x}}$$

$$= \frac{1}{1+\frac{2x}{x-1}}$$

$$= \frac{1}{\frac{(x-1)+2x}{x-1}}$$

$$= \frac{1}{\frac{3x-1}{x-1}}$$

$$= \frac{x-1}{3x-1}$$

74.
$$\frac{\frac{x^2+5x-6}{4x+3}}{\frac{(x+6)^2}{8x+6}} = \frac{(x-1)(x+6)}{4x+3} \cdot \frac{2(4x+3)}{(x+6)^2}$$

$$= \frac{2(x-1)}{x+6}$$

75.
$$\frac{\frac{x-3}{x+3}+\frac{x+3}{x-3}}{\frac{x-3}{x+3}-\frac{x+3}{x-3}} = \frac{\left(\frac{x-3}{x+3}+\frac{x+3}{x-3}\right)(x+3)(x-3)}{\left(\frac{x-3}{x+3}-\frac{x+3}{x-3}\right)(x+3)(x-3)}$$

$$= \frac{(x-3)^2+(x+3)^2}{(x-3)^2-(x+3)^2}$$

$$= \frac{(x^2-6x+9)+(x^2+6x+9)}{(x^2-6x+9)-(x^2+6x+9)}$$

$$= \frac{2x^2+18}{x^2-6x+9-x^2-6x-9}$$

$$= \frac{2x^2+18}{-12x}$$

$$= \frac{2(x^2+9)}{2(-6x)}$$

$$= -\frac{x^2+9}{6x}$$

76.
$$\frac{\frac{3}{x-1}-\frac{2}{1-x}}{\frac{2}{x-1}-\frac{2}{x}} = \frac{\left(\frac{3}{x-1}-\frac{2}{1-x}\right)[x(x-1)]}{\left(\frac{2}{x-1}-\frac{2}{x}\right)[x(x-1)]}$$

$$= \frac{3x+2x}{2x-2(x-1)}$$

$$\frac{5x}{2x-2x+2}$$

$$= \frac{5x}{2}$$

77. $f(x) = \dfrac{3}{x}$

   a. $f(a+h) = \dfrac{3}{a+h}$

   b. $f(a) = \dfrac{3}{a}$

   c. $\dfrac{f(a+h)-f(a)}{h} = \dfrac{\frac{3}{a+h}-\frac{3}{a}}{h}$

d. $\dfrac{\frac{3}{a+h}-\frac{3}{a}}{h}=\dfrac{\left(\frac{3}{a+h}-\frac{3}{a}\right)[a(a+h)]}{h[a(a+h)]}$

$\qquad = \dfrac{3a-3(a+h)}{ah(a+h)}$

$\qquad = \dfrac{3a-3a-3h}{ah(a+h)}$

$\qquad = \dfrac{-3h}{ah(a+h)}$

$\qquad = \dfrac{-3}{a(a+h)}$

78. $\dfrac{3x^5yb^9}{9xy^7}=\dfrac{x^{5-1}b^9}{3y^{7-1}}=\dfrac{x^4b^9}{3y^6}$

79. $\dfrac{-9xb^4z^3}{-4axb^2}=\dfrac{9x^{1-1}b^{4-2}z^3}{4a}=\dfrac{9b^2z^3}{4a}$

80. $\dfrac{4xy+2x^2-9}{4xy}=\dfrac{4xy}{4xy}+\dfrac{2x^2}{4xy}-\dfrac{9}{4xy}$

$\qquad = 1+\dfrac{x^{2-1}}{2y}-\dfrac{9}{4xy}$

$\qquad = 1+\dfrac{x}{2y}-\dfrac{9}{4xy}$

81. $\dfrac{12xb^2+16xb^4}{4xb^3}$

$\qquad = \dfrac{12xb^2}{4xb^3}+\dfrac{16xb^4}{4xb^3}$

$\qquad = \dfrac{3\cdot 4x^{1-1}}{4b^{3-2}}+\dfrac{4\cdot 4x^{1-1}b^{4-3}}{4}$

$\qquad = \dfrac{3}{b}+4b$

82.

$$
\begin{array}{r}
3x^3+\ 9x^2+2x+\ 6 \\
x-3\,\overline{\smash{\big)}\,3x^4+0x^3-25x^2+0x-20} \\
\underline{3x^4-9x^3} \\
9x^3-25x^2 \\
\underline{9x^3-27x^2} \\
2x^2+0x \\
\underline{2x^2-6x} \\
6x-20 \\
\underline{6x-18} \\
-\ 2
\end{array}
$$

Answer: $3x^3+9x^2+2x+6-\dfrac{2}{x-3}$

83.

$$
\begin{array}{r}
2x^3-4x^2+7x-\ 9 \\
x+2\,\overline{\smash{\big)}\,2x^4+0x^3-\ x^2+\ 5x-12} \\
\underline{2x^4+4x^3} \\
-4x^3-\ x^2 \\
\underline{-4x^3-8x^2} \\
7x^2+\ 5x \\
\underline{7x^2+14x} \\
-\ 9x-12 \\
\underline{-\ 9x-18} \\
6
\end{array}
$$

Answer: $2x^3-4x^2+7x-9+\dfrac{6}{x+2}$

84.

$$
\begin{array}{r}
2x^3+2x-2 \\
x-\tfrac{1}{2}\,\overline{\smash{\big)}\,2x^4-x^3+2x^2-3x+1} \\
\underline{2x^4-x^3} \\
2x^2-3x \\
\underline{2x^2-\ x} \\
-2x+1 \\
\underline{-2x+1} \\
0
\end{array}
$$

Answer: $2x^3+2x-2$

85.
$$x+\tfrac{3}{2}\overline{)\,2x^3+3x^2-2x+2\,}$$

$$\begin{array}{r} 2x^2 -2 \\ \underline{2x^3+3x^2} \\ -2x+2 \\ \underline{-2x-3} \\ 5 \end{array}$$

Answer: $2x^2-2+\dfrac{5}{x+\frac{3}{2}}$

86.
$$x^2+x+2\overline{)\,3x^4+5x^3+7x^2+3x-2\,}$$

$$\begin{array}{r} 3x^2+2x-1 \\ \underline{3x^4+3x^3+6x^2} \\ 2x^3+x^2+3x \\ \underline{2x^3+2x^2+4x} \\ -x^2-x-2 \\ \underline{-x^2-x-2} \\ 0 \end{array}$$

Answer: $3x^2+2x-1$

87.
$$3x^2-2x-5\overline{)\,9x^4-6x^3+3x^2-12x-30\,}$$

$$\begin{array}{r} 3x^2+6 \\ \underline{9x^4-6x^3-15x^2} \\ 18x^2-12x-30 \\ \underline{18x^2-12x-30} \\ 0 \end{array}$$

Answer: $3x^2+6$

88.

$$\begin{array}{r|rrrr} 2 & 3 & 0 & 12 & -4 \\ & & 6 & 12 & 48 \\ \hline & 3 & 6 & 24 & 44 \end{array}$$

$(3x^3+12x-4)\div(x-2)$

$=3x^2+6x+24+\dfrac{44}{x-2}$

89.

$$\begin{array}{r|rrrr} -\tfrac{3}{2} & 3 & 2 & -4 & -1 \\ & & -\tfrac{9}{2} & \tfrac{15}{4} & \tfrac{3}{8} \\ \hline & 3 & -\tfrac{5}{2} & -\tfrac{1}{4} & -\tfrac{5}{8} \end{array}$$

$(3x^3+2x^2-4x-1)\div\left(x+\dfrac{3}{2}\right)$

$=3x^2-\dfrac{5}{2}x-\dfrac{1}{4}-\dfrac{5}{8\left(x+\frac{3}{2}\right)}$

90.

$$\begin{array}{r|rrrrrr} -1 & 1 & 0 & 0 & 0 & 0 & -1 \\ & & -1 & 1 & -1 & 1 & -1 \\ \hline & 1 & -1 & 1 & -1 & 1 & -2 \end{array}$$

$(x^5-1)\div(x+1)$

$=x^4-x^3+x^2-x+1-\dfrac{2}{x+1}$

91.

$$\begin{array}{r|rrrr} 3 & 1 & 0 & 0 & -81 \\ & & 3 & 9 & 27 \\ \hline & 1 & 3 & 9 & -54 \end{array}$$

$(x^3-81)\div(x-3)$

$=x^2+3x+9-\dfrac{54}{x-3}$

92.

$$\begin{array}{r|rrrrr} 4 & 3 & 1 & -1 & 0 & -2 \\ & & 12 & 52 & 204 & 816 \\ \hline & 3 & 13 & 51 & 204 & 814 \end{array}$$

$(x^3-x^2+3x^4-2)\div(x-4)$

$=3x^3+13x^2+51x+204+\dfrac{814}{x-4}$

93.

$$\begin{array}{r|rrrrr} -2 & 3 & 0 & -2 & 0 & 10 \\ & & -6 & 12 & -20 & 40 \\ \hline & 3 & -6 & 10 & -20 & 50 \end{array}$$

$(3x^4-2x^2+10)\div(x+2)$

$=3x^3-6x^2+10x-20+\dfrac{50}{x+2}$

94.

$$\begin{array}{r|rrrrrr} 4 & 3 & 0 & 0 & 0 & -9 & 7 \\ & & 12 & 48 & 192 & 768 & 3036 \\ \hline & 3 & 12 & 48 & 192 & 759 & 3043 \end{array}$$

$P(4) = 3043$

95.

$$\begin{array}{r|rrrrrr} -5 & 3 & 0 & 0 & 0 & -9 & 7 \\ & & -15 & 75 & -375 & 1875 & -9330 \\ \hline & 3 & -15 & 75 & -375 & 1866 & -9323 \end{array}$$

$P(-5) = -9323$

96.

$$\begin{array}{r|rrrrrr} \frac{2}{3} & 3 & 0 & 0 & 0 & -9 & 7 \\ & & 2 & \frac{4}{3} & \frac{8}{9} & \frac{16}{27} & -\frac{454}{81} \\ \hline & 3 & 2 & \frac{4}{3} & \frac{8}{9} & -\frac{227}{27} & \frac{113}{81} \end{array}$$

$P\left(\dfrac{2}{3}\right) = \dfrac{113}{81}$

97.

$$\begin{array}{r|rrrrrr} -\frac{1}{2} & 3 & 0 & 0 & 0 & -9 & 7 \\ & & -\frac{3}{2} & \frac{3}{4} & -\frac{3}{8} & \frac{3}{16} & \frac{141}{32} \\ \hline & 3 & -\frac{3}{2} & \frac{3}{4} & -\frac{3}{8} & -\frac{141}{16} & \frac{365}{32} \end{array}$$

$P\left(-\dfrac{1}{2}\right) = \dfrac{365}{32}$

98.

$$\begin{array}{r|rrrrr} 3 & 1 & -1 & -6 & -6 & 18 \\ & & 3 & 6 & 0 & -18 \\ \hline & 1 & 2 & 0 & -6 & 0 \end{array}$$

The length of the rectangle is $(x^3 + 2x^2 - 6)$ miles.

99.

$$\frac{2}{5} = \frac{x}{15}$$

$$15\left(\frac{2}{5}\right) = 15\left(\frac{x}{15}\right)$$

$$6 = x$$

The solution is 6.

100.

$$\frac{3}{x} + \frac{1}{3} = \frac{5}{x}$$

$$\left(\frac{3}{x} + \frac{1}{3}\right)3x = \left(\frac{5}{x}\right)3x$$

$$9 + x = 15$$

$$x = 6$$

The solution is 6.

101.

$$4 + \frac{8}{x} = 8$$

$$\frac{8}{x} = 4$$

$$x\left(4 + \frac{8}{x}\right) = x \cdot 8$$

$$4x + 8 = 8x$$

$$8 = 4x$$

$$2 = x$$

The solution is 2.

102.

$$\frac{2x + 3}{5x - 9} = \frac{3}{2}$$

$$2(5x - 9)\left(\frac{2x + 3}{5x - 9}\right) = 2(5x - 9)\left(\frac{3}{2}\right)$$

$$2(2x + 3) = 3(5x - 9)$$

$$4x + 6 = 15x - 27$$

$$33 = 11x$$

$$3 = x$$

The solution is 3.

103.
$$\frac{1}{x-2} - \frac{3x}{x^2-4} = \frac{2}{x+2}$$

$$(x-2)(x+2)\left[\frac{1}{x-2} - \frac{3x}{(x+2)(x-2)}\right] = (x-2)(x+2)\left(\frac{2}{x+2}\right)$$

$$(x+2) - 3x = 2(x-2)$$
$$-2x + 2 = 2x - 4$$
$$6 = 4x$$
$$\frac{6}{4} = x$$
$$\frac{3}{2} = x$$

The solution is $\frac{3}{2}$.

104.
$$\frac{7}{x} - \frac{x}{7} = 0$$

$$7x\left(\frac{7}{x} - \frac{x}{7}\right) = 7x \cdot 0$$
$$49 - x^2 = 0$$
$$(7-x)(7+x) = 0$$
$$7 - x = 0 \quad \text{or} \quad 7 + x = 0$$
$$7 = x \quad \text{or} \qquad x = -7$$

The solutions are -7 and 7.

105.
$$\frac{x-2}{x^2-7x+10} = \frac{1}{5x-10} - \frac{1}{x-5}$$

$$\frac{x-2}{(x-2)(x-5)} = \frac{1}{5(x-2)} - \frac{1}{x-5}$$

$$5(x-2)(x-5)\left[\frac{x-2}{(x-2)(x-5)}\right] = 5(x-2)(x-5)\left[\frac{1}{5(x-2)} - \frac{1}{x-5}\right]$$

$$5(x-2) = (x-5) - 5(x-2)$$
$$5x - 10 = x - 5 - 5x + 10$$
$$5x - 10 = -4x + 5$$
$$9x = 15$$
$$x = \frac{15}{9} = \frac{5}{3}$$

The solution is $\frac{5}{3}$.

106. $\dfrac{5}{x^2-7x} + \dfrac{4}{2x-14} = \dfrac{5}{x(x-7)} + \dfrac{4}{2(x-7)} = \dfrac{10}{2x(x-7)} + \dfrac{4x}{2x(x-7)} = \dfrac{4x+10}{2x(x-7)} = \dfrac{2(2x+5)}{2x(x-7)} = \dfrac{2x+5}{x(x-7)}$

107.
$$3 - \frac{5}{x} - \frac{2}{x^2} = 0$$
$$x^2 \left(3 - \frac{5}{x} - \frac{2}{x^2} \right) = x^2 \cdot 0$$
$$3x^2 - 5x - 2 = 0$$
$$(3x + 1)(x - 2) = 0$$
$$3x + 1 = 0 \quad \text{or} \quad x - 2 = 0$$
$$3x = -1 \quad \text{or} \quad x = 2$$
$$x = -\frac{1}{3} \quad \text{or} \quad x = 2$$

The solutions are $-\frac{1}{3}$ and 2.

108.
$$\frac{4}{3 - x} - \frac{7}{2x - 6} + \frac{5}{x}$$
$$= \frac{-4}{x - 3} - \frac{7}{2(x - 3)} + \frac{5}{x}$$
$$= \frac{-8x}{2x(x - 3)} - \frac{7x}{2x(x - 3)} + \frac{10(x - 3)}{2x(x - 3)}$$
$$= \frac{-8x - 7x + 10x - 30}{2x(x - 3)}$$
$$= \frac{-5x - 30}{2x(x - 3)}$$

109.
$$A = \frac{h(a + b)}{2}$$
$$2A = ha + hb$$
$$2A - hb = ha$$
$$\frac{2A - hb}{h} = a$$

110.
$$\frac{1}{R} = \frac{1}{R_1} + \frac{1}{R_2}$$
$$(RR_1R_2)\frac{1}{R} = \left(\frac{1}{R_1} + \frac{1}{R_2} \right)(RR_1R_2)$$
$$R_1R_2 = RR_2 + RR_1$$
$$R_1R_2 - RR_2 = RR_1$$
$$R_2(R_1 - R) = RR_1$$
$$R_2 = \frac{RR_1}{R_1 - R}$$

111.
$$I = \frac{E}{R + r}$$
$$(R + r)I = \frac{E}{R + r}(R + r)$$
$$IR + Ir = E$$
$$IR = E - Ir$$
$$R = \frac{E - Ir}{I}$$

112.
$$A = P + Prt$$
$$A - P = Prt$$
$$\frac{A - P}{Pt} = r$$

113.
$$H = \frac{kA(T_1 - T_2)}{L}$$
$$HL = kA(T_1 - T_2)$$
$$\frac{HL}{k(T_1 - T_2)} = A$$

114. Let x = the number.
$$x + 2\left(\frac{1}{x} \right) = 3$$
$$\left(x + \frac{2}{x} \right)x = 3 \cdot x$$
$$x^2 + 2 = 3x$$
$$x^2 - 3x + 2 = 0$$
$$(x - 1)(x - 2) = 0$$
$$x - 1 = 0 \quad \text{or} \quad x - 2 = 0$$
$$x = 1 \quad \text{or} \quad x = 2$$
The numbers are 1 and 2.

115. Let x = the number added to the numerator, so $2x$ = the number added to the denominator.
$$\frac{3 + x}{7 + 2x} = \frac{10}{21}$$
$$21(7 + 2x)\left(\frac{3 + x}{7 + 2x} \right) = 21(7 + 2x)\left(\frac{10}{21} \right)$$
$$21(3 + x) = 10(7 + 2x)$$
$$63 + 21x = 70 + 20x$$
$$x = 7$$
7 is the number to be added to the numerator.

116. Let x = the numerator of the fraction, so $x + 2$ = the denominator of the fraction.

$$\frac{x-3}{(x+2)+5} = \frac{2}{3}$$

$$\frac{x-3}{x+7} = \frac{2}{3}$$

$$3(x+7)\left(\frac{x-3}{x+7}\right) = 3(x+7)\left(\frac{2}{3}\right)$$

$$3(x-3) = 2(x+7)$$

$$3x - 9 = 2x + 14$$

$$x = 23$$

The original fraction was $\frac{23}{25}$.

117. Let n and $n + 2$ represent two consecutive even integers.

$$\frac{1}{n} + \frac{1}{n+2} = -\frac{9}{40}$$

$$40n(n+2)\left(\frac{1}{n} + \frac{1}{n+2}\right) = 40n(n+2)\left(-\frac{9}{40}\right)$$

$$40(n+2) + 40n = -9n(n+2)$$

$$80n + 80 = -9n^2 - 18n$$

$$9n^2 + 98n + 80 = 0$$

$$(9n+8)(n+10) = 0$$

$$9n + 8 = 0 \quad \text{or} \quad n + 10 = 0$$

$$9n = -8 \quad \text{or} \quad n = -10$$

$$n = -\frac{8}{9} \quad \text{or} \quad n = -10$$

Discarding the non-integer solution, $-\frac{8}{9}$, we find the two consecutive even integers to be -10 and -8.

118. Let t = the time in hours to paint the fence together.

$$\frac{1}{4} + \frac{1}{5} + \frac{1}{6} = \frac{1}{t}$$

$$\left(\frac{1}{4} + \frac{1}{5} + \frac{1}{6}\right)60t = \left(\frac{1}{t}\right)60t$$

$$15t + 12t + 10t = 60$$

$$37t = 60$$

$$t = \frac{60}{37} = 1\frac{23}{37}$$

It will take all three boys $1\frac{23}{37}$ hours.

119. Let n = the number of hours required for Tom to type the mailing labels.

$$\frac{1}{6} + \frac{1}{n} = \frac{1}{4}$$

$$12n\left(\frac{1}{6} + \frac{1}{n}\right) = 12n\left(\frac{1}{4}\right)$$

$$2n + 12 = 3n$$

$$12 = n$$

Therefore, Tom can type the mailing labels in 12 hours.

120. Let t = the time in hours for the pipes to empty the tank together.

$$\frac{1}{2} - \frac{1}{2.5} = \frac{1}{t}$$

$$\left(\frac{1}{2} - \frac{1}{2.5}\right)10t = \frac{1}{t} \cdot 10t$$

$$5t - 4t = 10$$

$$t = 10$$

It will take 10 hours to empty the tank.

121. Let r = the speed of the car, so $r + 430$ = the speed of the jet.

Recall that $d = rt$ or $t = \dfrac{d}{r}$.

$$\frac{210}{r} = \frac{1715}{r+430}$$

$$r(r+430)\left(\frac{210}{r}\right) = r(r+430)\left(\frac{1715}{r+430}\right)$$

$$210(r+430) = 1715r$$

$$1715r = 210r + 90,300$$

$$1505r = 90,300$$

$$r = 60$$

$$r + 430 = 490$$

Thus, the speed of the jet is 490 miles per hour.

122.

$$\frac{1}{R} = \frac{1}{R_1} + \frac{1}{R_2}$$

$$\frac{1}{\frac{30}{11}} = \frac{1}{5} + \frac{1}{R_2}$$

$$\frac{11}{30} = \frac{1}{5} + \frac{1}{R_2}$$

$$\left(\frac{11}{30}\right)30R_2 = \left(\frac{1}{5} + \frac{1}{R_2}\right)30R_2$$

$$11R_2 = 6R_2 + 30$$

$$5R_2 = 30$$

$$R_2 = 6$$

The resistance of the other resistor is 6 ohms.

123. Let r = the speed of the current. Recall $d = rt$ or $t = \dfrac{d}{r}$.

$$\frac{72}{32 - r} = \frac{120}{32 + r}$$

$$(32 - r)(32 + r)\left(\frac{72}{32 - r}\right) = (32 - r)(32 + r)\left(\frac{120}{32 + r}\right)$$

$$72(32 + r) = 120(32 - r)$$

$$2304 + 72r = 3840 - 120r$$

$$192r = 1536$$

$$r = 8$$

Therefore, the speed of the current is 8 miles per hour.

124. Let x = the speed of the wind.

	Distance = Rate · Time $\left(\frac{d}{r}\right)$		
With the wind	445	$400 + x$	$\frac{445}{400+x}$
Against the wind	355	$400 - x$	$\frac{355}{400-x}$

$$\frac{445}{400 + x} = \frac{355}{400 - x}$$

$$(400 + x)(400 - x)\left(\frac{445}{400 + x}\right) = (400 + x)(400 - x)\left(\frac{355}{400 - x}\right)$$

$$445(400 - x) = 355(400 + x)$$

$$178,000 - 445x = 142,000 + 355x$$

$$36,000 = 800x$$

$$45 = x$$

The speed of the wind is 45 miles per hour.

125. Let r = the speed of the walker, so
$r + 3$ = the speed of the jogger.

Recall that $d = rt$ or $t = \dfrac{d}{r}$.

$$\frac{14}{r+3} = \frac{8}{r}$$

$$r(r+3)\left(\frac{14}{r+3}\right) = r(r+3)\left(\frac{8}{r}\right)$$

$$14r = 8(r+3)$$
$$14r = 8r + 24$$
$$6r = 24$$
$$r = 4$$

Thus, the speed of the walker is 4 miles per hour.

126. Let r = the speed of the faster train.

	Distance = Rate · Time $\left(\frac{d}{r}\right)$		
Faster train	378	r	$\frac{378}{r}$
Slower train	270	$r - 18$	$\frac{270}{r-18}$

$$\frac{378}{r} = \frac{270}{r-18}$$

$$r(r-18)\left(\frac{378}{r}\right) = r(r-18)\left(\frac{270}{r-18}\right)$$

$$378(r-18) = 270r$$
$$378r - 6804 = 270r$$
$$108r = 6804$$
$$r = 63$$
$$r - 18 = 63 - 18 = 45$$

The speeds of the trains are 63 miles per hour and 45 miles per hour.

127. $A = kB$
$6 = k(14)$
$$k = \frac{6}{14} = \frac{3}{7}$$
$$A = \frac{3}{7}B = \frac{3}{7}(21) = 3(3) = 9$$

128. $C = \dfrac{k}{D}$
$$12 = \frac{k}{8}$$
$$k = 96$$
$$C = \frac{96}{D} = \frac{96}{24} = 4$$

129. $P = \dfrac{k}{V}$
$$1250 = \frac{k}{2}$$
$$k = 2500$$
$$P = \frac{2500}{V}$$
$$800 = \frac{2500}{V}$$
$$800V = 2500$$
$$V = \frac{2500}{800} = 3.125$$

The volume is 3.125 cubic feet.

130. $A = kr^2$
$$36\pi = k(3)^2$$
$$36\pi = 9k$$
$$k = 4\pi$$
$$A = 4\pi r^2 = 4\pi(4)^2 = 64\pi$$

The surface area is 64π square inches.

Chapter 6 Test

1. $f(x) = \dfrac{5x^2}{1-x}$
Undefined when $1 - x = 0$.
$$1 - x = 0$$
$$1 = x$$
Domain: $\{x \mid x \text{ is a real number and } x \neq 1\}$

2. $g(x) = \dfrac{9x^2 - 9}{x^2 + 4x + 3}$

Undefined when $x^2 + 4x + 3 = 0$.

$$x^2 + 4x + 3 = 0$$
$$(x + 3)(x + 1) = 0$$
$$x + 3 = 0 \quad \text{or} \quad x + 1 = 0$$
$$x = -3 \quad \text{or} \quad x = -1$$

Domain:

$$\{x \mid x \text{ is a real number and } x \neq -3, \ x \neq -1\}$$

3. $\dfrac{5x^7}{3x^4} = \dfrac{5x^{7-4}}{3} = \dfrac{5x^3}{3}$

4. $\dfrac{7x - 21}{24 - 8x} = \dfrac{7(x - 3)}{8(3 - x)} = \dfrac{7(-1)}{8} = \dfrac{-7}{8} = -\dfrac{7}{8}$

5. $\dfrac{x^2 - 4x}{x^2 + 5x - 36} = \dfrac{x(x - 4)}{(x + 9)(x - 4)} = \dfrac{x}{x + 9}$

6. $\dfrac{x}{x - 2} \cdot \dfrac{x^2 - 4}{5x} = \dfrac{1}{x - 2} \cdot \dfrac{(x - 2)(x + 2)}{5}$

$$= \dfrac{x + 2}{5}$$

7. $\dfrac{2x^3 + 16}{6x^2 + 12x} \cdot \dfrac{5}{x^2 - 2x + 4}$

$$= \dfrac{2(x^3 + 8) \cdot 5}{6x(x + 2)(x^2 - 2x + 4)}$$

$$= \dfrac{5(x + 2)(x^2 - 2x + 4)}{3x(x + 2)(x^2 - 2x + 4)}$$

$$= \dfrac{5}{3x}$$

8. $\dfrac{26ab}{7c} \div \dfrac{13a^2 c^5}{14a^4 b^3} = \dfrac{26ab}{7c} \cdot \dfrac{14a^4 b^3}{13a^2 c^5}$

$$= \dfrac{2ab}{c} \cdot \dfrac{2a^4 b^3}{a^2 c^5}$$

$$= \dfrac{4a^5 b^4}{a^2 c^6}$$

$$= \dfrac{4a^3 b^4}{c^6}$$

9. $\dfrac{3x^2 - 12}{x^2 + 2x - 8} \div \dfrac{6x + 18}{x + 4}$

$$= \dfrac{3(x^2 - 4)}{(x + 4)(x - 2)} \cdot \dfrac{x + 4}{6(x + 3)}$$

$$= \dfrac{(x + 2)(x - 2)}{x - 2} \cdot \dfrac{1}{2(x + 3)}$$

$$= \dfrac{x + 2}{2(x + 3)}$$

10. $\dfrac{4x - 12}{2x - 9} \div \dfrac{3 - x}{4x^2 - 81} \cdot \dfrac{x + 3}{5x + 15}$

$$= \dfrac{4(x - 3)}{2x - 9} \cdot \dfrac{(2x + 9)(2x - 9)}{-(x - 3)} \cdot \dfrac{x + 3}{5(x + 3)}$$

$$= \dfrac{4}{1} \cdot \dfrac{2x + 9}{-1} \cdot \dfrac{1}{5}$$

$$= \dfrac{-4(2x + 9)}{5}$$

11. $\dfrac{5}{4x^3} + \dfrac{7}{4x^3} = \dfrac{5 + 7}{4x^3} = \dfrac{12}{4x^3} = \dfrac{3}{x^3}$

12. $\dfrac{3 + 2x}{10 - x} + \dfrac{13 + x}{x - 10} = \dfrac{(3 + 2x) - (13 + x)}{10 - x}$

$$= \dfrac{3 + 2x - 13 - x}{10 - x}$$

$$= \dfrac{-10 + x}{-(-10 + x)}$$

$$= \dfrac{1}{-1}$$

$$= -1$$

13. $\dfrac{3}{x^2 - x - 6} + \dfrac{2}{x^2 - 5x + 6}$

$$= \dfrac{3}{(x - 3)(x + 2)} + \dfrac{2}{(x - 3)(x - 2)}$$

$$= \dfrac{3(x - 2) + 2(x + 2)}{(x - 3)(x + 2)(x - 2)}$$

$$= \dfrac{3x - 6 + 2x + 4}{(x - 3)(x + 2)(x - 2)}$$

$$= \dfrac{5x - 2}{(x - 3)(x + 2)(x - 2)}$$

14. $\dfrac{5}{x-7} - \dfrac{2x}{3x-21} + \dfrac{x}{2x-14} = \dfrac{5}{x-7} - \dfrac{2x}{3(x-7)} + \dfrac{x}{2(x-7)} = \dfrac{6\cdot 5 - 2\cdot 2x + 3\cdot x}{6(x-7)} = \dfrac{30 - 4x + 3x}{6(x-7)} = \dfrac{30 - x}{6(x-7)}$

15. $\dfrac{3x}{5} \cdot \left(\dfrac{5}{x} - \dfrac{5}{2x} \right) = \dfrac{3x}{5}\left[5\left(\dfrac{1}{x} - \dfrac{1}{2x} \right) \right] = 3x\left(\dfrac{1}{x} - \dfrac{1}{2x} \right) = 3x \cdot \dfrac{2-1}{2x} = 3x\left(\dfrac{1}{2x} \right) = \dfrac{3}{2}$

16. $\dfrac{\frac{4x}{13}}{\frac{20x}{13}} = \dfrac{4x}{13} \cdot \dfrac{13}{20x} = \dfrac{4x}{20x} = \dfrac{1}{5}$

17. $\dfrac{\frac{5}{x} - \frac{7}{3x}}{\frac{9}{8x} - \frac{1}{x}} = \dfrac{\frac{15-7}{3x}}{\frac{9-8}{8x}} = \dfrac{\frac{8}{3x}}{\frac{1}{8x}} = \dfrac{8}{3x} \cdot \dfrac{8x}{1} = \dfrac{64}{3}$

18. $\dfrac{\frac{x^2-5x+6}{x+3}}{\frac{x^2-4x+4}{x^2-9}} = \dfrac{(x-3)(x-2)}{x+3} \cdot \dfrac{(x-3)(x+3)}{(x-2)(x-2)} = \dfrac{(x-3)^2}{x-2}$

19. $\dfrac{4x^2y + 9x + z}{3xz} = \dfrac{4x^2y}{3xz} + \dfrac{9x}{3xz} + \dfrac{z}{3xz} = \dfrac{4xy}{3z} + \dfrac{3}{z} + \dfrac{1}{3x}$

20.

$$
\begin{array}{r}
x^5 + 5x^4 + 8x^3 + 16x^2 + 33x + 63 \\
x-2\overline{\smash{\big)}\,x^6 + 3x^5 - 2x^4 + 0x^3 + x^2 - 3x + 2} \\
\underline{x^6 - 2x^5} \\
5x^5 - 2x^4 \\
\underline{5x^5 - 10x^4} \\
8x^4 + 0x^3 \\
\underline{8x^4 - 16x^3} \\
16x^3 + x^2 \\
\underline{16x^3 - 32x^2} \\
33x^2 - 3x \\
\underline{33x^2 - 66x} \\
63x + 2 \\
\underline{63x - 126} \\
128
\end{array}
$$

Answer: $x^5 + 5x^4 + 8x^3 + 16x^2 + 33x + 63 + \dfrac{128}{x-2}$

21.

$$
\begin{array}{r|rrrrr}
-3 & 4 & -3 & 2 & -1 & -1 \\
 & & -12 & 45 & -141 & 426 \\
\hline
 & 4 & -15 & 47 & -142 & 425
\end{array}
$$

$(4x^4 - 3x^3 + 2x^2 - x - 1) \div (x+3) = 4x^3 - 15x^2 + 47x - 142 + \dfrac{425}{x+3}$

22.

$$
\begin{array}{r|rrrrr}
-2 & 4 & 0 & 7 & -2 & -5 \\
   &   & -8 & 16 & -46 & 96 \\
\hline
   & 4 & -8 & 23 & -48 & 91
\end{array}
$$

$P(-2) = 91$

23.
$$\frac{5x+3}{3x-7} = \frac{19}{7}$$
$$7(3x-7)\left(\frac{5x+3}{3x-7}\right) = 7(3x-7)\left(\frac{19}{7}\right)$$
$$7(5x+3) = 19(3x-7)$$
$$35x + 21 = 57x - 133$$
$$154 = 22x$$
$$7 = x$$

The solution is 7.

24.
$$\frac{5}{x-5} + \frac{x}{x+5} = -\frac{29}{21}$$
$$21(x-5)(x+5)\left(\frac{5}{x-5} + \frac{x}{x+5}\right) = 21(x-5)(x+5)\left(-\frac{29}{21}\right)$$
$$21\cdot 5(x+5) + 21\cdot x(x-5) = -29(x-5)(x+5)$$
$$105x + 525 + 21x^2 - 105x = -29x^2 + 725$$
$$50x^2 - 200 = 0$$
$$x^2 - 4 = 0$$
$$(x+2)(x-2) = 0$$
$$x = -2 \text{ or } x = 2$$

The solutions are –2 and 2.

25.
$$\frac{x}{x-4} = 3 - \frac{4}{x-4}$$
$$(x-4)\left(\frac{x}{x-4}\right) = (x-4)\left(3 - \frac{4}{x-4}\right)$$
$$x = 3(x-4) - 4$$
$$x = 3x - 12 - 4$$
$$16 = 2x$$
$$8 = x$$

The solution is 8.

26.

$$\frac{x+b}{a} = \frac{4x-7a}{b}$$

$$ab\left(\frac{x+b}{a}\right) = ab\left(\frac{4x-7a}{b}\right)$$

$$b(x+b) = a(4x-7a)$$

$$bx + b^2 = 4ax - 7a^2$$

$$7a^2 + b^2 = 4ax - bx$$

$$7a^2 + b^2 = (4a-b)x$$

$$\frac{7a^2 + b^2}{4a-b} = x$$

27. Let $x =$ the number.

$$(x+1)\frac{2}{x} = \frac{12}{5}$$

$$\frac{2(x+1)}{x} = \frac{12}{5}$$

$$5x\left[\frac{2(x+1)}{x}\right] = 5x\left(\frac{12}{5}\right)$$

$$10(x+1) = 12x$$

$$10x + 10 = 12x$$

$$10 = 2x$$

$$5 = x$$

The number is 5.

28. Let $n =$ the number of hours it takes Jan and her husband to weed the garden together. Note that 1 hour and 30 minutes = 1.5 hours.

$$\frac{1}{2} + \frac{1}{1.5} = \frac{1}{n}$$

$$6n\left(\frac{1}{2} + \frac{1}{1.5}\right) = 6n\left(\frac{1}{n}\right)$$

$$3n + 4n = 6$$

$$7n = 6$$

$$n = \frac{6}{7}$$

It takes the two of them $\frac{6}{7}$ hour to weed the garden.

29.

$$W = \frac{k}{V}$$

$$20 = \frac{k}{12}$$

$$k = 240$$

$$W = \frac{240}{V} = \frac{240}{15} = 16$$

30.

$$Q = kRS^2$$

$$24 = k(3)4^2$$

$$24 = 48k$$

$$k = \frac{24}{48}$$

$$k = \frac{1}{2}$$

$$Q = \frac{1}{2}RS^2 = \frac{1}{2}(2)3^2 = 9$$

31.

$$S = k\sqrt{d}$$

$$160 = k\sqrt{400}$$

$$160 = 20k$$

$$k = \frac{160}{20}$$

$$k = 8$$

$$S = 8\sqrt{d}$$

$$128 = 8\sqrt{d}$$

$$16 = \sqrt{d}$$

$$256 = d$$

The height of the cliff is 256 feet.

Chapter 7

Exercise Set 7.1

1. $\sqrt{100} = 10$ because $10^2 = 100$.

3. $\sqrt{\dfrac{1}{4}} = \dfrac{1}{2}$ because $\left(\dfrac{1}{2}\right)^2 = \dfrac{1}{4}$.

5. $\sqrt{0.0001} = 0.01$ because $(0.01)^2 = 0.0001$.

7. $-\sqrt{36} = -6$ because $(6)^2 = 36$.

9. $\sqrt{x^{10}} = x^5$ because $(x^5)^2 = x^{10}$.

11. $\sqrt{16y^6} = \sqrt{16}\sqrt{y^6} = 4y^3$ because $(4y^3)^2 = 16y^6$

13. $\sqrt{7} \approx 2.646$
Since $4 < 7 < 9$, then $\sqrt{4} < \sqrt{7} < \sqrt{9}$, or $2 < \sqrt{7} < 3$. The approximation is between 2 and 3 and thus is reasonable.

15. $\sqrt{38} \approx 6.164$
Since $36 < 38 < 49$, then $\sqrt{36} < \sqrt{38} < \sqrt{49}$, or $6 < \sqrt{38} < 7$. The approximation is between 6 and 7 and thus is reasonable.

17. $\sqrt{200} \approx 14.142$
Since $196 < 200 < 225$, then $\sqrt{196} < \sqrt{200} < \sqrt{225}$, or $14 < \sqrt{200} < 15$. The approximation is between 14 and 15 and thus is reasonable.

19. $\sqrt[3]{64} = 4$ because $(4)^3 = 64$.

21. $\sqrt[3]{\dfrac{1}{8}} = \dfrac{1}{2}$ because $\left(\dfrac{1}{2}\right)^3 = \dfrac{1}{8}$.

23. $\sqrt[3]{-1} = -1$ because $(-1)^3 = -1$.

25. $\sqrt[3]{x^{12}} = x^4$ because $(x^4)^3 = x^{12}$.

27. $\sqrt[3]{-27x^9} = -3x^3$ because $(-3x^3)^3 = -27x^9$.

29. $-\sqrt[4]{16} = -2$ because $(2)^4 = 16$.

31. $\sqrt[4]{-16}$ is not a real number. There is no real number that, when raised to the fourth power, is -16.

33. $\sqrt[5]{-32} = -2$ because $(-2)^5 = -32$.

35. $\sqrt[5]{x^{20}} = x^4$ because $(x^4)^5 = x^{20}$.

37. $\sqrt[6]{64x^{12}} = 2x^2$ because $(2x^2)^6 = 64x^{12}$.

39. $\sqrt{81x^4} = 9x^2$ because $(9x^2)^2 = 81x^4$.

41. $\sqrt[4]{256x^8} = 4x^2$ because $(4x^2)^4 = 256x^8$.

43. $\sqrt{(-8)^2} = |-8| = 8$

45. $\sqrt[3]{(-8)^3} = -8$

47. $\sqrt{4x^2} = 2|x|$

49. $\sqrt[3]{x^3} = x$

51. $\sqrt{(x-5)^2} = |x-5|$

53. $\sqrt{x^2 + 4x + 4} = \sqrt{(x+2)^2} = |x+2|$

55. $-\sqrt{121} = -11$

57. $\sqrt[3]{8x^3} = 2x$

59. $\sqrt{y^{12}} = y^6$

61. $\sqrt{25a^2b^{20}} = 5ab^{10}$

63. $\sqrt[3]{-27x^{12}y^9} = -3x^4y^3$

65. $\sqrt[4]{a^{16}b^4} = a^4b$

67. $\sqrt[5]{-32x^{10}y^5} = \sqrt[5]{-32}\sqrt[5]{x^{10}}\sqrt[5]{y^5} = -2x^2y$

69. $\sqrt{\dfrac{25}{49}} = \dfrac{5}{7}$

71. $\sqrt{\dfrac{x^2}{4y^2}} = \dfrac{x}{2y}$

73. $-\sqrt[3]{\dfrac{z^{21}}{27x^3}} = -\dfrac{z^7}{3x}$

75. $\sqrt[4]{\dfrac{x^4}{16}} = \dfrac{x}{2}$

77. $f(x) = \sqrt{2x+3}$
$f(0) = \sqrt{2\cdot 0 + 3} = \sqrt{3}$

79. $g(x) = \sqrt[3]{x-8}$
$g(7) = \sqrt[3]{7-8} = \sqrt[3]{-1} = -1$

81. $g(x) = \sqrt[3]{x-8}$
$g(-19) = \sqrt[3]{-19-8} = \sqrt[3]{-27} = -3$

83. $f(x) = \sqrt{2x+3}$
$f(2) = \sqrt{2\cdot 2 + 3} = \sqrt{7}$

85. $f(x) = \sqrt{x} + 4$
Domain: $[0, \infty)$

87. $f(x) = \sqrt{x-3}$
Domain: $[3, \infty)$

x	$f(x) = \sqrt{x-3}$
3	$\sqrt{3-3} = \sqrt{0} = 0$
4	$\sqrt{4-3} = \sqrt{1} = 1$
7	$\sqrt{7-3} = \sqrt{4} = 2$
12	$\sqrt{12-3} = \sqrt{9} = 3$

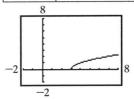

89. $f(x) = \sqrt{x} + 2$
Domain: $[0, \infty)$
y-intercept is $(0, 2)$, no x-intercept
Graph C

91. $f(x) = \sqrt{x-3}$
Domain: $[3, \infty)$
no y-intercept, x-intercept is $(3, 0)$
Graph D

93. $f(x) = \sqrt[3]{x} + 1$
Domain: All real numbers
y-intercept is $(0, 1)$, x-intercept is $(-1, 0)$
Graph A

95. $g(x) = \sqrt[3]{x-1}$
Domain: All real numbers
y-intercept is $(0, -1)$, x-intercept is $(1, 0)$
Graph B

97. $f(x) = \sqrt[3]{x} + 4$
Domain: $(-\infty, \infty)$

99. $g(x) = \sqrt[3]{x-1}$

Domain: $(-\infty, \infty)$

x	$g(x) = \sqrt[3]{x-1}$
1	$\sqrt[3]{1-1} = \sqrt[3]{0} = 0$
2	$\sqrt[3]{2-1} = \sqrt[3]{1} = 1$
0	$\sqrt[3]{0-1} = \sqrt[3]{-1} = -1$
9	$\sqrt[3]{9-1} = \sqrt[3]{8} = 2$
-7	$\sqrt[3]{-7-1} = \sqrt[3]{-8} = -2$

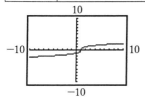

101. Answers may vary.

103. $(-2x^3y^2)^5 = (-2)^5 x^{3\cdot 5} y^{2\cdot 5} = -32x^{15}y^{10}$

105. $(-3x^2y^3z^5)(20x^5y^7)$
$= (-3)(20)x^{2+5}y^{3+7}z^5$
$= -60x^7y^{10}z^5$

107. $\dfrac{7x^{-1}y}{14(x^5y^2)^{-2}} = \dfrac{7x^{-1}y}{14x^{-10}y^{-4}} = \dfrac{x^9y^5}{2}$

Exercise Set 7.2

1. $49^{1/2} = \sqrt{49} = 7$

3. $27^{1/3} = \sqrt[3]{27} = 3$

5. $\left(\dfrac{1}{16}\right)^{1/4} = \sqrt[4]{\dfrac{1}{16}} = \dfrac{1}{2}$

7. $169^{1/2} = \sqrt{169} = 13$

9. $2m^{1/3} = 2\sqrt[3]{m}$

11. $(9x^4)^{1/2} = \sqrt{9x^4} = 3x^2$

13. $(-27)^{1/3} = \sqrt[3]{-27} = -3$

15. $-16^{1/4} = -\sqrt[4]{16} = -2$

17. $16^{3/4} = \left(\sqrt[4]{16}\right)^3 = 2^3 = 8$

19. $(-64)^{2/3} = \left(\sqrt[3]{-64}\right)^2 = (-4)^2 = 16$

21. $(-16)^{3/4} = \left(\sqrt[4]{-16}\right)^3$ is not a real number.

23. $(2x)^{3/5} = \sqrt[5]{(2x)^3}$ or $\left(\sqrt[5]{2x}\right)^3$

25. $(7x+2)^{2/3} = \sqrt[3]{(7x+2)^2}$ or $\left(\sqrt[3]{7x+2}\right)^2$

27. $\left(\dfrac{16}{9}\right)^{3/2} = \left(\sqrt{\dfrac{16}{9}}\right)^3 = \left(\dfrac{4}{3}\right)^3 = \dfrac{64}{27}$

29. $8^{-4/3} = \dfrac{1}{8^{4/3}} = \dfrac{1}{\left(\sqrt[3]{8}\right)^4} = \dfrac{1}{2^4} = \dfrac{1}{16}$

31. $(-64)^{-2/3} = \dfrac{1}{(-64)^{2/3}}$
$= \dfrac{1}{\left(\sqrt[3]{-64}\right)^2}$
$= \dfrac{1}{(-4)^2}$
$= \dfrac{1}{16}$

33. $(-4)^{-3/2} = \dfrac{1}{(-4)^{3/2}} = \dfrac{1}{\left(\sqrt{-4}\right)^3}$ is not a real number.

35. $x^{-1/4} = \dfrac{1}{x^{1/4}}$

37. $\dfrac{1}{a^{-2/3}} = a^{2/3}$

39. $\dfrac{5}{7x^{-3/4}} = \dfrac{5x^{3/4}}{7}$

41. Answers may vary.

43. $a^{2/3}a^{5/3} = a^{2/3+5/3} = a^{7/3}$

45. $x^{-2/5} \cdot x^{7/5} = x^{-2/5+7/5} = x^{5/5} = x$

47. $3^{1/4} \cdot 3^{3/8} = 3^{1/4+3/8} = 3^{2/8+3/8} = 3^{5/8}$

49. $\dfrac{y^{1/3}}{y^{1/6}} = y^{1/3-1/6} = y^{2/6-1/6} = y^{1/6}$

51. $(4u^2)^{3/2} = 4^{3/2}(u^2)^{3/2}$
$$= \left(\sqrt{4}\right)^3 \cdot u^3$$
$$= 2^3 u^3$$
$$= 8u^3$$

53. $\dfrac{b^{1/2}b^{3/4}}{-b^{1/4}} = -b^{1/2+3/4-1/4}$
$$= -b^{1/2+1/2}$$
$$= -b^1$$
$$= -b$$

55. $\dfrac{(3x^{1/4})^3}{x^{1/12}} = \dfrac{3^3 x^{3/4}}{x^{1/12}}$
$$= 27x^{3/4-1/12}$$
$$= 27x^{9/12-1/12}$$
$$= 27x^{8/12}$$
$$= 27x^{2/3}$$

57. $y^{1/2}(y^{1/2} - y^{2/3}) = y^{1/2}y^{1/2} - y^{1/2}y^{2/3}$
$$= y^{1/2+1/2} - y^{1/2+2/3}$$
$$= y^1 - y^{3/6+4/6}$$
$$= y - y^{7/6}$$

59. $x^{2/3}(2x-2) = 2x^{1+2/3} - 2x^{2/3}$
$$= 2x^{5/3} - 2x^{2/3}$$

61. $(2x^{1/3}+3)(2x^{1/3}-3) = (2x^{1/3})^2 - 3^2$
$$= 2^2(x^{1/3})^2 - 9$$
$$= 4x^{2/3} - 9$$

63. $x^{8/3} + x^{10/3} = x^{8/3} \cdot 1 + x^{8/3} \cdot x^{2/3}$
$$= x^{8/3}(1 + x^{2/3})$$

65. $x^{2/5} - 3x^{1/5} = x^{1/5} \cdot x^{1/5} - x^{1/5} \cdot 3$
$$= x^{1/5}(x^{1/5} - 3)$$

67. $5x^{-1/3} + x^{2/3} = x^{-1/3} \cdot 5 + x^{-1/3} \cdot x^1$
$$= x^{-1/3}(5 + x)$$

69. $\sqrt[6]{x^3} = x^{3/6} = x^{1/2} = \sqrt{x}$

71. $\sqrt[6]{4} = 4^{1/6} = (2^2)^{1/6} = 2^{2/6} = 2^{1/3} = \sqrt[3]{2}$

73. $\sqrt[4]{16x^2} = (16x^2)^{1/4}$
$$= 16^{1/4}x^{2/4}$$
$$= 2x^{1/2}$$
$$= 2\sqrt{x}$$

75. $\sqrt[8]{x^4 y^4} = (x^4 y^4)^{1/8}$
$$= x^{4/8}y^{4/8}$$
$$= x^{1/2}y^{1/2}$$
$$= (xy)^{1/2}$$
$$= \sqrt{xy}$$

77. $\sqrt[3]{y} \cdot \sqrt[5]{y^2} = y^{1/3} \cdot y^{2/5}$
$$= y^{5/15} \cdot y^{6/15}$$
$$= y^{11/15}$$
$$= \sqrt[15]{y^{11}}$$

79. $\dfrac{\sqrt[3]{b^2}}{\sqrt[4]{b}} = \dfrac{b^{2/3}}{b^{1/4}}$

$\qquad = b^{2/3-1/4}$

$\qquad = b^{8/12-3/12}$

$\qquad = b^{5/12}$

$\qquad = \sqrt[12]{b^5}$

81. $\dfrac{\sqrt[3]{a^2}}{\sqrt[6]{a}} = \dfrac{a^{2/3}}{a^{1/6}}$

$\qquad = a^{2/3-1/6}$

$\qquad = a^{4/6-1/6}$

$\qquad = a^{3/6}$

$\qquad = a^{1/2}$

$\qquad = \sqrt{a}$

83. $\sqrt{3} \cdot \sqrt[3]{4} = 3^{1/2} \cdot 4^{1/3}$

$\qquad = 3^{3/6} \cdot 4^{2/6}$

$\qquad = (3^3 \cdot 4^2)^{1/6}$

$\qquad = (432)^{1/6}$

$\qquad = \sqrt[6]{432}$

85. $\sqrt[5]{7} \cdot \sqrt[3]{y} = 7^{1/5} \cdot y^{1/3}$

$\qquad = 7^{3/15} \cdot y^{5/15}$

$\qquad = (7^3 \cdot y^5)^{1/15}$

$\qquad = (343y^5)^{1/15}$

$\qquad = \sqrt[15]{343y^5}$

87. $\dfrac{\sqrt{t}}{\sqrt{u}} = \dfrac{t^{1/2}}{u^{1/2}}$

89. $B(w) = 70w^{3/4}$

$\quad B(50) = 70(50)^{3/4} \approx 1316$

The BMR is 1316 calories.

91. a. $f(x) = 6550x^{43/50}$

$\qquad f(9) = 6550(9^{43/50}) = 43,340$

HP's net revenue in 1999 was
$43,340 million.

b. $f(14) = 6550(14^{43/50}) = 63,374$

HP's net revenue in 2004 will be
$63,374 million.

93. $x^{3/8}$

$x^{3/8} \cdot x^{1/8} = x^{4/8}$, or $x^{1/2}$

95. $y^{1/4}$

$\dfrac{y^{1/4}}{y^{-3/4}} = y^{1/4+3/4} = y^{4/4} = y$

97. $20^{1/5} \approx 1.8206$

99. $76^{5/7} \approx 22.0515$

101. $20 = 4 \cdot 5$, where $4 = 2^2$

103. $45 = 9 \cdot 5$, where $9 = 3^2$

105. $56 = 8 \cdot 7$, where $8 = 2^3$

107. $80 = 8 \cdot 10$, where $8 = 2^3$

Exercise Set 7.3

1. $\sqrt{7} \cdot \sqrt{2} = \sqrt{7 \cdot 2} = \sqrt{14}$

3. $\sqrt[4]{8} \cdot \sqrt[4]{2} = \sqrt[4]{8 \cdot 2} = \sqrt[4]{16} = 2$

5. $\sqrt[3]{4} \cdot \sqrt[3]{9} = \sqrt[3]{4 \cdot 9} = \sqrt[3]{36}$

7. $\sqrt{2} \cdot \sqrt{3x} = \sqrt{2 \cdot 3x} = \sqrt{6x}$

9. $\sqrt{\dfrac{7}{x}} \cdot \sqrt{\dfrac{2}{y}} = \sqrt{\dfrac{7 \cdot 2}{x \cdot y}} = \sqrt{\dfrac{14}{xy}}$

11. $\sqrt[4]{4x^3} \cdot \sqrt[4]{5} = \sqrt[4]{5 \cdot 4x^3} = \sqrt[4]{20x^3}$

13. $\sqrt{\dfrac{6}{49}} = \dfrac{\sqrt{6}}{\sqrt{49}} = \dfrac{\sqrt{6}}{7}$

15. $\sqrt{\dfrac{2}{49}} = \dfrac{\sqrt{2}}{\sqrt{49}} = \dfrac{\sqrt{2}}{7}$

17. $\sqrt[4]{\dfrac{x^3}{16}} = \dfrac{\sqrt[4]{x^3}}{\sqrt[4]{16}} = \dfrac{\sqrt[4]{x^3}}{2}$

19. $\sqrt[3]{\dfrac{4}{27}} = \dfrac{\sqrt[3]{4}}{\sqrt[3]{27}} = \dfrac{\sqrt[3]{4}}{3}$

21. $\sqrt[4]{\dfrac{8}{x^8}} = \dfrac{\sqrt[4]{8}}{\sqrt[4]{x^8}} = \dfrac{\sqrt[4]{8}}{x^{8/4}} = \dfrac{\sqrt[4]{8}}{x^2}$

23. $\sqrt[3]{\dfrac{2x}{81y^{12}}} = \dfrac{\sqrt[3]{2x}}{\sqrt[3]{81y^{12}}}$

$= \dfrac{\sqrt[3]{2x}}{\sqrt[3]{27y^{12}} \cdot \sqrt[3]{3}}$

$= \dfrac{\sqrt[3]{2x}}{3y^4 \sqrt[3]{3}}$

25. $\sqrt{\dfrac{x^2 y}{100}} = \dfrac{\sqrt{x^2 y}}{\sqrt{100}} = \dfrac{\sqrt{x^2} \cdot \sqrt{y}}{10} = \dfrac{x\sqrt{y}}{10}$

27. $\sqrt{\dfrac{5x^2}{4y^2}} = \dfrac{\sqrt{5x^2}}{\sqrt{4y^2}} = \dfrac{\sqrt{5}\sqrt{x^2}}{\sqrt{4}\sqrt{y^2}} = \dfrac{\sqrt{5}x}{2y}$

29. $-\sqrt[3]{\dfrac{z^7}{27x^3}} = \dfrac{-\sqrt[3]{z^7}}{\sqrt[3]{27x^3}}$

$= \dfrac{-\sqrt[3]{z^6 z}}{\sqrt[3]{27} \cdot \sqrt[3]{x^3}}$

$= \dfrac{-\sqrt[3]{z^6} \cdot \sqrt[3]{z}}{3x}$

$= \dfrac{-z^2 \sqrt[3]{z}}{3x}$

31. $\sqrt{32} = \sqrt{16(2)} = \sqrt{16} \cdot \sqrt{2} = 4\sqrt{2}$

33. $\sqrt[3]{192} = \sqrt[3]{64(3)} = \sqrt[3]{64} \cdot \sqrt[3]{3} = 4\sqrt[3]{3}$

35. $5\sqrt{75} = 5\sqrt{25(3)}$

$= 5\sqrt{25} \cdot \sqrt{3}$

$= 5(5)\sqrt{3}$

$= 25\sqrt{3}$

37. $\sqrt{24} = \sqrt{4 \cdot 6} = \sqrt{4} \cdot \sqrt{6} = 2\sqrt{6}$

39. $\sqrt{100x^5} = \sqrt{(100x^4)x}$

$= \sqrt{100x^4} \cdot \sqrt{x}$

$= 10x^2 \sqrt{x}$

41. $\sqrt[3]{16y^7} = \sqrt[3]{(8y^6)(2y)}$

$= \sqrt[3]{8y^6} \cdot \sqrt[3]{2y}$

$= 2y^2 \sqrt[3]{2y}$

43. $\sqrt[4]{a^8 b^7} = \sqrt[4]{a^8 b^4 b^3}$

$= \sqrt[4]{a^8 b^4} \cdot \sqrt[4]{b^3}$

$= a^2 b \sqrt[4]{b^3}$

45. $\sqrt{y^5} = \sqrt{y^4 y} = \sqrt{y^4} \cdot \sqrt{y} = y^2 \sqrt{y}$

47. $\sqrt{25a^2 b^3} = \sqrt{25a^2 b^2 \cdot b}$

$= \sqrt{25a^2 b^2} \cdot \sqrt{b}$

$= 5ab\sqrt{b}$

49. $\sqrt[5]{-32x^{10} y} = \sqrt[5]{-32x^{10}} \cdot \sqrt[5]{y} = -2x^2 \sqrt[5]{y}$

51. $\sqrt[3]{50x^{14}} = \sqrt[3]{x^{12}(50x^2)}$

$= \sqrt[3]{x^{12}} \cdot \sqrt[3]{50x^2}$

$= x^4 \sqrt[3]{50x^2}$

53. $-\sqrt{32a^8 b^7} = -\sqrt{16a^8 b^6 (2b)}$

$= -\sqrt{16a^8 b^6} \cdot \sqrt{2b}$

$= -4a^4 b^3 \sqrt{2b}$

55. $\sqrt{9x^7y^9} = \sqrt{9x^6y^8 \cdot xy}$
$\qquad\qquad = \sqrt{9x^6y^8} \cdot \sqrt{xy}$
$\qquad\qquad = 3x^3y^4\sqrt{xy}$

57. $\sqrt[3]{125r^9s^{12}} = 5r^3s^4$

59. $\dfrac{\sqrt{14}}{\sqrt{7}} = \sqrt{\dfrac{14}{7}} = \sqrt{2}$

61. $\dfrac{\sqrt[3]{24}}{\sqrt[3]{3}} = \sqrt[3]{\dfrac{24}{3}} = \sqrt[3]{8} = 2$

63. $\dfrac{5\sqrt[4]{48}}{\sqrt[4]{3}} = 5\sqrt[4]{\dfrac{48}{3}} = 5\sqrt[4]{16} = 5 \cdot 2 = 10$

65. $\dfrac{\sqrt{x^5y^3}}{\sqrt{xy}} = \sqrt{\dfrac{x^5y^3}{xy}}$
$\qquad\qquad = \sqrt{x^4y^2}$
$\qquad\qquad = x^2y$

67. $\dfrac{8\sqrt[3]{54m^7}}{\sqrt[3]{2m}} = 8\sqrt[3]{\dfrac{54m^7}{2m}}$
$\qquad\qquad = 8\sqrt[3]{27m^6}$
$\qquad\qquad = 8 \cdot 3m^2$
$\qquad\qquad = 24m^2$

69. $\dfrac{3\sqrt{100x^2}}{2\sqrt{2x^{-1}}} = \dfrac{3}{2}\sqrt{\dfrac{100x^2}{2x^{-1}}}$
$\qquad\qquad = \dfrac{3}{2}\sqrt{50x^3}$
$\qquad\qquad = \dfrac{3}{2}\sqrt{25x^2 \cdot 2x}$
$\qquad\qquad = \dfrac{3}{2} \cdot 5x\sqrt{2x}$
$\qquad\qquad = \dfrac{15x}{2}\sqrt{2x}$

71. $\dfrac{\sqrt[4]{96a^{10}b^3}}{\sqrt[4]{3a^2b^3}} = \sqrt[4]{\dfrac{96a^{10}b^3}{3a^2b^3}}$
$\qquad\qquad = \sqrt[4]{32a^8}$
$\qquad\qquad = \sqrt[4]{16a^8 \cdot 2}$
$\qquad\qquad = 2a^2\sqrt[4]{2}$

73. $A = \pi r\sqrt{r^2 + h^2}$

   a. $A = \pi \cdot 4\sqrt{4^2 + 3^2}$
$\qquad\quad = 4\pi\sqrt{16 + 9}$
$\qquad\quad = 4\pi\sqrt{25}$
$\qquad\quad = 4\pi(5) = 20\pi$
The surface area is 20π square centimeters.

   b. $A = \pi(6.8)\sqrt{(6.8)^2 + (7.2)^2}$
$\qquad\quad = 6.8\pi\sqrt{46.24 + 51.84}$
$\qquad\quad = 6.8\pi\sqrt{98.08} \approx 211.57$
The area is approximately 211.57 square feet.

75. $F(x) = 0.6\sqrt{49 - x^2}$

   a. $F(3) = 0.6\sqrt{49 - 3^2}$
$\qquad\quad = 0.6\sqrt{49 - 9}$
$\qquad\quad = 0.6\sqrt{40}$
$\qquad\quad \approx 3.8$
The demand is 3.8 times per week.

   b. $F(5) = 0.6\sqrt{49 - 5^2}$
$\qquad\quad = 0.6\sqrt{49 - 25}$
$\qquad\quad = 0.6\sqrt{24}$
$\qquad\quad \approx 2.9$
The demand is 2.9 times per week.

   c. Answers may vary.

77. $(6x)(8x) = (6)(8)x \cdot x = 48x^2$

79. $(2x+3)+(x-5)=2x+3+x-5$
$$=(2x+x)+(3-5)$$
$$=3x-2$$

81. $(9y^2)(-8y^2)=9(-8)y^{2+2}=-72y^4$

83. $-3+x+5=x+(-3+5)=x+2$

85. $(2x+1)^2=(2x)^2+2(2x)(1)+1^2$
$$=4x^2+4x+1$$

Section 7.4

Mental Math

1. $2\sqrt{3}+4\sqrt{3}=6\sqrt{3}$

2. $5\sqrt{7}+3\sqrt{7}=8\sqrt{7}$

3. $8\sqrt{x}-5\sqrt{x}=3\sqrt{x}$

4. $3\sqrt{y}+10\sqrt{y}=13\sqrt{y}$

5. $7\sqrt[3]{x}+5\sqrt[3]{x}=12\sqrt[3]{x}$

6. $8\sqrt[3]{z}-2\sqrt[3]{z}=6\sqrt[3]{z}$

Exercise Set 7.4

1. $\sqrt{8}-\sqrt{32}=\sqrt{4(2)}-\sqrt{16(2)}$
$$=\sqrt{4}\sqrt{2}-\sqrt{16}\sqrt{2}$$
$$=2\sqrt{2}-4\sqrt{2}$$
$$=-2\sqrt{2}$$

3. $2\sqrt{2x^3}+4x\sqrt{8x}$
$$=2\sqrt{x^2(2x)}+4x\sqrt{4(2x)}$$
$$=2\sqrt{x^2}\sqrt{2x}+4x\sqrt{4}\sqrt{2x}$$
$$=2x\sqrt{2x}+4x(2)\sqrt{2x}$$
$$=2x\sqrt{2x}+8x\sqrt{2x}$$
$$=10x\sqrt{2x}$$

5. $2\sqrt{50}-3\sqrt{125}+\sqrt{98}$
$$=2\sqrt{25(2)}-3\sqrt{25(5)}+\sqrt{49(2)}$$
$$=2\sqrt{25}\sqrt{2}-3\sqrt{25}\sqrt{5}+\sqrt{49}\sqrt{2}$$
$$=2(5)\sqrt{2}-3(5)\sqrt{5}+7\sqrt{2}$$
$$=10\sqrt{2}-15\sqrt{5}+7\sqrt{2}$$
$$=17\sqrt{2}-15\sqrt{5}$$

7. $\sqrt[3]{16x}-\sqrt[3]{54x}=\sqrt[3]{8(2x)}-\sqrt[3]{27(2x)}$
$$=\sqrt[3]{8}\sqrt[3]{2x}-\sqrt[3]{27}\sqrt[3]{2x}$$
$$=2\sqrt[3]{2x}-3\sqrt[3]{2x}$$
$$=-\sqrt[3]{2x}$$

9. $\sqrt{9b^3}-\sqrt{25b^3}+\sqrt{49b^3}$
$$=\sqrt{9b^2(b)}-\sqrt{25b^2(b)}+\sqrt{49b^2(b)}$$
$$=\sqrt{9b^2}\sqrt{b}-\sqrt{25b^2}\sqrt{b}+\sqrt{49b^2}\sqrt{b}$$
$$=3b\sqrt{b}-5b\sqrt{b}+7b\sqrt{b}$$
$$=5b\sqrt{b}$$

11. $\dfrac{5\sqrt{2}}{3}+\dfrac{2\sqrt{2}}{5}=\dfrac{5\left(5\sqrt{2}\right)+3\left(2\sqrt{2}\right)}{3(5)}$
$$=\dfrac{25\sqrt{2}+6\sqrt{2}}{15}$$
$$=\dfrac{31\sqrt{2}}{15}$$

13. $\sqrt[3]{\dfrac{11}{8}}-\dfrac{\sqrt[3]{11}}{6}=\dfrac{\sqrt[3]{11}}{\sqrt[3]{8}}-\dfrac{\sqrt[3]{11}}{6}$
$$=\dfrac{\sqrt[3]{11}}{2}-\dfrac{\sqrt[3]{11}}{6}$$
$$=\dfrac{3\sqrt[3]{11}-\sqrt[3]{11}}{6}$$
$$=\dfrac{2\sqrt[3]{11}}{6}$$
$$=\dfrac{\sqrt[3]{11}}{3}$$

15.
$$\frac{\sqrt{20x}}{9} + \sqrt{\frac{5x}{9}} = \frac{\sqrt{4(5x)}}{9} + \frac{\sqrt{5x}}{\sqrt{9}}$$
$$= \frac{\sqrt{4}\sqrt{5x}}{9} + \frac{\sqrt{5x}}{3}$$
$$= \frac{2\sqrt{5x} + 3\sqrt{5x}}{9}$$
$$= \frac{5\sqrt{5x}}{9}$$

17.
$$7\sqrt{9} - 7 + \sqrt{3} = 7(3) - 7 + \sqrt{3}$$
$$= 21 - 7 + \sqrt{3}$$
$$= 14 + \sqrt{3}$$

19.
$$2 + 3\sqrt{y^2} - 6\sqrt{y^2} + 5 = 7 - 3\sqrt{y^2}$$
$$= 7 - 3y$$

21.
$$3\sqrt{108} - 2\sqrt{18} - 3\sqrt{48}$$
$$= 3\sqrt{36}\sqrt{3} - 2\sqrt{9}\sqrt{2} - 3\sqrt{16}\sqrt{3}$$
$$= 3(6)\sqrt{3} - 2(3)\sqrt{2} - 3(4)\sqrt{3}$$
$$= 18\sqrt{3} - 6\sqrt{2} - 12\sqrt{3}$$
$$= 6\sqrt{3} - 6\sqrt{2}$$

23.
$$-5\sqrt[3]{625} + \sqrt[3]{40} = -5\sqrt[3]{125}\sqrt[3]{5} + \sqrt[3]{8}\sqrt[3]{5}$$
$$= -5(5)\sqrt[3]{5} + 2\sqrt[3]{5}$$
$$= -25\sqrt[3]{5} + 2\sqrt[3]{5}$$
$$= -23\sqrt[3]{5}$$

25.
$$\sqrt{9b^3} - \sqrt{25b^3} + \sqrt{16b^3}$$
$$= \sqrt{9b^2}\sqrt{b} - \sqrt{25b^2}\sqrt{b} + \sqrt{16b^2}\sqrt{b}$$
$$= 3b\sqrt{b} - 5b\sqrt{b} + 4b\sqrt{b}$$
$$= (3 - 5 + 4)b\sqrt{b}$$
$$= 2b\sqrt{b}$$

27.
$$5y\sqrt{8y} + 2\sqrt{50y^3}$$
$$= 5y\sqrt{4}\sqrt{2y} + 2\sqrt{25y^2}\sqrt{2y}$$
$$= 5y(2)\sqrt{2y} + 2(5y)\sqrt{2y}$$
$$= 10y\sqrt{2y} + 10y\sqrt{2y}$$
$$= 20y\sqrt{2y}$$

29.
$$\sqrt[3]{54xy^3} - 5\sqrt[3]{2xy^3} + y\sqrt[3]{128x}$$
$$= \sqrt[3]{27y^3}\sqrt[3]{2x} - 5\sqrt[3]{y^3}\sqrt[3]{2x} + y\sqrt[3]{64}\sqrt[3]{2x}$$
$$= 3y\sqrt[3]{2x} - 5y\sqrt[3]{2x} + y(4)\sqrt[3]{2x}$$
$$= -2y\sqrt[3]{2x} + 4y\sqrt[3]{2x}$$
$$= 2y\sqrt[3]{2x}$$

31. $6\sqrt[3]{11} + 8\sqrt{11} - 12\sqrt{11} = 6\sqrt[3]{11} - 4\sqrt{11}$

33.
$$-2\sqrt[4]{x^7} + 3\sqrt[4]{16x^7}$$
$$= -2\sqrt[4]{x^4}\sqrt[4]{x^3} + 3\sqrt[4]{16x^4}\sqrt[4]{x^3}$$
$$= -2x\sqrt[4]{x^3} + 3(2x)\sqrt[4]{x^3}$$
$$= -2x\sqrt[4]{x^3} + 6x\sqrt[4]{x^3}$$
$$= 4x\sqrt[4]{x^3}$$

35.
$$\frac{4\sqrt{3}}{3} - \frac{\sqrt{12}}{3} = \frac{4\sqrt{3}}{3} - \frac{\sqrt{4}\sqrt{3}}{3}$$
$$= \frac{4\sqrt{3} - 2\sqrt{3}}{3}$$
$$= \frac{2\sqrt{3}}{3}$$

37.
$$\frac{\sqrt[3]{8x^4}}{7} + \frac{3x\sqrt[3]{x}}{7} = \frac{\sqrt[3]{8x^3}\sqrt[3]{x} + 3x\sqrt[3]{x}}{7}$$
$$= \frac{2x\sqrt[3]{x} + 3x\sqrt[3]{x}}{7}$$
$$= \frac{5x\sqrt[3]{x}}{7}$$

39.
$$\sqrt{\frac{28}{x^2}} + \sqrt{\frac{7}{4x^2}} = \frac{\sqrt{28}}{\sqrt{x^2}} + \frac{\sqrt{7}}{\sqrt{4x^2}}$$
$$= \frac{\sqrt{4}\sqrt{7}}{x} + \frac{\sqrt{7}}{2x}$$
$$= \frac{2\sqrt{7}}{x} + \frac{\sqrt{7}}{2x}$$
$$= \frac{4\sqrt{7} + \sqrt{7}}{2x}$$
$$= \frac{5\sqrt{7}}{2x}$$

41. $\sqrt[3]{\dfrac{16}{27}} - \dfrac{\sqrt[3]{54}}{6} = \dfrac{\sqrt[3]{16}}{\sqrt[3]{27}} - \dfrac{\sqrt[3]{27}\sqrt[3]{2}}{6}$

$\qquad = \dfrac{\sqrt[3]{8}\sqrt[3]{2}}{3} - \dfrac{3\sqrt[3]{2}}{6}$

$\qquad = \dfrac{2(2)\sqrt[3]{2}}{6} - \dfrac{3\sqrt[3]{2}}{6}$

$\qquad = \dfrac{4\sqrt[3]{2} - 3\sqrt[3]{2}}{6}$

$\qquad = \dfrac{\sqrt[3]{2}}{6}$

43. $-\dfrac{\sqrt[3]{2x^4}}{9} + \sqrt[3]{\dfrac{250x^4}{27}}$

$\qquad = \dfrac{-\sqrt[3]{x^3}\sqrt[3]{2x}}{9} + \dfrac{\sqrt[3]{250x^4}}{\sqrt[3]{27}}$

$\qquad = \dfrac{-x\sqrt[3]{2x}}{9} + \dfrac{\sqrt[3]{125x^3}\sqrt[3]{2x}}{3}$

$\qquad = \dfrac{-x\sqrt[3]{2x}}{9} + \dfrac{5x\sqrt[3]{2x}}{3}$

$\qquad = \dfrac{-x\sqrt[3]{2x} + 15x\sqrt[3]{2x}}{9}$

$\qquad = \dfrac{14x\sqrt[3]{2x}}{9}$

45. $P = 2\sqrt{12} + \sqrt{12} + 2\sqrt{27} + 3\sqrt{3}$

$\qquad = 2\sqrt{4}\sqrt{3} + \sqrt{4}\sqrt{3} + 2\sqrt{9}\sqrt{3} + 3\sqrt{3}$

$\qquad = 2\cdot2\sqrt{3} + 2\sqrt{3} + 2\cdot3\sqrt{3} + 3\sqrt{3}$

$\qquad = (4 + 2 + 6 + 3)\sqrt{3}$

$\qquad = 15\sqrt{3}$

The perimeter of the trapezoid is
$15\sqrt{3}$ inches.

47. $\sqrt{7}\left(\sqrt{5} + \sqrt{3}\right) = \sqrt{7}\sqrt{5} + \sqrt{7}\sqrt{3}$

$\qquad\qquad = \sqrt{35} + \sqrt{21}$

49. $\left(\sqrt{5} - \sqrt{2}\right)^2 = \left(\sqrt{5}\right)^2 - 2\sqrt{5}\sqrt{2} + \left(\sqrt{2}\right)^2$

$\qquad\qquad = 5 - 2\sqrt{10} + 2$

$\qquad\qquad = 7 - 2\sqrt{10}$

51. $\sqrt{3x}\left(\sqrt{3} - \sqrt{x}\right) = \sqrt{3x}\sqrt{3} - \sqrt{3x}\sqrt{x}$

$\qquad\qquad = \sqrt{3\cdot3\cdot x} - \sqrt{3\cdot x\cdot x}$

$\qquad\qquad = 3\sqrt{x} - x\sqrt{3}$

53. $\left(2\sqrt{x} - 5\right)\left(3\sqrt{x} + 1\right)$

$\qquad = \left(2\sqrt{x}\right)\left(3\sqrt{x}\right) + \left(2\sqrt{x}\right)1 - 5\left(3\sqrt{x}\right) - 5\cdot1$

$\qquad = 6x + 2\sqrt{x} - 15\sqrt{x} - 5$

$\qquad = 6x - 13\sqrt{x} - 5$

55. $\left(\sqrt[3]{a} - 4\right)\left(\sqrt[3]{a} + 5\right)$

$\qquad = \left(\sqrt[3]{a}\right)^2 + 5\sqrt[3]{a} - 4\sqrt[3]{a} - 4\cdot5$

$\qquad = \sqrt[3]{a^2} + \sqrt[3]{a} - 20$

57. $6\left(\sqrt{2} - 2\right) = 6\sqrt{2} - 6\cdot2 = 6\sqrt{2} - 12$

59. $\sqrt{2}\left(\sqrt{2} + x\sqrt{6}\right) = \left(\sqrt{2}\right)^2 + \sqrt{2}\left(x\sqrt{6}\right)$

$\qquad\qquad = 2 + \sqrt{2}\left(x\sqrt{2}\sqrt{3}\right)$

$\qquad\qquad = 2 + 2x\sqrt{3}$

61. $\left(2\sqrt{7}+3\sqrt{5}\right)\left(\sqrt{7}-2\sqrt{5}\right)=2\left(\sqrt{7}\right)^2-\left(2\sqrt{7}\right)\left(2\sqrt{5}\right)+\left(3\sqrt{5}\right)\sqrt{7}-3\cdot2\left(\sqrt{5}\right)^2$

$$= 2\cdot7-4\sqrt{35}+3\sqrt{35}-6\cdot5$$
$$= 14-\sqrt{35}-30$$
$$= -16-\sqrt{35}$$

63. $\left(\sqrt{x}-y\right)\left(\sqrt{x}+y\right)=\left(\sqrt{x}\right)^2-y^2=x-y^2$

65. $\left(\sqrt{3}+x\right)^2=\left(\sqrt{3}\right)^2+2x\sqrt{3}+x^2=3+2x\sqrt{3}+x^2$

67. $\left(\sqrt{5x}-3\sqrt{2}\right)\left(\sqrt{5x}-3\sqrt{3}\right)=\left(\sqrt{5x}\right)^2-3\sqrt{3}\sqrt{5x}-3\sqrt{2}\sqrt{5x}+\left(3\sqrt{2}\right)\left(3\sqrt{3}\right)$

$$= 5x-3\sqrt{15x}-3\sqrt{10x}+9\sqrt{6}$$

69. $\left(\sqrt[3]{4}+2\right)\left(\sqrt[3]{2}-1\right)=\sqrt[3]{4}\sqrt[3]{2}-\sqrt[3]{4}\cdot1+2\sqrt[3]{2}-2\cdot1$

$$= \sqrt[3]{8}-\sqrt[3]{4}+2\sqrt[3]{2}-2$$
$$= 2-\sqrt[3]{4}+2\sqrt[3]{2}-2$$
$$= -\sqrt[3]{4}+2\sqrt[3]{2}$$

71. $\left(\sqrt[3]{x}+1\right)\left(\sqrt[3]{x}-4\sqrt{x}+7\right)=\sqrt[3]{x}\left(\sqrt[3]{x}-4\sqrt{x}+7\right)+1\left(\sqrt[3]{x}-4\sqrt{x}+7\right)$

$$= \left(\sqrt[3]{x}\right)^2-\sqrt[3]{x}\left(4\sqrt{x}\right)+\sqrt[3]{x}(7)+\sqrt[3]{x}-4\sqrt{x}+7$$
$$= \left(\sqrt[3]{x}\right)^2-x^{1/3}(4x^{1/2})+8\sqrt[3]{x}-4\sqrt{x}+7$$
$$= \sqrt[3]{x^2}-4x^{5/6}+8\sqrt[3]{x}-4\sqrt{x}+7$$
$$= \sqrt[3]{x^2}-4\sqrt[6]{x^5}+8\sqrt[3]{x}-4\sqrt{x}+7$$

73. a. $P=2\left(3\sqrt{20}\right)+2\sqrt{125}$

$$= 6\sqrt{4}\sqrt{5}+2\sqrt{25}\sqrt{5}$$
$$= 6(2)\sqrt{5}+2(5)\sqrt{5}$$
$$= 12\sqrt{5}+10\sqrt{5}=22\sqrt{5}$$

The perimeter is $22\sqrt{5}$ feet.

b. $A=3\sqrt{20}\cdot\sqrt{125}$

$$= 3\sqrt{4}\sqrt{5}\sqrt{5}\sqrt{25}$$
$$= 3\cdot2\cdot5\cdot5=150$$

The area is 150 square feet.

75. Answers may vary.

77. $\dfrac{2x-14}{2}=\dfrac{2(x-7)}{2}=x-7$

79. $\dfrac{7x-7y}{x^2-y^2}=\dfrac{7(x-y)}{(x-y)(x+y)}=\dfrac{7}{x+y}$

81. $\dfrac{6a^2b-9ab}{3ab}=\dfrac{3ab(2a-3)}{3ab}=2a-3$

83. $\dfrac{-4+2\sqrt{3}}{6}=\dfrac{2\left(-2+\sqrt{3}\right)}{2\cdot3}=\dfrac{-2+\sqrt{3}}{3}$

Section 7.5

Mental Math

1. The conjugate of $\sqrt{2} + x$ is $\sqrt{2} - x$.

2. The conjugate of $\sqrt{3} + y$ is $\sqrt{3} - y$.

3. The conjugate of $5 - \sqrt{a}$ is $5 + \sqrt{a}$.

4. The conjugate of $6 - \sqrt{b}$ is $6 + \sqrt{b}$.

5. The conjugate of $7\sqrt{5} + 8\sqrt{x}$ is $7\sqrt{5} - 8\sqrt{x}$.

6. The conjugate of $9\sqrt{2} - 6\sqrt{y}$ is $9\sqrt{2} + 6\sqrt{y}$.

Exercise Set 7.5

1. $\dfrac{\sqrt{2}}{\sqrt{7}} = \dfrac{\sqrt{2} \cdot \sqrt{7}}{\sqrt{7} \cdot \sqrt{7}} = \dfrac{\sqrt{14}}{7}$

3. $\sqrt{\dfrac{1}{5}} = \dfrac{\sqrt{1}}{\sqrt{5}} = \dfrac{1}{\sqrt{5}} = \dfrac{1 \cdot \sqrt{5}}{\sqrt{5} \cdot \sqrt{5}} = \dfrac{\sqrt{5}}{5}$

5. $\sqrt[3]{\dfrac{3}{4}} = \dfrac{\sqrt[3]{3}}{\sqrt[3]{4}} = \dfrac{\sqrt[3]{3} \cdot \sqrt[3]{2}}{\sqrt[3]{4} \cdot \sqrt[3]{2}} = \dfrac{\sqrt[3]{6}}{\sqrt[3]{8}} = \dfrac{\sqrt[3]{6}}{2}$

7. $\dfrac{4}{\sqrt[3]{3}} = \dfrac{4}{\sqrt[3]{3}} \cdot \dfrac{\sqrt[3]{9}}{\sqrt[3]{9}} = \dfrac{4\sqrt[3]{9}}{\sqrt[3]{27}} = \dfrac{4\sqrt[3]{9}}{3}$

9. $\dfrac{3}{\sqrt{8x}} = \dfrac{3}{\sqrt{8x}} \cdot \dfrac{\sqrt{2x}}{\sqrt{2x}} = \dfrac{3\sqrt{2x}}{\sqrt{16x^2}} = \dfrac{3\sqrt{2x}}{4x}$

11. $\dfrac{3}{\sqrt[3]{4x^2}} = \dfrac{3}{\sqrt[3]{4x^2}} \cdot \dfrac{\sqrt[3]{2x}}{\sqrt[3]{2x}} = \dfrac{3\sqrt[3]{2x}}{\sqrt[3]{8x^3}} = \dfrac{3\sqrt[3]{2x}}{2x}$

13. $\sqrt{\dfrac{4}{x}} = \dfrac{\sqrt{4}}{\sqrt{x}} = \dfrac{2}{\sqrt{x}} = \dfrac{2 \cdot \sqrt{x}}{\sqrt{x} \cdot \sqrt{x}} = \dfrac{2\sqrt{x}}{x}$

15. $\dfrac{9}{\sqrt{3a} \cdot} = \dfrac{9 \cdot \sqrt{3a}}{\sqrt{3a} \cdot \sqrt{3a}} = \dfrac{9\sqrt{3a}}{3a} = \dfrac{3\sqrt{3a}}{a}$

17. $\dfrac{3}{\sqrt[3]{2}} = \dfrac{3 \cdot \sqrt[3]{4}}{\sqrt[3]{2} \cdot \sqrt[3]{4}} = \dfrac{3\sqrt[3]{4}}{\sqrt[3]{8}} = \dfrac{3\sqrt[3]{4}}{2}$

19. $\dfrac{2\sqrt{3}}{\sqrt{7}} = \dfrac{2\sqrt{3} \cdot \sqrt{7}}{\sqrt{7} \cdot \sqrt{7}} = \dfrac{2\sqrt{21}}{7}$

21. $\sqrt{\dfrac{2x}{5y}} = \dfrac{\sqrt{2x}}{\sqrt{5y}} = \dfrac{\sqrt{2x} \cdot \sqrt{5y}}{\sqrt{5y} \cdot \sqrt{5y}} = \dfrac{\sqrt{10xy}}{5y}$

23. $\sqrt[4]{\dfrac{81}{8}} = \dfrac{\sqrt[4]{81}}{\sqrt[4]{8}} = \dfrac{3}{\sqrt[4]{8}} = \dfrac{3 \cdot \sqrt[4]{2}}{\sqrt[4]{8} \cdot \sqrt[4]{2}} = \dfrac{3\sqrt[4]{2}}{\sqrt[4]{16}} = \dfrac{3\sqrt[4]{2}}{2}$

25. $\sqrt[4]{\dfrac{16}{9x^7}} = \dfrac{\sqrt[4]{16}}{\sqrt[4]{9x^7}}$

$= \dfrac{2}{x\sqrt[4]{9x^3}}$

$= \dfrac{2 \cdot \sqrt[4]{9x}}{x\sqrt[4]{9x^3} \cdot \sqrt[4]{9x}}$

$= \dfrac{2\sqrt[4]{9x}}{x\sqrt[4]{81x^4}}$

$= \dfrac{2\sqrt[4]{9x}}{3x^2}$

27. $\dfrac{5a}{\sqrt[5]{8a^9b^{11}}} = \dfrac{5a}{ab^2\sqrt[5]{8a^4b}}$

$= \dfrac{5a \cdot \sqrt[5]{4ab^4}}{ab^2\sqrt[5]{8a^4b} \cdot \sqrt[5]{4ab^4}}$

$= \dfrac{5a\sqrt[5]{4ab^4}}{ab^2\sqrt[5]{32a^5b^5}}$

$= \dfrac{5a\sqrt[5]{4ab^4}}{ab^2(2ab)}$

$= \dfrac{5a\sqrt[5]{4ab^4}}{2a^2b^3}$

$= \dfrac{5\sqrt[5]{4ab^4}}{2ab^3}$

29. $\sqrt{\dfrac{5}{3}} = \dfrac{\sqrt{5}}{\sqrt{3}} = \dfrac{\sqrt{5} \cdot \sqrt{5}}{\sqrt{3} \cdot \sqrt{5}} = \dfrac{5}{\sqrt{15}}$

31.
$$\sqrt{\frac{18}{5}} = \frac{\sqrt{18}}{\sqrt{5}}$$
$$= \frac{\sqrt{9}\sqrt{2}}{\sqrt{5}}$$
$$= \frac{3\sqrt{2}}{\sqrt{5}}$$
$$= \frac{3\sqrt{2} \cdot \sqrt{2}}{\sqrt{5} \cdot \sqrt{2}}$$
$$= \frac{3(2)}{\sqrt{10}}$$
$$= \frac{6}{\sqrt{10}}$$

33.
$$\frac{\sqrt{4x}}{7} = \frac{\sqrt{4}\sqrt{x}}{7} = \frac{2\sqrt{x}}{7} = \frac{2\sqrt{x} \cdot \sqrt{x}}{7 \cdot \sqrt{x}} = \frac{2x}{7\sqrt{x}}$$

35.
$$\frac{\sqrt[3]{5y^2}}{\sqrt[3]{4x}} = \frac{\sqrt[3]{5y^2} \cdot \sqrt[3]{5^2 y}}{\sqrt[3]{4x} \cdot \sqrt[3]{5^2 y}}$$
$$= \frac{\sqrt[3]{5^3 y^3}}{\sqrt[3]{4(5^2)xy}}$$
$$= \frac{5y}{\sqrt[3]{100xy}}$$

37.
$$\sqrt{\frac{2}{5}} = \frac{\sqrt{2}}{\sqrt{5}} = \frac{\sqrt{2} \cdot \sqrt{2}}{\sqrt{5} \cdot \sqrt{2}} = \frac{2}{\sqrt{10}}$$

39.
$$\frac{\sqrt{2x}}{11} = \frac{\sqrt{2x} \cdot \sqrt{2x}}{11 \cdot \sqrt{2x}} = \frac{2x}{11\sqrt{2x}}$$

41.
$$\sqrt[3]{\frac{7}{8}} = \frac{\sqrt[3]{7}}{\sqrt[3]{8}}$$
$$= \frac{\sqrt[3]{7}}{2}$$
$$= \frac{\sqrt[3]{7} \cdot \sqrt[3]{7^2}}{2 \cdot \sqrt[3]{7^2}}$$
$$= \frac{\sqrt[3]{7^3}}{2\sqrt[3]{7^2}}$$
$$= \frac{7}{2\sqrt[3]{49}}$$

43.
$$\frac{\sqrt[3]{3x^5}}{10} = \frac{\sqrt[3]{3x^5} \cdot \sqrt[3]{3^2 x}}{10 \cdot \sqrt[3]{3^2 x}}$$
$$= \frac{\sqrt[3]{3^3 x^6}}{10\sqrt[3]{3^2 x}}$$
$$= \frac{3x^2}{10\sqrt[3]{9x}}$$

45.
$$\sqrt{\frac{18x^4 y^6}{3z}} = \sqrt{\frac{6x^4 y^6}{z}}$$
$$= \frac{\sqrt{6x^4 y^6}}{\sqrt{z}}$$
$$= \frac{x^2 y^3 \sqrt{6}}{\sqrt{z}}$$
$$= \frac{x^2 y^3 \sqrt{6} \cdot \sqrt{6}}{\sqrt{z} \cdot \sqrt{6}}$$
$$= \frac{6x^2 y^3}{\sqrt{6z}}$$

47. Answers may vary.

49.
$$\frac{6}{2 - \sqrt{7}} = \frac{6(2 + \sqrt{7})}{(2 - \sqrt{7})(2 + \sqrt{7})}$$
$$= \frac{6(2 + \sqrt{7})}{2^2 - (\sqrt{7})^2}$$
$$= \frac{6(2 + \sqrt{7})}{4 - 7}$$
$$= \frac{6(2 + \sqrt{7})}{-3}$$
$$= -2(2 + \sqrt{7})$$

51. $\dfrac{-7}{\sqrt{x}-3} = \dfrac{-7(\sqrt{x}+3)}{(\sqrt{x}-3)(\sqrt{x}+3)}$

$\qquad = \dfrac{-7(\sqrt{x}+3)}{(\sqrt{x})^2 - 3^2}$

$\qquad = \dfrac{-7(\sqrt{x}+3)}{x-9}$

$\qquad = \dfrac{7(\sqrt{x}+3)}{9-x}$

53. $\dfrac{\sqrt{2}-\sqrt{3}}{\sqrt{2}+\sqrt{3}} = \dfrac{(\sqrt{2}-\sqrt{3})(\sqrt{2}-\sqrt{3})}{(\sqrt{2}+\sqrt{3})(\sqrt{2}-\sqrt{3})}$

$\qquad = \dfrac{(\sqrt{2})^2 - 2\sqrt{2}\sqrt{3} + (\sqrt{3})^2}{(\sqrt{2})^2 - (\sqrt{3})^2}$

$\qquad = \dfrac{2 - 2\sqrt{6} + 3}{2-3}$

$\qquad = \dfrac{5 - 2\sqrt{6}}{-1}$

$\qquad = -5 + 2\sqrt{6}$

55. $\dfrac{\sqrt{a}+1}{2\sqrt{a}-\sqrt{b}}$

$\qquad = \dfrac{(\sqrt{a}+1)(2\sqrt{a}+\sqrt{b})}{(2\sqrt{a}-\sqrt{b})(2\sqrt{a}+\sqrt{b})}$

$\qquad = \dfrac{\sqrt{a}(2\sqrt{a}) + \sqrt{a}\sqrt{b} + 1(2\sqrt{a}) + 1\sqrt{b}}{(2\sqrt{a})^2 - (\sqrt{b})^2}$

$\qquad = \dfrac{2a + \sqrt{ab} + 2\sqrt{a} + \sqrt{b}}{4a-b}$

57. $\dfrac{8}{1+\sqrt{10}} = \dfrac{8(1-\sqrt{10})}{(1+\sqrt{10})(1-\sqrt{10})}$

$\qquad = \dfrac{8(1-\sqrt{10})}{1^2 - (\sqrt{10})^2}$

$\qquad = \dfrac{8(1-\sqrt{10})}{1-10}$

$\qquad = \dfrac{8(1-\sqrt{10})}{-9}$

$\qquad = -\dfrac{8(1-\sqrt{10})}{9}$

59. $\dfrac{\sqrt{x}}{\sqrt{x}+\sqrt{y}} = \dfrac{\sqrt{x}(\sqrt{x}-\sqrt{y})}{(\sqrt{x}+\sqrt{y})(\sqrt{x}-\sqrt{y})}$

$\qquad = \dfrac{\sqrt{x}\sqrt{x} - \sqrt{x}\sqrt{y}}{(\sqrt{x})^2 - (\sqrt{y})^2}$

$\qquad = \dfrac{x - \sqrt{xy}}{x-y}$

61. $\dfrac{2\sqrt{3}+\sqrt{6}}{4\sqrt{3}-\sqrt{6}} = \dfrac{(2\sqrt{3}+\sqrt{6})(4\sqrt{3}+\sqrt{6})}{(4\sqrt{3}-\sqrt{6})(4\sqrt{3}+\sqrt{6})}$

$\qquad = \dfrac{8\cdot 3 + 2\sqrt{18} + 4\sqrt{18} + 6}{16\cdot 3 - 6}$

$\qquad = \dfrac{24 + 6\sqrt{18} + 6}{48-6}$

$\qquad = \dfrac{30 + 6\sqrt{9}\sqrt{2}}{42}$

$\qquad = \dfrac{30 + 18\sqrt{2}}{42}$

$\qquad = \dfrac{6(5 + 3\sqrt{2})}{6\cdot 7}$

$\qquad = \dfrac{5 + 3\sqrt{2}}{7}$

63. $\dfrac{2-\sqrt{11}}{6} = \dfrac{\left(2-\sqrt{11}\right)\left(2+\sqrt{11}\right)}{6\left(2+\sqrt{11}\right)}$

$= \dfrac{2^2 - \left(\sqrt{11}\right)^2}{6\left(2+\sqrt{11}\right)}$

$= \dfrac{4-11}{6\left(2+\sqrt{11}\right)}$

$= \dfrac{-7}{6\left(2+\sqrt{11}\right)}$

65. $\dfrac{2-\sqrt{7}}{-5} = \dfrac{\left(2-\sqrt{7}\right)\left(2+\sqrt{7}\right)}{-5\left(2+\sqrt{7}\right)}$

$= \dfrac{4-7}{-5\left(2+\sqrt{7}\right)}$

$= \dfrac{-3}{-5\left(2+\sqrt{7}\right)}$

$= \dfrac{-1(3)}{-5\left(2+\sqrt{7}\right)}$

$= \dfrac{3}{5\left(2+\sqrt{7}\right)}$

67. $\dfrac{\sqrt{x}+3}{\sqrt{x}} = \dfrac{\left(\sqrt{x}+3\right)\left(\sqrt{x}-3\right)}{\sqrt{x}\left(\sqrt{x}-3\right)}$

$= \dfrac{x-9}{x-3\sqrt{x}}$

69. $\dfrac{\sqrt{2}-1}{\sqrt{2}+1} = \dfrac{\left(\sqrt{2}-1\right)\left(\sqrt{2}+1\right)}{\left(\sqrt{2}+1\right)\left(\sqrt{2}+1\right)}$

$= \dfrac{2-1}{2+2\sqrt{2}+1}$

$= \dfrac{1}{3+2\sqrt{2}}$

71. $\dfrac{\sqrt{x}+1}{\sqrt{x}-1} = \dfrac{\left(\sqrt{x}+1\right)\left(\sqrt{x}-1\right)}{\left(\sqrt{x}-1\right)\left(\sqrt{x}-1\right)}$

$= \dfrac{x-1}{x-2\sqrt{x}+1}$

73. $r = \sqrt{\dfrac{A}{4\pi}}$

$= \dfrac{\sqrt{A}}{\sqrt{4\pi}}$

$= \dfrac{\sqrt{A}}{2\sqrt{\pi}}$

$= \dfrac{\sqrt{A}\sqrt{\pi}}{2\sqrt{\pi}\sqrt{\pi}}$

$= \dfrac{\sqrt{A\pi}}{2\pi}$

75. Answers may vary.

77. $2x - 7 = 3(x-4)$
$2x - 7 = 3x - 12$
$12 - 7 = 3x - 2x$
$5 = x$
The solution is 5.

79. $(x-6)(2x+1) = 0$
$x - 6 = 0$ or $2x + 1 = 0$
$x = 6$ or $x = -\dfrac{1}{2}$

The solutions are $-\dfrac{1}{2}$ and 6.

81. $x^2 - 8x = -12$
$x^2 - 8x + 12 = 0$
$(x-6)(x-2) = 0$
$x - 6 = 0$ or $x - 2 = 0$
$x = 6$ or $x = 2$
The solutions are 2 and 6.

Exercise Set 7.6

1. $\sqrt{2x} = 4$
$2x = 4^2$
$2x = 16$
$x = 8$
The solution is 8.
Check: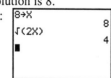

3. $\sqrt{x-3} = 2$

$x - 3 = 2^2$

$x - 3 = 4$

$x = 7$

The solution is 7.

Check:

```
7→X
                  7
√(X−3)
                  2
■
```

5. $\sqrt{2x} = -4$

The solution is $\varnothing$ since a principle square root does not yield a negative number.

7. $\sqrt{4x-3} - 5 = 0$

$\sqrt{4x-3} = 5$

$4x - 3 = 5^2$

$4x - 3 = 25$

$4x = 28$

$x = 7$

The solution is 7.

Check:

```
7→X
                  7
√(4X−3)−5
                  0
■
```

9. $\sqrt{2x-3} - 2 = 1$

$\sqrt{2x-3} = 3$

$2x - 3 = 3^2$

$2x - 3 = 9$

$2x = 12$

$x = 6$

The solution is 6.

Check:

```
6→X
                  6
√(2X−3)−2
                  1
■
```

11. $\sqrt[3]{6x} = -3$

$6x = (-3)^3$

$6x = -27$

$x = \dfrac{-27}{6} = -\dfrac{9}{2}$

The solution is $-\dfrac{9}{2}$.

Check: $y_1 = \sqrt[3]{6x}, y_2 = -3$

```
               5
  −10 |           | 5
     Intersection
     X=-4.5    Y=-3
              −10
```

The intersection of the two graphs has an x-value of -4.5 or $-\dfrac{9}{2}$, the solution of the equation.

13. $\sqrt[3]{x-2} - 3 = 0$

$\sqrt[3]{x-2} = 3$

$x - 2 = 3^3$

$x - 2 = 27$

$x = 29$

The solution is 29.

Check: $y_1 = \sqrt[3]{x-2} - 3, y_2 = 0$

```
               5
  −15 |           | 55
     Intersection
     X=29      Y=0
              −10
```

The intersection of the two graphs has an x-value of 29, the solution of the equation.

15. $\sqrt{13-x} = x-1$

$y_1 = \sqrt{13-x}, y_2 = x-1$

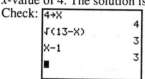

The intersection of the two graphs has an x-value of 4. The solution is 4.

Check:

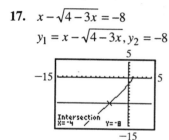

17. $x - \sqrt{4-3x} = -8$

$y_1 = x - \sqrt{4-3x}, y_2 = -8$

The intersection of the two graphs has an x-value of -4. The solution is -4.

Check:

19. $\sqrt{y+5} = 2 - \sqrt{y-4}$

$y_1 = \sqrt{x+5}, y_2 = 2 - \sqrt{x-4}$

The graphs do not intersect. The solution is $\varnothing$. Since there is no solution, we can not check the result numerically.

Check algebraically:

$$\sqrt{y+5} = 2 - \sqrt{y-4}$$

$$y+5 = \left(2 - \sqrt{y-4}\right)^2$$

$$y+5 = 4 - 4\sqrt{y-4} + (y-4)$$

$$y+5 = y - 4\sqrt{y-4}$$

$$5 = -4\sqrt{y-4}$$

$$25 = 16(y-4)$$

$$\frac{25}{16} = y-4$$

$$y = \frac{89}{16} = 5\frac{9}{16}$$

which we discard as extraneous.

21. $\sqrt{x-3} + \sqrt{x+2} = 5$

$y_1 = \sqrt{x-3} + \sqrt{x+2}, y_2 = 5$

The intersection of the two graphs has an x-value of 7. The solution is 7.

Check:

23. $\sqrt{3x-2} = 5$

$$3x - 2 = 5^2$$
$$3x - 2 = 25$$
$$3x = 27$$
$$x = 9$$

The solution is 9.

Check: $y_1 = \sqrt{3x-2}, y_2 = 5$

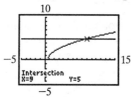

The intersection of the two graphs has an x-value of 9, the solution of the equation.

25. $-\sqrt{2x} + 4 = -6$

$$10 = \sqrt{2x}$$
$$10^2 = 2x$$
$$100 = 2x$$
$$x = 50$$

The solution is 50.

Check: $y_1 = -\sqrt{2x} + 4, y_2 = -6$

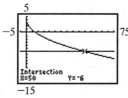

The intersection of the two graphs has an x-value of 50, the solution of the equation.

27. $\sqrt{3x+1} + 2 = 0$

$$\sqrt{3x+1} = -2$$

The solution is $\varnothing$ since a principle square root does not yield a negative number.

Check: $y_1 = \sqrt{3x+1} + 2, y_2 = 0$

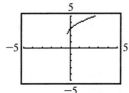

The graphs do not intersect.

29. $\sqrt[4]{4x+1} - 2 = 0$

$$\sqrt[4]{4x+1} = 2$$
$$4x + 1 = 2^4$$
$$4x + 1 = 16$$
$$4x = 15$$
$$x = \frac{15}{4}$$

The solution is $\frac{15}{4}$.

Check: $y_1 = \sqrt[4]{4x+1} - 2, y_2 = 0$

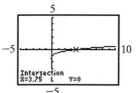

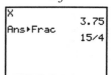

The intersection of the two graphs has an x-value of 3.75 or $\frac{15}{4}$, the solution of the equation.

31. $\sqrt{4x-3} = 7$

$$4x - 3 = 7^2$$
$$4x - 3 = 49$$
$$4x = 52$$
$$x = 13$$

The solution is 13.

Check: $y_1 = \sqrt{4x-3}, y_2 = 7$

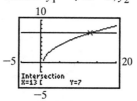

The intersection of the two graphs has an x-value of 13, the solution of the equation.

33. $\sqrt[3]{6x-3} - 3 = 0$

$\qquad \sqrt[3]{6x-3} = 3$

$\qquad 6x - 3 = 3^3$

$\qquad 6x - 3 = 27$

$\qquad 6x = 30$

$\qquad x = 5$

The solution is 5.

Check: $y_1 = \sqrt[3]{6x-3} - 3, y_2 = 0$

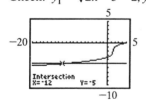

The intersection of the two graphs has an x-value of 5, the solution of the equation.

35. $\sqrt[3]{2x-3} - 2 = -5$

$\qquad \sqrt[3]{2x-3} = -3$

$\qquad 2x - 3 = (-3)^3$

$\qquad 2x - 3 = -27$

$\qquad 2x = -24$

$\qquad x = -12$

The solution is -12.

Check: $y_1 = \sqrt[3]{2x-3} - 2, y_2 = -5$

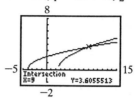

The intersection of the two graphs has an x-value of 12, the solution of the equation.

37. $\sqrt{x+4} = \sqrt{2x-5}$

$\qquad x + 4 = 2x - 5$

$\qquad 9 = x$

$\qquad x = 9$

The solution is 9.

Check: $y_1 = \sqrt{x+4}, y_2 = \sqrt{2x-5}$

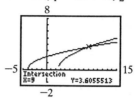

The intersection of the two graphs has an x-value of 9, the solution of the equation.

39. $\qquad x - \sqrt{1-x} = -5$

$\qquad x + 5 = \sqrt{1-x}$

$\qquad (x+5)^2 = 1 - x$

$\qquad x^2 + 10x + 25 = 1 - x$

$\qquad x^2 + 11x + 24 = 0$

$\qquad (x+3)(x+8) = 0$

$\qquad x + 3 = 0 \quad$ or $\quad x + 8 = 0$

$\qquad x = -3 \quad$ or $\qquad x = -8$

We discard -8 as extraneous, leaving -3 as the only solution.

Check: $y_1 = x - \sqrt{1-x}, y_2 = -5$

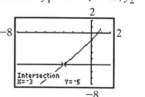

The intersection of the two graphs has an x-value of -3, the solution of the equation.

41. $\sqrt[3]{-6x-1} = \sqrt[3]{-2x-5}$
$$-6x-1 = -2x-5$$
$$4 = 4x$$
$$x = 1$$
The solution is 1.
Check: $y_1 = \sqrt[3]{-6x-1}, y_2 = \sqrt[3]{-2x-5}$

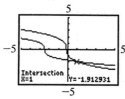

The intersection of the two graphs has an
x-value of 1, the solution of the equation.

43. $\sqrt{5x-1} - \sqrt{x+2} = 3$
$$\sqrt{5x-1} = \sqrt{x}+1$$
$$5x-1 = x+2\sqrt{x}+1$$
$$4x-2 = 2\sqrt{x}$$
$$2x-1 = \sqrt{x}$$
$$4x^2 - 4x + 1 = x$$
$$4x^2 - 5x + 1 = 0$$
$$(4x-1)(x-1) = 0$$
$$4x-1 = 0 \quad \text{or} \quad x-1 = 0$$
$$4x = 1 \quad \text{or} \quad x = 1$$
$$x = \frac{1}{4}$$

We discard $\frac{1}{4}$ as extraneous, leaving 1 as
the only solution.
Check: $y_1 = \sqrt{5x-1} - \sqrt{x+2}, y_2 = 3$

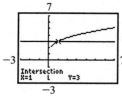

The intersection of the two graphs has an
x-value of 1, the solution of the equation.

45. $\sqrt{2x-1} = \sqrt{1-2x}$
$$2x-1 = 1-2x$$
$$4x = 2$$
$$x = \frac{1}{2}$$
The solution is $\frac{1}{2}$.
Check: $y_1 = \sqrt{2x-1}, y_2 = \sqrt{1-2x}$

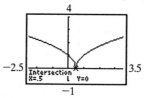

The intersection of the two graphs has an
x-value of 0.5 or $\frac{1}{2}$, the solution of
the equation.

47.
$$\sqrt{3x+4} - 1 = \sqrt{2x+1}$$
$$(3x+4) - 2\sqrt{3x+4} + 1 = 2x+1$$
$$x+4 = 2\sqrt{3x+4}$$
$$x^2 + 8x + 16 = 4(3x+4)$$
$$x^2 - 4x = 0$$
$$x(x-4) = 0$$
$$x = 0 \quad \text{or} \quad x = 4$$
The solutions are 0 and 4.
Check : $y_1 = \sqrt{3x+4} - 1, y_2 = \sqrt{2x+1}$

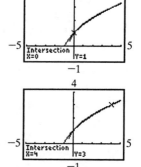

The intersections of the two graphs have
x-values of 0 and 4, the solutions of the
equation.

49. $\sqrt{y+3} - \sqrt{y-3} = 1$

$\sqrt{y+3} = 1 + \sqrt{y-3}$

$\left(\sqrt{y+3}\right)^2 = \left(1+\sqrt{y-3}\right)^2$

$y+3 = 1 + 2\sqrt{y-3} + (y-3)$

$5 = 2\sqrt{y-3}$

$25 = 4(y-3)$

$\dfrac{25}{4} = y - 3$

$\dfrac{25}{4} + \dfrac{12}{4} = y$

$\dfrac{37}{4} = y$

The solution is $\dfrac{37}{4}$.

Check : $y_1 = \sqrt{x+3} - \sqrt{x-3},\ y_2 = 1$

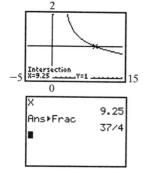

The intersection of the two graphs has an x-value of 9.25 or $\dfrac{37}{4}$, the solution of the equation.

51. Answers may vary. Possible answer: The left side of the equation on the third line should be

$(2x+5) + 2\sqrt{(2x+5)(4-x)} + 4 - x.$

53. Let c = the length of the hypotenuse of the right triangle. By the Pythagorean theorem,

$c^2 = 6^2 + 3^2 = 36 + 9 = 45$

or $c = \sqrt{45} = \sqrt{9}\sqrt{5} = 3\sqrt{5}$

The length is $3\sqrt{5}$ feet.

55. Let b = the length of the unknown leg of the right triangle. By the Pythagorean theorem,

$7^2 = 3^2 + b^2$

$49 = 9 + b^2$

$b^2 = 40$

$b = \sqrt{40} = \sqrt{4}\sqrt{10} = 2\sqrt{10}$

The length is $2\sqrt{10}$ meters.

57. Let b = the length of the unknown leg of the right triangle. By the Pythagorean theorem,

$\left(11\sqrt{5}\right)^2 = 9^2 + b^2$

$605 = 81 + b^2$

$b^2 = 524$

$b = \sqrt{524}$

$= \sqrt{4}\sqrt{131}$

$= 2\sqrt{131} \approx 22.9$

The length is $2\sqrt{131}$ meters

≈ 22.9 meters.

59. Let c = the length of the hypotenuse of the right triangle. By the Pythagorean theorem,

$c^2 = 7^2 + (7.2)^2 = 100.84.$

$c = \sqrt{100.84} \approx 10.0$

The length is $\sqrt{100.84}$ millimeters

≈ 10.0 millimeters.

61. Let c = the length of the hypotenuse of the right triangle as shown in the figure in the text. By the Pythagorean theorem,

$c^2 = 8^2 + 15^2 = 64 + 225 = 289.$

So, $c = \sqrt{289} = 17$ feet.

Thus, 17 feet of cable are needed.

63. Let c = the length of the hypotenuse of the right triangle as shown in the figure in the text. By the Pythagorean theorem,

$c^2 = 12^2 + 5^2 = 144 + 25 = 169.$

So, $x = \sqrt{169} = 13$ feet.

Thus, a 13-foot ladder is needed.

65.
$$r = \sqrt{\frac{A}{4\pi}}$$
$$1080 = \sqrt{\frac{A}{4\pi}}$$
$$(1080)^2 = \left(\sqrt{\frac{A}{4\pi}}\right)^2$$
$$1,166,400 = \frac{A}{4\pi}$$
$$14,657,415 \approx A$$
The surface area is 14,657,415 square miles.

67.
$$v = \sqrt{2gh}$$
$$80 = \sqrt{2(32)h}$$
$$(80)^2 = \left(\sqrt{2(32)h}\right)^2$$
$$6400 = 2(32)h$$
$$100 = h$$
The object fell 100 feet.

69.
$$\sqrt{\sqrt{x+3} + \sqrt{x}} = \sqrt{3}$$
$$\left(\sqrt{\sqrt{x+3} + \sqrt{x}}\right)^2 = \left(\sqrt{3}\right)^2$$
$$\sqrt{x+3} + \sqrt{x} = 3$$
$$\sqrt{x+3} = 3 - \sqrt{x}$$
$$\left(\sqrt{x+3}\right)^2 = \left(3 - \sqrt{x}\right)^2$$
$$x + 3 = 9 - 6\sqrt{x} + x$$
$$6\sqrt{x} = 6$$
$$\sqrt{x} = 1$$
$$x = 1$$
The solution is 1.

71. We need to solve the equation:
$$80\sqrt[3]{x} + 500 = 1620$$
$$80\sqrt[3]{x} = 1120$$
$$\sqrt[3]{x} = 14$$
$$x = 14^3$$
$$x = 2744$$
Thus, the company needs to make fewer than 2744 deliveries per day, or 2743 deliveries.

73. Answers may vary.

75. It is not a function since the vertical line, $x = 1$, intersects the graph in more than one point.

77. It is not a function since it is a vertical line.

79. It is a function since no vertical line intersects the graph in more than one point.

81.
$$\frac{\frac{1}{y} + \frac{4}{5}}{-\frac{3}{20}} = \frac{20y\left(\frac{1}{y} + \frac{4}{5}\right)}{20y\left(-\frac{3}{20}\right)}$$
$$= \frac{20y\left(\frac{1}{y}\right) + 20y\left(\frac{4}{5}\right)}{20y\left(-\frac{3}{20}\right)}$$
$$= \frac{20 + 16y}{-3y}$$
$$= -\frac{20 + 16y}{3y}$$

83.
$$\frac{\frac{1}{y} + \frac{1}{x}}{\frac{1}{y} - \frac{1}{x}} = \frac{xy\left(\frac{1}{y} + \frac{1}{x}\right)}{xy\left(\frac{1}{y} - \frac{1}{x}\right)}$$
$$= \frac{xy\left(\frac{1}{y}\right) + xy\left(\frac{1}{x}\right)}{xy\left(\frac{1}{y}\right) - xy\left(\frac{1}{x}\right)}$$
$$= \frac{x + y}{x - y}$$

85. Let $u = x^2 - x$.

$$\sqrt{u+7} = 2u - 1$$

$$\left(\sqrt{u+7}\right)^2 = (2u-1)^2$$

$$u + 7 = 4u^2 - 4u + 1$$

$$0 = 4u^2 - 5u - 6$$

$$0 = (4u+3)(u-2)$$

$$4u + 3 = 0 \quad \text{or} \quad u - 2 = 0$$

$$4u = -3 \qquad\qquad u = 2$$

$$u = -\frac{3}{4} \qquad\qquad u = 2$$

$u = -\dfrac{3}{4}$: Replace u with $x^2 - x$.

$$x^2 - x = -\frac{3}{4}$$

$$4(x^2 - x) = 4\left(-\frac{3}{4}\right)$$

$$4x^2 - 4x = -3$$

$$4x^2 - 4x + 3 = 0$$

$4x^2 - 4x + 3$ won't factor

$u = 2$: Replace u with $x^2 - x$.

$$x^2 - x = 2$$

$$x^2 - x - 2 = 0$$

$$(x-2)(x+1) = 0$$

$$x - 2 = 0 \quad \text{or} \quad x + 1 = 0$$

$$x = 2 \qquad\qquad x = -1$$

The solutions are -1 and 2.

87. Let $u = x^2 + 6x$.

$$u = 4\sqrt{u}$$

$$(u)^2 = \left(4\sqrt{u}\right)^2$$

$$u^2 = 16u$$

$$u^2 - 16u = 0$$

$$u(u-16) = 0$$

$$u = 0 \quad \text{or} \quad u - 16 = 0$$

$$u = 16$$

$u = 0$: Replace u with $x^2 + 6x$.

$$x^2 + 6x = 0$$

$$x(x+6) = 0$$

$$x = 0 \quad \text{or} \quad x + 6 = 0$$

$$x = -6$$

$u = 16$: Replace u with $x^2 + 6x$.

$$x^2 + 6x = 16$$

$$x^2 + 6x - 16 = 0$$

$$(x+8)(x-2) = 0$$

$$x + 8 = 0 \quad \text{or} \quad x - 2 = 0$$

$$x = -8 \qquad\qquad x = 2$$

The solutions are -8, -6, 0, and 2.

Section 7.7

Mental Math

1. $\sqrt{-81} = \sqrt{81}\sqrt{-1} = 9i$

2. $\sqrt{-49} = \sqrt{49}\sqrt{-1} = 7i$

3. $\sqrt{-7} = \sqrt{-1}\sqrt{7} = i\sqrt{7}$

4. $\sqrt{-3} = \sqrt{-1}\sqrt{3} = i\sqrt{3}$

5. $-\sqrt{16} = -4$

6. $-\sqrt{4} = -2$

7. $\sqrt{-64} = \sqrt{64}\sqrt{-1} = 8i$

8. $\sqrt{-100} = \sqrt{100}\sqrt{-1} = 10i$

Exercise Set 7.7

1. $\sqrt{-24} = \sqrt{-1}\sqrt{24} = i\sqrt{4}\sqrt{6} = 2i\sqrt{6}$

3. $-\sqrt{-36} = -\sqrt{36}\sqrt{-1} = -6i$

5. $8\sqrt{-63} = 8\sqrt{-1}\sqrt{63}$
$\quad\quad = 8i\sqrt{9}\sqrt{7}$
$\quad\quad = 8\cdot 3i\sqrt{7}$
$\quad\quad = 24i\sqrt{7}$

7. $-\sqrt{54} = -\sqrt{9}\sqrt{6} = -3\sqrt{6}$

9. $\sqrt{-2}\sqrt{-7} = \left(i\sqrt{2}\right)\left(i\sqrt{7}\right)$
$\quad\quad = i^2\sqrt{14}$
$\quad\quad = (-1)\sqrt{14}$
$\quad\quad = -\sqrt{14}$

11. $\sqrt{-5}\sqrt{-10} = \left(i\sqrt{5}\right)\left(i\sqrt{10}\right)$
$\quad\quad = i^2\sqrt{50}$
$\quad\quad = (-1)\sqrt{25}\sqrt{2}$
$\quad\quad = -5\sqrt{2}$

13. $\sqrt{16}\sqrt{-1} = 4i$

15. $\dfrac{\sqrt{-9}}{\sqrt{3}} = \dfrac{3i}{\sqrt{3}} = \dfrac{3i\cdot\sqrt{3}}{\sqrt{3}\cdot\sqrt{3}} = \dfrac{3i\sqrt{3}}{3} = i\sqrt{3}$

17. $\dfrac{\sqrt{-80}}{\sqrt{-10}} = \dfrac{i\sqrt{80}}{i\sqrt{10}}$
$\quad\quad = \sqrt{\dfrac{80}{10}}$
$\quad\quad = \sqrt{8}$
$\quad\quad = \sqrt{4}\sqrt{2}$
$\quad\quad = 2\sqrt{2}$

19. $(4-7i)+(2+3i) = (4+2)+(-7+3)i$
$\quad\quad\quad\quad\quad\quad\quad = 6-4i$

21. $(6+5i)-(8-i) = 6+5i-8+i$
$\quad\quad\quad\quad\quad\quad\quad = (6-8)+(5+1)i$
$\quad\quad\quad\quad\quad\quad\quad = -2+6i$

23. $6-(8+4i) = 6-8-4i = -2-4i$

25. $6i(2-3i) = 6i(2)-6i(3i)$
$\quad\quad\quad\quad = 12i-18i^2$
$\quad\quad\quad\quad = 12i-18(-1)$
$\quad\quad\quad\quad = 18+12i$

27. $\left(\sqrt{3}+2i\right)\left(\sqrt{3}-2i\right) = \left(\sqrt{3}\right)^2 - (2i)^2$
$\quad\quad\quad\quad\quad\quad\quad\quad\quad\quad = 3-4i^2$
$\quad\quad\quad\quad\quad\quad\quad\quad\quad\quad = 3+4$
$\quad\quad\quad\quad\quad\quad\quad\quad\quad\quad = 7$

29. $(4-2i)^2 = 4^2 - 2(4)(2i)+(2i)^2$
$\quad\quad\quad\quad = 16-16i+4i^2$
$\quad\quad\quad\quad = 16-16i+4(-1)$
$\quad\quad\quad\quad = 16-4-16i$
$\quad\quad\quad\quad = 12-16i$

31. $\dfrac{4}{i} = \dfrac{4\cdot(-i)}{i\cdot(-i)} = \dfrac{-4i}{-i^2} = \dfrac{-4i}{-(-1)} = \dfrac{-4i}{1} = -4i$

33. $\dfrac{7}{4+3i} = \dfrac{7(4-3i)}{(4+3i)(4-3i)}$
$\quad\quad\quad = \dfrac{28-21i}{4^2-(3i)^2}$
$\quad\quad\quad = \dfrac{28-21i}{16+9}$
$\quad\quad\quad = \dfrac{28-21i}{25}$
$\quad\quad\quad = \dfrac{28}{25} - \dfrac{21}{25}i$

35. $\dfrac{3+5i}{1+i} = \dfrac{(3+5i)(1-i)}{(1+i)(1-i)}$

$= \dfrac{3-3i+5i-5i^2}{1^2-i^2}$

$= \dfrac{3+2i-5(-1)}{1+1}$

$= \dfrac{3+5+2i}{2}$

$= \dfrac{8+2i}{2}$

$= 4+i$

37. $\dfrac{5-i}{3-2i} = \dfrac{(5-i)(3+2i)}{(3-2i)(3+2i)}$

$= \dfrac{15+10i-3i-2i^2}{3^2-(2i)^2}$

$= \dfrac{15+7i-2(-1)}{9+4}$

$= \dfrac{15+2+7i}{13}$

$= \dfrac{17+7i}{13}$

$= \dfrac{17}{13}+\dfrac{7}{13}i$

39. $(7i)(-9i) = -63i^2 = -63(-1) = 63$

41. $(6-3i)-(4-2i) = 6-3i-4+2i = 2-i$

43. $(6-2i)(3+i) = 18+6i-6i-2i^2$

$= 18-2(-1)$

$= 20$

45. $(8-3i)+(2+3i) = 8-3i+2+3i$

$= 10$

47. $(1-i)(1+i) = 1^2-i^2 = 1+1 = 2$

49. $\dfrac{16+15i}{-3i} = \dfrac{(16+5i)\cdot i}{-3i\cdot i}$

$= \dfrac{16i+15i^2}{-3i^2}$

$= \dfrac{16i+15(-1)}{-3(-1)}$

$= \dfrac{-15+16i}{3}$

$= \dfrac{-15}{3}+\dfrac{16}{3}i$

$= -5+\dfrac{16}{3}i$

51. $(9+8i)^2 = 9^2+2(9)(8i)+(8i)^2$

$= 81+144i+64i^2$

$= 81+144i+64(-1)$

$= 81-64+144i$

$= 17+144i$

53. $\dfrac{2}{3+i} = \dfrac{2(3-i)}{(3+i)(3-i)}$

$= \dfrac{2(3-i)}{3^2-i^2}$

$= \dfrac{6-2i}{9+1}$

$= \dfrac{6}{10}-\dfrac{2}{10}i$

$= \dfrac{3}{5}-\dfrac{1}{5}i$

55. $(5-6i)-4i = 5-6i-4i = 5-10i$

57. $\dfrac{2-3i}{2+i} = \dfrac{(2-3i)(2-i)}{(2+i)(2-i)}$

$= \dfrac{4-2i-6i+3i^2}{2^2-i^2}$

$= \dfrac{4-8i+3(-1)}{4+1}$

$= \dfrac{4-3-8i}{5}$

$= \dfrac{1-8i}{5}$

$= \dfrac{1}{5}-\dfrac{8}{5}i$

59. $(2+4i)+(6-5i) = (2+6)+(4-5)i$
$$= 8-i$$

61. $i^8 = (i^4)^2 = 1^2 = 1$

63. $i^{21} = i^{20}i = (i^4)^5 i = 1^5 i = 1i = i$

65. $i^{11} = i^{10}i = (i^2)^5 i = (-1)^5 i = (-1)i = -i$

67. $i^{-6} = (i^2)^{-3} = (-1)^{-3} = \dfrac{1}{(-1)^3} = \dfrac{1}{-1} = -1$

69. $i^3 + i^4 = -i + 1 = 1 - i$

71. $i^6 + i^8 = (i^4)(i^2) + (i^4)^2$
$$= 1(-1) + 1^2$$
$$= -1 + 1$$
$$= 0$$

73. $2 + \sqrt{-9} = 2 + \sqrt{9}\sqrt{-1} = 2 + 3i$

75. $\dfrac{6 + \sqrt{-18}}{3} = \dfrac{6 + 3i\sqrt{2}}{3}$
$$= \dfrac{6}{3} + \dfrac{3i\sqrt{2}}{3}$$
$$= 2 + i\sqrt{2}$$

77. $\dfrac{5 - \sqrt{-75}}{10} = \dfrac{5 - 5i\sqrt{3}}{10}$
$$= \dfrac{5}{10} - \dfrac{5i\sqrt{3}}{10}$$
$$= \dfrac{1}{2} - \dfrac{\sqrt{3}}{2}i$$

79. Answers may vary.

81. $\left(8 - \sqrt{-4}\right) - \left(2 + \sqrt{-16}\right)$
$$= (8 - 2i) - (2 + 4i)$$
$$= 8 - 2i - 2 - 4i$$
$$= (8 - 2) + (-2 - 4)i$$
$$= 6 - 6i$$

83.
$$x^2 + 2x = -2$$
$$(-1+i)^2 + 2(-1+i) \stackrel{?}{=} -2$$
$$1 - 2i + i^2 - 2 + 2i \stackrel{?}{=} -2$$
$$1 - 2i - 1 - 2 + 2i \stackrel{?}{=} -2$$
$$-2 = -2$$
Yes, $-1 + i$ is a solution.

85. $x + 57 + 90 = 180$
$$x + 147 = 180$$
$$x = 33$$
The unknown angle is 33°.

87.
$$
\begin{array}{r|rrrrr}
-2 & 5 & 0 & -3 & 0 & 2 \\
   &   & -10 & 20 & -34 & 68 \\
\hline
   & 5 & -10 & 17 & -34 & 70
\end{array}
$$
$(5x^4 - 3x^2 + 2) \div (x + 2)$
$$= 5x^3 - 10x^2 + 17x - 34 + \dfrac{70}{x+2}$$

89. 5 people reported an average checking balance of $0 to $100.

91. $6 + 2 + 3 = 11$
11 people reported an average checking balance of $301 or more.

93. $\dfrac{5}{30} = 0.166\overline{6}$
16.7% of the people reprted an average checking balance of $0 to $100.

Chapter 7 Review

1. $\sqrt{81} = 9$ because $9^2 = 81$.

2. $\sqrt[4]{81} = 3$ because $3^4 = 81$.

3. $\sqrt[3]{-8} = -2$ because $(-2)^3 = -8$.

4. $\sqrt[4]{-16}$ is not a real number since there is no real number whose 4th power is negative.

5. $-\sqrt{\dfrac{1}{49}} = -\dfrac{1}{7}$ because $\left(\dfrac{1}{7}\right)^2 = \dfrac{1}{49}$.

6. $\sqrt{x^{64}} = x^{32}$

because $(x^{32})^2 = x^{32 \cdot 2} = x^{64}$.

7. $-\sqrt{36} = -6$ because $(6)^2 = 36$.

8. $\sqrt[3]{64} = 4$ because $4^3 = 64$.

9. $\sqrt[3]{-a^6 b^9} = \sqrt[3]{-1} \sqrt[3]{a^6} \sqrt[3]{b^9}$

$= (-1)a^2 b^3$

$= -a^2 b^3$

10. $\sqrt{16a^4 b^{12}} = \sqrt{16}\sqrt{a^4}\sqrt{b^{12}} = 4a^2 b^6$

11. $\sqrt[5]{32a^5 b^{10}} = \sqrt[5]{32}\sqrt[5]{a^5}\sqrt[5]{b^{10}} = 2ab^2$

12. $\sqrt[5]{-32x^{15} y^{20}} = \sqrt[5]{-32}\sqrt[5]{x^{15}}\sqrt[5]{y^{20}}$

$= -2x^3 y^4$

13. $\sqrt{\dfrac{x^{12}}{36y^2}} = \dfrac{\sqrt{x^{12}}}{\sqrt{36y^2}} = \dfrac{x^6}{\sqrt{36}\sqrt{y^2}} = \dfrac{x^6}{6y}$

14. $\sqrt[3]{\dfrac{27y^3}{z^{12}}} = \dfrac{\sqrt[3]{27y^3}}{\sqrt[3]{z^{12}}} = \dfrac{\sqrt[3]{27}\sqrt[3]{y^3}}{z^4} = \dfrac{3y}{z^4}$

15. $\sqrt{(-x)^2} = |-x| = |x|$

16. $\sqrt[4]{(x^2 - 4)^4} = |x^2 - 4|$

17. $\sqrt[3]{(-27)^3} = -27$

18. $\sqrt[5]{(-5)^5} = -5$

19. $-\sqrt[5]{x^5} = -x$

20. $\sqrt[4]{16(2y + z)^{12}} = \sqrt[4]{16}\sqrt[4]{(2y + z)^{12}}$

$= 2|(2y + z)^3|$

21. $\sqrt{25(x - y)^{10}} = \sqrt{25}\sqrt{(x - y)^{10}}$

$= 5|(x - y)^5|$

22. $\sqrt[5]{-y^5} = \sqrt[5]{-1}\sqrt[5]{y^5} = (-1)y = -y$

23. $\sqrt[9]{-x^9} = \sqrt[9]{-1}\sqrt[9]{x^9} = -x$

24. $f(x) = \sqrt{x} + 3$

Domain: $[0, \infty)$

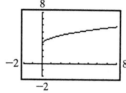

25. $g(x) = \sqrt[3]{x} - 3$

Domain: $(-\infty, \infty)$

x	-5	2	3	4	11
$g(x)$	-2	-1	0	1	2

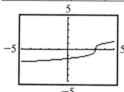

26. $\left(\dfrac{1}{81}\right)^{1/4} = \sqrt[4]{\dfrac{1}{81}} = \dfrac{\sqrt[4]{1}}{\sqrt[4]{81}} = \dfrac{1}{3}$

27. $\left(-\dfrac{1}{27}\right)^{1/3} = \sqrt[3]{\dfrac{-1}{27}} = \dfrac{\sqrt[3]{-1}}{\sqrt[3]{27}} = \dfrac{-1}{3} = -\dfrac{1}{3}$

28. $(-27)^{-1/3} = \dfrac{1}{(-27)^{1/3}} = \dfrac{1}{\sqrt[3]{-27}} = \dfrac{1}{-3} = -\dfrac{1}{3}$

29. $(-64)^{-1/3} = \dfrac{1}{(-64)^{1/3}} = \dfrac{1}{\sqrt[3]{-64}} = \dfrac{1}{-4} = -\dfrac{1}{4}$

30. $-9^{3/2} = -\left(\sqrt{9}\right)^3 = -3^3 = -27$

31. $64^{-1/3} = \dfrac{1}{(64)^{1/3}} = \dfrac{1}{\sqrt[3]{64}} = \dfrac{1}{4}$

32. $(-25)^{5/2} = \left(\sqrt{-25}\right)^5$ is not a real number, since there is no real number whose square is negative.

33. $\left(\dfrac{25}{49}\right)^{-3/2} = \dfrac{1}{\left(\frac{25}{49}\right)^{3/2}}$

$= \dfrac{1}{\left(\sqrt{\frac{25}{49}}\right)^3}$

$= \dfrac{1}{\left(\frac{5}{7}\right)^3}$

$= \dfrac{1}{\frac{125}{343}}$

$= \dfrac{343}{125}$

34. $\left(\dfrac{8}{27}\right)^{-2/3} = \dfrac{1}{\left(\frac{8}{27}\right)^{2/3}}$

$= \dfrac{1}{\left(\sqrt[3]{\frac{8}{27}}\right)^2}$

$= \dfrac{1}{\left(\frac{2}{3}\right)^2}$

$= \dfrac{1}{\frac{4}{9}}$

$= \dfrac{9}{4}$

35. $\left(-\dfrac{1}{36}\right)^{-1/4} = \dfrac{1}{\left(-\frac{1}{36}\right)^{1/4}}$ is not a real number, since there is no real number whose 4th power is negative.

36. $\sqrt[3]{x^2} = x^{2/3}$

37. $\sqrt[5]{5x^2y^3} = (5x^2y^3)^{1/5}$

$= 5^{1/5}(x^2)^{1/5}(y^3)^{1/5}$

$= 5^{1/5}x^{2/5}y^{3/5}$

38. $y^{4/5} = \sqrt[5]{y^4}$

39. $5(xy^2z^5)^{1/3} = 5\sqrt[3]{xy^2z^5}$

40. $(x+2y)^{-1/2} = \dfrac{1}{(x+2y)^{1/2}} = \dfrac{1}{\sqrt{x+2y}}$

41. $a^{1/3}a^{4/3}a^{1/2} = a^{(1/3+4/3+1/2)}$

$= a^{(5/3+1/2)}$

$= a^{(10/6+3/6)}$

$= a^{13/6}$

42. $\dfrac{b^{1/3}}{b^{4/3}} = b^{1/3-4/3} = b^{-1} = \dfrac{1}{b}$

43. $(a^{1/2}a^{-2})^3 = (a^{1/2-2})^3$

$= (a^{-3/2})^3$

$= a^{-9/2}$

$= \dfrac{1}{a^{9/2}}$

44. $(x^{-3}y^6)^{1/3} = (x^{-3})^{1/3}(y^6)^{1/3}$

$= x^{-1}y^2$

$= \dfrac{y^2}{x}$

45. $\left(\dfrac{b^{3/4}}{a^{-1/2}}\right)^8 = \dfrac{(b^{3/4})^8}{(a^{-1/2})^8} = \dfrac{b^6}{a^{-4}} = a^4b^6$

46. $\dfrac{x^{1/4}x^{-1/2}}{x^{2/3}} = \dfrac{x^{1/4-1/2}}{x^{2/3}}$

$= \dfrac{x^{-1/4}}{x^{2/3}}$

$= \dfrac{1}{x^{2/3-(-1/4)}}$

$= \dfrac{1}{x^{8/12+3/12}}$

$= \dfrac{1}{x^{11/12}}$

47. $\left(\dfrac{49c^{5/3}}{a^{-1/4}b^{5/6}}\right)^{-1} = \dfrac{49^{-1}c^{-5/3}}{a^{1/4}b^{-5/6}}$

$= \dfrac{b^{5/6}}{49a^{1/4}c^{5/3}}$

48. $a^{-1/4}(a^{5/4}-a^{9/4})$

$= a^{-1/4+5/4}-a^{-1/4+9/4}$

$= a - a^2$

49. $\sqrt{20} \approx 4.472$

50. $\sqrt[3]{-39} \approx -3.391$

51. $\sqrt[4]{726} \approx 5.191$

52. $56^{1/3} \approx 3.826$

53. $-78^{3/4} \approx -26.246$

54. $105^{-2/3} \approx 0.045$

55. $\sqrt[3]{2}\cdot\sqrt{7} = 2^{1/3}\cdot 7^{1/2}$

$= 2^{2/6}\cdot 7^{3/6}$

$= (2^2)^{1/6}\cdot(7^3)^{1/6}$

$= 4^{1/6}\cdot 343^{1/6}$

$= (4\cdot 343)^{1/6}$

$= \sqrt[6]{1372}$

56. $\sqrt[3]{3}\cdot\sqrt[4]{x} = 3^{1/3}\cdot x^{1/4}$

$= 3^{4/12}\cdot x^{3/12}$

$= (3^4)^{1/12}(x^3)^{1/12}$

$= (81)^{1/12}(x^3)^{1/12}$

$= (81x^3)^{1/12}$

$= \sqrt[12]{81x^3}$

57. $\sqrt{3}\cdot\sqrt{8} = \sqrt{24} = \sqrt{4\cdot 6} = 2\sqrt{6}$

58. $\sqrt[3]{7y}\cdot\sqrt[3]{x^2z} = \sqrt[3]{7x^2yz}$

59. $\dfrac{\sqrt{44x^3}}{\sqrt{11x}} = \sqrt{\dfrac{44x^3}{11x}} = \sqrt{4x^2} = 2x$

60. $\dfrac{\sqrt[4]{a^6b^{13}}}{\sqrt[4]{a^2b}} = \sqrt[4]{\dfrac{a^6b^{13}}{a^2b}}$

$= \sqrt[4]{a^{6-2}b^{13-1}}$

$= \sqrt[4]{a^4b^{12}}$

$= ab^3$

61. $\sqrt{60} = \sqrt{4\cdot 15} = 2\sqrt{15}$

62. $-\sqrt{75} = -\sqrt{25\cdot 3} = -5\sqrt{3}$

63. $\sqrt[3]{162} = \sqrt[3]{27\cdot 6} = 3\sqrt[3]{6}$

64. $\sqrt[3]{-32} = \sqrt[3]{(-8)(4)} = -2\sqrt[3]{4}$

65. $\sqrt{36x^7} = \sqrt{36x^6\cdot x} = 6x^3\sqrt{x}$

66. $\sqrt[3]{24a^5b^7} = \sqrt[3]{8a^3b^6\cdot 3a^2b} = 2ab^2\sqrt[3]{3a^2b}$

67. $\sqrt{\dfrac{p^{17}}{121}} = \dfrac{\sqrt{p^{16}\cdot p}}{\sqrt{121}} = \dfrac{p^8\sqrt{p}}{11}$

68. $\sqrt[3]{\dfrac{y^5}{27x^6}} = \dfrac{\sqrt[3]{y^3\cdot y^2}}{\sqrt[3]{27x^6}} = \dfrac{y\sqrt[3]{y^2}}{3x^2}$

69. $\sqrt[4]{\dfrac{xy^6}{81}} = \dfrac{\sqrt[4]{y^4 \cdot xy^2}}{\sqrt[4]{81}} = \dfrac{y\sqrt[4]{xy^2}}{3}$

70. $\sqrt{\dfrac{2x^3}{49y^4}} = \dfrac{\sqrt{x^2 \cdot 2x}}{\sqrt{49y^4}} = \dfrac{x\sqrt{2x}}{7y^2}$

71. $r = \sqrt{\dfrac{A}{\pi}}$

   a. $r = \sqrt{\dfrac{25}{\pi}} = \dfrac{\sqrt{25}}{\sqrt{\pi}} = \dfrac{5}{\sqrt{\pi}}$

      The radius is $\dfrac{5}{\sqrt{\pi}}$ meters or

      $\dfrac{5\sqrt{\pi}}{\pi}$ meters.

   b. $r = \sqrt{\dfrac{104}{\pi}} \approx 5.75$

      The radius is $\sqrt{\dfrac{104}{\pi}} \approx 5.75$ inches.

72. $x\sqrt{75xy} - \sqrt{27x^3y}$
$= x\sqrt{25}\sqrt{3xy} - \sqrt{9x^2}\sqrt{3xy}$
$= 5x\sqrt{3xy} - 3x\sqrt{3xy}$
$= 2x\sqrt{3xy}$

73. $2\sqrt{32x^2y^3} - xy\sqrt{98y}$
$= 2\sqrt{16x^2y^2 \cdot 2y} - xy\sqrt{49 \cdot 2y}$
$= 2\sqrt{16x^2y^2}\sqrt{2y} - xy\sqrt{49}\sqrt{2y}$
$= 8xy\sqrt{2y} - 7xy\sqrt{2y}$
$= xy\sqrt{2y}$

74. $\sqrt[3]{128} + \sqrt[3]{250} = \sqrt[3]{64 \cdot 2} + \sqrt[3]{125 \cdot 2}$
$= \sqrt[3]{64}\sqrt[3]{2} + \sqrt[3]{125}\sqrt[3]{2}$
$= 4\sqrt[3]{2} + 5\sqrt[3]{2}$
$= 9\sqrt[3]{2}$

75. $3\sqrt[4]{32a^5} - a\sqrt[4]{162a}$
$= 3\sqrt[4]{16a^4 \cdot 2a} - a\sqrt[4]{81 \cdot 2a}$
$= 3\sqrt[4]{16a^4}\sqrt[4]{2a} - a\sqrt[4]{81}\sqrt[4]{2a}$
$= 6a\sqrt[4]{2a} - 3a\sqrt[4]{2a}$
$= 3a\sqrt[4]{2a}$

76. $\dfrac{5}{\sqrt{4}} + \dfrac{\sqrt{3}}{3} = \dfrac{5}{2} + \dfrac{\sqrt{3}}{3}$
$= \dfrac{5(3) + 2\sqrt{3}}{2 \cdot 3} = \dfrac{15 + 2\sqrt{3}}{6}$

77. $\sqrt{\dfrac{8}{x^2}} - \sqrt{\dfrac{50}{16x^2}} = \dfrac{2\sqrt{2}}{x} - \dfrac{5\sqrt{2}}{4x}$
$= \dfrac{8\sqrt{2} - 5\sqrt{2}}{4x}$
$= \dfrac{3\sqrt{2}}{4x}$

78. $2\sqrt{50} - 3\sqrt{125} + \sqrt{98}$
$= 2\sqrt{25}\sqrt{2} - 3\sqrt{25}\sqrt{5} + \sqrt{49}\sqrt{2}$
$= 2(5)\sqrt{2} - 3(5)\sqrt{5} + 7\sqrt{2}$
$= 10\sqrt{2} - 15\sqrt{5} + 7\sqrt{2}$
$= 17\sqrt{2} - 15\sqrt{5}$

79. $2a\sqrt[4]{32b^5} - 3b\sqrt[4]{162a^4b} + \sqrt[4]{2a^4b^5} = 2a\sqrt[4]{16b^4}\sqrt[4]{2b} - 3b\sqrt[4]{81a^4}\sqrt[4]{2b} + \sqrt[4]{a^4b^4}\sqrt[4]{2b}$
$$= [(2a)(2b) - (3b)(3a) + ab]\sqrt[4]{2b}$$
$$= (4ab - 9ab + ab)\sqrt[4]{2b}$$
$$= -4ab\sqrt[4]{2b}$$

80. $\sqrt{3}\left(\sqrt{27} - \sqrt{3}\right) = \sqrt{81} - \sqrt{9} = 9 - 3 = 6$

81. $(\sqrt{x} - 3)^2 = \left(\sqrt{x}\right)^2 - 2(3)\sqrt{x} + 3^2 = x - 6\sqrt{x} + 9$

82. $\left(\sqrt{5} - 5\right)\left(2\sqrt{5} + 2\right) = 2\sqrt{25} + 2\sqrt{5} - 10\sqrt{5} - 10 = 10 - 8\sqrt{5} - 10 = -8\sqrt{5}$

83. $\left(2\sqrt{x} - 3\sqrt{y}\right)\left(2\sqrt{x} + 3\sqrt{y}\right) = \left(2\sqrt{x}\right)^2 - \left(3\sqrt{y}\right)^2 = 2^2\sqrt{x^2} - 3^2\sqrt{y^2} = 4x - 9y$

84. $\left(\sqrt{a} + 3\right)\left(\sqrt{a} - 3\right) = \left(\sqrt{a}\right)^2 - 3^2 = a - 9$

85. $\left(\sqrt[3]{a} + 2\right)^2 = \left(\sqrt[3]{a}\right)^2 + 2(2)\sqrt[3]{a} + 2^2 = \sqrt[3]{a^2} + 4\sqrt[3]{a} + 4$

86. $\left(\sqrt[3]{5x} + 9\right)\left(\sqrt[3]{5x} - 9\right) = \left(\sqrt[3]{5x}\right)^2 - 9^2 = \sqrt[3]{25x^2} - 81$

87. $\left(\sqrt[3]{a} + 4\right)\left(\sqrt[3]{a^2} - 4\sqrt[3]{a} + 16\right) = \sqrt[3]{a^3} - 4\sqrt[3]{a^2} + 16\sqrt[3]{a} + 4\sqrt[3]{a^2} - 16\sqrt[3]{a} + 64 = a + 64$

88. $\dfrac{3}{\sqrt{7}} = \dfrac{3\sqrt{7}}{\sqrt{7} \cdot \sqrt{7}} = \dfrac{3\sqrt{7}}{7}$

89. $\sqrt{\dfrac{x}{12}} = \dfrac{\sqrt{x}}{\sqrt{12}} = \dfrac{\sqrt{x}}{2\sqrt{3}} = \dfrac{\sqrt{x} \cdot \sqrt{3}}{2\sqrt{3} \cdot \sqrt{3}} = \dfrac{\sqrt{3x}}{2 \cdot 3} = \dfrac{\sqrt{3x}}{6}$

90. $\dfrac{5}{\sqrt[3]{4}} = \dfrac{5 \cdot \sqrt[3]{2}}{\sqrt[3]{4} \cdot \sqrt[3]{2}} = \dfrac{5\sqrt[3]{2}}{\sqrt[3]{8}} = \dfrac{5\sqrt[3]{2}}{2}$

91. $\sqrt{\dfrac{24x^5}{3y^2}} = \sqrt{\dfrac{8x^5}{y^2}} = \dfrac{\sqrt{4x^4 \cdot 2x}}{\sqrt{y^2}} = \dfrac{2x^2\sqrt{2x}}{y}$

92. $\sqrt[3]{\dfrac{15x^6y^7}{z^2}} = \dfrac{\sqrt[3]{15x^6y^7}}{\sqrt[3]{z^2}} = \dfrac{\sqrt[3]{x^6y^6 \cdot 15y} \; \sqrt[3]{z}}{\sqrt[3]{z^2} \cdot \sqrt[3]{z}} = \dfrac{x^2y^2\sqrt[3]{15yz}}{z}$

93. $\dfrac{5}{2 - \sqrt{7}} = \dfrac{5\left(2 + \sqrt{7}\right)}{\left(2 - \sqrt{7}\right)\left(2 + \sqrt{7}\right)} = \dfrac{5\left(2 + \sqrt{7}\right)}{2^2 - \left(\sqrt{7}\right)^2} = \dfrac{5\left(2 + \sqrt{7}\right)}{4 - 7} = \dfrac{5\left(2 + \sqrt{7}\right)}{-3} = -\dfrac{5\left(2 + \sqrt{7}\right)}{3}$

94.
$$\frac{3}{\sqrt{y}-2} = \frac{3\left(\sqrt{y}+2\right)}{\left(\sqrt{y}-2\right)\left(\sqrt{y}+2\right)}$$
$$= \frac{3\left(\sqrt{y}+2\right)}{\left(\sqrt{y}\right)^2 - 2^2}$$
$$= \frac{3\left(\sqrt{y}+2\right)}{y-4}$$

95.
$$\frac{\sqrt{2}-\sqrt{3}}{\sqrt{2}+\sqrt{3}} = \frac{\left(\sqrt{2}-\sqrt{3}\right)\left(\sqrt{2}-\sqrt{3}\right)}{\left(\sqrt{2}+\sqrt{3}\right)\left(\sqrt{2}-\sqrt{3}\right)}$$
$$= \frac{\left(\sqrt{2}\right)^2 - 2\sqrt{2}\sqrt{3} + \left(\sqrt{3}\right)^2}{\left(\sqrt{2}\right)^2 - \left(\sqrt{3}\right)^2}$$
$$= \frac{2 - 2\sqrt{6} + 3}{2-3}$$
$$= \frac{5 - 2\sqrt{6}}{-1}$$
$$= -5 + 2\sqrt{6}$$

96.
$$\frac{\sqrt{11}}{3} = \frac{\sqrt{11}\cdot\sqrt{11}}{3\cdot\sqrt{11}} = \frac{11}{3\sqrt{11}}$$

97.
$$\sqrt{\frac{18}{y}} = \frac{\sqrt{18}}{\sqrt{y}}$$
$$= \frac{3\sqrt{2}}{\sqrt{y}}$$
$$= \frac{3\sqrt{2}\cdot\sqrt{2}}{\sqrt{y}\cdot\sqrt{2}}$$
$$= \frac{3\cdot 2}{\sqrt{2y}}$$
$$= \frac{6}{\sqrt{2y}}$$

98.
$$\frac{\sqrt[3]{9}}{7} = \frac{\sqrt[3]{9}\cdot\sqrt[3]{3}}{7\cdot\sqrt[3]{3}} = \frac{3}{7\sqrt[3]{3}}$$

99.
$$\sqrt{\frac{24x^5}{3y^2}} = \sqrt{\frac{8x^5}{y^2}}$$
$$= \frac{\sqrt{4x^4\cdot 2x}}{\sqrt{y^2}}$$
$$= \frac{2x^2\sqrt{2x}}{y}$$
$$= \frac{2x^2\sqrt{2x}\cdot\sqrt{2x}}{y\sqrt{2x}}$$
$$= \frac{2x^2(2x)}{y\sqrt{2x}}$$
$$= \frac{4x^3}{y\sqrt{2x}}$$

100.
$$\sqrt[3]{\frac{xy^2}{10z}} = \frac{\sqrt[3]{xy^2}\cdot\sqrt[3]{x^2y}}{\sqrt[3]{10z}\cdot\sqrt[3]{x^2y}} = \frac{xy}{\sqrt[3]{10x^2yz}}$$

101.
$$\frac{\sqrt{x}+5}{-3} = \frac{\left(\sqrt{x}+5\right)\left(\sqrt{x}-5\right)}{-3\left(\sqrt{x}-5\right)} = \frac{x-25}{-3\left(\sqrt{x}-5\right)}$$

102. $\sqrt{y-7} = 5$
$$y - 7 = 25$$
$$y = 32$$
The solution is 32.
Check: $y_1 = \sqrt{x-7},\, y_2 = 5$

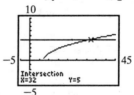

The intersection of the two graphs has an
x-value of 32, the solution of the equation.

103. $\sqrt{2x} + 10 = -4$

$\qquad \sqrt{2x} = -14$

The solution is $\varnothing$ since the principle square root of a number is not negative.

Check: $y_1 = \sqrt{2x} + 10, y_2 = -14$

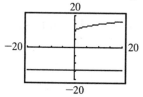

The graphs do not intersect.

104. $\sqrt[3]{2x - 6} = 4$

$\qquad \left(\sqrt[3]{2x - 6}\right)^3 = (4)^3$

$\qquad 2x - 6 = 64$

$\qquad 2x = 70$

$\qquad x = 35$

The solution is 35.

Check: $y_1 = \sqrt[3]{2x - 6}, y_2 = 4$

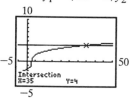

The intersection of the two graphs has an x-value of 35, the solution of the equation.

105. $\sqrt{x + 6} = \sqrt{x + 2}$

$\qquad x + 6 = x + 2$

$\qquad 6 = 2$

Since the last equation is never satisfied, the solution is $\varnothing$.

Check: $y_1 = \sqrt{x + 6}, y_2 = \sqrt{x + 2}$

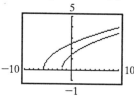

The graphs do not intersect.

106. $\qquad 2x - 5\sqrt{x} = 3$

$\qquad 2x - 3 = 5\sqrt{x}$

$\qquad (2x - 3)^2 = \left(5\sqrt{x}\right)^2$

$\qquad 4x^2 - 12x + 9 = 25x$

$\qquad 4x^2 - 37x + 9 = 0$

$\qquad (4x - 1)(x - 9) = 0$

$\qquad 4x - 1 = 0 \quad$ or $\quad x - 9 = 0$

$\qquad\quad 4x = 1 \quad$ or $\quad\quad x = 9$

$\qquad\quad x = \dfrac{1}{4}$

When $x = \dfrac{1}{4}$,

$2x - 5\sqrt{x} = 2\left(\dfrac{1}{4}\right) - 5\sqrt{\dfrac{1}{4}}$

$\qquad\qquad = \dfrac{1}{2} - \dfrac{5}{2}$

$\qquad\qquad = -2 \neq 3.$

When $x = 9$,

$2x - 5\sqrt{x} = 2(9) - 5\sqrt{9}$

$\qquad\qquad = 18 - 5$

$\qquad\qquad = 3.$

The solution is 9.

Check: $y_1 = 2x - 5\sqrt{x}, y_2 = 3$

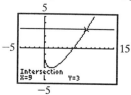

The intersection of the two graphs has an x-value of 9, the solution of the equation.

107. $\sqrt{x+9} = 2 + \sqrt{x-7}$

$x + 9 = 4 + 4\sqrt{x-7} + x - 7$

$12 = 4\sqrt{x-7}$

$3 = \sqrt{x-7}$

$9 = x - 7$

$16 = x$

The solution is 16.

Check: $y_1 = \sqrt{x+9}, y_2 = 2 + \sqrt{x-7}$

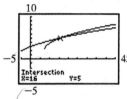

The intersection of the two graphs has an x-value of 16, the solution of the equation.

108. Let c = the length of the hypotenuse of the right triangle. By the Pythagorean theorem,

$c^2 = 3^2 + 3^2 = 9 + 9 = 18$

so $c = \sqrt{18} = \sqrt{9}\sqrt{2} = 3\sqrt{2}.$

109. Let c = the length of the hypotenuse of the right triangle. By the Pythagorean theorem,

$c^2 = 7^2 + \left(8\sqrt{3}\right)^2$

$= 49 + 192$

$= 241$

so $c = \sqrt{241}.$

110. $c^2 = a^2 + b^2$

$(65)^2 = (40)^2 + b^2$

$4225 - 1600 = b^2$

$2625 = b^2$

$b = \sqrt{2625} \approx 51.235$

The lake is 51.2 feet wide.

111. $c^2 = 3^2 + 3^2 = 9 + 9 = 18$

$c = \sqrt{18} \approx 4.243$

The length is 4.24 feet.

112. $\sqrt{-8} = \sqrt{4}\sqrt{2}\sqrt{-1} = 2i\sqrt{2}$

113. $-\sqrt{-6} = -\sqrt{6}\sqrt{-1} = -i\sqrt{6}$

114. $\sqrt{-4} + \sqrt{-16} = 2i + 4i = 6i$

115. $\sqrt{-2}\sqrt{-5} = i\sqrt{2} \cdot i\sqrt{5} = i^2\sqrt{2 \cdot 5} = -\sqrt{10}$

116. $(12 - 6i) + (3 + 2i) = (12 + 3) + (-6 + 2)i$

$= 15 - 4i$

117. $(-8 - 7i) - (5 - 4i) = -8 - 7i - 5 + 4i$

$= (-8 - 5) + (-7 + 4)i$

$= -13 - 3i$

118. $\left(\sqrt{3} + \sqrt{2}\right) + \left(3\sqrt{2} - \sqrt{-8}\right)$

$= \sqrt{3} + \sqrt{2} + 3\sqrt{2} - \sqrt{4}\sqrt{2}\sqrt{-1}$

$= \left(\sqrt{3} + 4\sqrt{2}\right) - 2i\sqrt{2}$

119. $2i(2 - 5i) = 4i - 10i^2$

$= 4i - 10(-1)$

$= 10 + 4i$

120. $-3i(6 - 4i) = -18i + 12i^2$

$= -18i + 12(-1)$

$= -12 - 18i$

121. $(3 + 2i)(1 + i) = 3 + 3i + 2i + 2i^2$

$= 3 + 5i + 2(-1)$

$= (3 - 2) + 5i$

$= 1 + 5i$

122. $(2 - 3i)^2 = (2)^2 - 2(2)(3i) + (3i)^2$

$= 4 - 12i + 9i^2$

$= 4 - 12i - 9$

$= -5 - 12i$

123. $\left(\sqrt{6} - 9i\right)\left(\sqrt{6} + 9i\right) = \left(\sqrt{6}\right)^2 - (9i)^2$

$= 6 - 81i^2$

$= 6 + 81$

$= 87$

124.
$$\frac{2+3i}{2i} = \frac{(2+3i)i}{2i^2}$$
$$= \frac{2i+3i^2}{2(-1)}$$
$$= \frac{2i+3(-1)}{-2}$$
$$= \frac{-3+2i}{-2}$$
$$= \frac{3}{2} - i$$

125.
$$\frac{1+i}{-3i} = \frac{(1+i)i}{-3i^2}$$
$$= \frac{i+i^2}{-3(-1)}$$
$$= \frac{i-1}{3}$$
$$= -\frac{1}{3} + \frac{1}{3}i$$

Chapter 7 Test

1. $\sqrt{216} = \sqrt{36 \cdot 6} = 6\sqrt{6}$

2. $-\sqrt[4]{x^{64}} = -(x^{64})^{1/4} = -(x^{64/4}) = -x^{16}$

3. $\left(\dfrac{1}{125}\right)^{1/3} = \sqrt[3]{\dfrac{1}{125}} = \dfrac{\sqrt[3]{1}}{\sqrt[3]{125}} = \dfrac{1}{5}$

4. $\left(\dfrac{1}{125}\right)^{-1/3} = \left[\left(\dfrac{1}{125}\right)^{-1}\right]^{1/3}$
$$= 125^{1/3}$$
$$= \sqrt[3]{125}$$
$$= 5$$

5.
$$\left(\frac{8x^3}{27}\right)^{2/3} = \frac{8^{2/3}(x^3)^{2/3}}{27^{2/3}}$$
$$= \frac{\left(\sqrt[3]{8}\right)^2 x^2}{\left(\sqrt[3]{27}\right)^2}$$
$$= \frac{2^2 x^2}{3^2}$$
$$= \frac{4x^2}{9}$$

6.
$$\sqrt[3]{-a^{18}b^9} = \sqrt[3]{-1}\sqrt[3]{a^{18}}\sqrt[3]{b^9}$$
$$= (-1)a^6 b^3$$
$$= -a^6 b^3$$

7.
$$\left(\frac{64c^{4/3}}{a^{-2/3}b^{5/6}}\right)^{1/2} = \left(\frac{64a^{2/3}c^{4/3}}{b^{5/6}}\right)^{1/2}$$
$$= \frac{64^{1/2}(a^{2/3})^{1/2}(c^{4/3})^{1/2}}{(b^{5/6})^{1/2}}$$
$$= \frac{8a^{1/3}c^{2/3}}{b^{5/12}}$$

8.
$$a^{-2/3}(a^{5/4} - a^3) = a^{-2/3+5/4} - a^{-2/3+3}$$
$$= a^{(-8+15)/12} - a^{(-2+9)/3}$$
$$= a^{7/12} - a^{7/3}$$

9. $\sqrt[4]{(4xy)^4} = |4xy|$

10. $\sqrt[3]{(-27)^3} = -27$

11. $\sqrt{\dfrac{9}{y}} = \dfrac{\sqrt{9}}{\sqrt{y}} = \dfrac{3}{\sqrt{y}} = \dfrac{3 \cdot \sqrt{y}}{\sqrt{y} \cdot \sqrt{y}} = \dfrac{3\sqrt{y}}{y}$

12.
$$\frac{4-\sqrt{x}}{4+2\sqrt{x}} = \frac{\left(4-\sqrt{x}\right)\left(4-2\sqrt{x}\right)}{\left(4+2\sqrt{x}\right)\left(4-2\sqrt{x}\right)}$$
$$= \frac{16-8\sqrt{x}-4\sqrt{x}+2x}{16-4x}$$
$$= \frac{16-12\sqrt{x}+2x}{16-4x}$$
$$= \frac{8-6\sqrt{x}+x}{8-2x}$$

13.
$$\frac{\sqrt[3]{ab}}{\sqrt[3]{ab^2}} = \sqrt[3]{\frac{ab}{ab^2}}$$
$$= \sqrt[3]{\frac{1}{b^1}}$$
$$= \frac{1}{\sqrt[3]{b}}$$
$$= \frac{1 \cdot \sqrt[3]{b^2}}{\sqrt[3]{b} \cdot \sqrt[3]{b^2}}$$
$$= \frac{\sqrt[3]{b^2}}{\sqrt[3]{b^3}}$$
$$= \frac{\sqrt[3]{b^2}}{b}$$

14.
$$\frac{\sqrt{6}+x}{8} = \frac{\left(\sqrt{6}+x\right)\left(\sqrt{6}-x\right)}{8\left(\sqrt{6}-x\right)}$$
$$= \frac{\left(\sqrt{6}\right)^2 - x^2}{8\left(\sqrt{6}-x\right)}$$
$$= \frac{6-x^2}{8\left(\sqrt{6}-x\right)}$$

15.
$$\sqrt{125x^3} - 3\sqrt{20x^3}$$
$$= \sqrt{25x^2}\sqrt{5x} - 3\sqrt{4x^2}\sqrt{5x}$$
$$= 5x\sqrt{5x} - 3(2x)\sqrt{5x}$$
$$= (5x-6x)\sqrt{5x}$$
$$= -x\sqrt{5x}$$

16.
$$\sqrt{3}\left(\sqrt{16}-\sqrt{2}\right) = \sqrt{3}\left(4-\sqrt{2}\right)$$
$$= 4\sqrt{3} - \sqrt{3}\sqrt{2}$$
$$= 4\sqrt{3} - \sqrt{6}$$

17. $\left(\sqrt{x}+1\right)^2 = \left(\sqrt{x}\right)^2 + 2\sqrt{x} + 1 = x + 2\sqrt{x} + 1$

18.
$$\left(\sqrt{2}-4\right)\left(\sqrt{3}+1\right) = \sqrt{2}\sqrt{3} + \sqrt{2} - 4\sqrt{3} - 4$$
$$= \sqrt{6} + \sqrt{2} - 4\sqrt{3} - 4$$

19.
$$\left(\sqrt{5}+5\right)\left(\sqrt{5}-5\right) = \left(\sqrt{5}\right)^2 - 5^2$$
$$= 5 - 25$$
$$= -20$$

20. $\sqrt{561} = 23.685$

21. $386^{-2/3} = 0.019$

22.
$$x = \sqrt{x-2} + 2$$
$$x - 2 = \sqrt{x-2}$$
$$(x-2)^2 = x-2$$
$$x^2 - 4x + 4 = x - 2$$
$$x^2 - 5x + 6 = 0$$
$$(x-2)(x-3) = 0$$
$$x - 2 = 0 \quad \text{or} \quad x - 3 = 0$$
$$x = 2 \quad \text{or} \quad x = 3$$
The solutions are 2 and 3.
Check: $y_1 = x, y_2 = \sqrt{x-2} + 2$

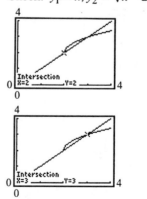

The intersections of the two graphs have x-values of 2 and 3, the solutions of the equation.

23. $\sqrt{x^2 - 7} + 3 = 0$

$\sqrt{x^2 - 7} = -3$

The solution is $\varnothing$ since a principle square root is never negative.

Check: $y_1 = \sqrt{x^2 - 7} + 3, y_2 = 0$

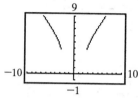

The graphs do not intersect.

24. $\sqrt{x+5} = \sqrt{2x-1}$

$x + 5 = 2x - 1$

$6 = x$

$x = 6$

The solution is 6.

Check: $y_1 = \sqrt{x+5}, y_2 = \sqrt{2x-1}$

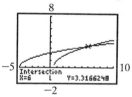

The intersection of the two graphs has an x-value of 6, the solution of the equation.

25. $\sqrt{-2} = \sqrt{-1}\sqrt{2} = i\sqrt{2}$

26. $-\sqrt{-8} = -\sqrt{4}\sqrt{2}\sqrt{-1} = -2i\sqrt{2}$

27. $(12 - 6i) - (12 - 3i) = 12 - 6i - 12 + 3i$
$= -3i$

28. $(6 - 2i)(6 + 2i) = 6^2 - (2i)^2 = 36 + 4 = 40$

29. $(4 + 3i)^2 = 4^2 + 2(4)(3i) + (3i)^2$

$= 16 + 24i + 9i^2$

$= 16 + 24i + 9(-1)$

$= (16 - 9) + 24i$

$= 7 + 24i$

30. $\dfrac{1 + 4i}{1 - i} = \dfrac{(1 + 4i)(1 + i)}{(1 - i)(1 + i)}$

$= \dfrac{1 + i + 4i + 4i^2}{1 - i^2}$

$= \dfrac{1 + 5i + 4(-1)}{1 - (-1)}$

$= \dfrac{(1 - 4) + 5i}{2}$

$= -\dfrac{3}{2} + \dfrac{5}{2}i$

31. By the Pythagorean theorem,

$x^2 + x^2 = 5^2$

$2x^2 = 25$

$x^2 = \dfrac{25}{2}$

$x = \sqrt{\dfrac{25}{2}} = \dfrac{\sqrt{25}}{\sqrt{2}} = \dfrac{5}{\sqrt{2}} \cdot \dfrac{\sqrt{2}}{\sqrt{2}} = \dfrac{5\sqrt{2}}{2}$

32. $g(x) = \sqrt{x+2}$

Domain: $[-2, \infty)$

x	-2	-1	2	7
$g(x)$	0	1	2	3

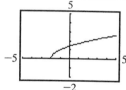

33. $V(300) = \sqrt{2.5(300)} \approx 27$

The maximum safe speed is 27 miles per hour.

34. $V(r) = \sqrt{2.5r}$

$30 = \sqrt{2.5r}$

$900 = 2.5r$

$360 = r$

The radius of curvature is 360 feet.

Chapter 8

1. $x^2 = 16$

$x = \pm\sqrt{16}$

$x = \pm 4$

Check: $y_1 = x^2$, $y_2 = 16$

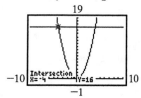

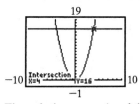

The solutions are -4 and 4.

3. $x^2 - 7 = 0$

$x^2 = 7$

$x = \pm\sqrt{7}$

Check: $y_1 = x^2 - 7$, $y_2 = 0$

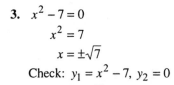

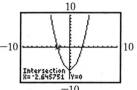

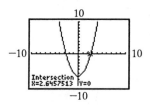

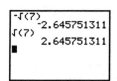

The approximate solutions are -2.646 and 2.646. The exact solutions are $-\sqrt{7}$ and $\sqrt{7}$.

5. $x^2 = 18$

$\quad x = \pm\sqrt{18}$

$\quad x = \pm 3\sqrt{2}$

Check: $y_1 = x^2$, $y_2 = 18$

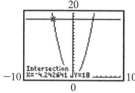

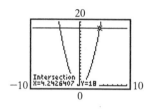

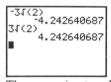

The approximate solutions are
–4.243 and 4.243. The exact solutions
are $-3\sqrt{2}$ and $3\sqrt{2}$.

7. $3z^2 - 30 = 0$

$\quad 3z^2 = 30$

$\quad z^2 = 10$

$\quad z = \pm\sqrt{10}$

Check: $y_1 = 3x^2 - 30$, $y_2 = 0$

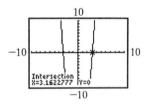

The approximate solutions are
–3.162 and 3.162. The exact solutions
are $-\sqrt{10}$ and $\sqrt{10}$.

9. $(x+5)^2 = 9$

$$x+5 = \pm\sqrt{9}$$
$$x+5 = \pm 3$$
$$x = -5 \pm 3$$
$$x = -8 \ \text{ or } \ x = -2$$

Check: $y_1 = (x+5)^2$, $y_2 = 9$

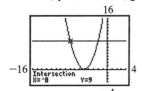

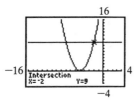

The solutions are –8 and –2.

11. $(z-6)^2 = 18$

$$z-6 = \pm\sqrt{18}$$
$$z-6 = \pm 3\sqrt{2}$$
$$z = 6 \pm 3\sqrt{2}$$

Check: $y_1 = (x-6)^2$, $y_2 = 18$

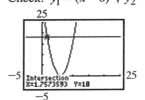

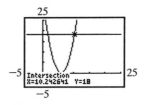

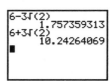

The approximate solutions are 1.757 and 10.243. The exact solutions are $6 - 3\sqrt{2}$ and $6 + 3\sqrt{2}$.

13. $(2x-3)^2 = 8$

$$2x - 3 = \pm\sqrt{8}$$
$$2x - 3 = \pm 2\sqrt{2}$$
$$2x = 3 \pm 2\sqrt{2}$$
$$x = \frac{3 \pm 2\sqrt{2}}{2}$$

Check: $y_1 = (2x-3)^2$, $y_2 = 8$

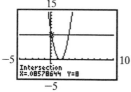

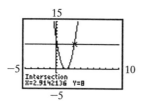

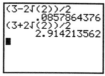

The approximate solutions are 0.086 and 2.914. The exact solutions are $\frac{3 - 2\sqrt{2}}{2}$ and $\frac{3 + 2\sqrt{2}}{2}$.

15. $x^2 + 9 = 0$

$$x^2 = -9$$
$$x = \pm\sqrt{-9}$$
$$x = \pm 3i$$

Check:

```
-3i→X:X²+9
                    0
3i→X:X²+9
                    0
■
```

The solutions are $-3i$ and $3i$.

17. $x^2 - 6 = 0$

$$x^2 = 6$$
$$x = \pm\sqrt{6}$$

Check:

```
-√(6)→X:X²-6
                    0
√(6)→X:X²-6
                    0
```

The solutions are $-\sqrt{6}$ and $\sqrt{6}$.

19. $2z^2 + 16 = 0$

$$2z^2 = -16$$
$$z^2 = -8$$
$$z = \pm\sqrt{-8}$$
$$z = \pm 2i\sqrt{2}$$

Check:

```
-2i√(2)→Z:2Z²+16
                    0
2i√(2)→Z:2Z²+16
                    0
```

The solutions are $-2i\sqrt{2}$ and $2i\sqrt{2}$.

21. $(x-1)^2 = -16$

$\qquad x-1 = \pm\sqrt{-16}$

$\qquad x-1 = \pm 4i$

$\qquad\quad x = 1 \pm 4i$

Check:

```
1-4i→X:(X-1)²
              -16
1+4i→X:(X-1)²
              -16
■
```

The solutions are $1 - 4i$ and $1 + 4i$.

23. $(z+7)^2 = 5$

$\qquad z+7 = \pm\sqrt{5}$

$\qquad\quad z = -7 \pm \sqrt{5}$

Check:

```
-7-√(5)→Z:(Z+7)²
                5
-7+√(5)→Z:(Z+7)²
                5
```

The solutions are $-7-\sqrt{5}$ and $-7+\sqrt{5}$.

25. $(x+3)^2 = -8$

$\qquad x+3 = \pm\sqrt{-8}$

$\qquad x+3 = \pm 2i\sqrt{2}$

$\qquad\quad x = -3 \pm 2i\sqrt{2}$

Check:

```
-3-2i√(2)→X:(X+3
)²
              -8
-3+2i√(2)→X:(X+3
)²
              -8
```

The solutions are
$-3-2i\sqrt{2}$ and $-3+2i\sqrt{2}$.

27. $x^2 + 16x + \left(\dfrac{16}{2}\right)^2 = x^2 + 16x + 64$

$\qquad\qquad\qquad = (x+8)^2$

29. $z^2 - 12z + \left(\dfrac{12}{2}\right)^2 = z^2 - 12z + 36$

$\qquad\qquad\qquad = (z-6)^2$

31. $p^2 + 9p + \left(\dfrac{9}{2}\right)^2 = p^2 + 9p + \dfrac{81}{4}$

$\qquad\qquad\qquad = \left(p + \dfrac{9}{2}\right)^2$

33. $x^2 + x + \left(\dfrac{1}{2}\right)^2 = x^2 + x + \dfrac{1}{4} = \left(x + \dfrac{1}{2}\right)^2$

35. $x^2 + \underline{\quad} + 16$

$\qquad \pm 2\sqrt{16} = \pm 2(4) = \pm 8$

$\qquad x^2 - 8x + 16 = (x-4)^2$

$\qquad x^2 + 8x + 16 = (x+4)^2$

The possible terms are $-8x$ and $8x$.

37. $z^2 + \underline{\quad} + \dfrac{25}{4}$

$\qquad \pm 2\sqrt{\dfrac{25}{4}} = \pm 2\left(\dfrac{5}{2}\right) = \pm 5$

$\qquad z^2 - 5z + \dfrac{25}{4} = \left(z - \dfrac{5}{2}\right)^2$

$\qquad z^2 + 5z + \dfrac{25}{4} = \left(z + \dfrac{5}{2}\right)^2$

The possible terms are $-5z$ and $5z$.

39.
$$x^2 + 8x = -15$$
$$x^2 + 8x + \left(\frac{8}{2}\right)^2 = -15 + 16$$
$$(x+4)^2 = 1$$
$$x + 4 = \pm\sqrt{1}$$
$$x = -4 \pm 1$$
$$x = -5 \text{ or } x = -3$$
Check: $y_1 = x^2 + 8x,\ y_2 = -15$

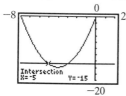

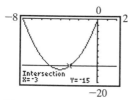

The solutions are -5 and -3.

41.
$$x^2 + 6x + 2 = 0$$
$$x^2 + 6x + \left(\frac{6}{2}\right)^2 = -2 + 9$$
$$(x+3)^2 = 7$$
$$x + 3 = \pm\sqrt{7}$$
$$x = -3 \pm \sqrt{7}$$
Check: $y_1 = x^2 + 6x + 2,\ y_2 = 0$

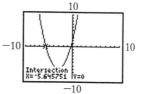

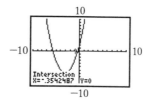

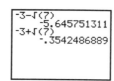

The approximate solutions are -5.646 and -0.354. The exact solutions are $-3 - \sqrt{7}$ and $-3 + \sqrt{7}$.

43. $x^2 + x - 1 = 0$

$$x^2 + x + \left(\frac{1}{2}\right)^2 = 1 + \left(\frac{1}{2}\right)^2$$

$$\left(x + \frac{1}{2}\right)^2 = \frac{5}{4}$$

$$x + \frac{1}{2} = \pm\sqrt{\frac{5}{4}}$$

$$x = -\frac{1}{2} \pm \frac{\sqrt{5}}{2}$$

$$x = \frac{-1 \pm \sqrt{5}}{2}$$

Check: $y_1 = x^2 + x - 1$, $y_2 = 0$

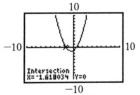

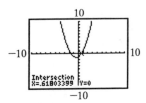

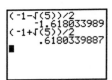

The approximate solutions are -1.618 and 0.618. The exact solutions are

$$\frac{-1 - \sqrt{5}}{2} \text{ and } \frac{-1 + \sqrt{5}}{2}.$$

45. $x^2 + 2x - 5 = 0$

$$x^2 + 2x + \left(\frac{2}{2}\right)^2 = 5 + 1$$

$$(x + 1)^2 = 6$$

$$x + 1 = \pm\sqrt{6}$$

$$x = -1 \pm \sqrt{6}$$

Check: $y_1 = x^2 + 2x - 5$, $y_2 = 0$

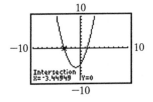

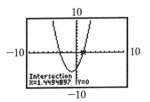

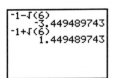

The approximate solutions are -3.449 and 1.449. The exact solutions are $-1 - \sqrt{6}$ and $-1 + \sqrt{6}$.

47.
$$3p^2 - 12p + 2 = 0$$
$$p^2 - 4p + \frac{2}{3} = 0$$
$$p^2 - 4p + \left(\frac{4}{2}\right)^2 = -\frac{2}{3} + 4$$
$$(p - 2)^2 = \frac{10}{3}$$
$$p - 2 = \pm\sqrt{\frac{10}{3}}$$
$$p - 2 = \pm\frac{\sqrt{30}}{3}$$
$$p = 2 \pm \frac{\sqrt{30}}{3}$$
$$p = \frac{6 \pm \sqrt{30}}{3}$$

Check: $y_1 = 3x^2 - 12x + 2$, $y_2 = 0$

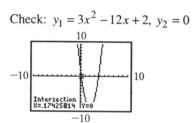

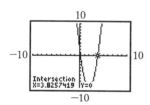

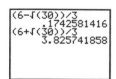

The approximate solutions are 0.174 and 3.826. The exact solutions are $\frac{6-\sqrt{30}}{3}$ and $\frac{6+\sqrt{30}}{3}$.

49.
$$4y^2 - 12y - 2 = 0$$
$$y^2 - 3y - \frac{1}{2} = 0$$
$$y^2 - 3y + \left(\frac{3}{2}\right)^2 = \frac{1}{2} + \frac{9}{4}$$
$$\left(y - \frac{3}{2}\right)^2 = \frac{11}{4}$$
$$y - \frac{3}{2} = \pm\sqrt{\frac{11}{4}}$$
$$y - \frac{3}{2} = \pm\frac{\sqrt{11}}{2}$$
$$y = \frac{3}{2} \pm \frac{\sqrt{11}}{2}$$
$$y = \frac{3 \pm \sqrt{11}}{2}$$

Check: $y_1 = 4x^2 - 12x - 2$, $y_2 = 0$

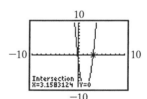

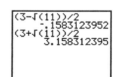

The approximate solutions are –0.158 and 3.158. The exact solutions are $\frac{3-\sqrt{11}}{2}$ and $\frac{3+\sqrt{11}}{2}$.

51.
$$2x^2 + 7x = 4$$
$$x^2 + \frac{7}{2}x = 2$$
$$x^2 + \frac{7}{2}x + \left(\frac{\frac{7}{2}}{2}\right)^2 = 2 + \frac{49}{16}$$
$$\left(x + \frac{7}{4}\right)^2 = \frac{81}{16}$$
$$x + \frac{7}{4} = \pm\sqrt{\frac{81}{16}}$$
$$x = -\frac{7}{4} \pm \frac{9}{4}$$
$$x = -4 \text{ or } x = \frac{1}{2}$$

Check: $y_1 = 2x^2 + 7x$, $y_2 = 4$

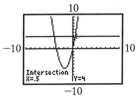

The solutions are -4 and $\frac{1}{2}$.

53.
$$x^2 - 4x - 5 = 0$$
$$x^2 - 4x = 5$$
$$x^2 - 4x + \left(\frac{4}{2}\right)^2 = 5 + 4$$
$$(x - 2)^2 = 9$$
$$x - 2 = \pm\sqrt{9}$$
$$x - 2 = \pm 3$$
$$x = 2 \pm 3$$
$$x = -1 \text{ or } x = 5$$

Check: $y_1 = x^2 - 4x - 5$, $y_2 = 0$

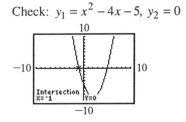

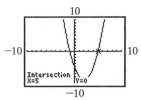

The solutions are -1 and 5.

55.
$$x^2 + 8x + 1 = 0$$
$$x^2 + 8x = -1$$
$$x^2 + 8x + \left(\frac{8}{2}\right)^2 = -1 + 16$$
$$(x+4)^2 = 15$$
$$x + 4 = \pm\sqrt{15}$$
$$x = -4 \pm \sqrt{15}$$

Check: $y_1 = x^2 + 8x + 1$, $y_2 = 0$

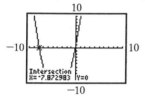

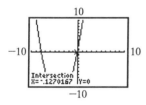

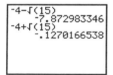

The approximate solutions are -7.873 and -0.127. The exact solutions are $-4 - \sqrt{15}$ and $-4 + \sqrt{15}$.

57.
$$3y^2 + 6y - 4 = 0$$
$$y^2 + 2y - \frac{4}{3} = 0$$
$$y^2 + 2y + \left(\frac{2}{2}\right)^2 = \frac{4}{3} + 1$$
$$(y+1)^2 = \frac{7}{3}$$
$$y + 1 = \pm\sqrt{\frac{7}{3}}$$
$$y + 1 = \pm\frac{\sqrt{21}}{3}$$
$$y = -1 \pm \frac{\sqrt{21}}{3}$$
$$y = \frac{-3 \pm \sqrt{21}}{3}$$

Check: $y_1 = 3x^2 + 6x - 4$, $y_2 = 0$

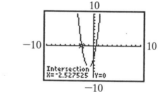

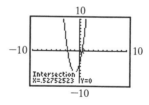

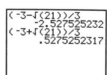

The approximate solutions are -2.528 and 0.528. The exact solutions are $\frac{-3 - \sqrt{21}}{3}$ and $\frac{-3 + \sqrt{21}}{3}$.

59. There are two real number solutions because the graph crosses the x-axis in two places.

61. There are no real number solutions because the graph does not cross the x-axis.

63.
$$y^2 + 2y + 2 = 0$$
$$y^2 + 2y + \left(\frac{2}{2}\right)^2 = -2 + 1$$
$$(y+1)^2 = -1$$
$$y + 1 = \pm\sqrt{-1}$$
$$y + 1 = \pm i$$
$$y = -1 \pm i$$

Check:

```
-1-i→X:X²+2X+2
              0
-1+i→X:X²+2X+2
              0
```

The solutions are $-1 - i$ and $-1 + i$.

65.
$$x^2 - 6x + 3 = 0$$
$$x^2 - 6x + \left(\frac{6}{2}\right)^2 = -3 + 9$$
$$(x-3)^2 = 6$$
$$x - 3 = \pm\sqrt{6}$$
$$x = 3 \pm \sqrt{6}$$

Check:

```
3-√(6)→X:X²-6X+3
              0
3+√(6)→X:X²-6X+3
              0
```

The solutions are $3 - \sqrt{6}$ and $3 + \sqrt{6}$.

67.
$$2a^2 + 8a = -12$$
$$a^2 + 4a = -6$$
$$a^2 + 4a + \left(\frac{4}{2}\right)^2 = -6 + 4$$
$$(a+2)^2 = -2$$
$$a + 2 = \pm\sqrt{-2}$$
$$a + 2 = \pm i\sqrt{2}$$
$$a = -2 \pm i\sqrt{2}$$

Check:

```
-2-i√(2)→X:2X²+8
X
              -12
-2+i√(2)→X:2X²+8
X
              -12
```

The solutions are $-2 - i\sqrt{2}$ and $-2 + i\sqrt{2}$.

69.
$$2x^2 - x + 6 = 0$$
$$x^2 - \frac{1}{2}x + 3 = 0$$
$$x^2 - \frac{1}{2}x + \left(\frac{\frac{1}{2}}{2}\right)^2 = -3 + \frac{1}{16}$$
$$\left(x - \frac{1}{4}\right)^2 = -\frac{47}{16}$$
$$x - \frac{1}{4} = \pm\sqrt{-\frac{47}{16}}$$
$$x - \frac{1}{4} = \pm\frac{i\sqrt{47}}{4}$$
$$x = \frac{1}{4} \pm \frac{i\sqrt{47}}{4}$$
$$x = \frac{1 \pm i\sqrt{47}}{4}$$

Check:

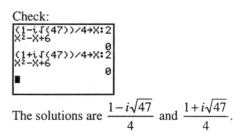

The solutions are $\dfrac{1 - i\sqrt{47}}{4}$ and $\dfrac{1 + i\sqrt{47}}{4}$.

71.
$$x^2 + 10x + 28 = 0$$
$$x^2 + 10x + \left(\frac{10}{2}\right)^2 = -28 + 25$$
$$(x+5)^2 = -3$$
$$x + 5 = \pm\sqrt{-3}$$
$$x + 5 = \pm i\sqrt{3}$$
$$x = -5 \pm i\sqrt{3}$$

Check:

```
-5-i√(3)→X:X²+10
X+28
                0
-5+i√(3)→X:X²+10
X+28
                0
```

The solutions are $-5 - i\sqrt{3}$ and $-5 + i\sqrt{3}$.

73.
$$z^2 + 3z - 4 = 0$$
$$z^2 + 3z + \left(\frac{3}{2}\right)^2 = 4 + \frac{9}{4}$$
$$\left(z + \frac{3}{2}\right)^2 = \frac{25}{4}$$
$$z + \frac{3}{2} = \pm\sqrt{\frac{25}{4}}$$
$$z + \frac{3}{2} = \pm\frac{5}{2}$$
$$z = -\frac{3}{2} \pm \frac{5}{2}$$
$$z = -4 \text{ or } z = 1$$

Check:

```
-4→X:X²+3X-4
                0
1→X:X²+3X-4
                0
```

The solutions are –4 and 1.

75.
$$2x^2 - 4x + 3 = 0$$
$$x^2 - 2x + \frac{3}{2} = 0$$
$$x^2 - 2x + \left(\frac{2}{2}\right)^2 = -\frac{3}{2} + 1$$
$$(x-1)^2 = -\frac{1}{2}$$
$$x - 1 = \pm\sqrt{-\frac{1}{2}}$$
$$x - 1 = \pm\frac{i\sqrt{2}}{2}$$
$$x = 1 \pm \frac{i\sqrt{2}}{2}$$
$$x = \frac{2 \pm i\sqrt{2}}{2}$$

Check:

```
(2-i√(2))/2→X:2X
²-4X+3
                0
(2+i√(2))/2→X:2X
²-4X+3
                0
■
```

The solutions are $\dfrac{2 - i\sqrt{2}}{2}$ and $\dfrac{2 + i\sqrt{2}}{2}$.

77.
$$3x^2 + 3x = 5$$
$$x^2 + x = \frac{5}{3}$$
$$x^2 + x + \left(\frac{1}{2}\right)^2 = \frac{5}{3} + \frac{1}{4}$$
$$\left(x + \frac{1}{2}\right)^2 = \frac{23}{12}$$
$$x + \frac{1}{2} = \pm\sqrt{\frac{23}{12}}$$
$$x + \frac{1}{2} = \pm\frac{\sqrt{69}}{6}$$
$$x = -\frac{1}{2} \pm \frac{\sqrt{69}}{6}$$
$$x = \frac{-3 \pm \sqrt{69}}{6}$$

Check:

```
(-3-√(69))/6→X:3
X²+3X
                5
(-3+√(69))/6→X:3
X²+3X
                5
■
```

The solutions are $\dfrac{-3 - \sqrt{69}}{6}$ and $\dfrac{-3 + \sqrt{69}}{6}$.

79.
$$A = P(1 + r)^t$$
$$4320 = 3000(1 + r)^2$$
$$\frac{4320}{3000} = (1 + r)^2$$
$$1.44 = (1 + r)^2$$
$$\pm\sqrt{1.44} = 1 + r$$
$$\pm 1.2 = 1 + r$$
$$-1 \pm 1.2 = r$$
$$r = -1 + 1.2 \quad \text{or} \quad r = -1 - 1.2$$
$$r = 0.2 \quad \quad \text{or} \quad r = -2.2$$
The rate cannot be negative, so the rate is $r = 0.2 = 20\%$.

81.
$$A = P(1 + r)^t$$
$$1000 = 810(1 + r)^2$$
$$\frac{1000}{810} = (1 + r)^2$$
$$\pm\sqrt{\frac{100}{81}} = 1 + r$$
$$\pm\frac{10}{9} = 1 + r$$
$$-1 \pm \frac{10}{9} = r$$
$$r = -1 + \frac{10}{9} \quad \text{or} \quad r = -1 - \frac{10}{9}$$
$$r = \frac{1}{9} \quad \quad \text{or} \quad r = -\frac{19}{9}$$
The rate cannot be negative, so
$r = \dfrac{1}{9}$, or about 11%.

83. Answers may vary.

85. Simple interest would be preferred because it would grow more slowly.

87.
$$s(t) = 16t^2$$
$$1053 = 16t^2$$
$$t = \pm\sqrt{\frac{1053}{16}}$$
$$t = 8.11 \text{ or } t = -8.11 \text{ (reject)}$$
It would take 8.11 seconds.

89.
$$s(t) = 16t^2$$
$$725 = 16t^2$$
$$t = \pm\sqrt{\frac{725}{16}}$$
$$t = 6.73 \text{ or } t = -6.73 \text{ (reject)}$$
It would take 6.73 seconds.

91.
$$A = \pi r^2$$
$$36\pi = \pi r^2$$
$$36 = r^2$$
$$\pm\sqrt{36} = r$$
$$\pm 6 = r$$
Reject the negative.
The radius of the circle is 6 inches.

93.
$$x^2 + x^2 = 27^2$$
$$2x^2 = 729$$
$$x^2 = \frac{729}{2}$$
$$x = \pm\sqrt{\frac{729}{2}}$$
$$x = \pm\frac{27\sqrt{2}}{2}$$

Reject the negative.

The side measures $\frac{27\sqrt{2}}{2}$ inches.

95.
$$p = -x^2 + 15$$
$$7 = -x^2 + 15$$
$$x^2 = 8$$
$$x = \pm\sqrt{8}$$
$$x \approx \pm 2.828$$

Demand cannot be negative, therefore the demand is approximately 2.828 thousand units.

97. $\dfrac{3}{5} + \sqrt{\dfrac{16}{25}} = \dfrac{3}{5} + \dfrac{4}{5} = \dfrac{7}{5}$

99. $\dfrac{9}{10} - \sqrt{\dfrac{49}{100}} = \dfrac{9}{10} - \dfrac{7}{10} = \dfrac{2}{10} = \dfrac{1}{5}$

101. $\dfrac{10 - 20\sqrt{3}}{2} = \dfrac{2\left(5 - 10\sqrt{3}\right)}{2} = 5 - 10\sqrt{3}$

103. $\dfrac{12 - 8\sqrt{7}}{16} = \dfrac{4\left(3 - 2\sqrt{7}\right)}{16} = \dfrac{3 - 2\sqrt{7}}{4}$

105.
$$\sqrt{b^2 - 4ac} = \sqrt{(6)^2 - 4(1)(2)}$$
$$= \sqrt{36 - 8}$$
$$= \sqrt{28}$$
$$= 2\sqrt{7}$$

107.
$$\sqrt{b^2 - 4ac} = \sqrt{(-3)^2 - 4(1)(-1)}$$
$$= \sqrt{9 + 4}$$
$$= \sqrt{13}$$

Exercise Set 8.2

1. $m^2 + 5m - 6 = 0$
$a = 1, b = 5, c = -6$
$$m = \frac{-5 \pm \sqrt{5^2 - 4(1)(-6)}}{2(1)}$$
$$= \frac{-5 \pm \sqrt{25 + 24}}{2}$$
$$= \frac{-5 \pm \sqrt{49}}{2}$$
$$= \frac{-5 \pm 7}{2}$$
$m = -6$ or $m = 1$
The solutions are -6 and 1.

3.
$$2y = 5y^2 - 3$$
$$5y^2 - 2y - 3 = 0$$
$a = 5, b = -2, c = -3$
$$y = \frac{2 \pm \sqrt{(-2)^2 - 4(5)(-3)}}{2(5)}$$
$$= \frac{2 \pm \sqrt{4 + 60}}{10}$$
$$= \frac{2 \pm \sqrt{64}}{10}$$
$$= \frac{2 \pm 8}{10}$$
$$y = -\frac{3}{5} \text{ or } y = 1$$

The solutions are $-\dfrac{3}{5}$ and 1.

5. $x^2 - 6x + 9 = 0$
$a = 1, b = -6, c = 9$
$$x = \frac{6 \pm \sqrt{(-6)^2 - 4(1)(9)}}{2(1)}$$
$$= \frac{6 \pm \sqrt{36 - 36}}{2}$$
$$= \frac{6 \pm \sqrt{0}}{2}$$
$$= 3$$
The solution is 3.

7. $x^2 + 7x + 4 = 0$

$a = 1, b = 7, c = 4$

$$x = \frac{-7 \pm \sqrt{7^2 - 4(1)(4)}}{2(1)}$$

$$= \frac{-7 \pm \sqrt{49 - 16}}{2}$$

$$= \frac{-7 \pm \sqrt{33}}{2}$$

The solutions are $\dfrac{-7 - \sqrt{33}}{2}$ and

$\dfrac{-7 + \sqrt{33}}{2}$.

9. $8m^2 - 2m = 7$

$8m^2 - 2m - 7 = 0$

$a = 8, b = -2, c = -7$

$$m = \frac{2 \pm \sqrt{(-2)^2 - 4(8)(-7)}}{2(8)}$$

$$= \frac{2 \pm \sqrt{4 + 224}}{16}$$

$$= \frac{2 \pm \sqrt{228}}{16}$$

$$= \frac{2 \pm 2\sqrt{57}}{16}$$

$$= \frac{1 \pm \sqrt{57}}{8}$$

The solutions are $\dfrac{1 - \sqrt{57}}{8}$ and $\dfrac{1 + \sqrt{57}}{8}$.

11. $3m^2 - 7m = 3$

$3m^2 - 7m - 3 = 0$

$a = 3, b = -7, c = -3$

$$m = \frac{7 \pm \sqrt{(-7)^2 - 4(3)(-3)}}{2(3)}$$

$$= \frac{7 \pm \sqrt{49 + 36}}{6}$$

$$= \frac{7 \pm \sqrt{85}}{6}$$

The solutions are $\dfrac{7 - \sqrt{85}}{6}$ and $\dfrac{7 + \sqrt{85}}{6}$.

13. $\dfrac{1}{2}x^2 - x - 1 = 0$

$x^2 - 2x - 2 = 0$

$a = 1, b = -2, c = -2$

$$x = \frac{2 \pm \sqrt{(-2)^2 - 4(1)(-2)}}{2(1)}$$

$$= \frac{2 \pm \sqrt{4 + 8}}{2}$$

$$= \frac{2 \pm \sqrt{12}}{2}$$

$$= \frac{2 \pm 2\sqrt{3}}{2}$$

$x = 1 \pm \sqrt{3}$

The solutions are $1 - \sqrt{3}$ and $1 + \sqrt{3}$.

15. $\dfrac{2}{5}y^2 + \dfrac{1}{5}y = \dfrac{3}{5}$

$2y^2 + y = 3$

$2y^2 + y - 3 = 0$

$a = 2, b = 1, c = -3$

$$y = \frac{-1 \pm \sqrt{1^2 - 4(2)(-3)}}{2(2)}$$

$$= \frac{-1 \pm \sqrt{1 + 24}}{4}$$

$$= \frac{-1 \pm \sqrt{25}}{4}$$

$$= \frac{-1 \pm 5}{4}$$

$y = -\dfrac{3}{2}$ or $y = 1$

The solutions are $-\dfrac{3}{2}$ and 1.

17. $\dfrac{1}{3}y^2 - y - \dfrac{1}{6} = 0$

$2y^2 - 6y - 1 = 0$

$a = 2,\ b = -6,\ c = -1$

$$y = \frac{6 \pm \sqrt{(-6)^2 - 4(2)(-1)}}{2(2)}$$

$$= \frac{6 \pm \sqrt{36 + 8}}{4}$$

$$= \frac{6 \pm \sqrt{44}}{4}$$

$$= \frac{6 \pm 2\sqrt{11}}{4}$$

$$= \frac{3 \pm \sqrt{11}}{2}$$

The solutions are $\dfrac{3 - \sqrt{11}}{2}$ and $\dfrac{3 + \sqrt{11}}{2}$.

19. Answers may vary.

21. $\qquad\qquad 6 = -4x^2 + 3x$

$4x^2 - 3x + 6 = 0$

$a = 4,\ b = -3,\ c = 6$

$$x = \frac{3 \pm \sqrt{(-3)^2 - 4(4)(6)}}{2(4)}$$

$$= \frac{3 \pm \sqrt{9 - 96}}{8}$$

$$= \frac{3 \pm \sqrt{-87}}{8}$$

$$= \frac{3 \pm i\sqrt{87}}{8}$$

Check:

```
(3+i√(87))/8→X: -
4X²+3X
              6
(3-i√(87))/8→X: -
4X²+3X
              6
■
```

The solutions are $\dfrac{3 - i\sqrt{87}}{8}$ and $\dfrac{3 + i\sqrt{87}}{8}$.

23. $(x + 5)(x - 1) = 2$

$x^2 + 4x - 5 = 2$

$x^2 + 4x - 7 = 0$

$a = 1,\ b = 4,\ c = -7$

$$x = \frac{-4 \pm \sqrt{4^2 - 4(1)(-7)}}{2(1)}$$

$$= \frac{-4 \pm \sqrt{16 + 28}}{2}$$

$$= \frac{-4 + \sqrt{44}}{2}$$

$$= \frac{-4 \pm 2\sqrt{11}}{2}$$

$$= -2 \pm \sqrt{11}$$

Check:

```
-2+√(11)→X:(X+5)
(X-1)
              2
-2-√(11)→X:(X+5)
(X-1)
              2
■
```

The solutions are $-2 - \sqrt{11}$ and $-2 + \sqrt{11}$.

25. $10y^2 + 10y + 3 = 0$

$a = 10, b = 10, c = 3$

$$y = \frac{-10 \pm \sqrt{10^2 - 4(10)(3)}}{2(10)}$$

$$= \frac{-10 \pm \sqrt{100 - 120}}{20}$$

$$= \frac{-10 \pm \sqrt{-20}}{20}$$

$$= \frac{-10 \pm 2i\sqrt{5}}{20}$$

$$= \frac{-5 \pm i\sqrt{5}}{10}$$

Check:
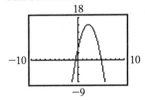

The solutions are $\dfrac{-5 - i\sqrt{5}}{10}$ and

$\dfrac{-5 + i\sqrt{5}}{10}$.

27. Answers may vary.

29. $9x - 2x^2 + 5 = 0$

$-2x^2 + 9x + 5 = 0$

$a = -2, b = 9, c = 5$

$b^2 - 4ac = 9^2 - 4(-2)(5)$

$\qquad = 81 + 40$

$\qquad = 121 > 0$

Therefore, there are two real solutions.
Confirm: Two x-intercepts indicate two real solutions.

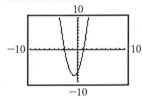

31. $4x^2 + 12x = -9$

$4x^2 + 12x + 9 = 0$

$a = 4, b = 12, c = 9$

$b^2 - 4ac = 12^2 - 4(4)(9)$

$\qquad = 144 - 144$

$\qquad = 0$

Therefore, there is 1 real solution.
Confirm: One x-intercept indicates one real solution.

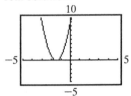

33. $3x = -2x^2 + 7$

$2x^2 + 3x - 7 = 0$

$a = 2, b = 3, c = -7$

$b^2 - 4ac = 3^2 - 4(2)(-7)$

$\qquad = 9 + 56$

$\qquad = 65 > 0$

Therefore, there are two real solutions.
Confirm: Two x-intercepts indicate two real solutions.

35.
$$6 = 4x - 5x^2$$
$$5x^2 - 4x + 6 = 0$$
$$a = 5,\ b = -4,\ c = 6$$
$$b^2 - 4ac = (-4)^2 - 4(5)(6)$$
$$= 16 - 120$$
$$= -104 < 0$$
Therefore, there are two complex but not real solutions.

Confirm: No x-intercepts indicate no real solutions, but two complex solutions.

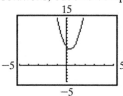

37.
$$x^2 + 5x = -2$$
$$x^2 + 5x + 2 = 0$$
$$a = 1,\ b = 5,\ c = 2$$
$$x = \frac{-5 \pm \sqrt{5^2 - 4(1)(2)}}{2(1)}$$
$$= \frac{-5 \pm \sqrt{25 - 8}}{2}$$
$$= \frac{-5 \pm \sqrt{17}}{2}$$

The solutions are $\dfrac{-5 - \sqrt{17}}{2}$ and $\dfrac{-5 + \sqrt{17}}{2}$.

39.
$$(m + 2)(2m - 6) = 5(m - 1) - 12$$
$$2m^2 - 2m - 12 = 5m - 5 - 12$$
$$2m^2 - 7m + 5 = 0$$
$$a = 2,\ b = -7,\ c = 5$$
$$m = \frac{7 \pm \sqrt{(-7)^2 - 4(2)(5)}}{2(2)}$$
$$= \frac{7 \pm \sqrt{49 - 40}}{4}$$
$$= \frac{7 \pm \sqrt{9}}{4}$$
$$= \frac{7 \pm 3}{4}$$
$$m = \frac{5}{2} \ \text{or}\ m = 1$$

The solutions are $\dfrac{5}{2}$ and 1.

41.
$$\frac{x^2}{3} - x = \frac{5}{3}$$
$$x^2 - 3x = 5$$
$$x^2 - 3x - 5 = 0$$
$$a = 1,\ b = -3,\ c = -5$$
$$x = \frac{3 \pm \sqrt{(-3)^2 - 4(1)(-5)}}{2(1)}$$
$$= \frac{3 \pm \sqrt{9 + 20}}{2}$$
$$= \frac{3 \pm \sqrt{29}}{2}$$

The solutions are $\dfrac{3 - \sqrt{29}}{2}$ and $\dfrac{3 + \sqrt{29}}{2}$.

43. $x(6x + 2) - 3 = 0$

$6x^2 + 2x - 3 = 0$

$a = 6,\ b = 2,\ c = -3$

$x = \dfrac{-2 \pm \sqrt{2^2 - 4(6)(-3)}}{2(6)}$

$= \dfrac{-2 \pm \sqrt{4 + 72}}{12}$

$= \dfrac{-2 \pm \sqrt{76}}{12}$

$= \dfrac{-2 \pm 2\sqrt{19}}{12}$

$= \dfrac{-1 \pm \sqrt{19}}{6}$

The solutions are $\dfrac{-1 - \sqrt{19}}{6}$ and

$\dfrac{-1 + \sqrt{19}}{6}$.

45. $x^2 + 6x + 13 = 0$

$a = 1,\ b = 6,\ c = 13$

$x = \dfrac{-6 \pm \sqrt{6^2 - 4(1)(13)}}{2(1)}$

$= \dfrac{-6 \pm \sqrt{36 - 52}}{2}$

$= \dfrac{-6 \pm \sqrt{-16}}{2}$

$= \dfrac{-6 \pm 4i}{2}$

$= -3 \pm 2i$

The solutions are $-3 - 2i$ and $-3 + 2i$.

47. $\dfrac{2}{5}y^2 + \dfrac{1}{5}y + \dfrac{3}{5} = 0$

$2y^2 + y + 3 = 0$

$a = 2,\ b = 1,\ c = 3$

$y = \dfrac{-1 \pm \sqrt{1^2 - 4(2)(3)}}{2(2)}$

$= \dfrac{-1 \pm \sqrt{1 - 24}}{4}$

$= \dfrac{-1 \pm \sqrt{-23}}{4}$

$= \dfrac{-1 \pm i\sqrt{23}}{4}$

The solutions are $\dfrac{-1 - i\sqrt{23}}{4}$ and

$\dfrac{-1 + i\sqrt{23}}{4}$.

49. $\dfrac{1}{2}y^2 = y - \dfrac{1}{2}$

$y^2 = 2y - 1$

$y^2 - 2y + 1 = 0$

$a = 1,\ b = -2,\ c = 1$

$y = \dfrac{2 \pm \sqrt{(-2)^2 - 4(1)(1)}}{2(1)}$

$= \dfrac{2 \pm \sqrt{4 - 4}}{2}$

$= \dfrac{2 \pm \sqrt{0}}{2}$

$= 1$

The solution is 1.

51. a. Two x-intercepts indicate two real solutions.

   b. One x-intercept indicates one real solution.

53.
$$(x+8)^2 + x^2 = 36^2$$
$$x^2 + 16x + 64 + x^2 = 1296$$
$$2x^2 + 16x - 1232 = 0$$
$$x^2 + 8x - 616 = 0$$
$$a = 1,\, b = 8,\, c = -616$$
$$x = \frac{-8 \pm \sqrt{8^2 - 4(1)(-616)}}{2(1)}$$
$$= \frac{-8 \pm \sqrt{2528}}{2}$$
$x \approx 21$ or $x \approx -29$ (disregard)
$$x + x + 8 = 2(21) + 8 = 50$$
$$50 - 36 = 14$$
They save about 14 feet of walking distance.

55. Let x = the length of each leg
$x + 2$ = the length of the hypotenuse
$$x^2 + x^2 = (x+2)^2$$
$$2x^2 = x^2 + 4x + 4$$
$$x^2 - 4x - 4 = 0$$
$$a = 1,\, b = -4,\, c = -4$$
$$x = \frac{4 \pm \sqrt{(-4)^2 - 4(1)(-4)}}{2(1)}$$
$$= \frac{4 \pm \sqrt{32}}{2}$$
$$= \frac{4 \pm 4\sqrt{2}}{2}$$
$$= 2 \pm 2\sqrt{2}$$
Reject the negative value.
The sides measure $2 + 2\sqrt{2}$ centimeters, $2 + 2\sqrt{2}$ centimeters, and $4 + 2\sqrt{2}$ centimeters.

57. Let x = the width, then $x + 10$ = the length.
Area = length $\cdot$ width
$$400 = (x+10)x$$
$$0 = x^2 + 10x - 400$$
$$a = 1,\, b = 10,\, c = -400$$
$$x = \frac{-10 \pm \sqrt{10^2 - 4(1)(-400)}}{2(1)}$$
$$= \frac{-10 \pm \sqrt{1700}}{2}$$
$$= \frac{-10 \pm 10\sqrt{17}}{2}$$
$$= -5 \pm 5\sqrt{17}$$
Reject the negative value.
The width is $-5 + 5\sqrt{17}$ feet and the length is $5 + 5\sqrt{17}$ feet.

59. a. Let x = the length.
$$x^2 + x^2 = 100^2$$
$$2x^2 - 10,000 = 0$$
$$a = 2,\, b = 0,\, c = -10,000$$
$$x = \frac{0 \pm \sqrt{0^2 - 4(2)(-10,000)}}{2(2)}$$
$$= \frac{\pm \sqrt{80,000}}{4}$$
$$= \frac{\pm 200\sqrt{2}}{4}$$
$$= \pm 50\sqrt{2}$$
Reject the negative value.
The sides measure $50\sqrt{2}$ meters.

b. Area $= s^2$
$$= \left(50\sqrt{2}\right)^2$$
$$= 50^2\left(\sqrt{2}\right)^2$$
$$= 2500 \cdot 2$$
$$= 5000$$
The area is 5000 square meters.

61. $\dfrac{x-1}{1} = \dfrac{1}{x}$

$x(x-1) = 1 \cdot 1$

$x^2 - x - 1 = 0$

$a = 1,\, b = -1,\, c = -1$

$x = \dfrac{1 \pm \sqrt{(-1)^2 - 4(1)(-1)}}{2(1)} = \dfrac{1 \pm \sqrt{5}}{2}$

Reject the negative value.

The value is $\dfrac{1 + \sqrt{5}}{2}$.

63. $h = -16t^2 + 20t + 1100$

$0 = -16t^2 + 20t + 1100$

$0 = -4t^2 + 5t + 275$

$a = -4,\, b = 5,\, c = 275$

$t = \dfrac{-5 \pm \sqrt{5^2 - 4(-4)(275)}}{2(-4)} = \dfrac{-5 \pm \sqrt{4475}}{-8}$

Reject the negative value.

$t \approx 8.9$

It will take about 8.9 seconds.

65. $h = -16t^2 - 20t + 180$

$0 = -16t^2 - 20t + 180$

$0 = -4(4t^2 + 5t - 45)$

$0 = 4t^2 + 5t - 45$

$a = 4,\, b = 5,\, c = -45$

$t = \dfrac{-5 \pm \sqrt{5^2 - 4(4)(-45)}}{2(4)} = \dfrac{-5 \pm \sqrt{745}}{8}$

Reject the negative value.

$t \approx 2.8$

The time until the ball strikes the ground is about 2.8 seconds.

67. From the graph, from Sunday to Monday shows the greatest decrease in the low temperature.

69. Wednesday had the lowest low temperature.

71. $f(4) = 3(4)^2 - 18(4) + 56 = 32$

This answer appears to agree with the graph.

73. a. $f(x) = 128.5x^2 - 69.5x + 2681$

$x = 1997 - 1995 = 2$

$f(2) = 128.5(2)^2 - 69.5(2) + 2681$

$\qquad = 3056$

Their net income in 1997 was $3056 million

b. $15,000 = 128.5x^2 - 69.5x + 2681$

$\qquad 0 = 128.5x^2 - 69.5x - 12,319$

$a = 128.5,\ b = -69.5,\ c = -12,319$

$$x = \frac{69.5 \pm \sqrt{(-69.5)^2 - 4(128.5)(-12,319)}}{2(128.5)}$$

Reject the negative value.

$x \approx 10$

$1995 + 10 = 2005$

10 years after 1995 or in 2005, the net income will be $15,000 million.

75. Exercise 63:

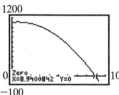

Exercise 65:

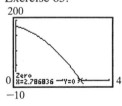

77.

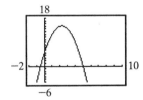

$y_1 = 9x - 2x^2 + 5$

There are 2 x-intercepts, thus two real solutions.

79. $\sqrt{5x - 2} = 3$

$\left(\sqrt{5x - 2}\right)^2 = 3^2$

$\qquad 5x - 2 = 9$

$\qquad\quad 5x = 11$

$\qquad\quad\ x = \dfrac{11}{5}$

The solution is $\dfrac{11}{5}$.

81. $\dfrac{1}{x} + \dfrac{2}{5} = \dfrac{7}{x}$

$5x \cdot \dfrac{1}{x} + 5x \cdot \dfrac{2}{5} = 5x \cdot \dfrac{7}{x}$

$\qquad\quad 5 + 2x = 35$

$\qquad\qquad 2x = 30$

$\qquad\qquad\ x = 15$

The solution is 15.

83. $x^4 + x^2 - 20 = (x^2 + 5)(x^2 - 4)$

$\qquad\qquad\qquad = (x^2 + 5)(x + 2)(x - 2)$

85. $z^4 - 13z^2 + 36 = (z^2 - 9)(z^2 - 4)$

$\qquad\qquad\qquad\ = (z + 3)(z - 3)(z + 2)(z - 2)$

87. $3x^2 - \sqrt{12}x + 1 = 0$

$a = 3,\ b = -\sqrt{12},\ c = 1$

$$x = \frac{\sqrt{12} \pm \sqrt{\left(-\sqrt{12}\right)^2 - 4(3)(1)}}{2(3)}$$

$$= \frac{\sqrt{12} \pm \sqrt{12 - 12}}{6}$$

$$= \frac{\sqrt{4}\sqrt{3} \pm \sqrt{0}}{6}$$

$$= \frac{2\sqrt{3}}{6}$$

$$= \frac{\sqrt{3}}{3}$$

The solution is $\dfrac{\sqrt{3}}{3}$.

89. $x^2 + \sqrt{2}x + 1 = 0$

$a = 1,\ b = \sqrt{2},\ c = 1$

$$x = \frac{-\sqrt{2} \pm \sqrt{\left(\sqrt{2}\right)^2 - 4(1)(1)}}{2(1)}$$

$$= \frac{-\sqrt{2} \pm \sqrt{2 - 4}}{2}$$

$$= \frac{-\sqrt{2} \pm \sqrt{-2}}{2}$$

$$= \frac{-\sqrt{2} \pm i\sqrt{2}}{2}$$

The solutions are $\dfrac{-\sqrt{2} - i\sqrt{2}}{2}$ and

$\dfrac{-\sqrt{2} + i\sqrt{2}}{2}$.

91. $2x^2 - \sqrt{3}x - 1 = 0$

$a = 2,\ b = -\sqrt{3},\ c = -1$

$$x = \frac{\sqrt{3} \pm \sqrt{\left(-\sqrt{3}\right)^2 - 4(2)(-1)}}{2(2)}$$

$$= \frac{\sqrt{3} \pm \sqrt{3 + 8}}{4}$$

$$= \frac{\sqrt{3} \pm \sqrt{11}}{4}$$

The solutions are $\dfrac{\sqrt{3} - \sqrt{11}}{4}$ and

$\dfrac{\sqrt{3} + \sqrt{11}}{4}$.

Exercise Set 8.3

1. 　　　　　　$2x = \sqrt{10 + 3x}$

　　　　　　$4x^2 = 10 + 3x$

$4x^2 - 3x - 10 = 0$

$(4x + 5)(x - 2) = 0$

$4x + 5 = 0$　　or　　$x - 2 = 0$

　　$x = -\dfrac{5}{4}$　　or　　　$x = 2$

Reject $x = -\dfrac{5}{4}$.

The solution is 2.

3. $x - 2\sqrt{x} = 8$

　　$2\sqrt{x} = x - 8$

　　　$4x = x^2 - 16x + 64$

　　　　$0 = x^2 - 20x + 64$

　　　　$0 = (x - 16)(x - 4)$

$x - 16 = 0$　　or　　$x - 4 = 0$

　　$x = 16$　or　　　$x = 4$

Reject $x = 4$.

The solution is 16.

5. $\sqrt{9x} = x + 2$

$9x = x^2 + 4x + 4$

$0 = x^2 - 5x + 4$

$0 = (x - 4)(x - 1)$

$x - 4 = 0$ or $x - 1 = 0$

$x = 4$ or $x = 1$

The solutions are 4 and 1.

7. $\dfrac{2}{x} + \dfrac{3}{x - 1} = 1$

$x(x - 1)\left(\dfrac{2}{x} + \dfrac{3}{x - 1}\right) = x(x - 1)(1)$

$2(x - 1) + 3x = x(x - 1)$

$2x - 2 + 3x = x^2 - x$

$5x - 2 = x^2 - x$

$x^2 - 6x + 2 = 0$

$x = \dfrac{6 \pm \sqrt{(-6)^2 - 4(1)(2)}}{2(1)} = \dfrac{6 \pm \sqrt{28}}{2} = \dfrac{6 \pm 2\sqrt{7}}{2} = 3 \pm \sqrt{7}$

The solutions are $3 - \sqrt{7}$ and $3 + \sqrt{7}$.

9. $\dfrac{3}{x} + \dfrac{4}{x + 2} = 2$

$x(x + 2)\left(\dfrac{3}{x} + \dfrac{4}{x + 2}\right) = x(x + 2)(2)$

$3(x + 2) + 4x = 2x(x + 2)$

$3x + 6 + 4x = 2x^2 + 4x$

$7x + 6 = 2x^2 + 4x$

$2x^2 - 3x - 6 = 0$

$x = \dfrac{3 \pm \sqrt{(-3)^2 - 4(2)(-6)}}{2(2)} = \dfrac{3 \pm \sqrt{57}}{4}$

The solutions are $\dfrac{3 - \sqrt{57}}{4}$ and $\dfrac{3 + \sqrt{57}}{4}$.

11.
$$\frac{7}{x^2 - 5x + 6} = \frac{2x}{x - 3} - \frac{x}{x - 2}$$
$$\frac{7}{(x - 3)(x - 2)} = \frac{2x}{x - 3} - \frac{x}{x - 2}$$
$$(x - 3)(x - 2)\left[\frac{7}{(x - 3)(x - 2)}\right] = (x - 3)(x - 2)\left(\frac{2x}{x - 3} - \frac{x}{x - 2}\right)$$
$$7 = 2x(x - 2) - x(x - 3)$$
$$7 = 2x^2 - 4x - x^2 + 3x$$
$$7 = x^2 - x$$
$$0 = x^2 - x - 7$$
$$x = \frac{1 \pm \sqrt{(-1)^2 - 4(1)(-7)}}{2(1)} = \frac{1 \pm \sqrt{29}}{2}$$

The solutions are $\dfrac{1 - \sqrt{29}}{2}$ and $\dfrac{1 + \sqrt{29}}{2}$.

13.
$$p^4 - 16 = 0$$
$$(p^2 + 4)(p^2 - 4) = 0$$
$$(p^2 + 4)(p + 2)(p - 2) = 0$$
$$p^2 + 4 = 0 \text{ or } p + 2 = 0 \text{ or } p - 2 = 0$$
$$p^2 = -4 \text{ or } p = -2 \text{ or } p = 2$$
$$p = \pm\sqrt{-4}$$
$$p = \pm 2i$$

The solutions are $-2i, 2i, -2,$ and 2.

15.
$$4x^4 + 11x^2 = 3$$
$$4x^4 + 11x^2 - 3 = 0$$
$$(4x^2 - 1)(x^2 + 3) = 0$$
$$(2x + 1)(2x - 1)(x^2 + 3) = 0$$
$$2x + 1 = 0 \quad \text{or} \quad 2x - 1 = 0 \quad \text{or} \quad x^2 + 3 = 0$$
$$2x = -1 \quad \text{or} \quad 2x = 1 \quad \text{or} \quad x^2 = -3$$
$$x = -\frac{1}{2} \quad \text{or} \quad x = \frac{1}{2} \quad \text{or} \quad x = \pm\sqrt{-3}$$
$$x = \pm i\sqrt{3}$$

The solutions are $-\dfrac{1}{2}, \dfrac{1}{2}, -i\sqrt{3},$ and $i\sqrt{3}$.

17.
$$z^4 - 13z^2 + 36 = 0$$
$$(z^2 - 9)(z^2 - 4) = 0$$
$$(z + 3)(z - 3)(z + 2)(z - 2) = 0$$
$$z + 3 = 0 \quad \text{or} \quad z - 3 = 0 \quad \text{or} \quad z + 2 = 0 \quad \text{or} \quad z - 2 = 0$$
$$z = -3 \quad \text{or} \quad z = 3 \quad \text{or} \quad z = -2 \quad \text{or} \quad z = 2$$
The solutions are $-3, 3, -2,$ and 2.

19. $x^{2/3} - 3x^{1/3} - 10 = 0$
Let $y = x^{1/3}$.
$$y^2 - 3y - 10 = 0$$
$$(y - 5)(y + 2) = 0$$
$$y - 5 = 0 \quad \text{or} \quad y + 2 = 0$$
$$y = 5 \quad \text{or} \quad y = -2$$
$$x^{1/3} = 5 \quad \text{or} \quad x^{1/3} = -2$$
$$x = 125 \quad \text{or} \quad x = -8$$
The solutions are 125 and -8.

21. $(5n + 1)^2 + 2(5n + 1) - 3 = 0$
Let $y = 5n + 1$.
$$y^2 + 2y - 3 = 0$$
$$(y + 3)(y - 1) = 0$$
$$y + 3 = 0 \quad \text{or} \quad y - 1 = 0$$
$$y = -3 \quad \text{or} \quad y = 1$$
$$5n + 1 = -3 \quad \text{or} \quad 5n + 1 = 1$$
$$5n = -4 \quad \text{or} \quad 5n = 0$$
$$n = -\frac{4}{5} \quad \text{or} \quad n = 0$$

The solutions are $-\frac{4}{5}$ and 0.

23. $2x^{2/3} - 5x^{1/3} = 3$
Let $y = x^{1/3}$.
$$2y^2 - 5y = 3$$
$$2y^2 - 5y - 3 = 0$$
$$(2y + 1)(y - 3) = 0$$
$$2y + 1 = 0 \quad \text{or} \quad y - 3 = 0$$
$$y = -\frac{1}{2} \quad \text{or} \quad y = 3$$
$$x^{1/3} = -\frac{1}{2} \quad \text{or} \quad x^{1/3} = 3$$
$$x = -\frac{1}{8} \quad \text{or} \quad x = 27$$
The solutions are $-\frac{1}{8}$ and 27.

25.
$$1 + \frac{2}{3t - 2} = \frac{8}{(3t - 2)^2}$$
$$(3t - 2)^2 + 2(3t - 2) = 8$$
Let $y = 3t - 2$.
$$y^2 + 2y = 8$$
$$y^2 + 2y - 8 = 0$$
$$(y + 4)(y - 2) = 0$$
$$y + 4 = 0 \quad \text{or} \quad y - 2 = 0$$
$$y = -4 \quad \text{or} \quad y = 2$$
$$3t - 2 = -4 \quad \text{or} \quad 3t - 2 = 2$$
$$3t = -2 \quad \text{or} \quad 3t = 4$$
$$t = -\frac{2}{3} \quad \text{or} \quad t = \frac{4}{3}$$
The solutions are $-\frac{2}{3}$ and $\frac{4}{3}$.

27. $20x^{2/3} - 6x^{1/3} - 2 = 0$

Let $y = x^{1/3}$.

$20y^2 - 6y - 2 = 0$

$(5y + 1)(4y - 2) = 0$

$$5y = -1 \quad \text{or} \quad 4y = 2$$
$$y = -\frac{1}{5} \quad \text{or} \quad y = \frac{1}{2}$$
$$x^{1/3} = -\frac{1}{5} \quad \text{or} \quad x^{1/3} = \frac{1}{2}$$
$$x = -\frac{1}{125} \quad \text{or} \quad x = \frac{1}{8}$$

The solutions are $-\dfrac{1}{125}$ and $\dfrac{1}{8}$.

29. $a^4 - 5a^2 + 6 = 0$

$(a^2 - 3)(a^2 - 2) = 0$

$$a^2 - 3 = 0 \quad \text{or} \quad a^2 - 2 = 0$$
$$a^2 = 3 \quad \text{or} \quad a^2 = 2$$
$$a = \pm\sqrt{3} \quad \text{or} \quad a = \pm\sqrt{2}$$

The solutions are $-\sqrt{3}$, $\sqrt{3}$, $-\sqrt{2}$, and $\sqrt{2}$.

31. $\dfrac{2x}{x-2} + \dfrac{x}{x+3} = -\dfrac{5}{x+3}$

$$\frac{2x}{x-2} = \frac{-x}{x+3} - \frac{5}{x+3}$$

$$\frac{2x}{x-2} = \frac{-x-5}{x+3}$$

$$2x(x+3) = (x-2)(-x-5)$$

$$2x^2 + 6x = -x^2 - 3x + 10$$

$3x^2 + 9x - 10 = 0$

$$x = \frac{-9 \pm \sqrt{9^2 - 4(3)(-10)}}{2(3)}$$

$$= \frac{-9 \pm \sqrt{201}}{6}$$

The solutions are $\dfrac{-9 - \sqrt{201}}{6}$ and $\dfrac{-9 + \sqrt{201}}{6}$.

33. $(p+2)^2 = 9(p+2) - 20$

$(p+2)^2 - 9(p+2) + 20 = 0$

Let $x = p + 2$.

$x^2 - 9x + 20 = 0$

$(x-5)(x-4) = 0$

$$x = 5 \quad \text{or} \quad x = 4$$
$$p + 2 = 5 \quad \text{or} \quad p + 2 = 4$$
$$p = 3 \quad \text{or} \quad p = 2$$

The solutions are 3 and 2.

35. $2x = \sqrt{11x + 3}$

$$4x^2 = 11x + 3$$

$4x^2 - 11x - 3 = 0$

$(4x + 1)(x - 3) = 0$

$$4x + 1 = 0 \quad \text{or} \quad x - 3 = 0$$
$$x = -\frac{1}{4} \quad \text{or} \quad x = 3$$

Reject $x = -\dfrac{1}{4}$.

The solution is 3.

37. $x^{2/3} - 8x^{1/3} + 15 = 0$

Let $y = x^{1/3}$.

$y^2 - 8y + 15 = 0$

$(y - 5)(y - 3) = 0$

$$y - 5 = 0 \quad \text{or} \quad y - 3 = 0$$
$$y = 5 \quad \text{or} \quad y = 3$$
$$x^{1/3} = 5 \quad \text{or} \quad x^{1/3} = 3$$
$$x = 125 \quad \text{or} \quad x = 27$$

The solutions are 125 and 27.

39. $y^3 + 9y - y^2 - 9 = 0$

$y(y^2 + 9) - 1(y^2 + 9) = 0$

$(y^2 + 9)(y - 1) = 0$

$$y^2 + 9 = 0 \quad \text{or} \quad y - 1 = 0$$
$$y^2 = -9 \quad \text{or} \quad y = 1$$
$$y = \pm\sqrt{-9}$$
$$y = \pm 3i$$

The solutions are $-3i$, $3i$, and 1.

41. $2x^{2/3} + 3x^{1/3} - 2 = 0$

Let $m = x^{1/3}$.

$2m^2 + 3m - 2 = 0$

$(2m - 1)(m + 2) = 0$

$2m - 1 = 0$ or $m + 2 = 0$

$m = \dfrac{1}{2}$ or $m = -2$

$x^{1/3} = \dfrac{1}{2}$ or $x^{1/3} = -2$

$x = \dfrac{1}{8}$ or $x = -8$

The solutions are $\dfrac{1}{8}$ and -8.

43. $x^{-2} - x^{-1} - 6 = 0$

Let $y = x^{-1}$.

$y^2 - y - 6 = 0$

$(y - 3)(y + 2) = 0$

$y - 3 = 0$ or $y + 2 = 0$

$y = 3$ or $y = -2$

$x^{-1} = 3$ or $x^{-1} = -2$

$x = \dfrac{1}{3}$ or $x = -\dfrac{1}{2}$

The solutions are $\dfrac{1}{3}$ and $-\dfrac{1}{2}$.

45. $x - \sqrt{x} = 2$

$x - 2 = \sqrt{x}$

$x^2 - 4x + 4 = x$

$x^2 - 5x + 4 = 0$

$(x - 4)(x - 1) = 0$

$x - 4 = 0$ or $x - 1 = 0$

$x = 4$ or $x = 1$

Reject $x = 1$.

The solution is 4.

47.
$$\frac{x}{x-1} + \frac{1}{x+1} = \frac{2}{x^2-1}$$

$$(x-1)(x+1)\left(\frac{x}{x-1} + \frac{1}{x+1}\right) = (x-1)(x+1)\left(\frac{2}{x^2-1}\right)$$

$$x(x+1) + 1(x-1) = 2$$

$$x^2 + x + x - 1 = 2$$

$$x^2 + 2x - 3 = 0$$

$$(x+3)(x-1) = 0$$

$$x + 3 = 0 \quad \text{or} \quad x - 1 = 0$$

$$x = -3 \quad \text{or} \quad x = 1$$

Reject $x = 1$.
The solution is -3.

49.
$$p^4 - p^2 - 20 = 0$$

$$(p^2 - 5)(p^2 + 4) = 0$$

$$p^2 - 5 = 0 \quad \text{or} \quad p^2 + 4 = 0$$

$$p^2 = 5 \quad \text{or} \quad p^2 = -4$$

$$p = \pm\sqrt{5} \quad \text{or} \quad p = \pm\sqrt{-4}$$

$$p = \pm 2i$$

The solutions are $-\sqrt{5}, \sqrt{5}, -2i,$ and $2i$.

51.
$$2x^3 = -54$$

$$x^3 = -27$$

$$x^3 + 27 = 0$$

$$(x+3)(x^2 - 3x + 9) = 0$$

$$x + 3 = 0 \quad \text{or} \quad x^2 - 3x + 9 = 0$$

$$x = -3 \quad \text{or} \quad x = \frac{3 \pm \sqrt{(-3)^2 - 4(1)(9)}}{2(1)} = \frac{3 \pm \sqrt{-27}}{2} = \frac{3 \pm 3i\sqrt{3}}{2}$$

The solutions are $-3, \dfrac{3 - 3i\sqrt{3}}{2},$ and $\dfrac{3 + 3i\sqrt{3}}{2}$.

53.

$$1 = \frac{4}{x-7} + \frac{5}{(x-7)^2}$$

$$(x-7)^2 = 4(x-7) + 5$$

Let $y = x - 7$.

$$y^2 - 4y - 5 = 0$$

$$(y-5)(y+1) = 0$$

$$y - 5 = 0 \quad \text{or} \quad y + 1 = 0$$

$$y = 5 \quad \text{or} \quad y = -1$$

$$x - 7 = 5 \quad \text{or} \quad x - 7 = -1$$

$$x = 12 \quad \text{or} \quad x = 6$$

The solutions are 6 and 12.

55.

$$27y^4 + 15y^2 = 2$$

$$27y^4 + 15y^2 - 2 = 0$$

$$(9y^2 - 1)(3y^2 + 2) = 0$$

$$(3y+1)(3y-1)(3y^2 + 2) = 0$$

$$3y + 1 = 0 \quad \text{or} \quad 3y - 1 = 0 \quad \text{or} \quad 3y^2 + 2 = 0$$

$$3y = -1 \quad \text{or} \quad 3y = 1 \quad \text{or} \quad 3y^2 = -2$$

$$y = -\frac{1}{3} \quad \text{or} \quad y = \frac{1}{3} \quad \text{or} \quad y^2 = -\frac{2}{3}$$

$$y = \pm\sqrt{-\frac{2}{3}}$$

$$y = \pm\frac{i\sqrt{6}}{3}$$

The solutions are $-\frac{1}{3}, \frac{1}{3}, -\frac{i\sqrt{6}}{3}$, and $\frac{i\sqrt{6}}{3}$.

57. Let x = The speed on the first part, then
$x - 1$ = the speed on the second part.

$$d = r \cdot t \text{ or } t = \frac{d}{r}, \quad 1\frac{3}{5} = \frac{8}{5}$$

$$\frac{3}{x} + \frac{4}{x-1} = \frac{8}{5}$$

$$3 \cdot 5(x-1) + 4 \cdot 5x = 8 \cdot x(x-1)$$

$$15x - 15 + 20x = 8x^2 - 8x$$

$$0 = 8x^2 - 43x + 15$$

$$0 = (8x - 3)(x - 5)$$

$$8x - 3 = 0 \quad \text{or} \quad x - 5 = 0$$

$$x = \frac{3}{8} \quad \text{or} \quad x = 5$$

Reject $x = \frac{3}{8}$, since $\frac{3}{8} - 1 = -\frac{5}{8}$.

So $x = 5$ and $x - 1 = 4$.

Her speed was 5 miles per hour on the first part, then 4 miles per hour on the second part.

59. Let x = the time for the hose alone, then
$x - 1$ = the time for the inlet pipe alone.

$$\frac{1}{x} + \frac{1}{x-1} = \frac{1}{8}$$

$$8(x-1) + 8x = x(x-1)$$

$$8x - 8 + 8x = x^2 - x$$

$$0 = x^2 - 17x + 8$$

$$x = \frac{17 \pm \sqrt{(-17)^2 - 4(1)(8)}}{2(1)} = \frac{17 \pm \sqrt{257}}{2}$$

$$x \approx 0.5 \quad \text{or} \quad x \approx 16.5$$

Reject $x \approx 0.5$ since $0.5 - 1 = -0.5$
So $x \approx 16.5$ and $x - 1 \approx 15.5$.
The inlet pipe can complete the job alone in 15.5 hours. The hose can complete the job alone in 16.5 hours.

61. Let x = the original speed, then
$x + 11$ = the return speed.

$$d = r \cdot t \text{ or } t = \frac{d}{r}$$

$$\frac{330}{x} - \frac{330}{x+11} = 1$$

$$330(x+11) - 330x = x(x+11)$$

$$330x + 3630 - 330x = x^2 + 11x$$

$$0 = x^2 + 11x - 3630$$

$$0 = (x - 55)(x + 66)$$

$$x = 55 \quad \text{or} \quad x = -66$$

Reject $x = -66$.
So $x = 55$ and $x + 11 = 66$.
Her original speed was 55 miles per hour. Her return speed was 66 miles per hour.

63.

	Hours	Part complete in one hour
Bill	$x - 1$	$\frac{1}{x-1}$
Billy	x	$\frac{1}{x}$
Together	4	$\frac{1}{4}$

$$\frac{1}{x-1} + \frac{1}{x} = \frac{1}{4}$$

$$4x \cdot (1) + 4(x-1) = x(x-1)$$

$$4x + 4x - 4 = x^2 - x$$

$$8x - 4 = x^2 - x$$

$$0 = x^2 - 9x + 4$$

$$x = \frac{9 \pm \sqrt{(-9)^2 - 4(1)(4)}}{2(1)}$$

$$x = \frac{9 - \sqrt{65}}{2} \quad \text{or} \quad x = \frac{9 + \sqrt{65}}{2}$$

$$x \approx 0.5 \quad \text{or} \quad x \approx 8.5$$

Reject $x \approx 0.5$ since $0.5 - 1 = -0.5$.
It takes his son about 8.5 hours.

65. Let x = the number.

$$x(x-4) = 96$$
$$x^2 - 4x = 96$$
$$x^2 - 4x - 96 = 0$$
$$(x-12)(x+8) = 0$$
$$x - 12 = 0 \quad \text{or} \quad x + 8 = 0$$
$$x = 12 \quad \text{or} \quad x = -8$$

The number is 12 or –8.

67. a. $x - 3 - 3 = x - 6$

The length is $(x-6)$ centimeters.

b. $V = l \cdot w \cdot h$

$$300 = (x-6)(x-6)(3)$$

c. $(x-6)(x-6) = 100$

$$x^2 - 12x + 36 = 100$$
$$x^2 - 12x - 64 = 0$$
$$(x-16)(x+4) = 0$$
$$x - 16 = 0 \quad \text{or} \quad x + 4 = 0$$
$$x = 16 \quad \text{or} \quad x = -4$$

Reject $x = -4$.

The sheet is 16 centimeters by 16 centimeters.

Check:

If the sheet of cardboard is 16 centimeters by 16 centimeters, then the length and width of the box are both $16 - 3 - 3 = 10$ centimeters.

$V = l \cdot w \cdot h = 10 \cdot 10 \cdot 3 = 300$

The volume is 300 cubic centimeters, as required.

69. a. Let x = Papis' fastest lap speed and $x + 3.8$ = Montoya's fastest lap speed.

$d = r \cdot t$, so $t = \dfrac{d}{r}$

$$\frac{7920}{x} = \frac{7920}{x+3.8} + 0.376$$
$$7920(x+3.8) = 7920x + 0.376x(x+3.8)$$
$$7920x + 30,096 = 7920x + 0.376x^2 + 1.4288x$$
$$0 = 0.376x^2 + 1.4288x - 30,096$$
$$x = \frac{-1.4288 \pm \sqrt{(1.4288)^2 - 4(0.376)(-30,096)}}{2(0.376)}$$

Using the positive square root, $x \approx 281.0$ feet per second.

Papis' fastest lap speed was 281.0 feet per second.

b. $x + 3.8 = 281.0 + 3.8 = 284.8$

Montoya's fastest lap speed was 284.8 feet per second.

c. There are 5280 feet in one mile, and 3600 seconds in one hour.

Papis: $\dfrac{281 \text{ feet}}{1 \text{ second}} \cdot \dfrac{3600 \text{ seconds}}{1 \text{ hour}} \cdot \dfrac{1 \text{ mile}}{5280 \text{ feet}} \approx 191.6$ miles per hour

Montoya: $\dfrac{284.8 \text{ feet}}{1 \text{ second}} \cdot \dfrac{3600 \text{ seconds}}{1 \text{ hour}} \cdot \dfrac{1 \text{ mile}}{5280 \text{ feet}} \approx 194.2$ miles per hour

71. Answers may vary.

73. $\dfrac{5x}{3} + 2 \le 7$

$\dfrac{5x}{3} \le 5$

$\dfrac{3}{5}\left(\dfrac{5x}{3}\right) \le \dfrac{3}{5}(5)$

$x \le 3$

The solution is $(-\infty, 3]$.

75. $\dfrac{y-1}{15} > -\dfrac{2}{5}$

$15\left(\dfrac{y-1}{15}\right) > 15\left(-\dfrac{2}{5}\right)$

$y - 1 > -6$

$y > -5$

The solution is $(-5, \infty)$.

77. Domain: $\{x \,|\, x \text{ is a real number}\}$

Range: $\{y \,|\, y \text{ is a real number}\}$

It is a function because it passes the vertical line test.

79. Domain: $\{x \,|\, x \text{ is a real number}\}$

Range: $\{y \,|\, y \ge -1\}$

It is a function because it passes the vertical line test.

Exercise Set 8.4

1. $(x + 1)(x + 5) > 0$
    $x + 1 = 0$ or $x + 5 = 0$
      $x = -1$ or $x = -5$

Region	Test Point	$(x + 1)(x + 5) > 0$	Result
A	-6	$(-6 + 1)(-6 + 5) > 0$	True
B	-2	$(-2 + 1)(-2 + 5) > 0$	False
C	0	$(0 + 1)(0 + 5) > 0$	True

The solution set is $(-\infty, -5) \cup (-1, \infty)$.

3. $(x - 3)(x + 4) \le 0$
    $x - 3 = 0$ or $x + 4 = 0$
      $x = 3$ or $x = -4$

Region	Test Point	$(x - 3)(x + 4) \le 0$	Result
A	-5	$(-5 - 3)(-5 + 4) \le 0$	False
B	0	$(0 - 3)(0 + 4) \le 0$	True
C	4	$(4 - 3)(4 + 4) \le 0$	False

The solution set is $[-4, 3]$.

5. $x^2 - 7x + 10 \le 0$

$(x - 5)(x - 2) \le 0$

$(x - 5)(x - 2) = 0$

$x - 5 = 0$　or　$x - 2 = 0$

$x = 5$　or　　$x = 2$

Region	Test Point	$(x - 5)(x - 2) \le 0$	Result
A	0	$(0 - 5)(0 - 2) \le 0$	False
B	3	$(3 - 5)(3 - 2) \le 0$	True
C	6	$(6 - 5)(6 - 2) \le 0$	False

The solution set is [2, 5].

7. $3x^2 + 16x < -5$

$3x^2 + 16x + 5 < 0$

$(3x + 1)(x + 5) < 0$

$(3x + 1)(x + 5) = 0$

$3x + 1 = 0$　or　$x + 5 = 0$

$3x = -1$　or　　$x = -5$

$x = -\dfrac{1}{3}$

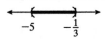

Region	Test Point	$(3x + 1)(x + 5) < 0$	Result
A	-6	$[3(-6) + 1](-6 + 5) < 0$	False
B	-1	$[3(-1) + 1](-1 + 5) < 0$	True
C	0	$[3(0) + 1](0 + 5) < 0$	False

The solution set is $\left(-5, -\dfrac{1}{3}\right)$.

9. $(x - 6)(x - 4)(x - 2) > 0$
$(x - 6)(x - 4)(x - 2) = 0$
$x - 6 = 0 \quad \text{or} \quad x - 4 = 0 \quad \text{or} \quad x - 2 = 0$
$\qquad x = 6 \quad \text{or} \qquad x = 4 \quad \text{or} \qquad x = 2$

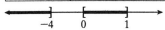

Region	Test Point	$(x-6)(x-4)(x-2)>0$	Result
A	1	$(1-6)(1-4)(1-2)>0$	False
B	3	$(3-6)(3-4)(3-2)>0$	True
C	5	$(5-6)(5-4)(5-2)>0$	False
D	7	$(7-6)(7-4)(7-2)>0$	True

The solution set is $(2, 4) \cup (6, \infty)$.

11. $x(x - 1)(x + 4) \le 0$
$x(x - 1)(x + 4) = 0$
$x = 0 \quad \text{or} \quad x - 1 = 0 \quad \text{or} \quad x + 4 = 0$
$x = 0 \quad \text{or} \qquad x = 1 \quad \text{or} \qquad x = -4$

Region	Test Point	$x(x-1)(x+4) \le 0$	Result
A	-5	$-5(-5-1)(-5+4) \le 0$	True
B	-1	$-1(-1-1)(-1+4) \le 0$	False
C	$\frac{1}{2}$	$\frac{1}{2}\left(\frac{1}{2}-1\right)\left(\frac{1}{2}+4\right) \le 0$	True
D	2	$2(2-1)(2+4) \le 0$	False

The solution set is $(-\infty, -4] \cup [0, 1]$.

13. $\qquad\qquad (x^2 - 9)(x^2 - 4) \le 0$
$(x + 3)(x - 3)(x + 2)(x - 2) \le 0$
$(x + 3)(x - 3)(x + 2)(x - 2) = 0$
$x + 3 = 0 \quad \text{or} \quad x - 3 = 0 \quad \text{or} \quad x + 2 = 0 \quad \text{or} \quad x - 2 = 0$
$\quad x = -3 \quad \text{or} \qquad x = 3 \quad \text{or} \qquad x = -2 \quad \text{or} \qquad x = 2$
The solution set is found from the portion of the graph on or below the x-axis.
The solution set is $\left[-3, -2\right] \cup \left[2, 3\right]$.

15. $\dfrac{x+7}{x-2} < 0$

$x + 7 = 0 \quad$ or $\quad x - 2 = 0$

$\quad\quad x = -7 \quad$ or $\quad\quad x = 2$

$$\overset{\text{A} \quad\quad \text{B} \quad\quad \text{C}}{\underset{-7 \quad\quad\quad 2}{\longleftrightarrow\!\!\!\!\!\!\mid\!\!\!\!\!\!\mid\!\!\!\!\!\!\longrightarrow}}$$

Choose -8 from region A.   Choose 0 from region B.   Choose 3 from region C.

$$\dfrac{x+7}{x-2} < 0 \quad\quad\quad \dfrac{x+7}{x-2} < 0 \quad\quad\quad \dfrac{x+7}{x-2} < 0$$

$$\dfrac{-8+7}{-8-2} < 0 \quad\quad\quad \dfrac{0+7}{0-2} < 0 \quad\quad\quad \dfrac{3+7}{3-2} < 0$$

$$\dfrac{-1}{-10} < 0 \quad\quad\quad\quad \dfrac{7}{-2} < 0 \quad\quad\quad\quad \dfrac{10}{1} < 0$$

$$\dfrac{1}{10} < 0 \quad \text{False} \quad\quad -\dfrac{7}{2} < 0 \quad \text{True} \quad\quad 10 < 0 \quad \text{False}$$

$$\underset{-7 \quad\quad\quad 2}{\longleftarrow\!\!\!\!\!\!(\!\!-\!\!-\!\!)\!\!\!\!\!\!\longrightarrow}$$

The solution set is $(-7, 2)$.

17. $\dfrac{5}{x+1} > 0$

$x + 1 = 0$

$\quad\quad x = -1$

$$\overset{\text{A} \quad\quad \text{B}}{\underset{-1}{\longleftrightarrow\!\!\!\!\!\!\mid\!\!\!\!\!\!\longrightarrow}}$$

Choose -2 from region A.   Choose 0 from region B.

$$\dfrac{5}{x+1} > 0 \quad\quad\quad\quad \dfrac{5}{x+1} > 0$$

$$\dfrac{5}{-2+1} > 0 \quad\quad\quad\quad \dfrac{5}{0+1} > 0$$

$$\dfrac{5}{-1} > 0 \quad\quad\quad\quad\quad \dfrac{5}{1} > 0$$

$$-5 > 0 \quad \text{False} \quad\quad\quad 5 > 0 \quad \text{True}$$

$$\underset{-1}{\longleftarrow\!\!\!\!\!\!(\!\!-\!\!-\!\!-\!\!-\!\!\longrightarrow}$$

The solution set is $(-1, \infty)$.

19. $\dfrac{x+1}{x-4} \geq 0$

$x+1=0$ or $x-4=0$

$\quad x=-1$ or $x=4$

The solution set is found from the portion of the graph on or above the x-axis.

The solution set is $(-\infty,\,-1] \cup (4,\,\infty)$.

21. Answers may vary.

23. $\qquad\qquad \dfrac{3}{x-2} < 4$

$\qquad\qquad \dfrac{3}{x-2}-4 < 0$

$\qquad\qquad \dfrac{3-4(x-2)}{x-2} < 0$

$\qquad\qquad \dfrac{3-4x+8}{x-2} < 0$

$\qquad\qquad \dfrac{11-4x}{x-2} < 0$

$11-4x=0$ and $x-2=0$

$\quad 11=4x$ and $x=2$

$\qquad x=\dfrac{11}{4}$

Choose 1 from region A.

$\dfrac{11-4x}{x-2} < 0$

$\dfrac{11-4(1)}{1-2} < 0$

$\dfrac{11--4}{1-2} < 0$

$\dfrac{7}{-1} < 0$

$-7 < 0$ True

Choose $\dfrac{5}{2}$ from region B.

$\dfrac{11-4x}{x-2} < 0$

$\dfrac{11-4\left(\frac{5}{2}\right)}{\frac{5}{2}-2} < 0$

$\dfrac{11-10}{\frac{5}{2}-\frac{4}{2}} < 0$

$\dfrac{1}{\frac{1}{2}} < 0$

$2 < 0$ False

Choose 3 from region C.

$\dfrac{11-4x}{x-2} < 0$

$\dfrac{11-4(3)}{3-2} < 0$

$\dfrac{11-12}{3-2} < 0$

$\dfrac{-1}{1} < 0$

$-1 < 0$ True

The solution set is $(-\infty,\, 2) \cup \left(\dfrac{11}{4},\, \infty\right)$.

25.

$$\frac{x^2+6}{5x} \geq 1$$

$$\frac{x^2+6}{5x} - 1 \geq 0$$

$$\frac{x^2+6-5x}{5x} \geq 0$$

$$\frac{(x-2)(x-3)}{5x} \geq 0$$

$$x-2=0 \quad \text{or} \quad x-3=0 \quad \text{or} \quad 5x=0$$

$$x=2 \quad \text{or} \quad x=3 \quad \text{or} \quad x=0$$

Choose -1 from region A.

$$\frac{(x-2)(x-3)}{5x} \geq 0$$

$$\frac{(-1-2)(-1-3)}{5(-1)} \geq 0$$

$$\frac{(-3)(-4)}{-5} \geq 0$$

$$\frac{12}{-5} \geq 0$$

$$-\frac{12}{5} \geq 0 \quad \text{False}$$

Choose 1 from region B.

$$\frac{(x-2)(x-3)}{5x} \geq 0$$

$$\frac{(1-2)(1-3)}{5(1)} \geq 0$$

$$\frac{(-1)(-2)}{5} \geq 0$$

$$\frac{2}{5} \geq 0 \quad \text{True}$$

Choose $\frac{5}{2}$ from region C.

$$\frac{(x-2)(x-3)}{5x} \geq 0$$

$$\frac{\left(\frac{5}{2}-2\right)\left(\frac{5}{2}-3\right)}{5\left(\frac{5}{2}\right)} \geq 0$$

$$\frac{\left(\frac{1}{2}\right)\left(-\frac{1}{2}\right)}{\frac{25}{2}} \geq 0$$

$$\frac{-\frac{1}{4}}{\frac{25}{2}} \geq 0$$

$$-\frac{1}{50} \geq 0 \quad \text{False}$$

Choose 4 from region D.

$$\frac{(x-2)(x-3)}{5x} \geq 0$$

$$\frac{(4-2)(4-3)}{5(4)} \geq 0$$

$$\frac{(2)(1)}{20} \geq 0$$

$$\frac{2}{20} \geq 0$$

$$\frac{1}{10} \geq 0 \quad \text{True}$$

The solution set is $(0, \ 2] \cup [3, \ \infty)$.

27. $(x-8)(x+7) > 0$

$(x-8)(x+7) = 0$

$x-8 = 0$ or $x+7 = 0$

$x = 8$ or $x = -7$

Region	Test Point	$(x-8)(x+7) > 0$	Result
A	-8	$(-8-8)(-8+7) > 0$	True
B	0	$(0-8)(0+7) > 0$	False
C	9	$(9-8)(9+7) > 0$	True

The solution set is $(-\infty, -7) \cup (8, \infty)$.

29. $(2x-3)(4x+5) \le 0$

$(2x-3)(4x+5) = 0$

$2x-3 = 0$ or $4x+5 = 0$

$2x = 3$ or $4x = -5$

$x = \dfrac{3}{2}$ or $x = -\dfrac{5}{4}$

Region	Test Point	$(2x-3)(4x+5) \le 0$	Result
A	-2	$[2(-2)-3][4(-2)+5] \le 0$	False
B	0	$[2(0)-3][4(0)+5] \le 0$	True
C	2	$[2(2)-3][4(2)+5] \le 0$	False

The solution set is $\left[-\dfrac{5}{4}, \dfrac{3}{2}\right]$.

31. $x^2 > x$

$x^2 - x > 0$

$x(x-1) > 0$

$x(x-1) = 0$

$x = 0$ or $x - 1 = 0$

$x = 1$

Region	Test Point	$x(x-1) > 0$	Result
A	-1	$(-1)(-1-1) > 0$	True
B	$\frac{1}{2}$	$\left(\frac{1}{2}\right)\left(\frac{1}{2}-1\right) > 0$	False
C	2	$(2)(2-1) > 0$	True

The solution set is $(-\infty, 0) \cup (1, \infty)$.

33. $(2x - 8)(x + 4)(x - 6) \le 0$

$(2x - 8)(x + 4)(x - 6) = 0$

$2x - 8 = 0$ or $x + 4 = 0$ or $x - 6 = 0$

$2x = 8$ or $x = -4$ or $x = 6$

$x = 4$

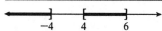

Region	Test Point	$(2x+8)(x+4)(x-6) \le 0$	Result
A	-5	$[2(-5) - 8](-5 + 4)(-5 - 6) \le 0$	True
B	0	$[2(0) - 8](0 + 4)(0 - 6) \le 0$	False
C	5	$[2(5) - 8](5 + 4)(5 - 6) \le 0$	True
D	7	$[2(7) - 8)](7 + 4)(7 - 6) \le 0$	False

The solution set is $(-\infty, -4] \cup [4, 6]$.

35.
$$6x^2 - 5x \geq 6$$
$$6x^2 - 5x - 6 \geq 0$$
$$(3x + 2)(2x - 3) \geq 0$$
$$(3x + 2)(2x - 3) = 0$$
$$3x + 2 = 0 \quad \text{or} \quad 2x - 3 = 0$$
$$3x = -2 \quad \text{or} \quad 2x = 3$$
$$x = -\frac{2}{3} \quad \text{or} \quad x = \frac{3}{2}$$

Region	Test Point	$(3x + 2)(2x - 3) \geq 0$	Result
A	-1	$[3(-1) + 2][2(-1) - 3] \geq 0$	True
B	0	$[3(0) + 2][2(0) - 3] \geq 0$	False
C	2	$[3(2) + 2][2(2) - 3] \geq 0$	True

The solution set is $\left(-\infty, -\frac{2}{3}\right] \cup \left[\frac{3}{2}, \infty\right)$.

37.
$$4x^3 + 16x^2 - 9x - 36 > 0$$
$$4x^2(x + 4) - 9(x + 4) > 0$$
$$(4x^2 - 9)(x + 4) > 0$$
$$(2x - 3)(2x + 3)(x + 4) > 0$$
$$2x - 3 = 0 \quad \text{or} \quad 2x + 3 = 0 \quad \text{or} \quad x + 4 = 0$$
$$x = \frac{3}{2} \quad \text{or} \quad x = -\frac{3}{2} \quad \text{or} \quad x = -4$$

Region	Test Point	$(2x - 3)(2x + 3)(x + 4) > 0$	Result
A	-5	$[2(-5) - 3][2(-5) + 3](-5 + 4) > 0$	False
B	-3	$[2(-3) - 3][2(-3) + 3](-3 + 4) > 0$	True
C	0	$[2(0) - 3][2(0) + 3](0 + 4) > 0$	False
D	2	$[2(2) - 3][2(2) + 3](2 + 4) > 0$	True

The solution set is $\left(-4, -\frac{3}{2}\right) \cup \left(\frac{3}{2}, \infty\right)$.

39.

$$x^4 - 26x^2 + 25 \geq 0$$

$$(x^2 - 25)(x^2 - 1) \geq 0$$

$$(x - 5)(x + 5)(x - 1)(x + 1) \geq 0$$

$$(x - 5)(x + 5)(x - 1)(x + 1) = 0$$

$x - 5 = 0$ or $x + 5 = 0$ or $x - 1 = 0$ or $x + 1 = 0$

$\quad x = 5$ or $\quad x = -5$ or $\quad x = 1$ or $\quad x = -1$

Region	Test Point	$(x-5)(x+5)(x-1)(x+1) \geq 0$	Result
A	–6	$(-6-5)(-6+5)(-6-1)(-6+1) \geq 0$	True
B	–3	$(-3-5)(-3+5)(-3-1)(-3+1) \geq 0$	False
C	0	$(0-5)(0+5)(0-1)(0+1) \geq 0$	True
D	3	$(3-5)(3+5)(3-1)(3+1) \geq 0$	False
E	6	$(6-5)(6+5)(6-1)(6+1) \geq 0$	True

The solution set is $(-\infty, -5] \cup [-1, 1] \cup [5, \infty)$.

41. $(2x - 7)(3x + 5) > 0$

$(2x - 7)(3x + 5) = 0$

$2x - 7 = 0$ or $3x + 5 = 0$

$\quad 2x = 7$ or $\quad 3x = -5$

$\quad x = \dfrac{7}{2}$ or $\quad x = -\dfrac{5}{3}$

Region	Test Point	$(2x - 7)(3x + 5) > 0$	Result
A	–2	$[2(-2) - 7][3(-2) + 5] > 0$	True
B	0	$[2(0) - 7][3(0) + 5] > 0$	False
C	4	$[2(4) - 7][2(4) + 5] > 0$	True

The solution set is $\left(-\infty, -\dfrac{5}{3}\right) \cup \left(\dfrac{7}{2}, \infty\right)$.

43. $\dfrac{x}{x-10} < 0$

$x = 0$ or $x - 10 = 0$

                            $x = 10$

$$\overset{\text{A} \quad\quad \text{B} \quad\quad \text{C}}{\underset{0 \quad\quad\quad 10}{\longleftarrow\!\!+\!\!-\!\!-\!\!-\!\!+\!\!\longrightarrow}}$$

Choose −1 from region A.	Choose 1 from region B.	Choose 11 from region C.
$\dfrac{x}{x-10} < 0$	$\dfrac{x}{x-10} < 0$	$\dfrac{x}{x-10} < 0$
$\dfrac{-1}{-1-10} < 0$	$\dfrac{1}{1-10} < 0$	$\dfrac{11}{11-10} < 0$
$\dfrac{-1}{-11} < 0$	$\dfrac{1}{-9} < 0$	$\dfrac{11}{1} < 0$
$\dfrac{1}{11} < 0$ False	$-\dfrac{1}{9} < 0$ True	$11 < 0$ False

$$\underset{0 \quad\quad\quad 10}{\longleftarrow\!\!(\!\!=\!\!=\!\!=\!\!)\!\!\longrightarrow}$$

The solution set is $(0,\ 10)$.

45. $\dfrac{x-5}{x+4} \geq 0$

$x - 5 = 0$ or $x + 4 = 0$

    $x = 5$ or $x = -4$

$$\overset{\text{A} \quad\quad \text{B} \quad\quad \text{C}}{\underset{-4 \quad\quad\quad 5}{\longleftarrow\!\!+\!\!-\!\!-\!\!-\!\!+\!\!\longrightarrow}}$$

Choose −5 from region A.	Choose 0 from region B.	Choose 6 from region C.
$\dfrac{x-5}{x+4} \geq 0$	$\dfrac{x-5}{x+4} \geq 0$	$\dfrac{x-5}{x+4} \geq 0$
$\dfrac{-5-5}{-5+4} \geq 0$	$\dfrac{0-5}{0+4} \geq 0$	$\dfrac{6-5}{6+4} \geq 0$
$\dfrac{-10}{-1} \geq 0$	$\dfrac{-5}{4} \geq 0$	$\dfrac{1}{10} \geq 0$ True
$10 \geq 0$ True	$-\dfrac{5}{4} \geq 0$ False	

$$\underset{-4 \quad\quad\quad 5}{\longleftarrow\!\!-\!\!)\!\!-\!\!-\!\!-\!\![\!\!-\!\!\longrightarrow}$$

The solution set is $(-\infty,\ -4) \cup [5,\ \infty)$.

47. $\dfrac{x(x+6)}{(x-7)(x+1)} \geq 0$

$x = 0$ or $x+6 = 0$ or $x-7 = 0$ or $x+1 = 0$

$x = 0$ or $x = -6$ or $x = 7$ or $x = -1$

$$
\begin{array}{ccccc}
\text{A} & \text{B} & \text{C} & \text{D} & \text{E} \\
\end{array}
$$

Choose –7 from region A.

$\dfrac{x(x+6)}{(x-7)(x+1)} \geq 0$

$\dfrac{(-7)(-7+6)}{(-7-7)(-7+1)} \geq 0$

$\dfrac{(-7)(-1)}{(-14)(-6)} \geq 0$

$\dfrac{7}{84} \geq 0$

$\dfrac{1}{12} \geq 0$ True

Choose –2 from region B.

$\dfrac{x(x+6)}{(x-7)(x+1)} \geq 0$

$\dfrac{(-2)(-2+6)}{(-2-7)(-2+1)} \geq 0$

$\dfrac{(-2)(-4)}{(-9)(-1)} \geq 0$

$\dfrac{-8}{9} \geq 0$

$-\dfrac{8}{9} \geq 0$ False

Choose $-\dfrac{1}{2}$ from region C.

$\dfrac{x(x+6)}{(x-7)(x+1)} \geq 0$

$\dfrac{\left(-\frac{1}{2}\right)\left(-\frac{1}{2}+6\right)}{\left(-\frac{1}{2}-7\right)\left(-\frac{1}{2}+1\right)} \geq 0$

$\dfrac{\left(-\frac{1}{2}\right)\left(\frac{11}{2}\right)}{\left(-\frac{15}{2}\right)\left(\frac{1}{2}\right)} \geq 0$

$\dfrac{\left(-\frac{11}{4}\right)}{\left(-\frac{15}{4}\right)} \geq 0$

$\dfrac{11}{15} \geq 0$ True

Choose 1 from region D.

$\dfrac{x(x+6)}{(x-7)(x+1)} \geq 0$

$\dfrac{(1)(1+6)}{(1-7)(1+1)} \geq 0$

$\dfrac{(1)(7)}{(-6)(2)} \geq 0$

$\dfrac{7}{-12} \geq 0$

$-\dfrac{7}{12} \geq 0$ False

Choose 8 from region E.

$\dfrac{x(x+6)}{(x-7)(x+1)} \geq 0$

$\dfrac{(8)(8+6)}{(8-7)(8+1)} \geq 0$

$\dfrac{(8)(14)}{(1)(9)} \geq 0$

$\dfrac{112}{9} \geq 0$ True

The solution set is $(-\infty, -6] \cup (-1, 0] \cup (7, \infty)$.

49.

$$\frac{-1}{x-1} > -1$$

$$-\frac{1}{x-1} + 1 > 0$$

$$\frac{-1 + (x-1)}{x-1} > 0$$

$$\frac{x-2}{x-1} > 0$$

$$x - 2 = 0 \quad \text{or} \quad x - 1 = 0$$

$$x = 2 \quad \text{or} \quad x = 1$$

Choose 0 from region A.   Choose $\frac{3}{2}$ from region B.   Choose 3 from region C.

$$\frac{x-2}{x-1} > 0 \qquad\qquad \frac{x-2}{x-1} > 0 \qquad\qquad \frac{x-2}{x-1} > 0$$

$$\frac{0-2}{0-1} > 0 \qquad\qquad \frac{\frac{3}{2}-2}{\frac{3}{2}-1} > 0 \qquad\qquad \frac{3-2}{3-1} > 0$$

$$\frac{-2}{-1} > 0 \qquad\qquad \frac{-\frac{1}{2}}{\frac{1}{2}} > 0 \qquad\qquad \frac{1}{2} > 0 \quad \text{True}$$

$$2 > 0 \quad \text{True} \qquad\qquad -1 > 0 \quad \text{False}$$

The solution set is $(-\infty, 1) \cup (2, \infty)$.

51.

$$\frac{x}{x+4} \leq 2$$

$$\frac{x}{x+4} - 2 \leq 0$$

$$\frac{x - 2(x+4)}{x+4} \leq 0$$

$$\frac{-x - 8}{x+4} \leq 0$$

$$-x - 8 = 0 \quad \text{or} \quad x + 4 = 0$$

$$-x = 8 \quad \text{or} \qquad x = -4$$

$$x = -8$$

A   B   C

───┼───┼───>

  −8   −4

Choose −9 from region A.

$$\frac{-x - 8}{x + 4} \leq 0$$

$$\frac{-(-9) - 8}{-9 + 4} \leq 0$$

$$\frac{9 - 8}{-9 + 4} \leq 0$$

$$\frac{1}{-5} \leq 0$$

$$-\frac{1}{5} \leq 0 \quad \text{True}$$

Choose −5 from region B.

$$\frac{-x - 8}{x + 4} \leq 0$$

$$\frac{-(-5) - 8}{-5 + 4} \leq 0$$

$$\frac{5 - 8}{-5 + 4} \leq 0$$

$$\frac{-3}{-1} \leq 0$$

$$3 \leq 0 \quad \text{False}$$

Choose 0 from region C.

$$\frac{-x - 8}{x + 4} \leq 0$$

$$\frac{0 - 8}{0 + 4} \leq 0$$

$$\frac{-8}{4} \leq 0$$

$$-2 \leq 0 \quad \text{True}$$

<───┼───────(───────>

  −8   −4

The solution set is $(-\infty,\ -8] \cup (-4,\ \infty)$.

53.

$$\frac{z}{z-5} \geq 2z$$

$$\frac{z}{z-5} - 2z \geq 0$$

$$\frac{z - 2z(z-5)}{z-5} \geq 0$$

$$\frac{z - 2z^2 + 10z}{z-5} \geq 0$$

$$\frac{11z - 2z^2}{z-5} \geq 0$$

$$\frac{z(11-2z)}{z-5} \geq 0$$

$$z = 0 \quad \text{or} \quad 11 - 2z = 0 \quad \text{or} \quad z - 5 = 0$$

$$11 = 2z \quad \text{or} \qquad z = 5$$

$$\frac{11}{2} = z$$

```
      A     B     C     D
  <---+-----+-----+-----+--->
      0     5    11/2
```

Choose -1 from region A.

$$\frac{z(11-2z)}{z-5} \geq 0$$

$$\frac{(-1)[11 - 2(-1)]}{-1-5} \geq 0$$

$$\frac{(-1)(13)}{-6} \geq 0$$

$$\frac{-13}{-6} \geq 0$$

$$\frac{13}{6} \geq 0 \quad \text{True}$$

Choose 1 from region B.

$$\frac{z(11-2z)}{z-5} \geq 0$$

$$\frac{(1)[11 - 2(1)]}{1-5} \geq 0$$

$$\frac{(1)(9)}{-4} \geq 0$$

$$\frac{9}{-4} \geq 0$$

$$-\frac{9}{4} \geq 0 \quad \text{False}$$

Choose $\frac{21}{4}$ from region C.

$$\frac{z(11-2z)}{z-5} \geq 0$$

$$\frac{\left(\frac{21}{4}\right)\left[11 - 2\left(\frac{21}{4}\right)\right]}{\frac{21}{4} - 5} \geq 0$$

$$\frac{\left(\frac{21}{4}\right)\left(\frac{1}{2}\right)}{\frac{1}{4}} \geq 0$$

$$\frac{\frac{21}{8}}{\frac{1}{4}} \geq 0$$

$$\frac{21}{2} \geq 0 \quad \text{True}$$

Choose 6 from region D.

$$\frac{z(11-2z)}{z-5} \geq 0$$

$$\frac{(6)[11 - 2(6)]}{6-5} \geq 0$$

$$\frac{(6)(-1)}{1} \geq 0$$

$$\frac{-6}{1} \geq 0$$

$$-6 \geq 0 \quad \text{False}$$

```
  <===+=====+=====(=====+--->
      0     5    11/2
```

The solution set is $(-\infty,\ 0] \cup \left(5,\ \frac{11}{2}\right]$.

55. $\dfrac{(x+1)^2}{5x} > 0$

$(x+1)^2 = 0$ or $5x = 0$

$x+1 = 0$ or $x = 0$

$x = -1$

$$\overset{\text{A} \quad\ \text{B} \quad\ \text{C}}{\underset{-1 \qquad\ 0}{\longleftrightarrow}}$$

Choose -2 from region A.   Choose $-\dfrac{1}{2}$ from region B.   Choose 1 from region C.

$\dfrac{(x+1)^2}{5x} > 0$ $\dfrac{(x+1)^2}{5x} > 0$ $\dfrac{(x+1)^2}{5x} > 0$

$\dfrac{(-2+1)^2}{5(-2)} > 0$ $\dfrac{\left(-\frac{1}{2}+1\right)^2}{5\left(-\frac{1}{2}\right)} > 0$ $\dfrac{(1+1)^2}{5(1)} > 0$

$\dfrac{(-1)^2}{-10} > 0$ $\dfrac{\left(\frac{1}{2}\right)^2}{-\frac{5}{2}} > 0$ $\dfrac{(2)^2}{5} > 0$

$\dfrac{1}{-10} > 0$ $\dfrac{\frac{1}{4}}{-\frac{5}{2}} > 0$ $\dfrac{4}{5} > 0$ True

$-\dfrac{1}{10} > 0$ False $-\dfrac{5}{2} > 0$ False

$$\overset{}{\underset{0}{\longleftarrow\hspace{-0.3em}\blacktriangleright\hspace{-1em}\longrightarrow}}$$

The solution set is $(0, \infty)$.

57. Let $x =$ the number, then

$\dfrac{1}{x} =$ the reciprocal of the number.

$$x - \dfrac{1}{x} < 0$$

$$\dfrac{x^2 - 1}{x} < 0$$

$$\dfrac{(x+1)(x-1)}{x} < 0$$

$x + 1 = 0 \quad$ or $\quad x - 1 = 0 \quad$ or $\quad x = 0$

$x = -1 \quad$ or $\qquad x = 1 \quad$ or $\quad x = 0$

A   B   C   D

$\xleftarrow{\quad\;\;|\qquad|\qquad|\quad}\rightarrow$
      $-1\quad\;\; 0\quad\;\; 1$

Choose -2 from region A.

$$\dfrac{(x+1)(x-1)}{x} < 0$$

$$\dfrac{(-2+1)(-2-1)}{-2} < 0$$

$$\dfrac{(-1)(-3)}{-2} < 0$$

$$\dfrac{3}{-2} < 0$$

$$-\dfrac{3}{2} < 0 \quad \text{True}$$

Choose $-\dfrac{1}{2}$ from region B.

$$\dfrac{(x+1)(x-1)}{x} < 0$$

$$\dfrac{\left(-\frac{1}{2}+1\right)\left(-\frac{1}{2}-1\right)}{-\frac{1}{2}} < 0$$

$$\dfrac{\left(\frac{1}{2}\right)\left(-\frac{3}{2}\right)}{-\frac{1}{2}} < 0$$

$$\dfrac{-\frac{3}{4}}{-\frac{1}{2}} < 0$$

$$\dfrac{3}{2} < 0 \quad \text{False}$$

Choose $\dfrac{1}{2}$ from region C.

$$\dfrac{(x+1)(x-1)}{x} < 0$$

$$\dfrac{\left(\frac{1}{2}+1\right)\left(\frac{1}{2}-1\right)}{\frac{1}{2}} < 0$$

$$\dfrac{\left(\frac{3}{2}\right)\left(-\frac{1}{2}\right)}{\frac{1}{2}} < 0$$

$$\dfrac{-\frac{3}{4}}{\frac{1}{2}} < 0$$

$$-\dfrac{3}{2} < 0 \quad \text{True}$$

Choose 2 from region D.

$$\dfrac{(x+1)(x-1)}{x} < 0$$

$$\dfrac{(2+1)(2-1)}{2} < 0$$

$$\dfrac{(3)(1)}{2} < 0$$

$$\dfrac{3}{2} < 0 \quad \text{False}$$

The solution set is $(-\infty, \, -1) \cup (0, \, 1)$. The number is any number less than -1 or between 0 and 1.

59. $P(x) = -2x^2 + 26x - 44$

$-2x^2 + 26x - 44 > 0$

$-2(x^2 - 13x + 22) > 0$

$-2(x - 11)(x - 2) > 0$

$x - 11 = 0$　or　$x - 2 = 0$

　　$x = 11$　or　　$x = 2$

Region	Test Point	$-2(x-11)(x-2) > 0$	Result
A	0	$-2(0-11)(0-2) > 0$	False
B	3	$-2(3-11)(3-2) > 0$	True
C	12	$-2(12-11)(12-2) > 0$	False

The solution set is (2, 11).

The company makes a profit when x is between 2 and 11.

61. $g(x) = |x| + 2$

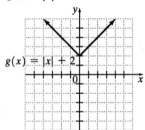

63. $F(x) = |x| - 1$

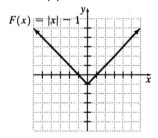

65. $F(x) = x^2 - 3$

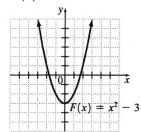

67. $H(x) = x^2 + 1$

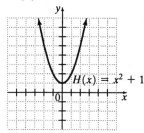

Section 8.5

Mental Math

1. $f(x) = x^2$; vertex: (0, 0)

2. $f(x) = -5x^2$; vertex: (0, 0)

3. $g(x) = (x - 2)^2$; vertex: (2, 0)

4. $g(x) = (x + 5)^2$; vertex: (–5, 0)

5. $f(x) = 2x^2 + 3$; vertex: (0, 3)

6. $h(x) = x^2 - 1$; vertex: (0, –1)

7. $g(x) = (x + 1)^2 + 5$; vertex: (–1, 5)

8. $h(x) = (x - 10)^2 - 7$; vertex: (10, –7)

Exercise Set 8.5

1. $f(x) = x^2 - 1$

The graph of $f(x) = x^2 - 1$ is obtained by shifting the graph of $y = x^2$ downward 1 unit.

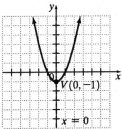

3. $h(x) = x^2 + 5$

The graph of $h(x) = x^2 + 5$ is obtained by shifting the graph of $y = x^2$ upward 5 units.

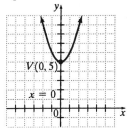

5. $g(x) = x^2 + 7$

The graph of $g(x) = x^2 + 7$ is obtained by shifting the graph of $y = x^2$ upward 7 units.

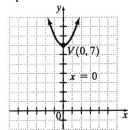

7. $f(x) = (x-5)^2$

The graph of $f(x) = (x-5)^2$ is obtained by shifting the graph of $y = x^2$ to the right 5 units.

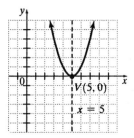

9. $h(x) = (x+2)^2$

The graph of $h(x) = (x+2)^2$ is obtained by shifting the graph of $y = x^2$ to the left 2 units.

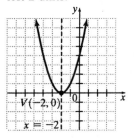

11. $G(x) = (x+3)^2$

The graph of $G(x) = (x+3)^2$ is obtained by shifting the graph of $y = x^2$ to the left 3 units.

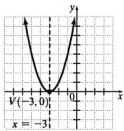

13. $f(x) = (x-2)^2 + 5$

The graph of $f(x) = (x-2)^2 + 5$ is the graph of $y = x^2$ shifted 2 units to the right and 5 units up.

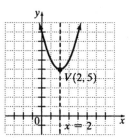

15. $h(x) = (x+1)^2 + 4$

The graph of $h(x) = (x+1)^2 + 4$ is the graph of $y = x^2$ shifted 1 unit to the left and 4 units up.

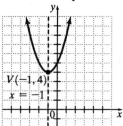

17. $g(x) = (x+2)^2 - 5$

The graph of $g(x) = (x+2)^2 - 5$ is the graph of $y = x^2$ shifted 2 units to the left and 5 units down.

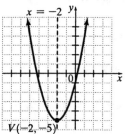

19. $g(x) = -x^2$

Because $a = -1$, this parabola opens downward.

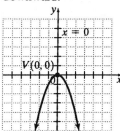

21. $h(x) = \dfrac{1}{3}x^2$

Because $a = \dfrac{1}{3}$ and $\dfrac{1}{3} < 1$, the parabola is wider than the graph of $y = x^2$.

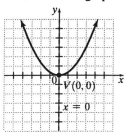

23. $H(x) = 2x^2$

Because $a = 2$ and $2 > 1$, the parabola is narrower than the graph of $y = x^2$.

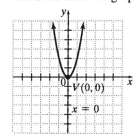

25. $f(x) = 2(x-1)^2 + 3$

This graph is similar to the graph of $y = x^2$ shifted 1 unit to the right and 3 units up, and it is narrower because a is 2.

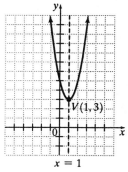

27. $h(x) = -3(x+3)^2 + 1$

This graph is similar to the graph of $y = x^2$ shifted 3 units to the left and 1 unit up, and it opens downward and is narrower because a is -3.

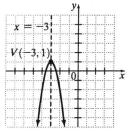

29. $H(x) = \dfrac{1}{2}(x-6)^2 - 3$

This graph is similar to the graph of $y = x^2$ shifted 6 units to the right and 3 units down, and it is wider because a is $\dfrac{1}{2}$.

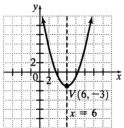

31. $f(x) = -(x-2)^2$

This graph is the same as the graph of $y = -x^2$ shifted 2 units to the right.

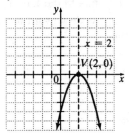

33. $F(x) = -x^2 + 4$

This graph is the same as the graph of $y = -x^2$ shifted 4 units up.

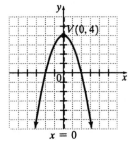

35. $F(x) = 2x^2 - 5$

This graph is similar to the graph of $y = x^2$ shifted 5 units down and it is narrower because a is 2.

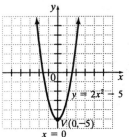

37. $h(x) = (x-6)^2 + 4$

This graph is the same as the graph of $y = x^2$ shifted 6 units to the right and 4 units up.

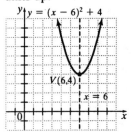

39. $F(x) = \left(x + \dfrac{1}{2}\right)^2 - 2$

This graph is the same as the graph of $y = x^2$ shifted $\dfrac{1}{2}$ unit to the left and 2 units down.

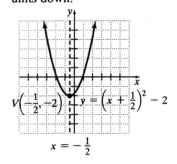

41. $F(x) = \dfrac{3}{2}(x+7)^2 + 1$

This graph is similar to the graph of $y = x^2$ shifted 7 units to the left and 1 unit up, and it is narrower because a is $\dfrac{3}{2}$.

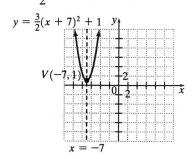

$y = \frac{3}{2}(x+7)^2 + 1$

$V(-7, 1)$

$x = -7$

43. $f(x) = \dfrac{1}{4}x^2 - 9$

This graph is similar to the graph of $y = x^2$ shifted 9 units down and it is wider because a is $\dfrac{1}{4}$.

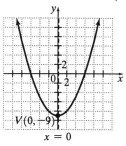

$V(0, -9)$

$x = 0$

45. $G(x) = 5\left(x + \dfrac{1}{2}\right)^2$

This graph is similar to the graph of $y = x^2$ shifted $\dfrac{1}{2}$ unit to the left and it is narrower because a is 5.

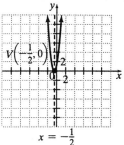

$V\left(-\dfrac{1}{2}, 0\right)$

$x = -\dfrac{1}{2}$

47. $f(x) = -(x-1)^2 - 1$

This graph is the same as the graph of $y = -x^2$ shifted 1 unit to the right and 1 unit down.

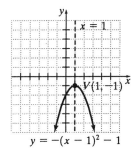

$x = 1$

$V(1, -1)$

$y = -(x-1)^2 - 1$

49. $g(x) = \sqrt{3}(x+5)^2 + \dfrac{3}{4}$

This graph is similar to the graph of

$y = x^2$ shifted 5 units to the left and

$\dfrac{3}{4}$ unit up and it is narrower because

a is $\sqrt{3}$.

$y = \sqrt{3}(x+5)^2 + \dfrac{3}{4}$

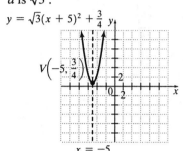

51. $h(x) = 10(x+4)^2 - 6$

This graph is similar to the graph of

$y = x^2$ shifted 4 units to the left and

6 units down, and it is narrower because

a is 10 .

$y = 10(x+4)^2 - 6$

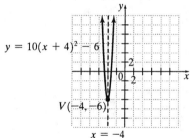

53. $f(x) = -2(x-4)^2 + 5$

This graph is similar to the graph of

$y = x^2$ shifted 4 units to the right and 5

units up, and it opens downward and is

narrower because a is -2 .

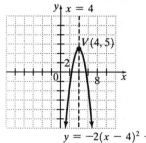

$y = -2(x-4)^2 + 5$

55. $f(x) = 5(x-2)^2 + 3$

57. $f(x) = 5[x-(-3)]^2 + 6$

$f(x) = 5(x+3)^2 + 6$

59. $F(x) = \sqrt{x};\ G(x) = \sqrt{x} + 1$

First, graph $y_1 = \sqrt{x}$.

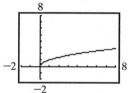

The graph of $G(x) = \sqrt{x} + 1$ is obtained

by shifting the graph of $y_1 = \sqrt{x}$ upward

1 unit.

Now, graph $y_1 = \sqrt{x}$ and $y_2 = \sqrt{x} + 1$.

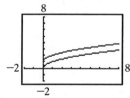

61. $H(x) = |x|; \ f(x) = |x - 5|$

First, graph $y_1 = |x|$.

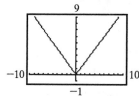

The graph of $f(x) = |x - 5|$ is obtained by shifting the graph of $y_1 = |x|$ to the right 5 units.

Now, graph $y_1 = |x|$ and $y_2 = |x - 5|$.

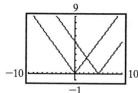

63. $f(x) = |x + 4|; \ F(x) = |x + 4| + 3$

First, graph $y_1 = |x + 4|$.

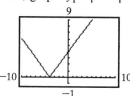

The graph of $F(x) = |x + 4| + 3$ is obtained by shifting the graph of $y_1 = |x + 4|$ upward 3 units.

Now, graph

$y_1 = |x + 4|$ and $y_2 = |x + 4| + 3$.

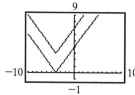

65. The graph of $y = f(x) + 1$ is obtained by shifting the graph of $y = f(x)$ upward 1 unit.

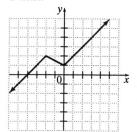

67. The graph of $y = f(x - 3)$ is obtained by shifting the graph of $y = f(x)$ to the right 3 units.

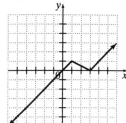

69. The graph of $y = f(x + 2) + 2$ is obtained by shifting the graph of $y = f(x)$ to the left 2 units and upward 2 units.

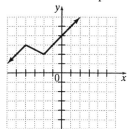

71. $x^2 + 8x$

$\left[\dfrac{1}{2}(8)\right]^2 = 4^2 = 16$

$x^2 + 8x + 16$

73. $z^2 - 16z$

$$\left[\frac{1}{2}(-16)\right]^2 = (-8)^2 = 64$$

$$z^2 - 16z + 64$$

75. $y^2 + y$

$$\left[\frac{1}{2}(1)\right]^2 = \left(\frac{1}{2}\right)^2 = \frac{1}{4}$$

$$y^2 + y + \frac{1}{4}$$

77. $x^2 + 4x = 12$

$$x^2 + 4x + 4 = 12 + 4$$

$$(x+2)^2 = 16$$

$$x + 2 = \pm\sqrt{16}$$

$$x + 2 = \pm 4$$

$$x = -2 \pm 4$$

$$x = -2 + 4 \quad \text{or} \quad x = -2 - 4$$

$$x = 2 \qquad \text{or} \quad x = -6$$

The solutions are 2 and –6.

79. $z^2 + 10z - 1 = 0$

$$z^2 + 10z = 1$$

$$z^2 + 10z + 25 = 1 + 25$$

$$(z+5)^2 = 26$$

$$z + 5 = \pm\sqrt{26}$$

$$z = -5 \pm \sqrt{26}$$

$$z = -5 + \sqrt{26} \text{ or } z = -5 - \sqrt{26}$$

The solutions are $-5 + \sqrt{26}$ and $-5 - \sqrt{26}$.

81. $z^2 - 8z = 2$

$$z^2 - 8z + 16 = 2 + 16$$

$$(z - 4)^2 = 18$$

$$z - 4 = \pm\sqrt{18}$$

$$z - 4 = \pm 3\sqrt{2}$$

$$z = 4 \pm 3\sqrt{2}$$

$$z = 4 + 3\sqrt{2} \text{ or } z = 4 - 3\sqrt{2}$$

The solutions are $4 + 3\sqrt{2}$ and $4 - 3\sqrt{2}$.

Exercise Set 8.6

1. $f(x) = x^2 + 8x + 7$

$$\frac{-b}{2a} = \frac{-8}{2(1)} = -4 \text{ and}$$

$$f(-4) = (-4)^2 + 8(-4) + 7$$

$$f(-4) = 16 - 32 + 7 = -9$$

Thus, the vertex is (–4, –9).

3. $f(x) = -x^2 + 10x + 5$

$$\frac{-b}{2a} = \frac{-10}{2(-1)} = 5 \text{ and}$$

$$f(5) = -5^2 + 10(5) + 5$$

$$f(5) = -25 + 50 + 5 = 30$$

Thus, the vertex is (5, 30).

5. $f(x) = 5x^2 - 10x + 3$

$$\frac{-b}{2a} = \frac{-(-10)}{2(5)} = 1 \text{ and}$$

$$f(1) = 5(1)^2 - 10(1) + 3$$

$$f(1) = 5 - 10 + 3 = -2$$

Thus, the vertex is (1, –2).

7. $f(x) = -x^2 + x + 1$

$$\frac{-b}{2a} = \frac{-1}{2(-1)} = \frac{1}{2} \text{ and}$$

$$f\left(\frac{1}{2}\right) = -\left(\frac{1}{2}\right)^2 + \frac{1}{2} + 1$$

$$f\left(\frac{1}{2}\right) = -\frac{1}{4} + \frac{1}{2} + 1 = \frac{5}{4}$$

Thus, the vertex is $\left(\frac{1}{2}, \frac{5}{4}\right)$.

9. $f(x) = x^2 - 4x + 3$

$\dfrac{-b}{2a} = \dfrac{-(-4)}{2(1)} = 2$ and

$f(2) = 2^2 - 4(2) + 3 = -1$

The vertex is (2, –1), so Graph D.

11. $f(x) = x^2 - 2x - 3$

$\dfrac{-b}{2a} = \dfrac{-(-2)}{2(1)} = 1$ and

$f(1) = 1^2 - 2(1) - 3 = -4$

The vertex is (1, –4), so Graph B.

13. $f(x) = x^2 + 4x - 5$

$\dfrac{-b}{2a} = \dfrac{-4}{2(1)} = -2$ and

$f(-2) = (-2)^2 + 4(-2) - 5$
$f(-2) = 4 - 8 - 5 = -9$

Thus, the vertex is (–2, –9).
The graph opens upward since $a = 1 > 0$.

$x^2 + 4x - 5 = 0$
$(x + 5)(x - 1) = 0$
$x + 5 = 0$ or $x - 1 = 0$
  $x = -5$ or $x = 1$

The x-intercepts are (–5, 0) and (1, 0).
$f(0) = -5$, so the y-intercept is (0, –5).

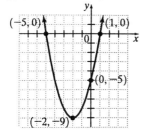

15. $f(x) = -x^2 + 2x - 1$

$\dfrac{-b}{2a} = \dfrac{-2}{2(-1)} = 1$ and

$f(1) = -1^2 + 2(1) - 1$
$f(1) = -1 + 2 - 1 = 0$

Thus, the vertex is (1, 0).
The graph opens downward since
$a = -1 < 0$.

$-x^2 + 2x - 1 = 0$
$-(x^2 - 2x + 1) = 0$
$-(x - 1)^2 = 0$
$x - 1 = 0$
$x = 1$

The x-intercept is (1, 0).
$f(0) = -1$, so the y-intercept is (0, –1).

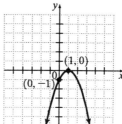

17. $f(x) = x^2 - 4$

$\dfrac{-b}{2a} = \dfrac{-0}{2(1)} = 0$ and

$f(0) = (0)^2 - 4$

$f(0) = 0 - 4 = -4$

Thus, the vertex is $(0, -4)$.

The graph opens upward since $a = 1 > 0$.

$$x^2 - 4 = 0$$
$$(x+2)(x-2) = 0$$
$$x + 2 = 0 \quad \text{or} \quad x - 2 = 0$$
$$x = -2 \quad \text{or} \quad x = 2$$

The x-intercepts are $(-2, 0)$ and $(2, 0)$.

$f(0) = -4$, so the y-intercept is $(0, -4)$.

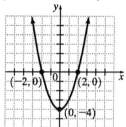

19. $f(x) = 4x^2 + 4x - 3$

$\dfrac{-b}{2a} = \dfrac{-4}{2(4)} = -\dfrac{1}{2}$ and

$f\left(-\dfrac{1}{2}\right) = 4\left(-\dfrac{1}{2}\right)^2 + 4\left(-\dfrac{1}{2}\right) - 3$

$f\left(-\dfrac{1}{2}\right) = 1 - 2 - 3 = -4$

Thus, the vertex is $\left(-\dfrac{1}{2}, -4\right)$.

The graph opens upward since $a = 4 > 0$.

$$4x^2 + 4x - 3 = 0$$
$$(2x-1)(2x+3) = 0$$
$$2x - 1 = 0 \quad \text{or} \quad 2x + 3 = 0$$
$$2x = 1 \quad \text{or} \quad 2x = -3$$
$$x = \dfrac{1}{2} \quad \text{or} \quad x = -\dfrac{3}{2}$$

The x-intercepts are $\left(\dfrac{1}{2}, 0\right)$ and $\left(-\dfrac{3}{2}, 0\right)$.

$f(0) = -3$, so the y-intercept is $(0, -3)$.

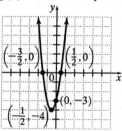

21.
$$y = x^2 + 8x + 15$$
$$y - 15 = x^2 + 8x$$
$$y - 15 + 16 = x^2 + 8x + 16$$
$$y + 1 = (x + 4)^2$$
$$f(x) = (x + 4)^2 - 1$$
Thus, the vertex is (–4, –1).
The graph opens upward since $a = 1 > 0$.
$$x^2 + 8x + 15 = 0$$
$$(x + 3)(x + 5) = 0$$
$$x + 3 = 0 \quad \text{or} \quad x + 5 = 0$$
$$x = -3 \quad \text{or} \qquad x = -5$$
The x-intercepts are (–3, 0) and (–5, 0).
$f(0) = 15$, so the y-intercept is (0, 15).

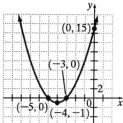

23.
$$y = x^2 - 6x + 5$$
$$y - 5 = x^2 - 6x$$
$$y - 5 + 9 = x^2 - 6x + 9$$
$$y + 4 = (x - 3)^2$$
$$f(x) = (x - 3)^2 - 4$$
Thus, the vertex is (3, –4).
The graph opens upward since $a = 1 > 0$.
$$x^2 - 6x + 5 = 0$$
$$(x - 1)(x - 5) = 0$$
$$x - 1 = 0 \quad \text{or} \quad x - 5 = 0$$
$$x = 1 \quad \text{or} \qquad x = 5$$
The x-intercepts are (1, 0) and (5, 0).
$f(0) = 5$, so the y-intercept is (0, 5).

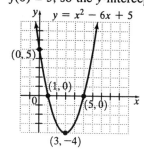

25.
$$y = x^2 - 4x + 5$$
$$y - 5 = x^2 - 4x$$
$$y - 5 + 4 = x^2 - 4x + 4$$
$$y - 1 = (x - 2)^2$$
$$f(x) = (x - 2)^2 + 1$$
Thus, the vertex is (2, 1).
The graph opens upward since $a = 1 > 0$.
$$x^2 - 4x + 5 = 0$$
$$(x - 2)^2 + 1 = 0$$
$$(x - 2)^2 = -1$$
Hence, there are no x-intercepts.
$f(0) = 5$, so the y-intercept is (0, 5).

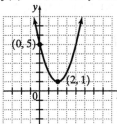

27.
$$y = 2x^2 + 4x + 5$$
$$y - 5 = 2(x^2 + 2x)$$
$$y - 5 + 2 = 2(x^2 + 2x + 1)$$
$$y - 3 = 2(x + 1)^2$$
$$f(x) = 2(x + 1)^2 + 3$$
Thus, the vertex is (–1, 3).
The graph opens upward since $a = 2 > 0$.
$$2x^2 + 4x + 5 = 0$$
$$2(x + 1)^2 + 3 = 0$$
$$2(x + 1)^2 = -3$$
Hence, there are no x-intercepts.
$f(0) = 5$, so the y-intercept is (0, 5).

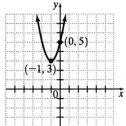

29.
$$y = -2x^2 + 12x$$
$$y = -2(x^2 - 6x)$$
$$y - 18 = -2(x^2 - 6x + 9)$$
$$y - 18 = -2(x - 3)^2$$
$$f(x) = -2(x - 3)^2 + 18$$
Thus, the vertex is (3, 18).
The graph opens downward since
$a = -2 < 0$.
$$-2x^2 + 12x = 0$$
$$-2x(x - 6) = 0$$
$$-2x = 0 \quad \text{or} \quad x - 6 = 0$$
$$x = 0 \quad \text{or} \qquad x = 6$$
The x-intercepts are (0, 0) and (6, 0).
$f(0) = 0$, so the y-intercept is (0, 0).

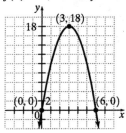

31. $f(x) = x^2 + 1$
$$\frac{-b}{2a} = \frac{-0}{2(1)} = 0 \text{ and}$$
$$f(0) = 0^2 + 1$$
$$f(0) = 0 + 1 = 1$$
Thus, the vertex is (0, 1).
The graph opens upward since $a = 1 > 0$.
$$x^2 + 1 = 0$$
$$x^2 = -1$$
Hence, there are no x-intercepts.
$f(0) = 1$, so the y-intercept is (0, 1).

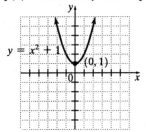

33. $f(x) = x^2 - 2x - 15$

$\dfrac{-b}{2a} = \dfrac{-(-2)}{2(1)} = 1$ and

$f(1) = 1^2 - 2(1) - 15$

$f(1) = 1 - 2 - 15 = -16$

Thus, the vertex is $(1, -16)$.

The graph opens upward since $a = 1 > 0$.

$x^2 - 2x - 15 = 0$

$(x - 5)(x + 3) = 0$

$x - 5 = 0$ or $x + 3 = 0$

$\quad x = 5$ or $\quad\quad x = -3$

The x-intercepts are $(5, 0)$ and $(-3, 0)$.

$f(0) = -15$, so the y-intercept is $(0, -15)$.

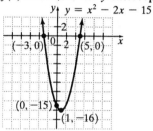

35. $f(x) = -5x^2 + 5x$

$\dfrac{-b}{2a} = \dfrac{-5}{2(-5)} = \dfrac{1}{2}$ and

$f\left(\dfrac{1}{2}\right) = -5\left(\dfrac{1}{2}\right)^2 + 5\left(\dfrac{1}{2}\right)$

$f\left(\dfrac{1}{2}\right) = -\dfrac{5}{4} + \dfrac{5}{2} = \dfrac{5}{4}$

Thus, the vertex is $\left(\dfrac{1}{2}, \dfrac{5}{4}\right)$.

The graph opens downward since $a = -5 < 0$.

$-5x^2 + 5x = 0$

$-5x(x - 1) = 0$

$-5x = 0$ or $x - 1 = 0$

$\quad x = 0$ or $\quad\quad x = 1$

The x-intercepts are $(0, 0)$ and $(1, 0)$.

$f(0) = 0$, so the y-intercept is $(0, 0)$.

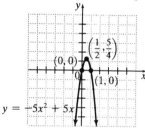

37. $f(x) = -x^2 + 2x - 12$

$\dfrac{-b}{2a} = \dfrac{-2}{2(-1)} = 1$ and

$f(1) = -1^2 + 2(1) - 12$

$f(1) = -1 + 2 - 12 = -11$

Thus, the vertex is $(1, -11)$.
The graph opens downward since
$a = -1 < 0$.

$-x^2 + 2x - 12 = 0$

$x^2 - 2x = -12$

$x^2 - 2x + 1 = -12 + 1$

$(x - 1)^2 = -11$

Hence, there are no x-intercepts.
$f(0) = -12$, so the y-intercept is $(0, -12)$.

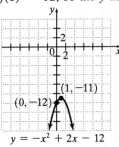

$y = -x^2 + 2x - 12$

39. $f(x) = 3x^2 - 12x + 15$

$\dfrac{-b}{2a} = \dfrac{-(-12)}{2(3)} = 2$ and

$f(2) = 3(2)^2 - 12(2) + 15$

$f(2) = 12 - 24 + 15 = 3$

Thus, the vertex is $(2, 3)$.
The graph opens upward since $a = 3 > 0$.

$3x^2 - 12x + 15 = 0$

$x^2 - 4x + 5 = 0$

$x^2 - 4x + 4 = -5 + 4$

$(x - 2)^2 = -1$

Hence, there are no x-intercepts.
$f(0) = 15$, so the y-intercept is $(0, 15)$.

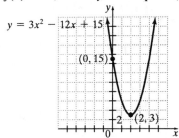

$y = 3x^2 - 12x + 15$

41. $f(x) = x^2 + x - 6$

$$\frac{-b}{2a} = \frac{-1}{2(1)} = -\frac{1}{2} \text{ and}$$

$$f\left(-\frac{1}{2}\right) = \left(-\frac{1}{2}\right)^2 + \left(-\frac{1}{2}\right) - 6$$

$$f\left(-\frac{1}{2}\right) = \frac{1}{4} - \frac{1}{2} - 6 = -\frac{25}{4}$$

Thus, the vertex is $\left(-\frac{1}{2}, -\frac{25}{4}\right)$.

The graph opens upward since $a = 1 > 0$.

$$x^2 + x - 6 = 0$$
$$(x + 3)(x - 2) = 0$$
$$x + 3 = 0 \quad \text{or} \quad x - 2 = 0$$
$$x = -3 \quad \text{or} \quad x = 2$$

The x-intercepts are $(-3, 0)$ and $(2, 0)$.

$f(0) = -6$, so the y-intercept is $(0, -6)$.

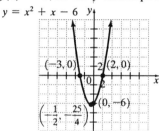

43. $f(x) = -2x^2 - 3x + 35$

$$\frac{-b}{2a} = \frac{-(-3)}{2(-2)} = -\frac{3}{4} \text{ and}$$

$$f\left(-\frac{3}{4}\right) = -2\left(-\frac{3}{4}\right)^2 - 3\left(-\frac{3}{4}\right) + 35$$

$$f\left(-\frac{3}{4}\right) = -\frac{9}{8} + \frac{9}{4} + 35 = \frac{289}{8}$$

Thus, the vertex is $\left(-\frac{3}{4}, \frac{289}{8}\right)$.

The graph opens downward since $a = -2 < 0$.

$$-2x^2 - 3x + 35 = 0$$
$$2x^2 + 3x - 35 = 0$$
$$(2x - 7)(x + 5) = 0$$
$$2x - 7 = 0 \quad \text{or} \quad x + 5 = 0$$
$$2x = 7 \quad \text{or} \quad x = -5$$
$$x = \frac{7}{2}$$

The x-intercepts are $\left(\frac{7}{2}, 0\right)$ and $(-5, 0)$.

$f(0) = 35$, so the y-intercept is $(0, 35)$.

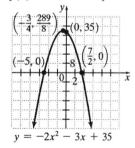

$$y = -2x^2 - 3x + 35$$

45. $y = -x^2 + 6x + 5$

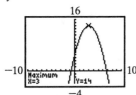

The vertex is $(3, 14)$ and it is a maximum point.
The axis of symmetry is $x = 3$.

47. $y = x^2 + 4x - 12$

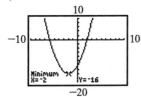

The vertex is $(-2, -16)$ and it is a minimum point.
The axis of symmetry is $x = -2$.

49. $C(x) = 2x^2 - 800x + 92,000$

a. $\dfrac{-b}{2a} = \dfrac{-(-800)}{2(2)} = 200$

200 bicycles must be manufactured to minimize the cost.

b. $C(200) = 2(200)^2 - 800(200) + 92,000$
$= 12,000$
The minimum cost is $12,000.

51. $h(t) = -16t^2 + 32t$
$\dfrac{-b}{2a} = \dfrac{-32}{2(-16)} = 1$
$h(1) = -16(1)^2 + 32(1)$
$h(1) = 16$
The maximum height of the ball is 16 feet.

53. Let $x =$ one number, then
$60 - x =$ the other number.
$f(x) = x(60 - x)$
$f(x) = 60x - x^2$
$\dfrac{-b}{2a} = \dfrac{-60}{2(-1)} = 30$
The maximum occurs at $x = 30$. The numbers are 30 and 30.

55. Let $x =$ one number, then
$10 + x =$ the other number.
$f(x) = x(10 + x)$
$f(x) = 10x + x^2$
$\dfrac{-b}{2a} = \dfrac{-10}{2(1)} = -5$
The minimum occurs at $x = -5$. The numbers are -5 and 5.

57. Let $x =$ the width, then
$40 - x =$ the length.
$f(x) = x(40 - x)$
$f(x) = 40x - x^2$
$\dfrac{-b}{2a} = \dfrac{-40}{2(-1)} = 20$
The maximum occurs at $x = 20$. The width is 20 units and the length is 20 units.

59. a. $f(14) = -0.74(14)^2 + 8.66(14) + 159.07$
≈ 135.27
In 2004, the emissions will be 135.27 million metric tons.

b. The maximum value occurs when
$x = \dfrac{-8.66}{2(-0.74)} \approx 6.$
Therefore emissions were at a maximum in 1996.

c. $f(6) = -0.74(6)^2 + 8.66(6) + 159.07$
≈ 184.39
The maximum methane emissions level was 184.39 million metric tons.

61. $f(x) = x^2 + 10x + 15$

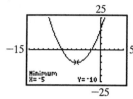

The vertex is $(-5, -10)$ and the graph opens upward.

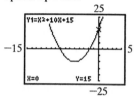

The y-intercept is $(0, 15)$.

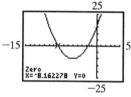

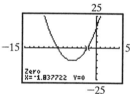

The x-intercepts are $(-8.2, 0)$ and $(-1.8, 0)$.

63. $f(x) = 3x^2 - 6x + 7$

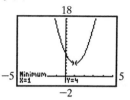

The vertex is $(1, 4)$ and the graph opens upward.

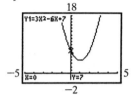

The y-intercept is $(0, 7)$ and there are no x-intercepts.

65. $f(x) = 2.3x^2 - 6.1x + 3.2$

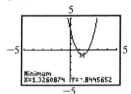

The minimum value is about -0.84.

67. $f(x) = -1.9x^2 + 5.6x - 2.7$

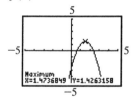

The maximum value is about 1.43.

69. This graph is the same as the graph of $y = x^2$ shifted 2 units up.

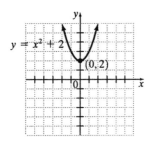

71. This graph is the same as the graph of $y = x$ shifted 2 units up.

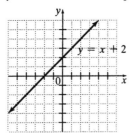

73. This graph is the same as the graph of $y = x^2$ shifted 5 units to the left and 2 units up.

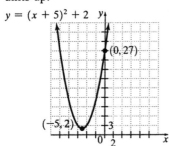

75. This graph is similar to the graph of $y = x^2$ shifted 4 units to the right and 1 unit up, and it is narrower because a is 3 and $3 > 1$.

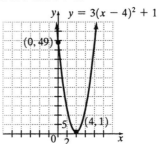

77. This graph is the same as the graph of $y = x^2$ shifted 4 units to the right and $\dfrac{3}{2}$ units up, and it opens downward because a is -1.

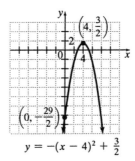

Exercise Set 8.7

1. Since the data points fall in approximately a straight line, a linear model may be best.

3. Since the data points do not appear to be in a straight line, nor in the shape of a parabola, neither model would be used.

5. Since the data points fall in approximately a straight line, a linear model may be best.

7. Since the vertical distance between the data points is increasing rather than constant, a quadratic model may be best.

9. Let $x = 0$ correspond to 1900.

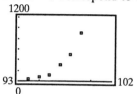

The graph shows that a quadratic model may be best.

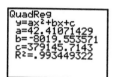

The equation is

$$y = 42.411x^2 - 8019.554x + 379,145.714.$$

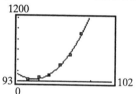

Graph of quadratic regression equation.
$x = 2004 - 1900 = 104$
To predict the gallons in 2004, find $Y_1(104)$.

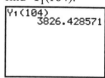

In 2004, 3826.4 million gallons of reclaimed water can be expected to be used.

11. Let $x = 0$ correspond to 1900.

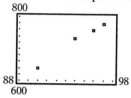

The graph shows that a linear model may be best.

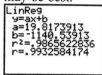

The equation is $y = 19.817x - 1140.539$.

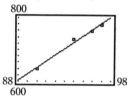

Graph of linear regression equation.
$x = 2007 - 1900 = 107$
To predict the earnings in 2007, find $Y_1(107)$.

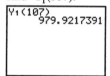

In 2007, the median weekly earnings of full-time wage and salary workers 25 years and older with a four-year degree is predicted to be \$979.92.

13. Let $x = 0$ correspond to 1900.

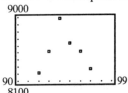

The graph shows that a quadratic model may be best.

The equation is

$$y = -91.5x^2 + 17,292.529x - 808,233.2.$$

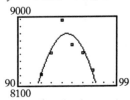

Graph of quadratic regression equation.
$x = 2001 - 1900 = 101$
To predict the sales in 2001, find $Y_1(101)$.

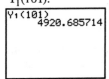

In 2001, 4920.7 thousand cars will be sold.

15. Let $x = 0$ correspond to 1900.

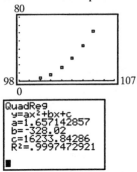

The quadratic regression equation is

$$y = 1.657x^2 - 328.02x + 16,233.843.$$

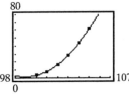

Graph of quadratic regression equation.
$x = 2006 - 1900 = 106$
To predict the number of households in 2006, find $Y_1(106)$.

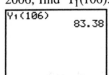

In 2006, 83.38 million households will be using personal rich media.

17. Let $x = 0$ correspond to 1900.

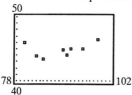

The graph shows that a quadratic model may be best.

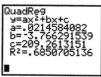

The equation is
$$y = 0.021x^2 - 3.766x + 209.261.$$

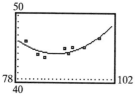

Graph of quadratic regression equation.
$x = 2005 - 1900 = 105$
To predict the percent in 2005,
find $Y_1(105)$.

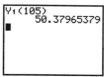

In 2005, 50.38% of Americans aged 60 to 64 are predicted to be in the work force.

19.

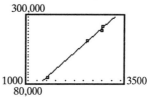

The graph shows that a linear model may be best.

```
LinReg
 y=ax+b
 a=114.4530447
 b=-81378.6754
 r²=.9937400738
 r=.9968651232
■
```

The equation is
$y = 114.453x - 81,378.675.$

Graph of linear regression equation.

21.
```
Y₁(2200)
    170418.0229
■
```

The predicted selling price of a house with 2200 square feet is $170,418.

23. $y = 0.0075x^2 - 29.372x + 28,770.797$

x (year)	Actual MPG	Predicted MPG	Difference
1990	21.0	21.3	0.3
1991	21.7	21.8	0.1
1992	21.7	22.3	0.6
1993	21.6	22.8	1.2

25.

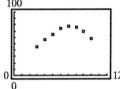

The graph shows that a quadratic model may be best.

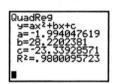

The equation is
$y = -1.994x^2 + 28.220x - 23.339.$

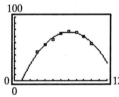

Graph of quadratic regression equation.
Answers may vary.

27. A quadratic model may be best.

29. A linear model may be best.

31. A quadratic model may be best.

33. A quadratic model may be best.

35. $(x + 3)(x - 5) = 0$
$\quad x + 3 = 0 \quad$ or $\quad x - 5 = 0$
$\quad\quad x = -3 \quad$ or $\quad\quad x = 5$
The solutions are -3 and 5.

37. $2x^2 - 7x - 15 = 0$
$\quad (2x + 3)(x - 5) = 0$
$\quad 2x + 3 = 0 \quad$ or $x - 5 = 0$
$\quad\quad x = -\dfrac{3}{2} \quad$ or $\quad x = 5$
The solutions are $-\dfrac{3}{2}$ and 5.

39. $3(x - 4) + 2 = 5(x - 6)$
$\quad 3x - 12 + 2 = 5x - 30$
$\quad\quad\quad -2x = -20$
$\quad\quad\quad\quad x = 10$
The solution is 10.

41. $f(x) = x^3 + 3x^2 - 5x - 8$
$\quad f(0) = 0^3 + 3(0)^2 - 5(0) - 8 = -8$
The y-intercept is $(0, -8)$.

43. $g(x) = x^2 - 3x + 5$
$\quad g(0) = 0^2 - 3(0) + 5 = 5$
The y-intercept is $(0, 5)$.

Chapter 8 Review

1. $x^2 - 15x + 14 = 0$
$(x - 14)(x - 1) = 0$
$x - 14 = 0$ or $x - 1 = 0$
$x = 14$ or $x = 1$
The solutions are 14 and 1.

2. $x^2 - x - 30 = 0$
$(x + 5)(x - 6) = 0$
$x + 5 = 0$ or $x - 6 = 0$
$x = -5$ or $x = 6$
The solutions are –5 and 6.

3. $10x^2 = 3x + 4$
$10x^2 - 3x - 4 = 0$
$(5x - 4)(2x + 1) = 0$
$5x - 4 = 0$ or $2x + 1 = 0$
$5x = 4$ or $2x = -1$
$x = \dfrac{4}{5}$ or $x = -\dfrac{1}{2}$
The solutions are $\dfrac{4}{5}$ and $-\dfrac{1}{2}$.

4. $7a^2 = 29a + 30$
$7a^2 - 29a - 30 = 0$
$(7a + 6)(a - 5) = 0$
$7a + 6 = 0$ or $a - 5 = 0$
$a = -\dfrac{6}{7}$ or $a = 5$

The solutions are $-\dfrac{6}{7}$ and 5.

5. $4m^2 = 196$
$m^2 = 49$
$m = \pm\sqrt{49}$
$m = \pm 7$
The solutions are 7 and –7.

6. $9y^2 = 36$
$y^2 = 4$
$y = \pm\sqrt{4}$
$y = \pm 2$
The solutions are 2 and –2.

7. $(9n + 1)^2 = 9$
$9n + 1 = \pm\sqrt{9}$
$9n + 1 = \pm 3$
$9n = -1 \pm 3$
$n = \dfrac{-1 \pm 3}{9}$
$n = \dfrac{2}{9}$ or $n = -\dfrac{4}{9}$
The solutions are $\dfrac{2}{9}$ and $-\dfrac{4}{9}$.

8. $(5x - 2)^2 = 2$
$5x - 2 = \pm\sqrt{2}$
$5x = 2 \pm \sqrt{2}$
$x = \dfrac{2 \pm \sqrt{2}}{5}$
The solutions are $\dfrac{2 - \sqrt{2}}{5}$ and $\dfrac{2 + \sqrt{2}}{5}$.

9. $z^2 + 3z + 1 = 0$
$z^2 + 3z = -1$
$z^2 + 3z + \left(\dfrac{3}{2}\right)^2 = -1 + \dfrac{9}{4}$
$\left(z + \dfrac{3}{2}\right)^2 = \dfrac{5}{4}$
$z + \dfrac{3}{2} = \pm\sqrt{\dfrac{5}{4}}$
$z + \dfrac{3}{2} = \dfrac{\pm\sqrt{5}}{2}$
$z = -\dfrac{3}{2} \pm \dfrac{\sqrt{5}}{2}$
$z = \dfrac{-3 \pm \sqrt{5}}{2}$
The solutions are $\dfrac{-3 - \sqrt{5}}{2}$ and $\dfrac{-3 + \sqrt{5}}{2}$.

10.
$$x^2 + x + 7 = 0$$
$$x^2 + x = -7$$
$$x^2 + x + \left(\frac{1}{2}\right)^2 = -7 + \frac{1}{4}$$
$$\left(x + \frac{1}{2}\right)^2 = -\frac{27}{4}$$
$$x + \frac{1}{2} = \pm\sqrt{\frac{-27}{4}}$$
$$x = -\frac{1}{2} \pm \frac{3i\sqrt{3}}{2}$$
$$x = \frac{-1 \pm 3i\sqrt{3}}{2}$$

The solutions are $\dfrac{-1 - 3i\sqrt{3}}{2}$ and $\dfrac{-1 + 3i\sqrt{3}}{2}$.

11.
$$(2x + 1)^2 = x$$
$$4x^2 + 4x + 1 = x$$
$$4x^2 + 3x = -1$$
$$x^2 + \frac{3}{4}x = -\frac{1}{4}$$
$$x^2 + \frac{3}{4}x + \left(\frac{\frac{3}{4}}{2}\right)^2 = -\frac{1}{4} + \frac{9}{64}$$
$$\left(x + \frac{3}{8}\right)^2 = -\frac{7}{64}$$
$$x + \frac{3}{8} = \pm\sqrt{\frac{-7}{64}}$$
$$x + \frac{3}{8} = \pm\frac{i\sqrt{7}}{8}$$
$$x = -\frac{3}{8} \pm \frac{i\sqrt{7}}{8}$$
$$x = \frac{-3 \pm i\sqrt{7}}{8}$$

The solutions are $\dfrac{-3 - i\sqrt{7}}{8}$ and $\dfrac{-3 + i\sqrt{7}}{8}$.

12.
$$(3x - 4)^2 = 10x$$
$$9x^2 - 24x + 16 - 10x = 0$$
$$9x^2 - 34x + 16 = 0$$
$$x^2 - \frac{34}{9}x = -\frac{16}{9}$$
$$x^2 - \frac{34}{9}x + \left(\frac{\frac{34}{9}}{2}\right)^2 = -\frac{16}{9} + \frac{289}{81}$$
$$\left(x - \frac{17}{9}\right)^2 = \frac{145}{81}$$
$$x - \frac{17}{9} = \pm\sqrt{\frac{145}{81}}$$
$$x = \frac{17 \pm \sqrt{145}}{9}$$

The solutions are $\dfrac{17 - \sqrt{145}}{9}$ and $\dfrac{17 + \sqrt{145}}{9}$.

13. In this problem, $P = 2500$, and $A = 2717$.
$$A = P(1 + r)^2$$
$$2717 = 2500(1 + r)^2$$
$$\frac{2717}{2500} = (1 + r)^2$$
$$1.0868 = (1 + r)^2$$
$$\pm\sqrt{1.0868} = 1 + r$$
$$-1 \pm \sqrt{1.0868} = r$$
Since the rate cannot be negative, the interest rate is $-1 + \sqrt{1.0868} \approx 0.0425$, or 4.25%.

14. $c^2 = a^2 + b^2$ where $a = b$.
$$c^2 = a^2 + a^2$$
$$(150)^2 = 2a^2$$
$$11,250 = a^2$$
$$\pm 75\sqrt{2} = a$$
Reject the negative value.
Each ship traveled $75\sqrt{2}$, or about 106.1, miles.

15. Two complex but not real solutions exist.

16. Two real solutions exist.

17. Two real solutions exist.

18. One real solution exists.

19. $x^2 - 16x + 64 = 0$
$a = 1, b = -16, c = 64$

$$x = \frac{16 \pm \sqrt{(-16)^2 - 4(1)(64)}}{2(1)}$$

$$= \frac{16 \pm \sqrt{256 - 256}}{2}$$

$$= \frac{16 \pm \sqrt{0}}{2}$$

$$= 8$$

The solution is 8.

20. $x^2 + 5x = 0$
$a = 1, b = 5, c = 0$

$$x = \frac{-5 \pm \sqrt{(5)^2 - 4(1)(0)}}{2(1)}$$

$$= \frac{-5 \pm \sqrt{25}}{2}$$

$$= \frac{-5 \pm 5}{2}$$

$$x = \frac{-5 + 5}{2} \quad \text{or} \quad x = \frac{-5 - 5}{2}$$

$$x = 0 \quad \text{or} \quad x = -5$$

The solutions are 0 and –5.

21. $x^2 + 11 = 0$
$a = 1, b = 0, c = 11$

$$x = \frac{-0 \pm \sqrt{0^2 - 4(1)(11)}}{2(1)}$$

$$= \frac{\pm \sqrt{-44}}{2}$$

$$= \frac{\pm 2i\sqrt{11}}{2}$$

$$= \pm i\sqrt{11}$$

The solutions are $i\sqrt{11}$ and $-i\sqrt{11}$.

22. $\quad 2x^2 + 3x = 5$
$2x^2 + 3x - 5 = 0$
$a = 2, b = 3, c = -5$

$$x = \frac{-3 \pm \sqrt{3^2 - 4(2)(-5)}}{2(2)}$$

$$= \frac{-3 \pm \sqrt{49}}{4}$$

$$= \frac{-3 \pm 7}{4}$$

$$x = \frac{-3 + 7}{4} \quad \text{or} \quad x = \frac{-3 - 7}{4}$$

$$x = 1 \quad\quad \text{or} \quad x = -\frac{5}{2}$$

The solutions are 1 and $-\frac{5}{2}$.

23. $\quad 6x^2 + 7 = 5x$
$6x^2 - 5x + 7 = 0$
$a = 6, b = -5, c = 7$

$$x = \frac{5 \pm \sqrt{(-5)^2 - 4(6)(7)}}{2(6)}$$

$$= \frac{5 \pm \sqrt{-143}}{12}$$

$$= \frac{5 \pm i\sqrt{143}}{12}$$

The solutions are $\frac{5 - i\sqrt{143}}{12}$ and $\frac{5 + i\sqrt{143}}{12}$.

24.
$$9a^2 + 4 = 2a$$
$$9a^2 - 2a + 4 = 0$$
$$a = 9,\ b = -2,\ c = 4$$
$$a = \frac{2 \pm \sqrt{(-2)^2 - 4(9)(4)}}{2(9)}$$
$$= \frac{2 \pm \sqrt{-140}}{18}$$
$$= \frac{2 \pm 2i\sqrt{35}}{18}$$
$$= \frac{1 \pm i\sqrt{35}}{9}$$

The solutions are $\dfrac{1 - i\sqrt{35}}{9}$ and $\dfrac{1 + i\sqrt{35}}{9}$.

25.
$$(5a - 2)^2 - a = 0$$
$$25a^2 - 20a + 4 - a = 0$$
$$25a^2 - 21a + 4 = 0$$
$$a = 25,\ b = -21,\ c = 4$$
$$a = \frac{21 \pm \sqrt{(-21)^2 - 4(25)(4)}}{2(25)}$$
$$= \frac{21 \pm \sqrt{41}}{50}$$

The solutions are $\dfrac{21 - \sqrt{41}}{50}$ and

$\dfrac{21 + \sqrt{41}}{50}$.

26.
$$(2x - 3)^2 = x$$
$$4x^2 - 12x + 9 - x = 0$$
$$4x^2 - 13x + 9 = 0$$
$$a = 4,\ b = -13,\ c = 9$$
$$x = \frac{13 \pm \sqrt{(-13)^2 - 4(4)(9)}}{2(4)}$$
$$= \frac{13 \pm \sqrt{25}}{8}$$
$$= \frac{13 \pm 5}{8}$$
$$x = \frac{13 + 5}{8} \quad \text{or} \quad x = \frac{13 - 5}{8}$$
$$x = \frac{9}{4} \quad\quad \text{or} \quad x = 1$$

The solutions are $\dfrac{9}{4}$ and 1.

27. $d(t) = -16t^2 + 30t + 6$

  a. $\quad d(1) = -16(1)^2 + 30(1) + 6$
$$= -16 + 30 + 6$$
$$= 20$$
The hat is 20 feet above the ground.

  b. $\quad -16t^2 + 30t + 6 = 0$
$$8t^2 - 15t - 3 = 0$$
$$a = 8,\ b = -15,\ c = -3$$
$$x = \frac{15 \pm \sqrt{(-15)^2 - 4(8)(-3)}}{2(8)}$$
$$= \frac{15 \pm \sqrt{321}}{16}$$

Rejecting the negative value as extraneous, we find that
$$x = \frac{15 + \sqrt{321}}{16} \approx 2.1.$$
It takes the hat $\dfrac{15 + \sqrt{321}}{16}$, or approximately 2.1, seconds to hit the ground.

28. Let x = the length of each leg, then
$x + 6$ = the length of the hypotenuse.

$$x^2 + x^2 = (x+6)^2$$
$$2x^2 = x^2 + 12x + 36$$
$$x^2 - 12x - 36 = 0$$
$$x = \frac{12 \pm \sqrt{(-12)^2 - 4(1)(-36)}}{2}$$
$$= \frac{12 \pm \sqrt{288}}{2}$$
$$= \frac{12 \pm 12\sqrt{2}}{2}$$
$$= 6 \pm 6\sqrt{2}$$

The length of each leg is
$\left(6 + 6\sqrt{2}\right)$ centimeters.

29.
$$x^3 = 27$$
$$x^3 - 27 = 0$$
$$(x-3)(x^2 + 3x + 9) = 0$$
$$x - 3 = 0 \quad \text{or} \quad x^2 + 3x + 9 = 0$$
$$x = 3$$
$$a = 1, b = 3, c = 9$$
$$x = \frac{-3 \pm \sqrt{3^2 - 4(1)(9)}}{2(1)}$$
$$= \frac{-3 \pm \sqrt{9 - 36}}{2}$$
$$= \frac{-3 \pm \sqrt{-27}}{2}$$
$$= \frac{-3 \pm 3i\sqrt{3}}{2}$$

The solutions are 3, $\dfrac{-3 + 3i\sqrt{3}}{2}$,

and $\dfrac{-3 - 3i\sqrt{3}}{2}$.

30.
$$y^3 = -64$$
$$y^3 + 64 = 0$$
$$(y+4)(y^2 - 4y + 16) = 0$$
$$y + 4 = 0 \quad \text{or} \quad y^2 - 4y + 16 = 0$$
$$y = -4 \quad \text{or} \quad y = \frac{4 \pm \sqrt{(-4)^2 - 4(1)(16)}}{2(1)}$$
$$= \frac{4 \pm \sqrt{-48}}{2}$$
$$= \frac{4 \pm 4i\sqrt{3}}{2}$$
$$= 2 \pm 2i\sqrt{3}$$

The solutions are -4, $2 - 2i\sqrt{3}$, and
$2 + 2i\sqrt{3}$.

31.
$$\frac{5}{x} + \frac{6}{x-2} = 3$$
$$x(x-2)\left(\frac{5}{x} + \frac{6}{x-2}\right) = 3x(x-2)$$
$$5(x-2) + 6x = 3x^2 - 6x$$
$$5x - 10 + 6x = 3x^2 - 6x$$
$$11x - 10 = 3x^2 - 6x$$
$$3x^2 - 17x + 10 = 0$$
$$(3x-2)(x-5) = 0$$
$$3x - 2 = 0 \quad \text{or} \quad x - 5 = 0$$
$$3x = 2 \quad \text{or} \quad x = 5$$
$$x = \frac{2}{3}$$

The solutions are $\dfrac{2}{3}$ and 5.

32.
$$\frac{7}{8} = \frac{8}{x^2}$$
$$7x^2 = 64$$
$$x^2 = \frac{64}{7}$$
$$x = \pm\sqrt{\frac{64}{7}} = \pm\frac{8\sqrt{7}}{7}$$

The solutions are $\dfrac{8\sqrt{7}}{7}$ and $-\dfrac{8\sqrt{7}}{7}$.

33. $x^4 - 21x^2 - 100 = 0$

$(x^2 - 25)(x^2 + 4) = 0$

$x^2 - 25 = 0$ or $x^2 + 4 = 0$
     $x^2 = 25$ or $x^2 = -4$
       $x = \pm\sqrt{25}$ or $x = \pm\sqrt{-4}$
        $x = \pm 5$ or $x = \pm 2i$

The solutions are 5, –5, 2*i*, and –2*i*.

34. $5(x + 3)^2 - 19(x + 3) = 4$

Let $u = x + 3$.

$5u^2 - 19u - 4 = 0$

$(5u + 1)(u - 4) = 0$

$5u + 1 = 0$ or $u - 4 = 0$
       $u = -\dfrac{1}{5}$ or $u = 4$

   $x + 3 = -\dfrac{1}{5}$ or $x + 3 = 4$

       $x = -\dfrac{16}{5}$ or $x = 1$

The solutions are $-\dfrac{16}{15}$ and 1.

35. $x^{2/3} - 6x^{1/3} + 5 = 0$

Let $m = x^{1/3}$.

$m^2 - 6m + 5 = 0$

$(m - 1)(m - 5) = 0$

$m - 1 = 0$ or $m - 5 = 0$
     $m = 1$ or $m = 5$
  $x^{1/3} = 1$ or $x^{1/3} = 5$
     $x = 1^3$ or $x = 5^3$
     $x = 1$ or $x = 125$

The solutions are 1 and 125.

36. $x^{2/3} - 6x^{1/3} = -8$

Let $m = x^{1/3}$.

$m^2 - 6m + 8 = 0$

$(m - 2)(m - 4) = 0$

$m - 2 = 0$ or $m - 4 = 0$
     $m = 2$ or $m = 4$
  $x^{1/3} = 2$ or $x^{1/3} = 4$
     $x = 2^3$ or $x = 4^3$
     $x = 8$ or $x = 64$

The solutions are 8 and 64.

37.
$$a^6 - a^2 = a^4 - 1$$
$$a^2(a^4 - 1) - (a^4 - 1) = 0$$
$$(a^2 - 1)(a^4 - 1) = 0$$
$$(a^2 - 1)(a^2 - 1)(a^2 + 1) = 0$$
$$(a^2 - 1)^2(a^2 + 1) = 0$$
$$[(a + 1)(a - 1)]^2(a^2 + 1) = 0$$
$$(a + 1)^2(a - 1)^2(a^2 + 1) = 0$$
$$(a + 1)^2 = 0 \quad \text{or} \quad (a - 1)^2 = 0 \quad \text{or} \quad a^2 + 1 = 0$$
$$a + 1 = 0 \quad \text{or} \quad a - 1 = 0 \quad \text{or} \quad a^2 = -1$$
$$a = -1 \quad \text{or} \quad a = 1 \quad \text{or} \quad a = \pm\sqrt{-1} = \pm i$$

The solutions are $1, -1, i$ and $-i$.

38.
$$y^{-2} + y^{-1} = 20$$
$$\frac{1}{y^2} + \frac{1}{y} = 20$$
$$y^2\left(\frac{1}{y^2} + \frac{1}{y}\right) = 20y^2$$
$$1 + y = 20y^2$$
$$0 = 20y^2 - y - 1$$
$$0 = (5y + 1)(4y - 1)$$
$$5y + 1 = 0 \quad \text{or} \quad 4y - 1 = 0$$
$$y = -\frac{1}{5} \quad \text{or} \quad y = \frac{1}{4}$$

The solutions are $-\dfrac{1}{5}$ and $\dfrac{1}{4}$.

39.

	Hours	Part complete in one hour
Jerome	x	$\frac{1}{x}$
Tim	$x - 1$	$\frac{1}{x-1}$
Together	5	$\frac{1}{5}$

$$\frac{1}{x} + \frac{1}{x - 1} = \frac{1}{5}$$
$$5x(x - 1)\left(\frac{1}{x} + \frac{1}{x - 1}\right) = 5x(x - 1)\left(\frac{1}{5}\right)$$
$$5(x - 1) + 5x = x(x - 1)$$
$$5x - 5 + 5x = x^2 - x$$
$$10x - 5 = x^2 - x$$
$$0 = x^2 - 11x + 5$$
$$a = 1, b = -11, c = 5$$
$$x = \frac{-(-11) \pm \sqrt{(-11)^2 - 4(1)(5)}}{2(1)}$$
$$= \frac{11 \pm \sqrt{101}}{2}$$
$$x \approx 0.475 \text{ or } 10.525$$

Reject $x \approx 0.475$ since $0.475 - 1 = -0.425$.
$x \approx 10.5, \ x - 1 \approx 9.5$
Jerome can sort the mail in 10.5 hours.
Tim can sort the mail in 9.5 hours.

40. Let x = the number, then $\dfrac{1}{x}$ = the reciprocal of the number.

$$x - \frac{1}{x} = -\frac{24}{5}$$

$$5x\left(x - \frac{1}{x}\right) = 5x\left(-\frac{24}{5}\right)$$

$$5x^2 - 5 = -24x$$

$$5x^2 + 24x - 5 = 0$$

$$(5x - 1)(x + 5) = 0$$

$$5x - 1 = 0 \quad \text{or} \quad x + 5 = 0$$

$$5x = 1 \quad \text{or} \quad x = -5$$

$$x = \frac{1}{5}$$

Since we were given that the number is negative, the number is –5.

41.

$$2x^2 - 50 \le 0$$

$$2(x^2 - 25) \le 0$$

$$2(x + 5)(x - 5) \le 0$$

$$2(x + 5)(x - 5) = 0$$

$$x + 5 = 0 \quad \text{or} \quad x - 5 = 0$$

$$x = -5 \quad \text{or} \quad x = 5$$

$$\begin{array}{ccc} \text{A} & \text{B} & \text{C} \\ \hline & -5 & 5 \end{array}$$

Region	Test Point	$2(x+5)(x-5) \le 0$	Result
A	–6	$2(-6+5)(-6-5) \le 0$	False
B	0	$2(0+5)(0-5) \le 0$	True
C	6	$2(6+5)(6-5) \le 0$	False

The solution set is [–5, 5].

42.

$$\frac{1}{4}x^2 < \frac{1}{16}$$

$$x^2 < \frac{1}{4}$$

$$x^2 - \frac{1}{4} < 0$$

$$\left(x + \frac{1}{2}\right)\left(x - \frac{1}{2}\right) < 0$$

$$\left(x + \frac{1}{2}\right)\left(x - \frac{1}{2}\right) = 0$$

$$x + \frac{1}{2} = 0 \quad \text{or} \quad x - \frac{1}{2} = 0$$

$$x = -\frac{1}{2} \quad \text{or} \quad x = \frac{1}{2}$$

Region	Test Point	$\left(x + \frac{1}{2}\right)\left(x - \frac{1}{2}\right) < 0$	Result
A	-1	$\left(-1 + \frac{1}{2}\right)\left(-1 - \frac{1}{2}\right) < 0$	False
B	0	$\left(0 + \frac{1}{2}\right)\left(0 - \frac{1}{2}\right) < 0$	True
C	1	$\left(1 + \frac{1}{2}\right)\left(1 - \frac{1}{2}\right) < 0$	False

The solution set is $\left(-\frac{1}{2}, \frac{1}{2}\right)$.

43. $(2x - 3)(4x + 5) \geq 0$

$(2x - 3)(4x + 5) = 0$

$2x - 3 = 0$ or $4x + 5 = 0$

$2x = 3$ or $4x = -5$

$x = \dfrac{3}{2}$ or $x = -\dfrac{5}{4}$

Region	Test Point	$(2x - 3)(4x + 5) \geq 0$	Result
A	-1	$[2(-2) - 3][4(-2) + 5] \geq 0$	True
B	0	$[2(0) - 3][4(0) + 5] \geq 0$	False
C	1	$[2(2) - 3][4(2) + 5] \geq 0$	True

The solution set is $\left(-\infty, \ -\dfrac{5}{4}\right] \cup \left[\dfrac{3}{2}, \ \infty\right)$.

44. $(x^2 - 16)(x^2 - 1) > 0$

$(x + 4)(x - 4)(x + 1)(x - 1) > 0$

$(x + 4)(x - 4)(x + 1)(x - 1) = 0$

$x + 4 = 0$ or $x - 4 = 0$ or $x + 1 = 0$ or $x - 1 = 0$

$x = -4$ or $x = 4$ or $x = -1$ or $x = 1$

Region	Test Point	$(x + 4)(x - 4)(x + 1)(x - 1) > 0$	Result
A	-5	$(-5 + 4)(-5 - 4)(-5 + 1)(-5 - 1) > 0$	True
B	-2	$(-2 + 4)(-2 - 4)(-2 + 1)(-2 - 1) > 0$	False
C	0	$(0 + 4)(0 - 4)(0 + 1)(0 - 1) > 0$	True
D	2	$(2 + 4)(2 - 4)(2 + 1)(2 - 1) > 0$	False
E	5	$(5 + 4)(5 - 4)(5 + 1)(5 - 1) > 0$	True

The solution set is $(-\infty, \ -4) \cup (-1, \ 1) \cup (4, \ \infty)$.

45. $\dfrac{x-5}{x-6} < 0$

$x - 5 = 0$ or $x - 6 = 0$

$\quad x = 5$ or $\quad\quad x = 6$

Choose 4 from region A.   Choose $\dfrac{11}{2}$ from region B.   Choose 7 from region C.

$$\dfrac{x-5}{x-6} < 0 \qquad\qquad \dfrac{x-5}{x-6} < 0 \qquad\qquad \dfrac{x-5}{x-6} < 0$$

$$\dfrac{4-5}{4-6} < 0 \qquad\qquad \dfrac{\frac{11}{2}-5}{\frac{11}{2}-6} < 0 \qquad\qquad \dfrac{7-5}{7-6} < 0$$

$$\dfrac{-1}{-2} < 0 \qquad\qquad \dfrac{\frac{1}{2}}{-\frac{1}{2}} < 0 \qquad\qquad \dfrac{2}{1} < 0$$

$$\dfrac{1}{2} < 0 \quad \text{False} \qquad\qquad -1 < 0 \quad \text{True} \qquad\qquad 2 < 0 \quad \text{False}$$

The solution set is $(5, 6)$.

46. $\dfrac{x(x+5)}{4x-3} \geq 0$

$x = 0$ or $x+5 = 0$ or $4x - 3 = 0$

$x = 0$ or $x = -5$ or $x = \dfrac{3}{4}$

```
      A     B        C       D
 ◄────┼─────┼────────┼───────────►
     −5     0        3
                     4
```

Choose −6 from region A.   Choose −1 from region B.

$$\dfrac{x(x+5)}{4x-3} \geq 0$$ $$\dfrac{x(x+5)}{4x-3} \geq 0$$

$$\dfrac{-6(-6+5)}{4(-6)-3} \geq 0$$ $$\dfrac{-1(-1+5)}{4(-1)-3} \geq 0$$

$$\dfrac{(-6)(-1)}{-24-3} \geq 0$$ $$\dfrac{(-1)(4)}{-4-3} \geq 0$$

$$\dfrac{6}{-27} \geq 0$$ $$\dfrac{-4}{-7} \geq 0$$

$$-\dfrac{2}{9} \geq 0 \quad \text{False}$$ $$\dfrac{4}{7} \geq 0 \quad \text{True}$$

Choose $\dfrac{1}{2}$ from region C.   Choose 1 from region D.

$$\dfrac{x(x+5)}{4x-3} \geq 0$$ $$\dfrac{x(x+5)}{4x-3} \geq 0$$

$$\dfrac{\left(\frac{1}{2}\right)\left(\frac{1}{2}+5\right)}{4\left(\frac{1}{2}\right)-3} \geq 0$$ $$\dfrac{(1)(1+5)}{4(1)-3} \geq 0$$

$$\dfrac{\left(\frac{1}{2}\right)\left(\frac{11}{2}\right)}{2-3} \geq 0$$ $$\dfrac{(1)(6)}{4-3} \geq 0$$

$$\dfrac{\frac{11}{4}}{-1} \geq 0$$ $$\dfrac{6}{1} \geq 0$$

$$-\dfrac{11}{4} \geq 0 \quad \text{False}$$ $$6 \geq 0 \quad \text{True}$$

```
 ◄────●─────●────────(───────────►
     −5     0        3
                     4
```

The solution set is $[-5,\, 0] \cup \left(\dfrac{3}{4},\, \infty\right)$.

47. $\dfrac{(4x+3)(x-5)}{x(x+6)} > 0$

$4x+3=0 \quad$ or $\quad x-5=0 \quad$ or $\quad x=0 \quad$ or $\quad x+6=0$

$x=-\dfrac{3}{4} \quad$ or $\qquad x=5 \quad$ or $\quad x=0 \quad$ or $\qquad x=-6$

$$
\begin{array}{ccccc}
A & B & C & D & E
\end{array}
$$

Number line with points at -6, $-\dfrac{3}{4}$, 0, 5.

Choose -7 from region A.

$\dfrac{(4x+3)(x-5)}{x(x+6)} > 0$

$\dfrac{[4(-7)+3](-7-5)}{(-7)(-7+6)} > 0$

$\dfrac{(-28+3)(-12)}{(-7)(-1)} > 0$

$\dfrac{(-25)(-12)}{7} > 0$

$\dfrac{300}{7} > 0 \quad$ True

Choose -1 from region B.

$\dfrac{(4x+3)(x-5)}{x(x+6)} > 0$

$\dfrac{[4(-1)+3](-1-5)}{(-1)(-1+6)} > 0$

$\dfrac{(-4+3)(-6)}{(-1)(5)} > 0$

$\dfrac{(-1)(-6)}{-5} > 0$

$\dfrac{6}{-5} > 0$

$-\dfrac{6}{5} > 0 \quad$ False

Choose $-\dfrac{1}{2}$ from region C.

$\dfrac{(4x+3)(x-5)}{x(x+6)} > 0$

$\dfrac{\left[4\left(-\frac{1}{2}\right)+3\right]\left(-\frac{1}{2}-5\right)}{\left(-\frac{1}{2}\right)\left(-\frac{1}{2}+6\right)} > 0$

$\dfrac{(-2+3)\left(-\frac{11}{2}\right)}{\left(-\frac{1}{2}\right)\left(\frac{11}{2}\right)} > 0$

$\dfrac{(1)\left(-\frac{11}{2}\right)}{-\frac{11}{4}} > 0$

$\dfrac{-\frac{11}{2}}{-\frac{11}{4}} > 0$

$2 > 0 \quad$ True

Choose 1 from region D.

$\dfrac{(4x+3)(x-5)}{x(x+6)} > 0$

$\dfrac{[4(1)+3](1-5)}{(1)(1+6)} > 0$

$\dfrac{(4+3)(-4)}{(1)(7)} > 0$

$\dfrac{(7)(-4)}{7} > 0$

$\dfrac{-28}{7} > 0$

$-4 > 0 \quad$ False

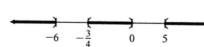

Choose 6 from region E.

$\dfrac{(4x+3)(x-5)}{x(x+6)} > 0$

$\dfrac{[4(6)+3](6-5)}{(6)(6+6)} > 0$

$\dfrac{(24+3)(1)}{(6)(12)} > 0$

$\dfrac{(27)(1)}{72} > 0$

$\dfrac{27}{72} > 0$

$\dfrac{3}{8} > 0 \quad$ True

The solution set is $(-\infty,\, -6) \cup \left(-\dfrac{3}{4},\, 0\right) \cup (5,\, \infty)$.

48. $(x+5)(x-6)(x+2) \le 0$

$x+5=0$　or　$x-6=0$　or　$x+2=0$

$x=-5$　or　　$x=6$　or　　$x=-2$

```
   A     B      C    D
 <──┼─────┼──────┼──────>
   -5    -2      6
```

Region	Test Point	$(x+5)(x-6)(x+2) \le 0$	Result
A	-6	$(-6+5)(-6-6)(-6+2) \le 0$	True
B	-3	$(-3+5)(-3-6)(-3+2) \le 0$	False
C	0	$(0+5)(0-6)(0+2) \le 0$	True
D	7	$(7+5)(7-6)(7+2) \le 0$	False

```
 ◄━━━━━┫━━━━━┣━━━━━┫━━━━━━►
      -5    -2     6
```

The solution set is $(-\infty, -5] \cup [-2, 6]$.

49. $x^3 + 3x^2 - 25x - 75 > 0$

$x^2(x+3) - 25(x+3) > 0$

$(x+3)(x^2 - 25) > 0$

$(x+3)(x+5)(x-5) > 0$

$x+3=0$　or　$x+5=0$　or　$x-5=0$

$x=-3$　or　　$x=-5$　or　　$x=5$

```
   A    B     C    D
 <──┼────┼─────┼──────>
   -5   -3     5
```

Region	Test Point	$(x+3)(x+5)(x-5) > 0$	Result
A	-6	$(-6+3)(-6+5)(-6-5) > 0$	False
B	-4	$(-4+3)(-4+5)(-4-5) > 0$	True
C	0	$(0+3)(0+5)(0-5) > 0$	False
D	6	$(6+3)(6+5)(6-5) > 0$	True

```
 ◄━━━━━┣━━━━━┫━━━━━┣━━━━━━►
      -5    -3     5
```

The solution set is $(-5, -3) \cup (5, \infty)$.

50. $\dfrac{x^2+4}{3x} \le 1$

$\dfrac{x^2+4}{3x} - 1 \le 0$

$\dfrac{x^2-3x+4}{3x} \le 0$

$x^2 - 3x + 4 = 0$ has no real solutions.

$3x = 0$

$x = 0$

Choose −1 from region A.   Choose 1 from region B.

$\dfrac{x^2-3x+4}{3x} \le 0$ $\dfrac{x^2-3x+4}{3x} \le 0$

$\dfrac{(-1)^2-3(-1)+4}{3(-1)} \le 0$ $\dfrac{(1)^2-3(1)+4}{3(1)} \le 0$

$\dfrac{1+3+4}{-3} \le 0$ $\dfrac{1-3+4}{3} \le 0$

$\dfrac{8}{-3} \le 0$ $\dfrac{2}{3} \le 0$ False

$-\dfrac{8}{3} \le 0$ True

The solution set is $(-\infty,\ 0)$.

51. $\dfrac{(5x+6)(x-3)}{x(6x-5)} < 0$

$5x+6=0 \quad$ or $\quad x-3=0 \quad$ or $\quad x=0 \quad$ or $\quad 6x-5=0$

$\qquad x=-\dfrac{6}{5} \quad$ or $\qquad x=3 \quad$ or $\quad x=0 \quad$ or $\qquad x=\dfrac{5}{6}$

A   B   C   D   E

$\begin{array}{ccccc} & & & & \\ -\dfrac{6}{5} & 0 & \dfrac{5}{6} & 3 & \end{array}$

Choose −2 from region A.

$\dfrac{(5x+6)(x-3)}{x(6x-5)} < 0$

$\dfrac{[5(-2)+6](-2-3)}{(-2)[6(-2)-5]} < 0$

$\dfrac{(-10+6)(-5)}{(-2)(-12-5)} < 0$

$\dfrac{(-4)(-5)}{(-2)(-17)} < 0$

$\dfrac{20}{34} < 0$

$\dfrac{10}{17} < 0 \quad$ False

Choose −1 from region B.

$\dfrac{(5x+6)(x-3)}{x(6x-5)} < 0$

$\dfrac{[5(-1)+6](-1-3)}{(-1)[6(-1)-5]} < 0$

$\dfrac{(-5+6)(-4)}{(-1)(-6-5)} < 0$

$\dfrac{(1)(-4)}{(-1)(-11)} < 0$

$\dfrac{-4}{11} < 0$

$-\dfrac{4}{11} < 0 \quad$ True

Choose $\dfrac{1}{2}$ from region C.

$\dfrac{(5x+6)(x-3)}{x(6x-5)} < 0$

$\dfrac{\left[5\left(\frac{1}{2}\right)+6\right]\left(\frac{1}{2}-3\right)}{\left(\frac{1}{2}\right)\left[6\left(\frac{1}{2}\right)-5\right]} < 0$

$\dfrac{\left(\frac{5}{2}+6\right)\left(-\frac{5}{2}\right)}{\left(\frac{1}{2}\right)(3-5)} < 0$

$\dfrac{\left(\frac{17}{2}\right)\left(-\frac{5}{2}\right)}{\left(\frac{1}{2}\right)(-2)} < 0$

$\dfrac{-\frac{85}{4}}{-1} < 0$

$\dfrac{85}{4} < 0 \quad$ False

Choose 1 from region D.

$\dfrac{(5x+6)(x-3)}{x(6x-5)} < 0$

$\dfrac{[5(1)+6](1-3)}{(1)[6(1)-5]} < 0$

$\dfrac{(5+6)(-2)}{(1)(6-5)} < 0$

$\dfrac{(11)(-2)}{(1)(1)} < 0$

$\dfrac{-22}{1} < 0$

$-22 < 0 \quad$ True

Choose 4 from region E.

$\dfrac{(5x+6)(x-3)}{x(6x-5)} < 0$

$\dfrac{[5(4)+6](4-3)}{(4)[6(4)-5]} < 0$

$\dfrac{(20+6)(1)}{(4)(24-5)} < 0$

$\dfrac{(26)(1)}{(4)(19)} < 0$

$\dfrac{26}{76} < 0$

$\dfrac{13}{38} < 0 \quad$ False

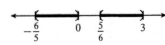

The solution set is $\left(-\dfrac{6}{5},\ 0\right) \cup \left(\dfrac{5}{6},\ 3\right)$.

52.
$$\frac{3}{x-2} > 2$$

$$\frac{3}{x-2} - 2 > 0$$

$$\frac{3 - 2(x-2)}{x-2} > 0$$

$$\frac{3 - 2x + 4}{x-2} > 0$$

$$\frac{7 - 2x}{x-2} > 0$$

$$7 - 2x = 0 \quad \text{or} \quad x - 2 = 0$$

$$x = \frac{7}{2} \quad \text{or} \quad x = 2$$

Choose 1 from region A.

$$\frac{7 - 2x}{x - 2} > 0$$

$$\frac{7 - 2(1)}{1 - 2} > 0$$

$$\frac{7 - 2}{-1} > 0$$

$$\frac{5}{-1} > 0$$

$$-5 > 0 \quad \text{False}$$

Choose 3 from region B.

$$\frac{7 - 2x}{x - 2} > 0$$

$$\frac{7 - 2(3)}{3 - 2} > 0$$

$$\frac{7 - 6}{1} > 0$$

$$\frac{1}{1} > 0$$

$$1 > 0 \quad \text{True}$$

Choose 4 from region C.

$$\frac{7 - 2x}{x - 2} > 0$$

$$\frac{7 - 2(4)}{4 - 2} > 0$$

$$\frac{7 - 8}{2} > 0$$

$$\frac{-1}{2} > 0$$

$$-\frac{1}{2} > 0 \quad \text{False}$$

The solution set is $\left(2, \dfrac{7}{2}\right)$.

53. $f(x) = x^2 - 4$

This graph is the same as the graph of $y = x^2$ shifted 4 units down.

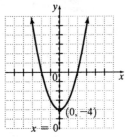

54. $g(x) = x^2 + 7$

This graph is the same as the graph of $y = x^2$ shifted 7 units up.

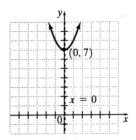

55. $H(x) = 2x^2$

Because $a = 2$ and $2 > 1$, the parabola is narrower than the graph of $y = x^2$.

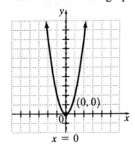

56. $h(x) = -\dfrac{1}{3}x^2$

Because $a = -\dfrac{1}{3}$, a negative value, this parabola opens downward. Since $\left|-\dfrac{1}{3}\right| = \dfrac{1}{3} < 1$, the parabola is wider than the graph of $y = x^2$.

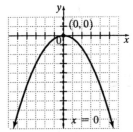

57. $F(x) = (x - 1)^2$

This graph is the same as the graph of $y = x^2$ shifted 1 unit to the right.

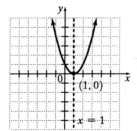

58. $G(x) = (x + 5)^2$

This graph is the same as the graph of $y = x^2$ shifted 5 units to the left.

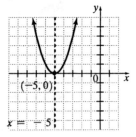

59. $f(x) = (x-4)^2 - 2$

This graph is the same as the graph of $y = x^2$ shifted 4 units to the right and 2 units down.

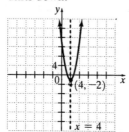

60. $f(x) = -3(x-1)^2 + 1$

This graph is similar to the graph of $y = x^2$ shifted 1 unit to the right and 1 unit down, and it opens downward and is narrower because $a = -3$.

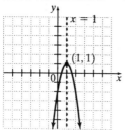

61. $f(x) = x^2 + 10x + 25$ or

$f(x) = (x+5)^2 + 0$

Vertex: $(-5, 0)$

$(x+5)^2 = 0$

$x + 5 = 0$

$x = -5$

x-intercept: $(-5, 0)$

$f(0) = 25$; y-intercept: $(0, 25)$

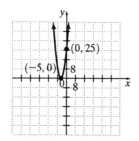

62. $f(x) = -x^2 + 6x - 9$

$$\frac{-b}{2a} = \frac{-6}{2(-1)} = 3$$

$f(3) = -(3)^2 + 6(3) - 9 = 0$

Vertex: $(3, 0)$

$f(0) = -9$

y-intercept: $(0, -9)$

$-x^2 + 6x - 9 = 0$

$-(x-3)^2 = 0$

$x - 3 = 0$

$x = 3$

x-intercept: $(3, 0)$

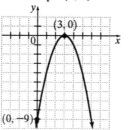

63. $f(x) = 4x^2 - 1$ or

$f(x) = 4(x-0)^2 - 1$

Vertex: $(0, -1)$

$4x^2 - 1 = 0$

$4x^2 = 1$

$x^2 = \dfrac{1}{4}$

$x = \pm\dfrac{1}{2}$

x-intercepts: $\left(\dfrac{1}{2}, 0\right), \left(-\dfrac{1}{2}, 0\right)$

$f(0) = -1$; y-intercept: $(0, -1)$

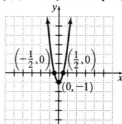

64. $f(x) = -5x^2 + 5$

$\dfrac{-b}{2a} = \dfrac{-0}{2(-5)} = 0$

$f(0) = 0 + 5$

Vertex: (0, 5)

y-intercept: (0, 5)

$-5x^2 + 5 = 0$

$-5(x^2 - 1) = 0$

$-5(x+1)(x-1) = 0$

$x + 1 = 0$ or $x - 1 = 0$

$x = -1$ or $x = 1$

x-intercepts: (−1, 0), (1, 0)

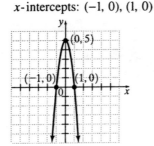

65. $f(x) = -3x^2 - 5x + 4$

$\dfrac{-b}{2a} = \dfrac{-(-5)}{2(-3)} = -\dfrac{5}{6}$

$f\left(\dfrac{5}{6}\right) = -3\left(-\dfrac{5}{6}\right)^2 - 5\left(-\dfrac{5}{6}\right) + 4 = \dfrac{73}{12}$

Vertex: $\left(-\dfrac{5}{6}, \dfrac{73}{12}\right)$

The graph opens downward because
$a = -3 < 0$.

y-intercept: (0, 4).

$x = \dfrac{-(-5) \pm \sqrt{(-5)^2 - 4(-3)(4)}}{2(-3)}$

$= \dfrac{5 \pm \sqrt{73}}{-6}$

The x-intercepts are approximately
(−2.3, 0) and (0.6, 0).

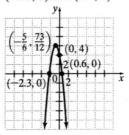

66. a. $350 = -16t^2 + 120t + 300$

$0 = -16t^2 + 120t - 50$

$0 = -8t^2 + 60t - 25$

$a = -8, b = 60, c = -25$

$t = \dfrac{-60 \pm \sqrt{(60)^2 - 4(-8)(-25)}}{2(-8)}$

$= \dfrac{-60 \pm \sqrt{2800}}{-16}$

$t \approx 0.4$ and 7.1

The object will reach a height of 350
feet at 0.4 second and 7.1 seconds.

b. Answers may vary.

67. Let x = one number, then
$420 - x$ = the other number.
Let $f(x)$ represent their product.
Thus,

$$f(x) = x(420 - x)$$

$$f(x) = -x^2 + 420x \;\cdot$$

$$\frac{-b}{2a} = \frac{-420}{2(-1)} = 210$$

$$420 - 210 = 210$$

The numbers are 210 and 210.

68. Since the vertex is at (–3, 7), the function
will have the form

$$y = a[x - (-3)]^2 + 7 \quad \text{or} \quad y = a(x + 3)^2 + 7.$$

Use the point (0, 0) to solve for a.

$$0 = a(0 + 3)^2 + 7$$

$$0 = 9a + 7$$

$$-7 = 9a$$

$$-\frac{7}{9} = a$$

The equation is $y = -\dfrac{7}{9}(x + 3)^2 + 7$.

69. Let $x = 0$ correspond to 1900.

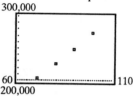

The graph shows that a linear model may
be best.

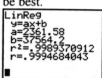

The equation is $y = 2361.58x + 37,564.2$.

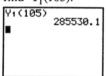

Graph of linear regression equation.
To predict the population in 2005,
find $Y_1(105)$.

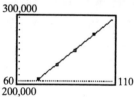

The predicted population in 2005 is
$285,530$ thousand.

70. Let $x = 0$ correspond to 1900.

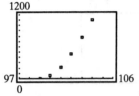

The graph shows that a quadratic model may be best.

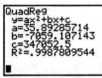

The equation is

$y = 35.893x^2 - 7059.107x + 347,052.5.$

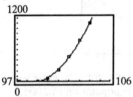

Graph of quadratic regression equation. To predict the sales in 2006, find $Y_1(106)$.

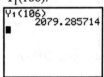

The predicted sales in 2006 is $2079.3 million.

Chapter 8 Test

1. $\quad 5x^2 - 2x = 7$

$5x^2 - 2x - 7 = 0$

$(5x - 7)(x + 1) = 0$

$5x - 7 = 0 \quad \text{or} \quad x + 1 = 0$

$\quad 5x = 7 \quad \text{or} \quad\quad x = -1$

$\quad\quad x = \dfrac{7}{5}$

The solutions are $\dfrac{7}{5}$ and -1.

2. $(x+1)^2 = 10$

$x+1 = \pm\sqrt{10}$

$x = -1 \pm \sqrt{10}$

The solutions are $-1+\sqrt{10}$ and $-1-\sqrt{10}$.

3. $m^2 - m + 8 = 0$

$a = 1, b = -1, c = 8$

$m = \dfrac{1 \pm \sqrt{(-1)^2 - 4(1)(8)}}{2(1)} = \dfrac{1 \pm \sqrt{1-32}}{2} = \dfrac{1 \pm \sqrt{-31}}{2} = \dfrac{1 \pm i\sqrt{31}}{2}$

The solutions are $\dfrac{1+i\sqrt{31}}{2}$ and $\dfrac{1-i\sqrt{31}}{2}$.

4. $u^2 - 6u + 2 = 0$

$a = 1, b = -6, c = 2$

$u = \dfrac{6 \pm \sqrt{(-6)^2 - 4(1)(2)}}{2(1)} = \dfrac{6 \pm \sqrt{36-8}}{2} = \dfrac{6 \pm \sqrt{28}}{2} = \dfrac{6 \pm 2\sqrt{7}}{2} = 3 \pm \sqrt{7}$

The solutions are $3+\sqrt{7}$ and $3-\sqrt{7}$.

5. $7x^2 + 8x + 1 = 0$

$(7x+1)(x+1) = 0$

$7x+1 = 0 \quad$ or $\quad x+1 = 0$

$7x = -1 \quad$ or $\qquad x = -1$

$x = -\dfrac{1}{7}$

The solutions are $-\dfrac{1}{7}$ and -1.

6. $a^2 - 3a = 5$

$a^2 - 3a - 5 = 0$

$a = 1, b = -3, c = -5$

$a = \dfrac{3 \pm \sqrt{(-3)^2 - 4(1)(-5)}}{2(1)} = \dfrac{3 \pm \sqrt{9+20}}{2} = \dfrac{3 \pm \sqrt{29}}{2}$

The solutions are $\dfrac{3+\sqrt{29}}{2}$ and $\dfrac{3-\sqrt{29}}{2}$.

7.
$$\frac{4}{x+2}+\frac{2x}{x-2}=\frac{6}{x^2-4}$$

$$\frac{4}{x+2}+\frac{2x}{x-2}=\frac{6}{(x+2)(x-2)}$$

$$(x+2)(x-2)\left(\frac{4}{x+2}+\frac{2x}{x-2}\right)=(x+2)(x-2)\frac{6}{(x+2)(x-2)}$$

$$4(x-2)+2x(x+2)=6$$

$$4x-8+2x^2+4x=6$$

$$2x^2+8x-8=6$$

$$2x^2+8x-14=0$$

$$x^2+4x-7=0$$

$a=1,\,b=4,\,c=-7$

$$x=\frac{-4\pm\sqrt{4^2-4(1)(-7)}}{2(1)}=\frac{-4\pm\sqrt{16+28}}{2}=\frac{-4\pm\sqrt{44}}{2}=\frac{-4\pm2\sqrt{11}}{2}=-2\pm\sqrt{11}$$

The solutions are $-2+\sqrt{11}$ and $-2-\sqrt{11}$.

8. $x^4-8x^2-9=0$

$(x^2-9)(x^2+1)=0$

$x^2-9=0$ or $x^2+1=0$

     $x^2=9$ or $x^2=-1$

       $x=\pm\sqrt{9}$ or $x=\pm\sqrt{-1}$

         $x=\pm3$ or $x=\pm i$

The solutions are $3,\,-3,\,i$ and $-i$.

9. $x^6+1=x^4+x^2$

         $x^6-x^4-x^2+1=0$

     $x^4(x^2-1)-(x^2-1)=0$

         $(x^4-1)(x^2-1)=0$

   $(x^2+1)(x^2-1)(x^2-1)=0$

       $(x^2+1)(x^2-1)^2=0$

  $(x^2+1)[(x+1)(x-1)]^2=0$

  $(x^2+1)(x+1)^2(x-1)^2=0$

$x^2+1=0$ or $(x+1)^2=0$ or $(x-1)^2=0$

     $x^2=-1$ or $x+1=0$ or $x-1=0$

       $x=\pm\sqrt{-1}$ or $x=-1$ or $x=1$

         $x=\pm i$

The solutions are $1,\,-1,\,i,$ and $-i$.

10. $(x+1)^2 - 15(x+1) + 56 = 0$

Let $u = x + 1$.

$u^2 - 15u + 56 = 0$

$(u - 7)(u - 8) = 0$

$u - 7 = 0$ or $u - 8 = 0$

$u = 7$ or $u = 8$

Substitute $x + 1$ for u.

$x + 1 = 7$ or $x + 1 = 8$

$x = 6$ or $x = 7$

The solutions are 6 and 7.

11.
$$x^2 - 6x = -2$$
$$x^2 - 6x + \left(\frac{6}{2}\right)^2 = -2 + 9$$
$$(x - 3)^2 = 7$$
$$x - 3 = \pm\sqrt{7}$$
$$x = 3 \pm \sqrt{7}$$

The solutions are $3 + \sqrt{7}$ and $3 - \sqrt{7}$.

12.
$$2a^2 + 5 = 4a$$
$$2a^2 - 4a = -5$$
$$a^2 - 2a = -\frac{5}{2}$$
$$a^2 - 2a + \left(\frac{2}{2}\right)^2 = -\frac{5}{2} + 1$$
$$(a - 1)^2 = -\frac{3}{2}$$
$$a - 1 = \pm\sqrt{\frac{-3}{2}}$$
$$a - 1 = \pm\frac{i\sqrt{3}}{\sqrt{2}}$$
$$a - 1 = \frac{\pm i\sqrt{6}}{2}$$
$$a = 1 \pm \frac{i\sqrt{6}}{2}$$
$$a = \frac{2 \pm i\sqrt{6}}{2}$$

The solutions are $\dfrac{2 + i\sqrt{6}}{2}$ and $\dfrac{2 - i\sqrt{6}}{2}$.

13. $2x^2 - 7x > 15$

$2x^2 - 7x - 15 > 0$

$(2x+3)(x-5) > 0$

$2x+3 = 0$ or $x-5 = 0$

$x = -\dfrac{3}{2}$ or $x = 5$

Region	Test Point	$(2x+3)(x-5) > 0$	Result
A	-2	$[2(-2)+3](-2-5) > 0$	True
B	0	$[2(0)+3](0-5) > 0$	False
C	6	$[2(6)+3](6-5) > 0$	True

The solution set is $\left(-\infty, \ -\dfrac{3}{2}\right) \cup (5, \ \infty)$.

14. $(x^2 - 16)(x^2 - 25) > 0$

$(x+4)(x-4)(x+5)(x-5) > 0$

$x = -4$ or $x = 4$ or $x = -5$ or $x = 5$

Region	Test Point	$(x+4)(x-4)(x+5)(x-5) > 0$	Result
A	-6	$(-6+4)(-6-4)(-6+5)(-6-5) > 0$	True
B	$-\dfrac{9}{2}$	$\left(-\dfrac{9}{2}+4\right)\left(-\dfrac{9}{2}-4\right)\left(-\dfrac{9}{2}+5\right)\left(-\dfrac{9}{2}-5\right) > 0$	False
C	0	$(0+4)(0-4)(0+5)(0-5) > 0$	True
D	$\dfrac{9}{2}$	$\left(\dfrac{9}{2}+4\right)\left(\dfrac{9}{2}-4\right)\left(\dfrac{9}{2}+5\right)\left(\dfrac{9}{2}-5\right) > 0$	False
E	6	$(6+4)(6-4)(6+5)(6-5) > 0$	True

The solution set is $(-\infty, \ -5) \cup (-4, \ 4) \cup (5, \ \infty)$.

15.

$$\frac{5}{x+3} < 1$$

$$\frac{5}{x+3} - \frac{x+3}{x+3} < 0$$

$$\frac{2-x}{x+3} < 0$$

$$2-x=0 \quad \text{or} \quad x+3=0$$

$$x=2 \quad \text{or} \quad x=-3$$

A   B   C

-3 2

Choose –4 from region A.   Choose 0 from region B.   Choose 3 from region C.

$$\frac{2-x}{x+3} < 0 \qquad\qquad \frac{2-x}{x+3} < 0 \qquad\qquad \frac{2-x}{x+3} < 0$$

$$\frac{2-(-4)}{-4+3} < 0 \qquad\qquad \frac{2-0}{0+3} < 0 \qquad\qquad \frac{2-3}{3+3} < 0$$

$$\frac{6}{-1} < 0 \qquad\qquad \frac{2}{3} < 0 \quad \text{False} \qquad\qquad \frac{-1}{6} < 0$$

$$-6 < 0 \quad \text{True} \qquad\qquad\qquad\qquad\qquad\qquad -\frac{1}{6} < 0 \quad \text{True}$$

-3 2

The solution set is $(-\infty,\ -3) \cup (2,\ \infty)$.

16.
$$\frac{7x - 14}{x^2 - 9} \le 0$$

$$\frac{7(x - 2)}{(x + 3)(x - 3)} \le 0$$

$x - 2 = 0$ or $x + 3 = 0$ or $x - 3 = 0$

    $x = 2$ or $x = -3$ or $x = 3$

Choose −4 from region A.   Choose 0 from region B.

$$\frac{7(x - 2)}{(x + 3)(x - 3)} \le 0 \qquad\qquad \frac{7(x - 2)}{(x + 3)(x - 3)} \le 0$$

$$\frac{7(-4 - 2)}{(-4 + 3)(-4 - 3)} \le 0 \qquad\qquad \frac{7(0 - 2)}{(0 + 3)(0 - 3)} \le 0$$

$$\frac{7(-6)}{(-1)(-7)} \le 0 \qquad\qquad\qquad \frac{(7)(-2)}{(3)(-3)} \le 0$$

$$\frac{-42}{7} \le 0 \qquad\qquad\qquad\qquad \frac{-14}{-9} \le 0$$

$$-6 \le 0 \quad \text{True} \qquad\qquad\qquad \frac{14}{9} \le 0 \qquad \text{False}$$

Choose $\frac{5}{2}$ from region C.   Choose 4 from region D.

$$\frac{7(x - 2)}{(x + 3)(x - 3)} \le 0 \qquad\qquad \frac{7(x - 2)}{(x + 3)(x - 3)} \le 0$$

$$\frac{7\left(\frac{5}{2} - 2\right)}{\left(\frac{5}{2} + 3\right)\left(\frac{5}{2} - 3\right)} \le 0 \qquad\qquad \frac{7(4 - 2)}{(4 + 3)(4 - 3)} \le 0$$

$$\frac{7\left(\frac{1}{2}\right)}{\left(\frac{11}{2}\right)\left(-\frac{1}{2}\right)} \le 0 \qquad\qquad \frac{(7)(2)}{(7)(1)} \le 0$$

$$\frac{\frac{7}{2}}{-\frac{11}{4}} \le 0 \qquad\qquad\qquad \frac{14}{7} \le 0$$

$$-\frac{14}{11} \le 0 \quad \text{True} \qquad\qquad 2 \le 0 \qquad \text{False}$$

The solution set is $(-\infty,\ -3) \cup [2,\ 3)$.

17. $f(x) = 3x^2$
vertex: (0, 0)

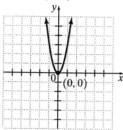

18. $G(x) = -2(x-1)^2 + 5$
vertex: (1, 5)

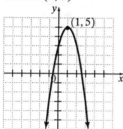

19. $h(x) = x^2 - 4x + 4$
$$\frac{-b}{2a} = \frac{-(-4)}{2(1)} = 2$$
$h(2) = 2^2 - 4(2) + 4 = 0$
vertex: (2, 0)
y-intercept: (0, 4)
$0 = x^2 - 4x + 4 = (x-2)^2$
x-intercept: (2, 0)

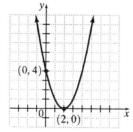

20. $F(x) = 2x^2 - 8x + 9$
$$\frac{-b}{2a} = \frac{-(-8)}{2(2)} = 2$$
$F(2) = 2(2)^2 - 8(2) + 9 = 1$
Vertex: (2, 1)
y-intercept: (0, 9)
There are no x-intercepts.

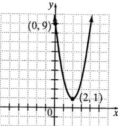

21. $c^2 = a^2 + b^2$
$(10)^2 = x^2 + (x-4)^2$
$100 = x^2 + x^2 - 8x + 16$
$0 = 2x^2 - 8x - 84$
$0 = x^2 - 4x - 42$
$a = 1,\ b = -4,\ c = -42$
$$x = \frac{-(-4) \pm \sqrt{(-4)^2 - 4(1)(-42)}}{2(1)}$$
$$= \frac{4 \pm \sqrt{16 + 168}}{2}$$
$$= \frac{4 \pm \sqrt{184}}{2}$$
$$= \frac{4 \pm 2\sqrt{46}}{2}$$
$$= 2 \pm \sqrt{46}$$
Reject the negative result. The top of the ladder is $\left(2 + \sqrt{46}\right)$, or about 8.8, feet from the ground.

22.

	Hours	Job complete in 1 hr
Dave	$x-2$	$\frac{1}{x-2}$
Sandy	x	$\frac{1}{x}$
Together	4	$\frac{1}{4}$

$$\frac{1}{x-2}+\frac{1}{x}=\frac{1}{4}$$

$$4x(x-2)\left(\frac{1}{x-2}+\frac{1}{x}\right)=4x(x-2)\left(\frac{1}{4}\right)$$

$$4x+4(x-2)=x(x-2)$$

$$4x+4x-8=x^2-2x$$

$$8x-8=x^2-2x$$

$$0=x^2-10x+8$$

$a=1,\,b=-10,\,c=8$

$$x=\frac{-(-10)\pm\sqrt{(-10)^2-4(1)(8)}}{2(1)}$$

$$=\frac{10\pm\sqrt{68}}{2}$$

$$=\frac{10\pm2\sqrt{17}}{2}$$

$$=5\pm\sqrt{17}$$

Since $x-2$ must be positive,

$x=5+\sqrt{17}$.

Sandy can paint the room in

$\left(5+\sqrt{17}\right)$, or about 9.12, hours.

23. a. $s(t)=-16t^2+32t+256$

$$\frac{-b}{2a}=\frac{-32}{2(-16)}=1$$

$$s(1)=-16(1)^2+32(1)+256=272$$

Vertex: $(1, 272)$

The maximum height is 272 feet.

b. $0=-16t^2+32t+256$

$0=-t^2+2t+16$

$a=-1,\,b=2,\,c=16$

$$x=\frac{-2\pm\sqrt{2^2-4(-1)(16)}}{2(-1)}$$

$$=\frac{-2\pm\sqrt{68}}{-2}$$

$$=\frac{-2\pm2\sqrt{17}}{-2}$$

$$=1\pm\sqrt{17}$$

$x\approx-3.12$ and 5.12

Reject the negative value.

The stone hits the water in about

5.12 seconds.

24. $\qquad(x+8)^2+x^2=20^2$

$$x^2+16x+64+x^2=400$$

$$2x^2+16x-336=0$$

$$x^2+8x-168=0$$

$a=1,\,b=8,\,c=-168$

$$x=\frac{-8\pm\sqrt{8^2-4(1)(-168)}}{2(1)}$$

$$=\frac{-8\pm\sqrt{736}}{2}$$

$$=-4\pm2\sqrt{46}$$

$x\approx9.56$ and -17.56.

Reject the negative value.

$(9.56+8)+9.56=27.12$

$27.12-20=7.12$

The person saves approximately 7 feet.

25. a. Let $x=0$ correspond to 1900.

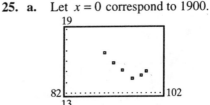

The graph shows that a quadratic
model may be best.

b.

The equation is $y = 0.052x^2 - 10.079x + 501.563$.

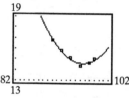

Graph of quadratic regression equation. To predict the rate for 2007, find $Y_1(107)$.

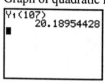

The predicted birth rate in 2007 is 20.2 births per 1000 population.

Chapter 9

Exercise Set 9.1

1. a. $(f+g)(x) = x - 7 + 2x + 1 = 3x - 6$

   b. $(f-g)(x) = x - 7 - (2x+1)$
$$= x - 7 - 2x - 1$$
$$= -x - 8$$

   c. $(f \cdot g)(x) = (x-7)(2x+1)$
$$= 2x^2 - 13x - 7$$

   d. $\left(\dfrac{f}{g}\right)(x) = \dfrac{x-7}{2x+1}$, where $x \neq -\dfrac{1}{2}$

3. a. $(f+g)(x) = x^2 + 1 + 5x = x^2 + 5x + 1$

   b. $(f-g)(x) = x^2 + 1 - 5x = x^2 - 5x + 1$

   c. $(f \cdot g)(x) = (x^2+1)(5x) = 5x^3 + 5x$

   d. $\left(\dfrac{f}{g}\right)(x) = \dfrac{x^2+1}{5x}$, where $x \neq 0$

5. a. $(f+g)(x) = \sqrt{x} + x + 5$

   b. $(f-g)(x) = \sqrt{x} - (x+5) = \sqrt{x} - x - 5$

   c. $(f \cdot g)(x) = \sqrt{x}(x+5) = x\sqrt{x} + 5\sqrt{x}$

   d. $\left(\dfrac{f}{g}\right)(x) = \dfrac{\sqrt{x}}{x+5}$, where $x \neq -5$

7. a. $(f+g)(x) = -3x + 5x^2 = 5x^2 - 3x$

   b. $(f-g)(x) = -3x - 5x^2 = -5x^2 - 3x$

   c. $(f \cdot g)(x) = -3x(5x^2) = -15x^3$

   d. $\left(\dfrac{f}{g}\right)(x) = \dfrac{-3x}{5x^2} = -\dfrac{3}{5x}$, where $x \neq 0$

9. $(f \circ g)(2) = f(g(2))$
$$= f(-4)$$
$$= (-4)^2 - 6(-4) + 2$$
$$= 16 + 24 + 2$$
$$= 42$$

11. $(g \circ f)(-1) = g(f(-1)) = g(9) = -2(9) = -18$

13. $(g \circ h)(0) = g(h(0)) = g(0) = -2(0) = 0$

15. $(f \circ g)(x) = f(g(x))$
$$= f(5x)$$
$$= (5x)^2 + 1$$
$$= 25x^2 + 1$$

$(g \circ f)(x) = g(f(x))$
$$= g(x^2 + 1)$$
$$= 5(x^2 + 1)$$
$$= 5x^2 + 5$$

17. $(f \circ g)(x) = f(g(x))$
$$= f(x+7)$$
$$= 2(x+7) - 3$$
$$= 2x + 14 - 3$$
$$= 2x + 11$$

$(g \circ f)(x) = g(f(x))$
$$= g(2x - 3)$$
$$= (2x - 3) + 7$$
$$= 2x + 4$$

19. $(f \circ g)(x) = f(g(x))$
$$= f(-2x)$$
$$= (-2x)^3 + (-2x) - 2$$
$$= -8x^3 - 2x - 2$$

$(g \circ f)(x) = g(f(x))$
$$= g(x^3 + x - 2)$$
$$= -2(x^3 + x - 2)$$
$$= -2x^3 - 2x + 4$$

21. $(f \circ g)(x) = f(g(x))$
$$= f(-5x + 2)$$
$$= \sqrt{-5x + 2}$$

$(g \circ f)(x) = g(f(x)) = g\left(\sqrt{x}\right) = -5\sqrt{x} + 2$

23. $H(x) = (g \circ h)(x)$
$$= g(h(x))$$
$$= g(x^2 + 2)$$
$$= \sqrt{x^2 + 2}$$

25. $F(x) = (h \circ f)(x)$
$$= h(f(x))$$
$$= h(3x)$$
$$= (3x)^2 + 2$$
$$= 9x^2 + 2$$

27. $G(x) = (f \circ g)(x) = f(g(x)) = f\left(\sqrt{x}\right) = 3\sqrt{x}$

29. Answers may vary. For example,
$g(x) = x + 2$ and $f(x) = x^2$ because
$(f \circ g)(x) = f(g(x))$
$$= f(x + 2)$$
$$= (x + 2)^2$$
$$= h(x)$$

31. Answers may vary. For example,
$g(x) = x + 5$ and $f(x) = \sqrt{x} + 2$ because
$(f \circ g)(x) = f(g(x))$
$$= f(x + 5)$$
$$= \sqrt{x + 5} + 2$$
$$= h(x)$$

33. Answers may vary. For example,
$g(x) = 2x - 3$ and $f(x) = \dfrac{1}{x}$ because
$(f \circ g)(x) = f(g(x))$
$$= f(2x + 3)$$
$$= \frac{1}{2x + 3}$$
$$= h(x)$$

35. $(f + g)(2) = 7 + (-1) = 6$

37. $(f \circ g)(2) = f(g(2)) = f(-1) = 4$

39. $(f \cdot g)(7) = 1 \cdot 4 = 4$

41. $\left(\dfrac{f}{g}\right)(-1) = \dfrac{4}{-4} = -1$

43. $P(x) = R(x) - C(x)$

45. $\quad x = y + 2$
$$x - 2 = y + 2 - 2$$
$$x - 2 = y$$
$$y = x - 2$$

47. $\quad x = 3y$
$$\frac{x}{3} = \frac{3y}{3}$$
$$\frac{x}{3} = y$$
$$y = \frac{x}{3}$$

49. $\quad x = -2y - 7$
$$x + 7 = -2y$$
$$\frac{x + 7}{-2} = y$$
$$y = -\frac{x + 7}{2}$$

Exercise Set 9.2

1. $f = \{(-1, -1), (1, 1), (0, 2), (2, 0)\}$ is a one-to-one function.
$f^{-1} = \{(-1, -1), (1, 1), (2, 0), (0, 2)\}$

3. $h = \{(10, 10)\}$ is a one-to-one function.
$h^{-1} = \{(10, 10)\}$

5. $f = \{(11, 12), (4, 3), (3, 4), (6, 6)\}$ is a one-to-one function.
$f^{-1} = \{(12, 11), (3, 4), (4, 3), (6, 6)\}$

7. This function is not one-to-one because there are two months with the same output: (January, 282) and (May, 282).

9. This function is one-to-one.

Rank in population (Input)	1	49	12	2	45
State (Output)	California	Vermont	Virginia	Texas	South Dakota

11. $f(x) = x^3 + 2$

    a. $f(1) = 1^3 + 2 = 3$

    b. $f^{-1}(3) = 1$

13. $f(x) = x^3 + 2$

    a. $f(-1) = (-1)^3 + 2 = 1$

    b. $f^{-1}(1) = -1$

15. The graph represents a one-to-one function because it passes the horizontal line test.

17. The graph does not represent a one-to-one function because it does not pass the horizontal line test.

19. The graph represents a one-to-one function because it passes the horizontal line test.

21. The graph does not represent a one-to-one function because it does not pass the horizontal line test.

23.
$$f(x) = x + 4$$
$$y = x + 4$$
$$x = y + 4$$
$$y = x - 4$$
$$f^{-1}(x) = x - 4$$

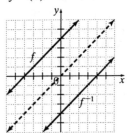

25.
$$f(x) = 2x - 3$$
$$y = 2x - 3$$
$$x = 2y - 3$$
$$2y = x + 3$$
$$y = \frac{x+3}{2}$$
$$f^{-1}(x) = \frac{x+3}{2}$$

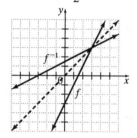

27.
$$f(x) = \frac{1}{2}x - 1$$
$$y = \frac{1}{2}x - 1$$
$$x = \frac{1}{2}y - 1$$
$$\frac{1}{2}y = x + 1$$
$$y = 2x + 2$$
$$f^{-1}(x) = 2x + 2$$

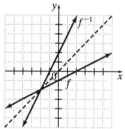

29.
$$f(x) = x^3$$
$$y = x^3$$
$$x = y^3$$
$$y = \sqrt[3]{x}$$
$$f^{-1}(x) = \sqrt[3]{x}$$

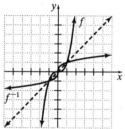

31.
$$f(x) = 5x + 2$$
$$y = 5x + 2$$
$$x = 5y + 2$$
$$5y = x - 2$$
$$y = \frac{x-2}{5}$$
$$f^{-1}(x) = \frac{x-2}{5}$$

33.
$$f(x) = \frac{x-2}{5}$$
$$y = \frac{x-2}{5}$$
$$x = \frac{y-2}{5}$$
$$5x = y - 2$$
$$y = 5x + 2$$
$$f^{-1}(x) = 5x + 2$$

35.
$$f(x) = \sqrt[3]{x}$$
$$y = \sqrt[3]{x}$$
$$x = \sqrt[3]{y}$$
$$x^3 = y$$
$$y = x^3$$
$$f^{-1}(x) = x^3$$

37. $f(x) = \dfrac{5}{3x+1}$

$y = \dfrac{5}{3x+1}$

$x = \dfrac{5}{3y+1}$

$3y+1 = \dfrac{5}{x}$

$3y = \dfrac{5}{x} - 1$

$3y = \dfrac{5-x}{x}$

$y = \dfrac{5-x}{3x}$

$f^{-1}(x) = \dfrac{5-x}{3x}$

39. $f(x) = (x+2)^3$

$y = (x+2)^3$

$x = (y+2)^3$

$\sqrt[3]{x} = y+2$

$y = \sqrt[3]{x} - 2$

$f^{-1}(x) = \sqrt[3]{x} - 2$

41.

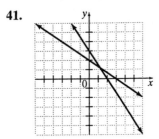

43.

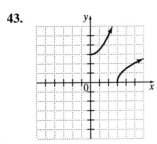

45.

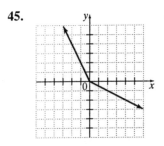

47. $(f \circ f^{-1})(x) = f(f^{-1}(x))$

$= f\left(\dfrac{x-1}{2}\right)$

$= 2\left(\dfrac{x-1}{2}\right) + 1$

$= x - 1 + 1$

$= x$

$(f^{-1} \circ f)(x) = f^{-1}(f(x))$

$= f^{-1}(2x+1)$

$= \dfrac{(2x+1)-1}{2}$

$= \dfrac{2x}{2}$

$= x$

49. $(f \circ f^{-1})(x) = f(f^{-1}(x))$

$= f\left(\sqrt[3]{x-6}\right)$

$= \left(\sqrt[3]{x-6}\right)^3 + 6$

$= x - 6 + 6$

$= x$

$(f^{-1} \circ f)(x) = f^{-1}(f(x))$

$= f^{-1}(x^3 + 6)$

$= \sqrt[3]{(x^3+6)-6}$

$= \sqrt[3]{x^3}$

$= x$

51. a. $\left(-2, \dfrac{1}{4}\right)$, $\left(-1, \dfrac{1}{2}\right)$, $(0, 1)$, $(1, 2)$, $(2, 5)$

b. $\left(\dfrac{1}{4}, -2\right)$, $\left(\dfrac{1}{2}, -1\right)$, $(1, 0)$, $(2, 1)$, $(5, 2)$

c.

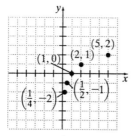

d.

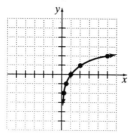

53. $f(x) = 3x + 1$ or $y = 3x + 1$
Find the inverse:
$$x = 3y + 1$$
$$3y = x - 1$$
$$y = \frac{x-1}{3}$$
$$f^{-1}(x) = \frac{x-1}{3}$$

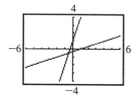

55. $f(x) = \sqrt[3]{x+1}$ or $y = \sqrt[3]{x+1}$
Find the inverse:
$$x = \sqrt[3]{y+1}$$
$$x^3 = y + 1$$
$$y = x^3 - 1$$
$$f^{-1}(x) = x^3 - 1$$

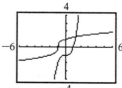

57. $25^{1/2} = \sqrt{25} = 5$

59. $16^{3/4} = (16^{1/4})^3 = \left(\sqrt[4]{16}\right)^3 = 2^3 = 8$

61. $9^{-3/2} = \dfrac{1}{9^{3/2}} = \dfrac{1}{\left(\sqrt{9}\right)^3} = \dfrac{1}{3^3} = \dfrac{1}{27}$

63. $f(x) = 3^x$
$$f(2) = 3^2 = 9$$

65. $f(x) = 3^x$
$$f\left(\frac{1}{2}\right) = 3^{1/2} \approx 1.73$$

Exercise Set 9.3

1. $y = 4^x$

x	-2	-1	0	1
y	$\frac{1}{16}$	$\frac{1}{4}$	1	4

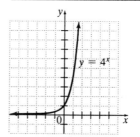

3. $y = 2^x + 1$

x	-2	-1	0	1	2
y	$\frac{5}{4}$	$\frac{3}{2}$	2	3	5

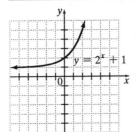

5. $y = \left(\frac{1}{4}\right)^x$

x	-1	0	1	2
y	4	1	$\frac{1}{4}$	$\frac{1}{16}$

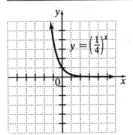

7. $y = \left(\frac{1}{2}\right)^x - 2$

x	-3	-2	-1	0	1
y	6	2	0	-1	$-\frac{3}{2}$

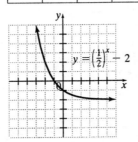

9. $y = -2^x$

x	-2	-1	0	1	2
y	$-\frac{1}{4}$	$-\frac{1}{2}$	-1	-2	-4

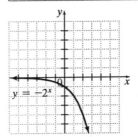

11. $y = -\left(\frac{1}{4}\right)^x$

x	-1	0	1	2
y	-4	-1	$-\frac{1}{4}$	$-\frac{1}{16}$

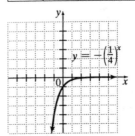

13. $f(x) = 2^{x+1}$

x	-2	-1	0	1	2
y	$\frac{1}{2}$	1	2	4	8

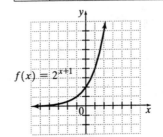

15. $f(x) = 4^{x-2}$

x	0	1	2	3
y	$\frac{1}{16}$	$\frac{1}{4}$	1	4

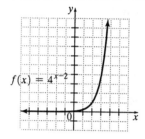

17. Answers may vary.

19. $3^x = 27$
$3^x = 3^3$
$x = 3$
The solution is 3.

21. $16^x = 8$
$(2^4)^x = 2^3$
$2^{4x} = 2^3$
$4x = 3$
$x = \frac{3}{4}$
The solution is $\frac{3}{4}$.

23. $32^{2x-3} = 2$
$(2^5)^{2x-3} = 2^1$
$2^{10x-15} = 2^1$
$10x - 15 = 1$
$10x = 16$
$x = \frac{8}{5}$
The solution is $\frac{8}{5}$.

25. $\dfrac{1}{4} = 2^{3x}$
$2^{-2} = 2^{3x}$
$3x = -2$
$x = -\dfrac{2}{3}$
The solution is $-\dfrac{2}{3}$.

27. $5^x = 625$
$5^x = 5^4$
$x = 4$
The solution is 4.

29. $4^x = 8$
$(2^2)^x = 2^3$
$2^{2x} = 2^3$
$2x = 3$
$x = \dfrac{3}{2}$
The solution is $\dfrac{3}{2}$.

31. $27^{x+1} = 9$
$(3^3)^{x+1} = 3^2$
$3^{3x+3} = 3^2$
$3x + 3 = 2$
$3x = -1$
$x = -\dfrac{1}{3}$
The solution is $-\dfrac{1}{3}$.

33. $81^{x-1} = 27^{2x}$
$(3^4)^{x-1} = (3^3)^{2x}$
$3^{4x-4} = 3^{6x}$
$4x - 4 = 6x$
$-4 = 2x$
$x = -2$
The solution is –2.

35. C; Check: $f(1) = \left(\dfrac{1}{2}\right)^1 = \dfrac{1}{2}$, point $\left(1, \ \dfrac{1}{2}\right)$

37. D; Check: $f(1) = \left(\frac{1}{4}\right)^1 = \frac{1}{4}$, point $\left(1, \frac{1}{4}\right)$

39. $y = 30(2.7)^{-0.004t}, \ t = 50$

$y = 30(2.7)^{-(0.004)(50)}$

$= 30(2.7)^{-0.2}$

≈ 24.6

Therefore, approximately 24.6 pounds of uranium will remain after 50 days.

41. $y = 200(2.7)^{0.08t}, \ t = 12$

$y = 200(2.7)^{0.08(12)}$

$= 200(2.7)^{0.96}$

≈ 519

There should be approximately 519 rats by next January.

43. $y = 5(2.7)^{-0.15t}, \ t = 10$

$y = 5(2.7)^{-0.15(10)}$

$= 5(2.7)^{-1.5}$

≈ 1.1

Approximately 1.1 grams of isotope remains.

45. $y = 15,525,000(2.7)^{0.007t}, \ t = 6$

$y = 15,525,000(2.7)^{0.007(6)}$

$= 15,525,000(2.7)^{0.042}$

$\approx 16,186,349$

Approximately 16,190,000 residents are living in the city.

47. Use $A = P\left(1 + \frac{r}{n}\right)^{nt}$ with $P = 6000$,

$r = 0.08, n = 12$, and $t = 3$.

$A = 6000\left(1 + \frac{0.08}{12}\right)^{12(3)}$

$= 6000\frac{151}{150}^{36}$

≈ 7621.42

Erica would owe $7621.42 after 3 years.

49. Use $A = P\left(1 + \frac{r}{n}\right)^{nt}$ with $P = 2000$,

$r = 0.06, n = 2$, and $t = 12$.

$A = 2000\left(1 + \frac{0.06}{2}\right)^{2(12)}$

$= 2000(1.03)^{24}$

≈ 4065.59

Janina has $4065.59 in her savings account.

51. $x = 2002 - 1900 = 12$

```
6.052*(1.378^12)
            283.719725
■
```

The prediction is that, in 2002, there will be approximately 284 million subscribers.

53. $t = 100$

```
30(2.7)^(-.004*1
00)
       20.16395637
```

After 100 days, there will be approximately 20.16 pounds of uranium available.

55. $t = 10$

```
75


                          1


6 X=10    .  .Y=50.409891 . 14
25
```

Approximately 50.41 grams of debris still remains after 10 days.

57. $y = 5,926,466,814(2.7)^{0.0132t}, \ t = 7$

$= 5,926,466,814(2.7)^{0.0132(7)}$

$\approx 6,496,117,455$

The predicted population is approximately 6,496,000,000 people.

59. $y = 358(1.004)^t$

   a. $t = 2004 - 1994 = 10$

      $y = 358(1.004)^{10} \approx 372.6$

      In 2004, the prediction is 372.6 parts per million by volume.

   b. $t = 2025 - 1994 = 31$

      $y = 358(1.004)^{31} \approx 405.2$

      In 2025, the prediction is 405.2 parts per million by volume.

61. $y = 6.052(1.378)^x$

   $t = 2008 - 1990 = 18$

   $y = 6.052(1.378)^{18} \approx 1943$

   There will be approximately 1943 million cellular phone users in 2008.

63. $x = 2006 - 1990 = 16$

```
Y₁(16)
        39872.88863
```

   The predicted number of mergers and acquisitions in 2006 is 39,873 billion.

65. $x = 2006 - 1990 = 16$

   $x = 2000 - 1990 = 10$

```
Y₁(16)
        39872.88863
Y₁(10)
        17401.3959
39872.9/17401.4
        2.291361615
```

   The predicted percent increase from the year 2000 to 2006 is approximately $(2.29 - 1)100 = 129\%$.

67. $3x - 7 = 11$

      $3x = 18$

       $x = 6$

   The solution is 6.

69. $2 - 6x = 6(1 - x)$

   $2 - 6x = 6 - 6x$

       $2 = 6$

   Inconsistent equation
The solution is $\varnothing$.

71.
$$18 = 11x - x^2$$
$$x^2 - 11x + 18 = 0$$
$$(x - 2)(x - 9) = 0$$
$$x - 2 = 0 \text{ or } x - 9 = 0$$
$$x = 2 \text{ or } \quad x = 9$$
The solutions are 2 and 9.

73. $3^x = 9$

   $3^x = 3^2$

    $x = 2$

   The solution is 2.

75. $4^x = 1$

   $4^x = 4^0$

    $x = 0$

   The solution is 0.

Exercise Set 9.4

1. $\log_6 36 = 2$

   $6^2 = 36$

3. $\log_3 \dfrac{1}{27} = -3$

   $3^{-3} = \dfrac{1}{27}$

5. $\log_{10} 1000 = 3$

   $10^3 = 1000$

7. $\log_e x = 4$

   $e^4 = x$

9. $\log_e \dfrac{1}{e^2} = -2$

   $e^{-2} = \dfrac{1}{e^2}$

11. $\log_7 \sqrt{7} = \dfrac{1}{2}$

   $7^{1/2} = \sqrt{7}$

13. $2^4 = 16$

   $\log_2 16 = 4$

15. $10^2 = 100$

$\log_{10} 100 = 2$

17. $e^3 = x$

$\log_e x = 3$

19. $10^{-1} = \dfrac{1}{10}$

$\log_{10} \dfrac{1}{10} = -1$

21. $4^{-2} = \dfrac{1}{16}$

$\log_4 \dfrac{1}{16} = -2$

23. $5^{1/2} = \sqrt{5}$

$\log_5 \sqrt{5} = \dfrac{1}{2}$

25. $\log_2 8 = 3$ because $2^3 = 8$.

27. $\log_3 \dfrac{1}{9} = -2$ because $3^{-2} = \dfrac{1}{9}$.

29. $\log_{25} 5 = \dfrac{1}{2}$ because $25^{1/2} = 5$.

31. $\log_{1/2} 2 = -1$ because $\left(\dfrac{1}{2}\right)^{-1} = 2$.

33. $\log_7 1 = 0$ because $7^0 = 1$.

35. $\log_2 2^4 = 4$ because $2^4 = 2^4$.

37. $\log_{10} 100 = 2$ because $10^2 = 100$.

39. $3^{\log_3 5} = 5$

41. $\log_3 81 = 4$ because $3^4 = 81$.

43. $\log_4 \left(\dfrac{1}{64}\right) = -3$ because $4^{-3} = \dfrac{1}{64}$.

45. Answers may vary.

47. $\log_3 9 = x$

$3^x = 9$

$3^x = 3^2$

$x = 2$

The solution is 2.

49. $\log_3 x = 4$

$x = 3^4 = 81$

The solution is 81.

51. $\log_x 49 = 2$

$x^2 = 49$

$x = \pm 7$

The base of a logarithm must be positive.

$x = 7$

The solution is 7.

53. $\log_2 \dfrac{1}{8} = x$

$2^x = \dfrac{1}{8}$

$2^x = 2^{-3}$

$x = -3$

The solution is -3.

55. $\log_3 \left(\dfrac{1}{27}\right) = x$

$3^x = \dfrac{1}{27}$

$3^x = 3^{-3}$

$x = -3$

The solution is -3.

57. $\log_8 x = \dfrac{1}{3}$

$x = 8^{1/3} = 2$

The solution is 2.

59. $\log_4 16 = x$

$4^x = 16$

$4^x = 4^2$

$x = 2$

The solution is 2.

61. $\log_{3/4} x = 3$

$$x = \left(\frac{3}{4}\right)^3 = \frac{3^3}{4^3} = \frac{27}{64}$$

The solution is $\dfrac{27}{64}$.

63. $\log_x 100 = 2$

$$x^2 = 100$$
$$x = \pm 10$$

The base of a logarithm must be positive.
$x = 10$
The solution is 10.

65. $\log_5 5^3 = 3$

67. $2^{\log_2 3} = 3$

69. $\log_9 9 = 1$

71. $y = \log_3 x$

$y = 0: 0 = \log_3 x$

$x = 3^0 = 1$ is the x-intercept.
$x = 0: y = \log_3 0$ which is not defined.
No y-intercept exists.

$x = 3^y$	y
$\frac{1}{3}$	-1
1	0
3	1

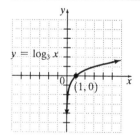

73. $f(x) = \log_{1/4} x$ or $y = \log_{1/4} x$

$y = 0: \ 0 = \log_{1/4} x$

$$x = \left(\frac{1}{4}\right)^0 = 1 \text{ is the } x\text{-intercept.}$$

$x = 0: \ y = \log_{1/4} 0$ which is not defined.
No y-intercept exists.

$x = \left(\frac{1}{4}\right)^y$	y
4	-1
1	0
$\frac{1}{4}$	1

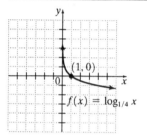

75. $f(x) = \log_5 x$ or $y = \log_5 x$

$y = 0: \ 0 = \log_5 x$

$x = 5^0 = 1$ is the x-intercept.
$x = 0: \ y = \log_5 0$ which is not defined.
No y-intercept exists.

$x = 5^y$	y
$\frac{1}{5}$	-1
1	0
5	1

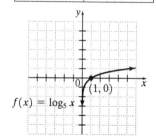

77. $f(x) = \log_{1/6} x$ or $y = \log_{1/6} x$

$y = 0$: $0 = \log_{1/6} x$

$x = \left(\dfrac{1}{6}\right)^0 = 1$ is the x-intercept.

$x = 0$: $y = \log_{1/6} 0$ which is not defined.

No y-intercept exists.

$x = \left(\frac{1}{6}\right)^y$	y
6	-1
1	0
$\frac{1}{6}$	1

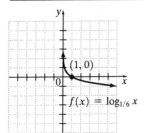

79. $y = 4^x$; $y = \log_4 x$

$x = 0$: $y = 4^0 = 1$ is the y-intercept of $y = 4^x$; hence it is the x-intercept of $y = \log_4 x$.

$y = 0$: $0 = 4^x$ has no solution so $y = 4^x$ has no x-intercept; hence $y = \log_4 x$ has no y-intercept.

x	$y = 4^x$	$y = 4^y$	y
-1	$\frac{1}{4}$	$\frac{1}{4}$	-1
0	1	1	0
1	4	4	1

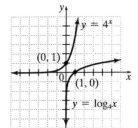

81. $y = \left(\frac{1}{3}\right)^x$; $y = \log_{1/3} x$

$x = 0$: $y = \left(\frac{1}{3}\right)^0 = 1$ is the y-intercept of

$y = \left(\frac{1}{3}\right)^x$; hence it is the x-intercept of

$y = \log_{1/3} x$.

$y = 0$: $0 = \left(\frac{1}{3}\right)^x$ has no solution so

$y = \left(\frac{1}{3}\right)^x$ has no x-intercept; hence

$y = \log_{1/3} x$ has no y-intercept.

x	$y = \left(\frac{1}{3}\right)^x$	$y = \left(\frac{1}{3}\right)^y$	y
-1	3	3	-1
0	1	1	0
1	$\frac{1}{3}$	$\frac{1}{3}$	1

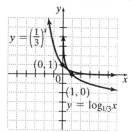

83. $\log_{10}(1-k) = \frac{-0.3}{H}$, $H = 8$

$\log_{10}(1-k) = \frac{-0.3}{8} = -0.0375$

$1 - k = 10^{-0.0375}$

$k = 1 - 10^{-0.0375}$

$k \approx 0.0827$

The rate of decay is 0.0827.

85. $\log_3 10$ is between 2 and 3 because $3^2 = 9$ and $3^3 = 27$.

87. $\frac{x-5}{5-x} = \frac{x-5}{-(x-5)} = \frac{1}{-1} = -1$

89. $\frac{x^2 - 3x - 10}{2 + x} = \frac{(x+2)(x-5)}{x+2} = x - 5$

91. $\frac{3x}{x+3} + \frac{9}{x+3} = \frac{3x+9}{x+3} = \frac{3(x+3)}{x+3} = 3$

93. $\frac{5}{y+1} - \frac{4}{y-1} = \frac{5(y-1) - 4(y+1)}{(y+1)(y-1)}$

$= \frac{5y - 5 - 4y - 4}{y^2 - 1}$

$= \frac{y-9}{y^2 - 1}$

Exercise Set 9.5

1. $\log_5 2 + \log_5 7 = \log_5 (2 \cdot 7) = \log_5 14$

3. $\log_4 9 + \log_4 x = \log_4 9x$

5. $\log_{10} 5 + \log_{10} 2 + \log_{10}(x^2 + 2)$

$= \log_{10}[5 \cdot 2(x^2 + 2)]$

$= \log_{10}(10x^2 + 20)$

7. $\log_5 12 - \log_5 4 = \log_5 \left(\frac{12}{4}\right) = \log_5 3$

9. $\log_2 x - \log_2 y = \log_2 \frac{x}{y}$

11. $\log_4 2 + \log_4 10 - \log_4 5$

$= \log_4 2 \cdot 10 - \log_4 5$

$= \log_4 \left(\frac{20}{5}\right)$

$= \log_4 4 = 1$

13. $\log_3 x^2 = 2 \log_3 x$

15. $\log_4 5^{-1} = -\log_4 5$

17. $\log_5 \sqrt{y} = \log_5 y^{1/2} = \frac{1}{2} \log_5 y$

19. $2\log_2 5 = \log_2 5^2 = \log_2 25$

21. $3\log_5 x + 6\log_5 z = \log_5 x^3 + \log_5 z^6$
$$= \log_5 x^3 z^6$$

23. $\log_{10} x - \log_{10}(x+1) + \log_{10}(x^2-2)$
$$= \log_{10}\frac{x}{x+1} + \log_{10}(x^2-2)$$
$$= \log_{10}\frac{x(x^2-2)}{x+1}$$
$$= \log_{10}\frac{x^3-2x}{x+1}$$

25. $\log_4 5 + \log_4 7 = \log_4(5\cdot 7) = \log_4 35$

27. $\log_3 8 - \log_3 2 = \log_3\left(\frac{8}{2}\right) = \log_3 4$

29. $\log_7 6 + \log_7 3 - \log_7 4$
$$= \log_7(6\cdot 3) - \log_7 4$$
$$= \log_7\left(\frac{18}{4}\right)$$
$$= \log_7\frac{9}{2}$$

31. $3\log_4 2 + \log_4 6 = \log_4 2^3 + \log_4 6$
$$= \log_4 8 + \log_4 6$$
$$= \log_4(8\cdot 6)$$
$$= \log_4 48$$

33. $3\log_2 x + \frac{1}{2}\log_2 x - 2\log_2(x+1)$
$$= \log_2 x^3 + \log_2 x^{1/2} - \log_2(x+1)^2$$
$$= \log_2(x^3 \cdot x^{1/2}) - \log_2(x+1)^2$$
$$= \log_2 x^{7/2} - \log_2(x+1)^2$$
$$= \log_2\frac{x^{7/2}}{(x+1)^2}$$

35. $2\log_8 x - \frac{2}{3}\log_8 x + 4\log_8 x$
$$= \left(2 - \frac{2}{3} + 4\right)\log_8 x$$
$$= \frac{16}{3}\log_8 x$$
$$= \log_8 x^{16/3}$$

37. $\log_2\frac{7\cdot 11}{3} = \log_2(7\cdot 11) - \log_2 3$
$$= \log_2 7 + \log_2 11 - \log_2 3$$

39. $\log_3\left(\frac{4y}{5}\right) = \log_3 4y - \log_3 5$
$$= \log_3 4 + \log_3 y - \log_3 5$$

41. $\log_2\left(\frac{x^3}{y}\right) = \log_2 x^3 - \log_2 y$
$$= 3\log_2 x - \log_2 y$$

43. $\log_b \sqrt{7x} = \log_b(7x)^{1/2}$
$$= \frac{1}{2}\log_b(7x)$$
$$= \frac{1}{2}(\log_b 7 + \log_b x)$$
$$= \frac{1}{2}\log_b 7 + \frac{1}{2}\log_b x$$

45. $\log_7\left(\frac{5x}{4}\right) = \log_7 5x - \log_7 4$
$$= \log_7 5 + \log_7 x - \log_7 4$$

47. $\log_5 x^3(x+1) = \log_5 x^3 + \log_5(x+1)$
$$= 3\log_5 x + \log_5(x+1)$$

49. $\log_6\frac{x^2}{x+3} = \log_6 x^2 - \log_6(x+3)$
$$= 2\log_6 x - \log_6(x+3)$$

51. $\log_b\left(\frac{5}{3}\right) = \log_b 5 - \log_b 3$
$$= 0.7 - 0.5 = 0.2$$

53. $\log_b 15 = \log_b(5 \cdot 3)$
$$= \log_b 5 + \log_b 3$$
$$= 0.7 + 0.5 = 1.2$$

55. $\log_b \sqrt[3]{5} = \log_b 5^{1/3}$
$$= \frac{1}{3}\log_b 5$$
$$= \frac{1}{3}(0.7)$$
$$\approx 0.233$$

57. $\log_2 x^3 = 3\log_2 x$
True

59. $\dfrac{\log_7 10}{\log_7 5} = \log_7 2$
False

61. $\dfrac{\log_7 x}{\log_7 y} = (\log_7 x) - (\log_7 y)$
False

63. $\log_b 8 = \log_b 2^3$
$$= 3\log_b 2$$
$$= 3(0.43)$$
$$= 1.29$$

65. $\log_b\left(\dfrac{3}{9}\right) = \log_b\left(\dfrac{1}{3}\right)$
$$= \log_b 3^{-1}$$
$$= (-1)\log_b 3$$
$$= -(0.68)$$
$$= -0.68$$

67. $\log_b \sqrt{\dfrac{2}{3}} = \log_b\left(\dfrac{2}{3}\right)^{1/2}$
$$= \frac{1}{2}\log_b \frac{2}{3}$$
$$= \frac{1}{2}(\log_b 2 - \log_b 3)$$
$$= \frac{1}{2}(0.43 - 0.68)$$
$$= \frac{1}{2}(-0.25)$$
$$= -0.125$$

69.

x	$y = 10^x$	$x = 10^y$	y
-1	$\dfrac{1}{10}$	$\dfrac{1}{10}$	-1
0	1	1	0
1	10	10	1

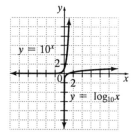

71. $\log_{10} \dfrac{1}{10} = -1$ because $10^{-1} = \dfrac{1}{10}$.

73. $\log_7 \sqrt{7} = \dfrac{1}{2}$ because $7^{1/2} = \sqrt{7}$.

Exercise Set 9.6

1.
```
log(8)
        .903089987
```
$\log 8 \approx 0.9031$

3.
```
log(2.31)
        .3636119799
```
$\log 2.31 \approx 0.3636$

5.
```
ln(2)
        .6931471806
■
```
$\ln 2 \approx 0.6931$

7.

$\ln 0.0716 \approx -2.6367$

9.

$\log 12.6 \approx 1.1004$

11.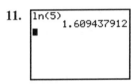

$\ln 5 \approx 1.6094$

13.

$\log 41.5 \approx 1.6180$

15. Answers may vary.

17. $\log 100 = \log 10^2 = 2$

19. $\log \dfrac{1}{1000} = \log 10^{-3} = -3$

21. $\ln e^2 = 2$

23. $\ln \sqrt[4]{e} = \ln e^{1/4} = \dfrac{1}{4}$

25. $\log 10^3 = 3$

27. $\ln e^2 = 2$

29. $\log 0.0001 = \log 10^{-4} = -4$

31. $\ln \sqrt{e} = \ln e^{1/2} = \dfrac{1}{2}$

33. ln 50 is larger.
Answers may vary.

35. $\log x = 1.3$
$x = 10^{1.3} \approx 19.9526$
The exact solution is $10^{1.3}$. To four decimal places, $x \approx 19.9526$.

37. $\log 2x = 1.1$
$2x = 10^{1.1}$
$x = \dfrac{10^{1.1}}{2} \approx 6.2946$

The exact solution is $\dfrac{10^{1.1}}{2}$. To four decimal places, $x \approx 6.2946$.

39. $\ln x = 1.4$
$x = e^{1.4} \approx 4.0552$
The exact solution is $e^{1.4}$. To four decimal places, $x \approx 4.0552$.

41. $\ln(3x - 4) = 2.3$
$3x - 4 = e^{2.3}$
$3x = 4 + e^{2.3}$
$x = \dfrac{4 + e^{2.3}}{3} \approx 4.6581$

The exact solution is $\dfrac{4 + e^{2.3}}{3}$. To four decimal places, $x \approx 4.6581$.

43. $\log x = 2.3$
$x = 10^{2.3} \approx 199.5262$
The exact solution is $10^{2.3}$. To four decimal places, $x \approx 199.5262$.

45. $\ln x = -2.3$
$x = e^{-2.3} \approx 0.1003$
The exact solution is $e^{-2.3}$. To four decimal places, $x \approx 0.1003$.

47. $\log(2x+1) = -0.5$

$2x + 1 = 10^{-0.5}$

$2x = 10^{-0.5} - 1$

$x = \dfrac{10^{-0.5} - 1}{2} \approx -0.3419$

The exact solution is $\dfrac{10^{-0.5} - 1}{2}$. To four decimal places, $x \approx -0.3419$.

49. $\ln 4x = 0.18$

$4x = e^{0.18}$

$x = \dfrac{e^{0.18}}{4} \approx 0.2993$

The exact solution is $\dfrac{e^{0.18}}{4}$. To four decimal places, $x \approx 0.2993$.

51. $\log_2 3 = \dfrac{\ln 3}{\ln 2} \approx 1.5850$

53. $\log_{1/2} 5 = \dfrac{\ln 5}{\ln\left(\frac{1}{2}\right)} \approx -2.3219$

55. $\log_4 9 = \dfrac{\ln 9}{\ln 4} \approx 1.5850$

57. $\log_3\left(\dfrac{1}{6}\right) = \log_3 6^{-1}$

$= (-1)\log_3 6$

$= -\dfrac{\ln 6}{\ln 3} \approx -1.6309$

59. $\log_8 6 = \dfrac{\ln 6}{\ln 8} \approx 0.8617$

61. $R = \log\left(\dfrac{a}{T}\right) + B$, $a = 200$, $T = 1.6$, and $B = 2.1$.

$R = \log\left(\dfrac{200}{1.6}\right) + 2.1 \approx 4.2$

The earthquake measures 4.2 on the Richter scale.

63. $R = \log\left(\dfrac{a}{T}\right) + B$, $a = 400$, $T = 2.6$, and $B = 3.1$.

$R = \log\left(\dfrac{400}{2.6}\right) + 3.1 \approx 5.3$

The earthquake measures 5.3 on the Richter scale.

65. $A = Pe^{rt}$, $t = 12$, $P = 1400$, and $r = 0.08$.

$A = 1400e^{(0.08)12} = 1400e^{0.96} \approx 3656.38$

Dana has \$3656.38 after 12 years.

67. $A = Pe^{rt}$, $t = 4$, $r = 0.06$, and $P = 2000$.

$A = 2000e^{(0.06)4} = 2000e^{0.24} \approx 2542.50$

Barbara owes \$2542.50 at the end of 4 years.

69.

x	$y = e^x$
-1	0.4
0	1
1	2.7

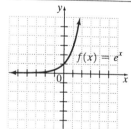

71.

x	$y = e^{-3x}$
−0.5	4.5
0	1
0.5	0.2

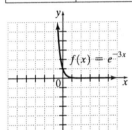

73.

x	$y = e^x + 2$
−1	2.4
0	3
1	4.7

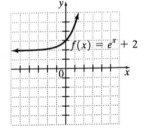

75.

x	$y = e^{x-1}$
0	0.4
1	1
2	2.7

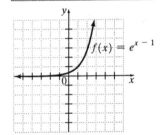

77.

x	$y = 3e^x$
−2	0.4
−1	1.1
0	3

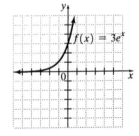

79.

x	$y = \ln x$
0.5	–0.7
1	0
6	1.8

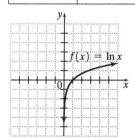

81.

x	$y = -2 \log x$
0.5	0.6
1	0
6	–1.6

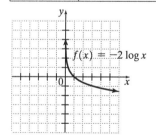

83.

x	$y = \log(x + 2)$
–1.5	–0.3
–1	0
6	0.9

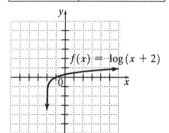

85.

x	$y = \ln x - 3$
0.5	–3.7
1	–3
6	–1.2

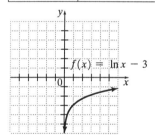

87. $f(x) = e^x$

$f(x) = e^x + 2$

$f(x) = e^x - 3$

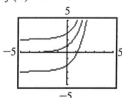

Answers may vary.

89. $6x - 3(2 - 5x) = 6$

$6x - 6 + 15x = 6$

$21x - 6 = 6$

$21x = 12$

$x = \dfrac{12}{21} = \dfrac{4}{7}$

The solution is $\dfrac{4}{7}$.

91. $2x + 3y = 6x$

$3y = 4x$

$x = \dfrac{3y}{4}$

The solution is $\dfrac{3y}{4}$.

93. $x^2 + 7x = -6$

$x^2 + 7x + 6 = 0$

$(x + 6)(x + 1) = 0$

$x + 6 = 0 \quad$ or $\quad x + 1 = 0$

$x = -6 \quad$ or $\quad x = -1$

The solutions are –6 and –1.

95. $\begin{cases} x + 2y = -4 \\ 3x - y = 9 \end{cases}$ or $\begin{cases} x + 2y = -4 \\ 6x - 2y = 18 \end{cases}$

Add.

$7x = 14$

$x = 2$

Substitute back.

$2 + 2y = -4$

$2y = -6$

$y = -3$

The solution is (2, –3).

Exercise Set 9.7

1. $3^x = 6$

$\log 3^x = \log 6$

$x \log 3 = \log 6$

$x = \dfrac{\log 6}{\log 3} \approx 1.6309$

The solution is $\dfrac{\log 6}{\log 3}$, or approximately 1.6309.

Check: $y_1 = 3^x$, $y_2 = 6$

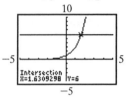

See that $x \approx 1.6309$.

3. $3^{2x} = 3.8$

$\log 3^{2x} = \log 3.8$

$2x \log 3 = \log 3.8$

$2x = \dfrac{\log 3.8}{\log 3}$

$x = \dfrac{\log 3.8}{2 \log 3} \approx 0.6076$

The solution is $\dfrac{\log 3.8}{2 \log 3}$, or approximately 0.6076.

Check: $y_1 = 3^{2x}$, $y_2 = 3.8$

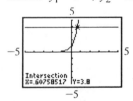

See that $x \approx 0.6076$.

5.
$$2^{x-3} = 5$$
$$\log 2^{x-3} = \log 5$$
$$(x-3)\log 2 = \log 5$$
$$x - 3 = \frac{\log 5}{\log 2}$$
$$x = 3 + \frac{\log 5}{\log 2} \approx 5.3219$$

The solution is $3 + \dfrac{\log 5}{\log 2}$, or

approximately 5.3219.

Check: $y_1 = 2^{x-3}$, $y_2 = 5$

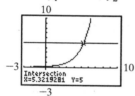

See that $x \approx 5.3219$.

7.
$$9^x = 5$$
$$\log 9^x = \log 5$$
$$x \log 9 = \log 5$$
$$x = \frac{\log 5}{\log 9} \approx 0.7325$$

The solution is $\dfrac{\log 5}{\log 9}$, or

approximately 0.7325.

Check: $y_1 = 9^x$, $y_2 = 5$

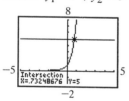

See that $x \approx 0.7325$.

9.
$$4^{x+7} = 3$$
$$\log 4^{x+7} = \log 3$$
$$(x+7)\log 4 = \log 3$$
$$x + 7 = \frac{\log 3}{\log 4}$$
$$x = \frac{\log 3}{\log 4} - 7 \approx -6.2075$$

The solution is $\dfrac{\log 3}{\log 4} - 7$, or

approximately −6.2075.

Check: $y_1 = 4^{x+7}$, $y_2 = 3$

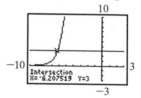

See that $x \approx -6.2075$.

11.
$$7^{3x-4} = 11$$
$$\log 7^{3x-4} = \log 11$$
$$(3x-4)\log 7 = \log 11$$
$$3x - 4 = \frac{\log 11}{\log 7}$$
$$3x = 4 + \frac{\log 11}{\log 7}$$
$$x = \frac{1}{3}\left(4 + \frac{\log 11}{\log 7}\right) \approx 1.7441$$

The solution is $\dfrac{1}{3}\left(4 + \dfrac{\log 11}{\log 7}\right)$, or

approximately 1.7441.

Check: $y_1 = 7^{3x-4}$, $y_2 = 11$

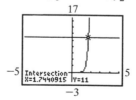

See that $x \approx 1.7441$.

13. $e^{6x} = 5$

$\ln e^{6x} = \ln 5$

$6x = \ln 5$

$x = \dfrac{\ln 5}{6} \approx 0.2682$

The solution is $\dfrac{\ln 5}{6}$, or approximately 0.2682.

Check: $y_1 = e^{6x}$, $y_2 = 5$

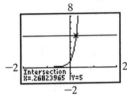

See that $x \approx 0.2682$.

15. $\log_2(x+5) = 4$

$x + 5 = 2^4$

$x + 5 = 16$

$x = 11$

The solution is 11.

17. $\log_3 x^2 = 4$

$x^2 = 3^4$

$x^2 = 81$

$x = \pm 9$

The solutions are –9 and 9.

19. $\log_4 2 + \log_4 x = 0$

$\log_4(2x) = 0$

$2x = 4^0$

$2x = 1$

$x = \dfrac{1}{2}$

The solution is $\dfrac{1}{2}$.

21. $\log_2 6 - \log_2 x = 3$

$\log_2\left(\dfrac{6}{x}\right) = 3$

$\dfrac{6}{x} = 2^3$

$\dfrac{6}{x} = 8$

$8x = 6$

$x = \dfrac{6}{8} = \dfrac{3}{4}$

The solution is $\dfrac{3}{4}$.

23. $\log_4 x + \log_4(x+6) = 2$

$\log_4 x(x+6) = 2$

$x(x+6) = 4^2$

$x^2 + 6x = 16$

$x^2 + 6x - 16 = 0$

$(x+8)(x-2) = 0$

$x + 8 = 0$ or $x - 2 = 0$

$x = -8$ or $x = 2$

We reject –8.

The solution is 2.

25. $\log_5(x+3) - \log_5 x = 2$

$\log_5\left(\dfrac{x+3}{x}\right) = 2$

$\dfrac{x+3}{x} = 5^2$

$\dfrac{x+3}{x} = 25$

$x + 3 = 25x$

$3 = 24x$

$x = \dfrac{3}{24} = \dfrac{1}{8}$

The solution is $\dfrac{1}{8}$.

27. $\log_3(x-2) = 2$

$$x-2 = 3^2$$
$$x-2 = 9$$
$$x = 11$$

The solution is 11.

29. $\log_4(x^2 - 3x) = 1$

$$x^2 - 3x = 4^1$$
$$x^2 - 3x = 4$$
$$x^2 - 3x - 4 = 0$$
$$(x-4)(x+1) = 0$$
$$x-4 = 0 \quad \text{or} \quad x+1 = 0$$
$$x = 4 \quad \text{or} \quad x = -1$$

The solutions are -1 and 4.

31. $\ln 5 + \ln x = 0$

$$\ln(5x) = 0$$
$$5x = e^0$$
$$5x = 1$$
$$x = \frac{1}{5}$$

The solution is $\frac{1}{5}$.

33. $3\log x - \log x^2 = 2$

$$3\log x - 2\log x = 2$$
$$\log x = 2$$
$$x = 10^2$$
$$x = 100$$

The solution is 100.

35. $\log_2 x + \log_2(x+5) = 1$

$$\log_2 x(x+5) = 1$$
$$x(x+5) = 2^1$$
$$x^2 + 5x = 2$$
$$x^2 + 5x - 2 = 0$$
$$a = 1, \, b = 5, \, c = -2$$
$$x = \frac{-5 \pm \sqrt{5^2 - 4(1)(-2)}}{2(1)}$$
$$= \frac{-5 \pm \sqrt{25 + 8}}{2}$$
$$= \frac{-5 \pm \sqrt{33}}{2}$$

We reject $\dfrac{-5 - \sqrt{33}}{2}$.

The solution is $\dfrac{-5 + \sqrt{33}}{2}$.

37. $\log_4 x - \log_4(2x-3) = 3$

$$\log_4\left(\frac{x}{2x-3}\right) = 3$$
$$\frac{x}{2x-3} = 4^3$$
$$\frac{x}{2x-3} = 64$$
$$x = 64(2x-3)$$
$$x = 128x - 192$$
$$192 = 127x$$
$$x = \frac{192}{127}$$

The solution is $\dfrac{192}{127}$.

39.
$$\log_2 x + \log_2(3x+1) = 1$$
$$\log_2 x(3x+1) = 1$$
$$x(3x+1) = 2^1$$
$$3x^2 + x = 2$$
$$3x^2 + x - 2 = 0$$
$$(3x-2)(x+1) = 0$$
$$3x - 2 = 0 \quad \text{or} \quad x+1 = 0$$
$$3x = 2 \quad \text{or} \quad x = -1$$
$$x = \frac{2}{3}$$

We reject -1.

The solution is $\frac{2}{3}$.

41. $y = y_0 e^{0.043t}$, $y_0 = 83$, and $t = 5$.
$$y = 83e^{0.043(5)} = 83e^{0.215} \approx 103$$
There should be 103 wolves in 5 years.

43. $y = y_0 e^{0.026t}$, $y_0 = 10,052,000$, and $t = 6$.
$$y = 10,052,000e^{0.026(6)} \approx 11,750,000$$
There will be approximately
11,750,000 inhabitants in 2005.

45. $y = y_0 e^{-0.005t}$, $y_0 = 146,394$,
and, $y = 120,000$.
$$120,000 = 146,394e^{-0.005t}$$
$$\frac{120,000}{146,394} = e^{-0.005t}$$
$$\ln\left(\frac{120,000}{146,394}\right) = \ln e^{-0.005t}$$
$$\ln\left(\frac{120,000}{146,394}\right) = -0.005t \ln e$$
$$t = \frac{\ln\left(\frac{120,000}{146,394}\right)}{-0.005} \approx 39.8$$
It will take approximately 39.8 years to
reach 120,000 thousand.

47. $A = P\left(1 + \frac{r}{n}\right)^{nt}$, $P = 600$,
$A = 2(600) = 1200$, $r = 0.07$, and $n = 12$.
$$1200 = 600\left(1 + \frac{0.07}{12}\right)^{12t}$$
$$2 = (1.0058\overline{3})^{12t}$$
$$\log 2 = \log\left(1.0058\overline{3}^{12t}\right)$$
$$\log 2 = 12t(\log 1.0058\overline{3})$$
$$t = \frac{\log 2}{12\log 1.0058\overline{3}} \approx 9.9$$
It would take approximately 9.9 years for
the \$600 to double.

49. $A = P\left(1 + \frac{r}{n}\right)^{nt}$, $P = 1200$,
$A = P + I = 1200 + 200 = 1400$,
$r = 0.09$, and $n = 4$.
$$1400 = 1200\left(1 + \frac{0.09}{4}\right)^{4t}$$
$$\frac{7}{6} = (1.0225)^{4t}$$
$$\log\frac{7}{6} = \log 1.0225^{4t}$$
$$\log\frac{7}{6} = 4t(\log 1.0225)$$
$$t = \frac{\log\frac{7}{6}}{4\log 1.0225} \approx 1.7$$
It would take the investment
approximately 1.7 years to earn \$200.

51. $A = P\left(1 + \frac{r}{n}\right)^{nt}$, $P = 1000$,
$A = 2(1000) = 2000$, $r = 0.08$, and $n = 2$.
$$2000 = 1000\left(1 + \frac{0.08}{2}\right)^{2t}$$
$$2 = (1.04)^{2t}$$
$$\log 2 = \log 1.04^{2t}$$
$$\log 2 = 2t \log 1.04$$
$$t = \frac{\log 2}{2\log 1.04} \approx 8.8$$
It would take approximately 8.8 years for
the \$1000 to double.

53. $w = 0.00185h^{2.67}$ and $h = 35$.

$w = 0.00185(35)^{2.67} \approx 24.5$

The expected weight of a boy 35 inches tall is 24.5 pounds.

55. $w = 0.00185h^{2.67}$ and $w = 85$.

$$85 = 0.00185h^{2.67}$$

$$\frac{85}{0.00185} = h^{2.67}$$

$$\left(h^{2.67}\right)^{1/2.67} = \left(\frac{85}{0.00185}\right)^{1/2.67}$$

$$h \approx 55.7$$

The expected height of the boy is approximately 55.7 inches.

57. $P = 14.7e^{-0.21x}$ and $x = 1$.

$P = 14.7e^{-0.21(1)}$

$= 14.7e^{-0.21}$

≈ 11.9

The average atmospheric pressure of Denver is approximately 11.9 pounds per square inch.

59. $P = 14.7e^{-0.21x}$ and $P = 7.5$.

$$7.5 = 14.7e^{-0.21x}$$

$$\frac{7.5}{14.7} = e^{-0.21x}$$

$$\ln\left(\frac{7.5}{14.7}\right) = \ln e^{-0.21x}$$

$$\ln\left(\frac{7.5}{14.7}\right) = -0.21x(\ln e)$$

$$-0.21x = \ln\left(\frac{7.5}{14.7}\right)$$

$$x = -\frac{1}{0.21}\ln\left(\frac{7.5}{14.7}\right) \approx 3.2$$

The elevation of the jet is approximately 3.2 miles.

61.

$t = \dfrac{1}{c}\ln\left(\dfrac{A}{A-N}\right)$, $N = 50$, $A = 75$, and $c = 0.09$.

$$t = \frac{1}{0.09}\ln\left(\frac{75}{75-50}\right) = \frac{1}{0.09}\ln(3) \approx 12.21$$

It should take about 12 weeks.

63.

$t = \dfrac{1}{c}\ln\left(\dfrac{A}{A-N}\right)$, $N = 150$, $A = 210$, and $c = 0.07$.

$$t = \frac{1}{0.07}\ln\left(\frac{210}{210-150}\right) = \frac{1}{0.07}\ln(3.5) \approx 17.9$$

It should take about 18 weeks.

65. a.
```
ExpReg
y=a*b^x
a=942.4201707
b=1.23528862
r²=.9155541739
r=.956845951
```

The exponential regression equation is $y = 942.42(1.235)^x$.

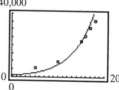

b. $x = 1987 - 1980 = 7$
```
Y1(7)
        4136.410746
```

In 1987, the number of shipments was approximately 4136 thousand.

c. $x = 2005 - 1980 = 25$

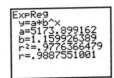

In 2005, the number of shipments is predicted to be approximately 185,548 thousand.

67. a.

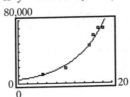

The exponential regression equation is $y = 5173.90(1.160)^x$.

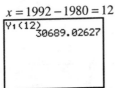

b. $x = 1992 - 1980 = 12$

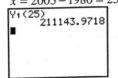

In 1992, the revenue was approximately \$30,689 million.

c. $x = 2005 - 1980 = 25$

In 2005, the revenue is predicted to be approximately \$211,144 million.

69. a.

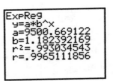

The exponential regression equation is $y = 9500.669(1.182)^x$.

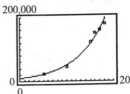

b. $x = 1992 - 1980 = 12$

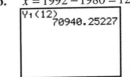

In 1992, the revenue was approximately \$70,940 million.

c. $x = 2005 - 1980 = 25$

In 2005, the revenue is predicted to be approximately \$626,315 million.

71. $\dfrac{x^2 - y + 2z}{3x} = \dfrac{(-2)^2 - (0) + 2(3)}{3(-2)}$

$$= \frac{4 + 6}{-6}$$

$$= \frac{-10}{6}$$

$$= -\frac{5}{3}$$

73. $\dfrac{3z-4x+y}{x+2z} = \dfrac{3(3)-4(-2)+0}{-2+2(3)}$

$\qquad\qquad = \dfrac{9+8}{-2+6}$

$\qquad\qquad = \dfrac{17}{4}$

75. $\quad f(x) = 5x+2$

$\qquad\quad y = 5x+2$

$\qquad\quad x = 5y+2$

$\qquad x-2 = 5y$

$\qquad \dfrac{x-2}{5} = y$

$\qquad f^{-1}(x) = \dfrac{x-2}{5}$

Chapter 9 Review

1. $(f+g)(x) = f(x)+g(x)$
$\qquad\qquad = (x-5)+(2x+1)$
$\qquad\qquad = x-5+2x+1$
$\qquad\qquad = 3x-4$

2. $(f-g)(x) = f(x)-g(x)$
$\qquad\qquad = (x-5)-(2x+1)$
$\qquad\qquad = x-5-2x-1$
$\qquad\qquad = -x-6$

3. $(f \cdot g)(x) = f(x) \cdot g(x)$
$\qquad\qquad = (x-5)(2x+1)$
$\qquad\qquad = 2x^2+x-10x-5$
$\qquad\qquad = 2x^2-9x-5$

4. $\left(\dfrac{g}{f}\right)(x) = \dfrac{g(x)}{f(x)} = \dfrac{2x+1}{x-5}, \ x \neq 5$

5. $(f \circ g)(x) = f(g(x))$
$\qquad\qquad = f(x+1)$
$\qquad\qquad = (x+1)^2-2$
$\qquad\qquad = x^2+2x+1-2$
$\qquad\qquad = x^2+2x-1$

6. $(g \circ f)(x) = g(f(x))$
$\qquad\qquad = g(x^2-2)$
$\qquad\qquad = x^2-2+1$
$\qquad\qquad = x^2-1$

7. $(h \circ g)(2) = h(g(2))$
$\qquad\qquad = h(3)$
$\qquad\qquad = 3^3-3^2$
$\qquad\qquad = 18$

8. $(f \circ f)(x) = f(f(x))$
$\qquad\qquad = f(x^2-2)$
$\qquad\qquad = (x^2-2)^2-2$
$\qquad\qquad = x^4-4x^2+4-2$
$\qquad\qquad = x^4-4x^2+2$

9. $(f \circ g)(-1) = f(g(-1))$
$\qquad\qquad = f(0)$
$\qquad\qquad = 0^2-2$
$\qquad\qquad = -2$

10. $(h \circ h)(2) = h(h(2))$
$\qquad\qquad = h(4)$
$\qquad\qquad = 4^3-4^2$
$\qquad\qquad = 48$

11. The function is one-to-one.
    $h^{-1} = \{(14, -9), (8, 6), (12, -11), (15, 15)\}$

12. The function is not one-to-one.

13. The function is one-to-one.

Rank in Automobile Thefts (Input)	2	4	1	3
U.S. Region (Output)	West	Midwest	South	Northeast

14. The function is not one-to-one.

15. $f(x) = \sqrt{x+2}$

    a. $f(7) = \sqrt{7+2} = \sqrt{9} = 3$

    b. $f^{-1}(3) = 7$

16. $f(x) = \sqrt{x+2}$

    a. $f(-1) = \sqrt{-1+2} = \sqrt{1} = 1$

    b. $f^{-1}(1) = -1$

17. The graph is not a one-to-one function since it fails the horizontal line test.

18. The graph is not a one-to-one function since it fails the horizontal line test.

19. The graph is not a one-to-one function since it fails the horizontal line test.

20. The graph is a one-to-one function since it passes the horizontal line test.

21. $f(x) = x - 9$
    $y = x - 9$
    $x = y - 9$
    $y = x + 9$
    $f^{-1}(x) = x + 9$

22. $f(x) = x + 8$
    $y = x + 8$
    $x = y + 8$
    $y = x - 8$
    $f^{-1}(x) = x - 8$

23. $f(x) = 6x + 11$
    $y = 6x + 11$
    $x = 6y + 11$
    $6y = x - 11$
    $y = \dfrac{x - 11}{6}$
    $f^{-1}(x) = \dfrac{x - 11}{6}$

24. $f(x) = 12x$
    $y = 12x$
    $x = 12y$
    $y = \dfrac{x}{12}$
    $f^{-1}(x) = \dfrac{x}{12}$

25. $f(x) = x^3 - 5$
    $y = x^3 - 5$
    $x = y^3 - 5$
    $y^3 = x + 5$
    $y = \sqrt[3]{x + 5}$
    $f^{-1}(x) = \sqrt[3]{x + 5}$

26.
$$f(x) = \sqrt[3]{x+2}$$
$$y = \sqrt[3]{x+2}$$
$$x = \sqrt[3]{y+2}$$
$$x^3 = y+2$$
$$y = x^3 - 2$$
$$f^{-1}(x) = x^3 - 2$$

27.
$$g(x) = \frac{12x-7}{6}$$
$$y = \frac{12x-7}{6}$$
$$x = \frac{12y-7}{6}$$
$$6x = 12y - 7$$
$$6x + 7 = 12y$$
$$\frac{6x+7}{12} = y$$
$$g^{-1}(x) = \frac{6x+7}{12}$$

28.
$$r(x) = \frac{13}{2}x - 4$$
$$y = \frac{13}{2}x - 4$$
$$x = \frac{13}{2}y - 4$$
$$x + 4 = \frac{13}{2}y$$
$$2(x+4) = 13y$$
$$\frac{2(x+4)}{13} = y$$
$$r^{-1}(x) = \frac{2(x+4)}{13}$$

29.
$$y = g(x) = \sqrt{x}$$
$$x = \sqrt{y}$$
$$x^2 = y = g^{-1}(x), \ \ x \geq 0$$

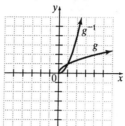

30. $y = h(x) = 5x - 5$
$$x = 5y - 5$$
$$5y = x + 5$$
$$y = \frac{x+5}{5}$$
$$h^{-1}(x) = \frac{x+5}{5}$$

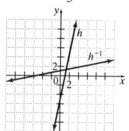

31. $y = f(x) = 2x - 3$
Find the inverse:
$$x = 2y - 3$$
$$2y = x + 3$$
$$y = \frac{x+3}{2}$$
$$f^{-1}(x) = \frac{x+3}{2}$$

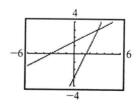

32. $4^x = 64$

$4^x = 4^3$

$x = 3$

The solution is 3.

33. $3^x = \dfrac{1}{9}$

$3^x = 3^{-2}$

$x = -2$

The solution is –2.

34. $2^{3x} = \dfrac{1}{16}$

$2^{3x} = 2^{-4}$

$3x = -4$

$x = -\dfrac{4}{3}$

The solution is $-\dfrac{4}{3}$.

35. $5^{2x} = 125$

$5^{2x} = 5^3$

$2x = 3$

$x = \dfrac{3}{2}$

The solution is $\dfrac{3}{2}$.

36. $9^{x+1} = 243$

$(3^2)^{x+1} = 3^5$

$3^{2x+2} = 3^5$

$2x + 2 = 5$

$2x = 3$

$x = \dfrac{3}{2}$

The solution is $\dfrac{3}{2}$.

37. $8^{3x-2} = 4$

$(2^3)^{3x-2} = 2^2$

$2^{9x-6} = 2^2$

$9x - 6 = 2$

$9x = 8$

$x = \dfrac{8}{9}$

The solution is $\dfrac{8}{9}$.

38.

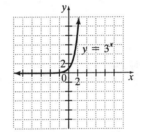

39.

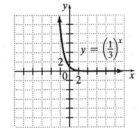

40.

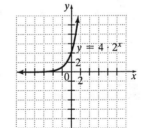

41.

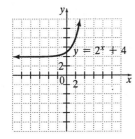

42. $A = P\left(1 + \dfrac{r}{n}\right)^{nt}$

$A = 1600\left(1 + \dfrac{0.09}{2}\right)^{(2)(7)} \approx 2963.11$

The amount accrued is \$2963.11.

43. $A = P\left(1 + \dfrac{r}{n}\right)^{nt}$

$A = 800\left(1 + \dfrac{0.07}{4}\right)^{4(5)} \approx 1131.82$

The certificate is worth \$1131.82 at the end of 5 years.

44.

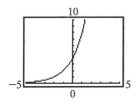

45. $7^2 = 49$

$\log_7 49 = 2$

46. $2^{-4} = \dfrac{1}{16}$

$\log_2 \dfrac{1}{16} = -4$

47. $\log_{1/2} 16 = -4$

$\left(\dfrac{1}{2}\right)^{-4} = 16$

48. $\log_{0.4} 0.064 = 3$

$0.4^3 = 0.064$

49. $\log_4 x = -3$

$x = 4^{-3} = \dfrac{1}{64}$

The solution is $\dfrac{1}{64}$.

50. $\log_3 x = 2$

$x = 3^2 = 9$

The solution is 9.

51. $\log_3 1 = x$

$3^x = 1$

$3^x = 3^0$

$x = 0$

The solution is 0.

52. $\log_4 64 = x$

$4^x = 64$

$4^x = 4^3$

$x = 3$

The solution is 3.

53. $\log_x 64 = 2$

$x^2 = 64$

$x = \pm\sqrt{64} = \pm 8$

The base of a logarithm must be positive.

$x = 8$

The solution is 8.

54. $\log_x 81 = 4$

$x^4 = 81$

$x = \pm 3$

The base of a logarithm must be positive.

$x = 3$

The solution is 3.

55. $\log_4 4^5 = x$

$4^x = 4^5$

$x = 5$

The solution is 5.

56. $\log_7 7^{-2} = x$
$$7^x = 7^{-2}$$
$$x = -2$$
The solution is -2.

57. $5^{\log_5 4} = x$
$$\log_5 x = \log_5 4$$
$$x = 4$$
The solution is 4.

58. $2^{\log_2 9} = x$
$$\log_2 x = \log_2 9$$
$$x = 9$$
The solution is 9.

59. $\log_2(3x - 1) = 4$
$$3x - 1 = 2^4$$
$$3x - 1 = 16$$
$$3x = 17$$
$$x = \frac{17}{3}$$
The solution is $\dfrac{17}{3}$.

60. $\log_3(2x + 5) = 2$
$$2x + 5 = 3^2$$
$$2x + 5 = 9$$
$$2x = 4$$
$$x = 2$$
The solution is 2.

61. $\log_4(x^2 - 3x) = 1$
$$x^2 - 3x = 4^1$$
$$x^2 - 3x = 4$$
$$x^2 - 3x - 4 = 0$$
$$(x + 1)(x - 4) = 0$$
$$x + 1 = 0 \quad \text{or} \quad x - 4 = 0$$
$$x = -1 \quad \text{or} \qquad x = 4$$
The solutions are -1 and 4.

62. $\log_8(x^2 + 7x) = 1$
$$x^2 + 7x = 8^1$$
$$x^2 + 7x = 8$$
$$x^2 + 7x - 8 = 0$$
$$(x + 8)(x - 1) = 0$$
$$x + 8 = 0 \quad \text{or} \quad x - 1 = 0$$
$$x = -8 \quad \text{or} \qquad x = 1$$
The solutions are -8 and 1.

63.

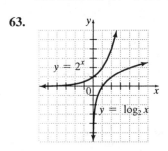

64.

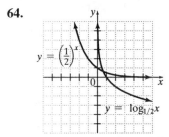

65. $\log_3 8 + \log_3 4 = \log_3(8 \cdot 4) = \log_3 32$

66. $\log_2 6 + \log_2 3 = \log_2(6 \cdot 3) = \log_2 18$

67. $\log_7 15 - \log_7 20 = \log_7 \dfrac{15}{20} = \log_7 \dfrac{3}{4}$

68. $\log 18 - \log 12 = \log \dfrac{18}{12} = \log \dfrac{3}{2}$

69. $\log_{11} 8 + \log_{11} 3 - \log_{11} 6$
$$= \log_{11}(8 \cdot 3) - \log_{11} 6$$
$$= \log_{11}\left(\frac{24}{6}\right)$$
$$= \log_{11} 4$$

70. $\log_5 14 + \log_5 3 - \log_5 21$
$= \log_5(14 \cdot 3) - \log_5 21$
$= \log_5\left(\dfrac{42}{21}\right)$
$= \log_5 2$

71. $2\log_5 x - 2\log_5(x+1) + \log_5 x$
$= \log_5 x^2 - \log_5(x+1)^2 + \log_5 x$
$= \log_5 \dfrac{(x^2)(x)}{(x+1)^2}$
$= \log_5 \dfrac{x^3}{(x+1)^2}$

72. $4\log_3 x - \log_3 x + \log_3(x+2)$
$= 3\log_3 x + \log_3(x+2)$
$= \log_3 x^3 + \log_3(x+2)$
$= \log_3[x^3(x+2)]$
$= \log_3(x^4 + 2x^3)$

73. $\log_3 \dfrac{x^3}{x+2} = \log_3 x^3 - \log_3(x+2)$
$= 3\log_3 x - \log_3(x+2)$

74. $\log_4 \dfrac{x+5}{x^2} = \log_4(x+5) - \log_4 x^2$
$= \log_4(x+5) - 2\log_4 x$

75. $\log_2 \dfrac{3x^2 y}{z}$
$= \log_2 3x^2 y - \log_2 z$
$= \log_2 3 + \log_2 x^2 + \log_2 y - \log_2 z$
$= \log_2 3 + 2\log_2 x + \log_2 y - \log_2 z$

76. $\log_7 \dfrac{yz^3}{x} = \log_7(yz^3) - \log_7 x$
$= \log_7 y + \log_7 z^3 - \log_7 x$
$= \log_7 y + 3\log_7 z - \log_7 x$

77. $\log_b 50 = \log_b(5 \cdot 5 \cdot 2)$
$= \log_b(5) + \log_b(5) + \log_b(2)$
$= 0.83 + 0.83 + 0.36$
$= 2.02$

78. $\log_b \dfrac{4}{5} = \log_b 4 - \log_b 5$
$= \log_b 2^2 - \log_b 5$
$= 2\log_b 2 - \log_b 5$
$= 2(0.36) - 0.83$
$= 0.72 - 0.83$
$= -0.11$

79. $\log 3.6 \approx 0.5563$

80. $\log 0.15 \approx -0.8239$

81. $\ln 1.25 \approx 0.2231$

82. $\ln 4.63 \approx 1.5326$

83. $\log 1000 = 3$

84. $\log \dfrac{1}{10} = \log 1 - \log 10 = 0 - 1 = -1$

85. $\ln \dfrac{1}{e} = \ln 1 - \ln e = 0 - 1 = -1$

86. $\ln e^4 = 4$

87. $\ln(2x) = 2$
$2x = e^2$
$x = \dfrac{e^2}{2}$
The solution is $\dfrac{e^2}{2}$.

88. $\ln(3x) = 1.6$
$3x = e^{1.6}$
$x = \dfrac{e^{1.6}}{3}$
The solution is $\dfrac{e^{1.6}}{3}$.

89. $\ln(2x-3) = -1$
$$2x - 3 = e^{-1}$$
$$2x = e^{-1} + 3$$
$$x = \frac{e^{-1} + 3}{2}$$
The solution is $\dfrac{e^{-1} + 3}{2}$.

90. $\ln(3x+1) = 2$
$$3x + 1 = e^2$$
$$3x = e^2 - 1$$
$$x = \frac{e^2 - 1}{3}$$
The solution is $\dfrac{e^2 - 1}{3}$.

91. $\ln\dfrac{I}{I_0} = -kx$
$$\ln\frac{0.03 I_0}{I_0} = -2.1x$$
$$\ln 0.03 = -2.1x$$
$$x = \frac{\ln 0.03}{-2.1} \approx 1.67$$
The depth is approximately 1.67 millimeters.

92. $\ln\left(\dfrac{I}{I_0}\right) = -kx$
$$\ln\frac{0.02 I_0}{I_0} = -3.2x$$
$$\ln 0.02 = -3.2x$$
$$x = \frac{\ln 0.02}{-3.2} \approx 1.22$$
The depth is approximately 1.22 millimeters.

93. $\log_5 1.6 = \dfrac{\log 1.6}{\log 5} = 0.2920$

94. $\log_3 4 = \dfrac{\log 4}{\log 3} \approx 1.2619$

95. $A = Pe^{rt}$
$A = 1450e^{(0.06)(5)} \approx 1957.30$
The amount accrued is $1957.30.

96. $A = Pe^{rt}$
$A = 940e^{0.11(3)} \approx 1307.51$
The $940 investment grows to $1307.51 in 3 years.

97. $3^{2x} = 7$
$$2x \log 3 = \log 7$$
$$x = \frac{\log 7}{2 \log 3} \approx 0.8856$$
The solution is $\dfrac{\log 7}{2 \log 3}$, or approximately 0.8856.

98. $6^{3x} = 5$
$$3x \log 6 = \log 5$$
$$x = \frac{\log 5}{3 \log 6} \approx 0.2994$$
The solution is $\dfrac{\log 5}{3 \log 6}$, or approximately 0.2994.

99. $3^{2x+1} = 6$
$$(2x+1)\log 3 = \log 6$$
$$2x + 1 = \frac{\log 6}{\log 3}$$
$$2x = \frac{\log 6}{\log 3} - 1$$
$$x = \frac{1}{2}\left(\frac{\log 6}{\log 3} - 1\right) \approx 0.3155$$
The solution is $\dfrac{1}{2}\left(\dfrac{\log 6}{\log 3} - 1\right)$, or approximately 0.3155.

100.
$$4^{3x+2} = 9$$
$$(3x+2)\log 4 = \log 9$$
$$3x+2 = \frac{\log 9}{\log 4}$$
$$3x = \frac{\log 9}{\log 4} - 2$$
$$x = \frac{1}{3}\left(\frac{\log 9}{\log 4} - 2\right) \approx -0.1383$$

The solution is $\frac{1}{3}\left(\dfrac{\log 9}{\log 4} - 2\right)$, or approximately -0.1383.

101.
$$5^{3x-5} = 4$$
$$(3x-5)\log 5 = \log 4$$
$$3x-5 = \frac{\log 4}{\log 5}$$
$$3x = \frac{\log 4}{\log 5} + 5$$
$$x = \frac{1}{3}\left(\frac{\log 4}{\log 5} + 5\right)$$
$$x \approx 1.9538$$

The solution is $\frac{1}{3}\left(\dfrac{\log 4}{\log 5} + 5\right)$, or approximately 1.9538.

102.
$$8^{4x-2} = 3$$
$$(4x-2)\log 8 = \log 3$$
$$4x-2 = \frac{\log 3}{\log 8}$$
$$4x = \frac{\log 3}{\log 8} + 2$$
$$x = \frac{1}{4}\left(\frac{\log 3}{\log 8} + 2\right)$$
$$x \approx 0.6321$$

The solution is $\frac{1}{4}\left(\dfrac{\log 3}{\log 8} + 2\right)$, or approximately 0.6321.

103.
$$5^{x-1} = 1$$
$$(x-1)\log 5 = \log 1$$
$$x-1 = \frac{\log 1}{\log 5}$$
$$x = \frac{\log 1}{\log 5} + 1$$
$$= 0 + 1$$
$$= 1$$

The solution is $\dfrac{\log 1}{\log 5} + 1$, or 1.

104.
$$4^{x+5} = 2$$
$$(x+5)\log 4 = \log 2$$
$$x+5 = \frac{\log 2}{\log 4}$$
$$x = \frac{\log 2}{\log 4} - 5$$
$$= 0.5 - 5$$
$$= -4.5$$

The solution is $\dfrac{\log 2}{\log 4} - 5$, or -4.5.

105.
$$\log_5 2 + \log_5 x = 2$$
$$\log_5(2x) = 2$$
$$2x = 5^2$$
$$2x = 25$$
$$x = \frac{25}{2}$$

The solution is $\dfrac{25}{2}$.

106.
$$\log_3 x + \log_3 10 = 2$$
$$\log_3(10x) = 2$$
$$10x = 3^2$$
$$10x = 9$$
$$x = \frac{9}{10}$$

The solution is $\dfrac{9}{10}$.

107. $\log(5x) - \log(x+1) = 4$

$$\log \frac{5x}{x+1} = 4$$

$$\frac{5x}{x+1} = 10^4$$

$$\frac{5x}{x+1} = 10,000$$

$$5x = 10,000x + 10,000$$

$$-10,000 = 9995x$$

$$-\frac{10,000}{9995} = x$$

We reject this solution because $5x$ must be positive. The solution is $\varnothing$.

108. $\ln(3x) - \ln(x-3) = 2$

$$\ln\left(\frac{3x}{x-3}\right) = 2$$

$$\frac{3x}{x-3} = e^2$$

$$3x = e^2(x-3)$$

$$3x = e^2 x - 3e^2$$

$$3x - e^2 x = -3e^2$$

$$(3 - e^2)x = -3e^2$$

$$x = \frac{-3e^2}{3 - e^2} = \frac{3e^2}{e^2 - 3}$$

The solution is $\dfrac{3e^2}{e^2 - 3}$.

109. $\log_2 x + \log_2 2x - 3 = 1$

$$\log_2(x \cdot 2x) = 4$$

$$2x^2 = 2^4$$

$$2x^2 = 16$$

$$x^2 = 8$$

$$x = \pm\sqrt{8}$$

$$x = \pm 2\sqrt{2}$$

We reject $-2\sqrt{2}$ because x must be positive. The solution is $2\sqrt{2}$.

110. $-\log_6(4x+7) + \log_6 x = 1$

$$\log_6 \frac{x}{4x+7} = 1$$

$$\frac{x}{4x+7} = 6^1$$

$$\frac{x}{4x+7} = 6$$

$$x = 6(4x+7)$$

$$x = 24x + 42$$

$$-42 = 23x$$

$$x = -\frac{42}{23}$$

We reject $x = -\dfrac{42}{23}$ because x must be positive. The solution is $\varnothing$.

111. $y = y_0 e^{kt}$

$y = 155,000 e^{(0.06)(4)} \approx 197,044$

The expected number of ducks is approximately 197,044.

112. $y = y_0 e^{kt}$

$y = 212,942,000 e^{0.015(8)}$

$\approx 240,091,435$

The expected population of Indonesia by the year 2006 is approximately 240,091,435.

113. $y = y_0 e^{kt}$

$140,000,000 = 125,932,000 e^{0.002t}$

$$t = \frac{\ln \frac{140,000,000}{125,932,000}}{0.002}$$

$$t \approx 53$$

It will take approximately 53 years.

114.

$$y = y_0 e^{kt}$$

$$2(30,675,000) = 30,675,000 e^{0.011t}$$

$$2 = e^{0.011t}$$

$$t = \frac{\ln 2}{0.011}$$

$$t \approx 63$$

It will take approximately 63 years.

115. $2(66,050,000) = 66,050,000e^{(0.019)t}$

$$2 = e^{0.019t}$$

$$t = \frac{\ln 2}{0.019} \approx 36$$

It will take approximately 36 years.

116.

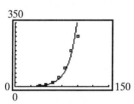

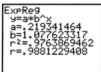

The exponential regression equation is

$y = 0.219(1.078)^x$.

$x = 2005 - 1900 = 105$

```
Y₁(105)
        562.5402481
■
```

The predicted cost is $562.54 billion for 2005.

117. $A = P\left(1 + \frac{r}{n}\right)^{nt}$

$$10,000 = 5,000\left(1 + \frac{0.08}{4}\right)^{4t}$$

$$2 = (1.02)^{4t}$$

$$\log 2 = 4t \log 1.02$$

$$t = \frac{\log 2}{4 \log 1.02} \approx 8.8$$

It will take approximately 8.8 years.

118. $A = P\left(1 + \frac{r}{n}\right)^{nt}$

$$10,000 = 6,000\left(1 + \frac{0.06}{12}\right)^{12t}$$

$$\frac{5}{3} = (1.005)^{12t}$$

$$\log\frac{5}{3} = 12t \log 1.005$$

$$t = \frac{\log\dfrac{5}{3}}{12 \log 1.005} \approx 8.5$$

It was invested for approximately 8.5 years.

119. $y_1 = e^x$, $y_2 = 2$

The solution is approximately 0.69.

120. $y_1 = 10^{0.3x}$, $y_2 = 7$

The solution is approximately 2.82.

Chapter 9 Test

1. $(f \circ h)(0) = f(h(0)) = f(5) = 5$

2. $(g \circ f)(x) = g(f(x)) = g(x) = x - 7$

3. $(g \circ h)(x) = g(h(x))$

$$= g(x^2 - 6x + 5)$$

$$= x^2 - 6x + 5 - 7$$

$$= x^2 - 6x - 2$$

4. $f(x) = 7x - 14$, so $f^{-1}(x) = \dfrac{x + 14}{7}$.

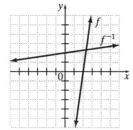

5. The graph is a one-to-one function.

6. The graph does not represent a function, hence does not represent a one-to-one function.

7. $y = 6 - 2x$ is one-to-one.

$$x = 6 - 2y$$
$$2y = -x + 6$$
$$y = \frac{-x + 6}{2}$$
$$f^{-1}(x) = \frac{-x + 6}{2}$$

8. $f = \{(0, 0), (2, 3), (-1, 5)\}$ is one-to-one.
$$f^{-1} = \{(0, 0), (3, 2), (5, -1)\}$$

9. This function is not one-to-one because there are at least 2 inputs that correspond to the same output (dog and desk correspond to d, and cat and circle correspond to c).

10. $\log_3 6 + \log_3 4 = \log_3(6 \cdot 4) = \log_3 24$

11. $\log_5 x + 3\log_5 x - \log_5(x + 1)$

$$= 4\log_5 x - \log_5(x + 1)$$
$$= \log_5 x^4 - \log_5(x + 1)$$
$$= \log_5 \frac{x^4}{x + 1}$$

12. $\log_6 \dfrac{2x}{y^3} = \log_6 2x - \log_6 y^3$

$$= \log_6 2 + \log_6 x - 3\log_6 y$$

13. $\log_b\left(\dfrac{3}{25}\right) = \log_b 3 - \log_b 25$

$$= \log_b 3 - \log_b 5^2$$
$$= \log_b 3 - 2\log_b 5$$
$$= 0.79 - 2(1.16)$$
$$= -1.53$$

14. $\log_7 8 = \dfrac{\ln 8}{\ln 7} \approx 1.0686$

15. $8^{x-1} = \dfrac{1}{64}$

$$8^{x-1} = 8^{-2}$$
$$x - 1 = -2$$
$$x = -1$$
The solution is -1.

16. $\quad 3^{2x+5} = 4$

$$(2x + 5)\log 3 = \log 4$$
$$2x + 5 = \frac{\log 4}{\log 3}$$
$$2x = \frac{\log 4}{\log 3} - 5$$
$$x = \frac{1}{2}\left(\frac{\log 4}{\log 3} - 5\right)$$
$$x \approx -1.8691$$

The solution is $\dfrac{1}{2}\left(\dfrac{\log 4}{\log 3} - 5\right)$, or approximately -1.8691.

17. $\log_3 x = -2$

$$x = 3^{-2}$$
$$x = \frac{1}{9}$$

The solution is $\dfrac{1}{9}$.

18. $\ln \sqrt{e} = x$

$$\ln e^{1/2} = x$$

$$e^x = e^{1/2}$$

$$x = \frac{1}{2}$$

The solution is $\frac{1}{2}$.

19. $\log_8(3x - 2) = 2$

$$3x - 2 = 8^2$$

$$3x - 2 = 64$$

$$3x = 66$$

$$x = \frac{66}{3} = 22$$

The solution is 22.

20. $\log_5 x + \log_5 3 = 2$

$$\log_5(3x) = 2$$

$$3x = 5^2$$

$$3x = 25$$

$$x = \frac{25}{3}$$

The solution is $\frac{25}{3}$.

21. $\log_4(x + 1) - \log_4(x - 2) = 3$

$$\log_4 \frac{x + 1}{x - 2} = 3$$

$$\frac{x + 1}{x - 2} = 4^3$$

$$x + 1 = 64(x - 2)$$

$$x + 1 = 64x - 128$$

$$129 = 63x$$

$$x = \frac{43}{21}$$

The solution is $\frac{43}{21}$.

22. $\ln(3x + 7) = 1.31$

$$3x + 7 = e^{1.31}$$

$$3x = e^{1.31} - 7$$

$$x = \frac{e^{1.31} - 7}{3} \approx -1.0979$$

The solution is -1.0979.

23. $y = \left(\dfrac{1}{2}\right)^x + 1$

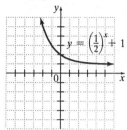

24. $y = 3^x$ and $y = \log_3 x$

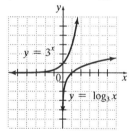

25. $A = P\left(1 + \dfrac{r}{n}\right)^{nt}$, $P = 4000$, $t = 3$, $r = 0.09$, and $n = 12$.

$$A = 4000\left(1 + \frac{0.09}{12}\right)^{12(3)}$$

$$= 4000\left(1 + \frac{0.09}{12}\right)^{12(3)}$$

$$= 4000(1.0075)^{36}$$

$$\approx 5234.58$$

$5234.58 will be in the account.

26. $A = P\left(1 + \dfrac{r}{n}\right)^{nt}$, $P = 2000$, $A = 3000$,

$r = 0.07$, and $n = 2$.

$$3000 = 2000\left(1 + \dfrac{0.07}{2}\right)^{2t}$$

$$1.5 = (1.035)^{2t}$$

$$\log 1.5 = 2t \log 1.035$$

$$t = \dfrac{\log 1.5}{2 \log 1.035} \approx 5.9$$

It would take 6 years for the investment to reach $3000.

27. $y = y_0 e^{kt}$, $y_0 = 57,000$, $k = 0.026$, and $t = 5$.

$$y = 57,000 e^{0.026(5)}$$

$$= 57,000 e^{0.13}$$

$$\approx 64,913$$

There will be approximately 64,913 prairie dogs 5 years from now.

28. $y = y_0 e^{kt}$, $y_0 = 400$, $y = 1000$, and $k = 0.062$.

$$1000 = 400 e^{0.062t}$$

$$2.5 = e^{0.062t}$$

$$0.062t = \ln 2.5$$

$$t = \dfrac{\ln 2.5}{0.062} \approx 14.8$$

It will take the naturalists approximately 15 years to reach their goal.

29. $\log(1 + k) = \dfrac{0.3}{D}$, $D = 56$

$$\log(1 + k) = \dfrac{0.3}{56}$$

$$1 + k = 10^{0.3/56}$$

$$k = -1 + 10^{0.3/56}$$

$$k \approx .012$$

The rate of population increase is approximately 1.2%.

30. $y_1 = e^{0.2x}$

$y_2 = e^{-0.4x} + 2$

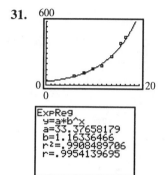

The solution is approximately 3.95.

31.

The exponential regression equation is $y = 33.377(1.163)^x$.

$x = 2000 - 1980 = 20$

Y₁(20)
688.270563

The predicted amount charged is $688.271 billion for 2000.

$x = 2005 - 1980 = 25$

Y₁(25)
1466.690672

The predicted amount charged is $1466.691 billion for 2005.

Chapter 10

Mental Math

1. $y = x^2 - 7x + 5$; upward

2. $y = -x^2 + 16$; downward

3. $x = -y^2 - y + 2$; to the left

4. $x = 3y^2 + 2y - 5$; to the right

5. $y = -x^2 + 2x + 1$; downward

6. $x = -y^2 + 2y - 6$; to the left

Exercise Set 10.1

1. $x = 3y^2$
   $x = 3(y - 0)^2$
   The vertex is (0, 0).

   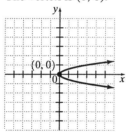

3. $x = (y - 2)^2 + 3$
   The vertex is (3, 2).

   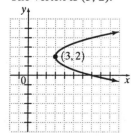

5. $y = 3(x - 1)^2 + 5$
   The vertex is (1, 5).

   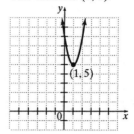

7. $x = y^2 + 6y + 8$
   $x = y^2 + 6y + 9 + 8 - 9$
   $x = (y + 3)^2 - 1$
   The vertex is (−1, −3).

   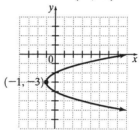

9. $y = x^2 + 10x + 20$
   $y = x^2 + 10x + 25 + 20 - 25$
   $y = (x + 5)^2 - 5$
   The vertex is (−5, −5).

   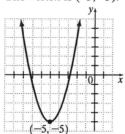

11. $x = -2y^2 + 4y + 6$

$x = -2(y^2 - 2y) + 6$

$x = -2(y^2 - 2y + 1) + 6 + 2$

$x = -2(y-1)^2 + 8$

The vertex is (8, 1).

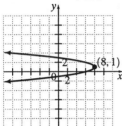

13. (5, 1), (8, 5)

$d = \sqrt{(8-5)^2 + (5-1)^2}$

$= \sqrt{9+16}$

$= \sqrt{25}$

$= 5$

The distance is 5 units.

15. (−3, 2), (1, −3)

$d = \sqrt{[1-(-3)]^2 + (-3-2)^2}$

$= \sqrt{16+25}$

$= \sqrt{41}$

The distance is $\sqrt{41}$ units.

17. (−9, 4), (−8, 1)

$d = \sqrt{[-8-(-9)]^2 + (1-4)^2}$

$= \sqrt{(-8+9)^2 + (-3)^2}$

$= \sqrt{1+9}$

$= \sqrt{10}$

The distance is $\sqrt{10}$ units.

19. $\left(0, -\sqrt{2}\right), \left(\sqrt{3}, 0\right)$

$d = \sqrt{\left(\sqrt{3}-0\right)^2 + \left[0-\left(-\sqrt{2}\right)\right]^2}$

$= \sqrt{\left(\sqrt{3}\right)^2 + \left(\sqrt{2}\right)^2}$

$= \sqrt{3+2}$

$= \sqrt{5}$

The distance is $\sqrt{5}$ units.

21. (1.7, −3.6), (−8.6, 5.7)

$d = \sqrt{(-8.6-1.7)^2 + [5.7-(-3.6)]^2}$

$= \sqrt{(-10.3)^2 + (9.3)^2}$

$= \sqrt{192.58} \approx 13.88$

The distance is exactly $\sqrt{192.58}$ units or approximately 13.88 units.

23. $\left(2\sqrt{3}, \sqrt{6}\right), \left(-\sqrt{3}, 4\sqrt{6}\right)$

$d = \sqrt{\left(-\sqrt{3}-2\sqrt{3}\right)^2 + \left(4\sqrt{6}-\sqrt{6}\right)^2}$

$= \sqrt{\left(-3\sqrt{3}\right)^2 + \left(3\sqrt{6}\right)^2}$

$= \sqrt{27+54}$

$= \sqrt{81}$

$= 9$

The distance is 9 units.

25. (6, −8), (2, 4)

$\left(\dfrac{6+2}{2}, \dfrac{-8+4}{2}\right)$

The midpoint of the segment is (4, −2).

27. (−2, −1), (−8, 6)

$\left(\dfrac{-2+(-8)}{2}, \dfrac{-1+6}{2}\right)$

The midpoint of the segment is $\left(-5, \dfrac{5}{2}\right)$.

29. (7, 3), (−1, −3)

$\left(\dfrac{7+(-1)}{2}, \dfrac{3+(-3)}{2}\right)$

The midpoint of the segment is (3, 0).

31. $\left(\dfrac{1}{2},\dfrac{3}{8}\right), \left(-\dfrac{3}{2},\dfrac{5}{8}\right)$

$\left(\dfrac{\frac{1}{2}+\left(-\frac{3}{2}\right)}{2},\dfrac{\frac{3}{8}+\frac{5}{8}}{2}\right)$

The midpoint of the segment is $\left(-\dfrac{1}{2},\dfrac{1}{2}\right)$.

33. $\left(\sqrt{2},3\sqrt{5}\right), \left(\sqrt{2},-2\sqrt{5}\right)$

$\left(\dfrac{\sqrt{2}+\sqrt{2}}{2},\dfrac{3\sqrt{5}-2\sqrt{5}}{2}\right)$

The midpoint of the segment is

$\left(\sqrt{2},\dfrac{\sqrt{5}}{2}\right)$.

35. $(4.6, -3.5), (7.8, -9.8)$

$\left(\dfrac{4.6+7.8}{2},\dfrac{-3.5+(-9.8)}{2}\right)$

The midpoint of the segment is
$(6.2, -6.65)$.

37. $x^2 + y^2 = 9$

$(x-0)^2 + (y-0)^2 = 3^2$

The center is $(0, 0)$ and the radius is 3.

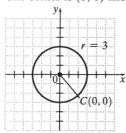

39. $x^2 + (y-2)^2 = 1$

$(x-0)^2 + (y-2)^2 = 1^2$

The center is $(0, 2)$ and the radius is 1.

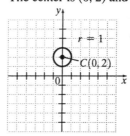

41. $(x-5)^2 + (y+2)^2 = 1$

$(x-5)^2 + (y+2)^2 = 1^2$

The center is $(5, -2)$ and the radius is 1.

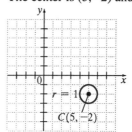

43. $x^2 + y^2 + 6y = 0$

$x^2 + y^2 + 6y + 9 = 9$

$(x-0)^2 + (y+3)^2 = 3^2$

The center is $(0, -3)$ and the radius is 3.

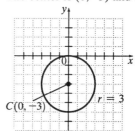

45.
$$x^2 + y^2 + 2x - 4y = 4$$
$$x^2 + 2x + 1 + y^2 - 4y + 4 = 4 + 1 + 4$$
$$(x+1)^2 + (y-2)^2 = 9$$
$$(x+1)^2 + (y-2)^2 = 3^2$$
The center is (–1, 2) and the radius is 3.

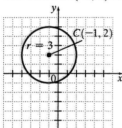

47.
$$x^2 + y^2 - 4x - 8y - 2 = 0$$
$$(x^2 - 4x + 4) + (y^2 - 8y + 16) = 2 + 4 + 16$$
$$(x-2)^2 + (y-4)^2 = 22$$
The center is (2, 4) and the radius is $\sqrt{22}$.

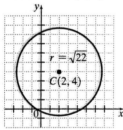

49. The center is (2, 3); the radius is 6.
$$(x-2)^2 + (y-3)^2 = 6^2$$
$$(x-2)^2 + (y-3)^2 = 36$$

51. The center is (0, 0); the radius is $\sqrt{3}$.
$$(x-0)^2 + (y-0)^2 = \left(\sqrt{3}\right)^2$$
$$x^2 + y^2 = 3$$

53. The center is (–5, 4); the radius is $3\sqrt{5}$.
$$[x-(-5)]^2 + (y-4)^2 = \left(3\sqrt{5}\right)^2$$
$$(x+5)^2 + (y-4)^2 = 45$$

55. Answers may vary.

57. $x = y^2 - 3$
The vertex is (–3, 0).

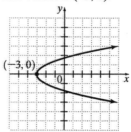

59. $y = (x-2)^2 - 2$
The vertex is (2, –2).

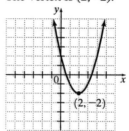

61. $x^2 + y^2 = 1$
The center is (0, 0); the radius is 1.

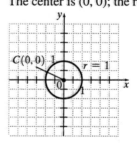

63. $x = (y+3)^2 - 1$
The vertex is (–1, –3).

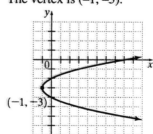

65. $(x-2)^2 + (y-2)^2 = 16$

The center is (2, 2); the radius is 4.

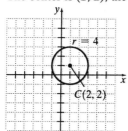

67. $x = -(y-1)^2$

The vertex is (0, 1).

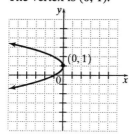

69. $(x-4)^2 + y^2 = 7$

The center is $(4,0)$; the radius is $\sqrt{7}$.

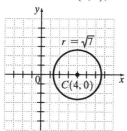

71. $y = 5(x+5)^2 + 3$

The vertex is (–5, 3).

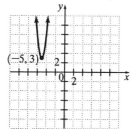

73. $\dfrac{x^2}{8} + \dfrac{y^2}{8} = 2$

$x^2 + y^2 = 16$

The center is (0, 0); the radius is 4.

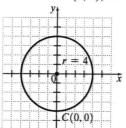

75. $y = x^2 + 7x + 6$

$y = x^2 + 7x + \dfrac{49}{4} + 6 - \dfrac{49}{4}$

$y = \left(x + \dfrac{7}{2}\right)^2 - \dfrac{25}{4}$

The vertex is $\left(-\dfrac{7}{2}, -\dfrac{25}{4}\right)$.

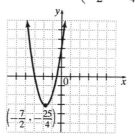

77. $\qquad x^2 + y^2 + 2x + 12y - 12 = 0$

$x^2 + 2x + 1 + y^2 + 12y + 36 = 12 + 1 + 36$

$\qquad (x+1)^2 + (y+6)^2 = 49$

The center is (–1, –6); the radius is 7.

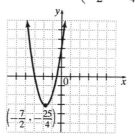

79. $x = y^2 + 8y - 4$

$x = y^2 + 8y + 16 - 4 - 16$

$x = (y + 4)^2 - 20$

The vertex is $(-20, -4)$.

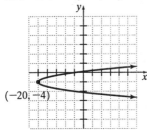

81. $x^2 - 10y + y^2 + 4 = 0$

$x^2 + y^2 - 10y + 25 = -4 + 25$

$x^2 + (y - 5)^2 = 21$

The center is $(0, 5)$; the radius is $\sqrt{21}$.

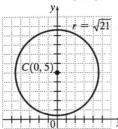

83. $x = -3y^2 + 30y$

$x = -3(y^2 - 10y)$

$x = -3(y^2 - 10y + 25) + 75$

$x = -3(y - 5)^2 + 75$

The vertex is $(75, 5)$.

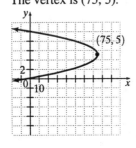

85. $5x^2 + 5y^2 = 25$

$x^2 + y^2 = 5$

The center is $(0, 0)$; the radius is $\sqrt{5}$.

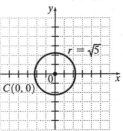

87. $y = 5x^2 - 20x + 16$

$y = 5(x^2 - 4x) + 16$

$y = 5(x^2 - 4x + 4) + 16 - 20$

$y = 5(x - 2)^2 - 4$

The vertex is $(2, -4)$.

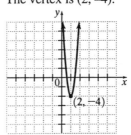

89. $x^2 + y^2 = 20$

Graph $y_1 = \sqrt{20 - x^2}$ and $y_2 = -y_1$

91. $x = -7 - 6y - y^2$

$y^2 + 6y + (x + 7) = 0$

$y = \dfrac{-6 \pm \sqrt{6^2 - 4(x + 7)}}{2}$

Set $y_1 = 36 - 4(x + 7)$

Graph $y_2 = \dfrac{-6 + \sqrt{y_1}}{2}$ and $y_3 = \dfrac{-6 - \sqrt{y_1}}{2}$

93. $(x + 3)^2 + (y - 1)^2 = 15$

Set $y_1 = \sqrt{15 - (x + 3)^2}$

Graph $y_2 = y_1 + 1$ and $y_3 = -y_1 + 1$

95. $x = 9y^2 - 6y + 4$

$9y^2 - 6y + (4 - x) = 0$

$y = \dfrac{6 \pm \sqrt{(-6)^2 - 4(9)(4 - x)}}{2(9)}$

Set $y_1 = \sqrt{36 - 36(4 - x)}$

Graph $y_2 = \dfrac{6 + y_1}{18}$ and $y_3 = \dfrac{6 - y_1}{18}$

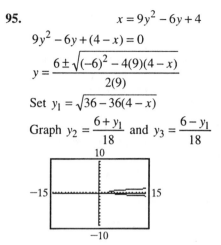

97. The distance between the points (5, 1) and (2, 6) is

$d = \sqrt{(5 - 2)^2 + (1 - 6)^2}$

$ = \sqrt{3^2 + (-5)^2}$

$ = \sqrt{9 + 25}$

$ = \sqrt{34}$

The distance between the points (5, 1) and (0, –2) is

$d = \sqrt{(5 - 0)^2 + [1 - (-2)]^2}$

$ = \sqrt{5^2 + 3^2}$

$ = \sqrt{25 + 9}$

$ = \sqrt{34}$

Therefore, the triangle with vertices (2, 6), (0, –2) and (5, 1) is an isosceles triangle.

99. Setting up a coordinate system with the axis of symmetry as the y-axis and the base of the bridge on the x-axis. The parabola would pass through the points $(-50, 0)$, $(0, 40)$, and $(50, 0)$.

Use the equation for a parabola:

$y = ax^2 + bx + c$.

Substituting the x- and y-coordinates for the known points yields a system of equations.

$$\begin{cases} 0 = a(-50)^2 + b(-50) + c \\ 40 = a(0)^2 + b(0) + c \\ 0 = a(50)^2 + b(50) + c \end{cases}$$

$$\begin{cases} 0 = 2500a - 50b + c \\ 40 = c \\ 0 = 2500a + 50b + c \end{cases}$$

Replace c with 40.

$$\begin{cases} 0 = 2500a - 50b + 40 \\ 0 = 2500a + 50b + 40 \end{cases}$$

Add the two equations.

$$0 = 5000a + 80$$

$$-80 = 5000a$$

$$\frac{-80}{5000} = a$$

$$-\frac{2}{125} = a$$

$$0 = 2500\left(-\frac{2}{125}\right) - 50b + 40$$

$$0 = -40 - 50b + 40$$

$$0 = -50b$$

$$0 = b$$

Thus, the equation of the parabola is

$$y = -\frac{2}{125}x^2 + 40.$$

101. $y = -3x + 3$

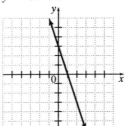

103. $x = -2$

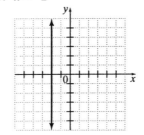

105.

$$\frac{\sqrt{5}}{\sqrt{8}} = \frac{\sqrt{5}}{2\sqrt{2}}$$

$$= \frac{\sqrt{5} \cdot \sqrt{2}}{2\sqrt{2} \cdot \sqrt{2}}$$

$$= \frac{\sqrt{10}}{2\sqrt{4}}$$

$$= \frac{\sqrt{10}}{2(2)}$$

$$= \frac{\sqrt{10}}{4}$$

107.

$$\frac{10}{\sqrt{5}} = \frac{10 \cdot \sqrt{5}}{\sqrt{5} \cdot \sqrt{5}}$$

$$= \frac{10\sqrt{5}}{\sqrt{25}}$$

$$= \frac{10\sqrt{5}}{5}$$

$$= 2\sqrt{5}$$

Exercise Set 10.2

1. $\dfrac{x^2}{4} + \dfrac{y^2}{25} = 1$

$\dfrac{x^2}{2^2} + \dfrac{y^2}{5^2} = 1$

The center is (0, 0).
x-intercepts: $(-2, 0)$ and $(2, 0)$
y-intercepts: $(0, -5)$ and $(5, 0)$

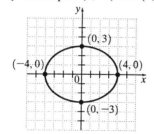

3. $\dfrac{x^2}{16} + \dfrac{y^2}{9} = 1$

$\dfrac{x^2}{4^2} + \dfrac{y^2}{3^2} = 1$

The center is (0, 0).
x-intercepts: $(-4, 0)$ and $(4, 0)$
y-intercepts: $(0, -3)$ and $(0, 3)$

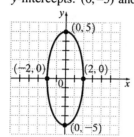

5. $9x^2 + 4y^2 = 36$

$\dfrac{x^2}{4} + \dfrac{y^2}{9} = 1$

$\dfrac{x^2}{2^2} + \dfrac{y^2}{3^2} = 1$

The center is (0, 0).
x-intercepts: $(-2, 0)$ and $(2, 0)$
y-intercepts: $(0, -3)$ and $(0, 3)$

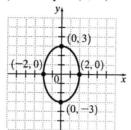

7. $4x^2 + 25y^2 = 100$

$\dfrac{x^2}{25} + \dfrac{y^2}{4} = 1$

$\dfrac{x^2}{5^2} + \dfrac{y^2}{2^2} = 1$

The center is (0, 0).
x-intercepts: $(-5, 0)$ and $(5, 0)$
y-intercepts: $(0, -2)$ and $(0, 2)$

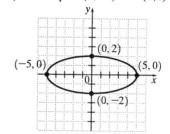

9. $\dfrac{(x+1)^2}{36} + \dfrac{(y-2)^2}{49} = 1$

$\dfrac{(x+1)^2}{6^2} + \dfrac{(y-2)^2}{7^2} = 1$

The center is $(-1, 2)$.
Other points:
$(-1 - 6, 2)$ or $(-7, 2)$
$(-1 + 6, 2)$ or $(5, 2)$
$(-1, 2 - 7)$ or $(-1, -5)$
$(-1, 2 + 7)$ or $(-1, 9)$

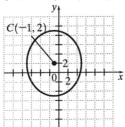

11. $\dfrac{(x-1)^2}{4} + \dfrac{(y-1)^2}{25} = 1$

$\dfrac{(x-1)^2}{2^2} + \dfrac{(y-1)^2}{5^2} = 1$

The center is $(1, 1)$.
Other points:
$(1 - 2, 1)$ or $(-1, 1)$
$(1 + 2, 1)$ or $(3, 1)$
$(1, 1 - 5)$ or $(1, -4)$
$(1, 1 + 5)$ or $(1, 6)$

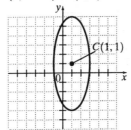

13. $\dfrac{x^2}{4} - \dfrac{y^2}{9} = 1$

$\dfrac{x^2}{2^2} - \dfrac{y^2}{3^2} = 1$

$a = 2, b = 3$

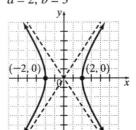

15. $\dfrac{y^2}{25} - \dfrac{x^2}{16} = 1$

$\dfrac{y^2}{5^2} - \dfrac{x^2}{4^2} = 1$

$a = 4, b = 5$

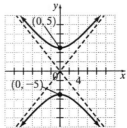

17. $x^2 - 4y^2 = 16$

$\dfrac{x^2}{16} - \dfrac{y^2}{4} = 1$

$\dfrac{x^2}{4^2} - \dfrac{y^2}{2^2} = 1$

$a = 4, b = 2$

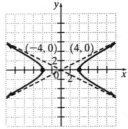

19. $16y^2 - x^2 = 16$

$$\frac{y^2}{1} - \frac{x^2}{16} = 1$$

$$\frac{y^2}{1^2} - \frac{x^2}{4^2} = 1$$

$a = 4, b = 1$

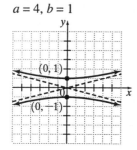

21. Answers may vary.

23. $y = x^2 + 4$

parabola: The vertex is $(0, 4)$.
opens upward

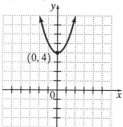

25. $\dfrac{x^2}{25} + \dfrac{y^2}{9} = 1$

ellipse: center $(0, 0)$
$a = 5, b = 3$
x-intercepts: $(5, 0)$ and $(-5, 0)$
y-intercepts: $(0, 3)$ and $(0, -3)$

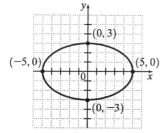

27. $\dfrac{x^2}{25} - \dfrac{y^2}{4} = 1$

hyperbola: center $(0, 0)$
$a = 5, b = 2$
x-intercepts: $(5, 0)$ and $(-5, 0)$

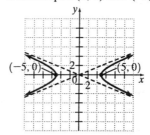

29. $x^2 + y^2 = 16$

circle: center $(0, 0)$
radius $= 4$

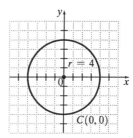

31. $x = -y^2 + 6y$

parabola: $\dfrac{-b}{2a} = \dfrac{-6}{2(-1)} = 3$

$x = -9 + 18 = 9$
The vertex is $(9, 3)$.
opens to the left

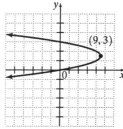

33. $4x^2 + 9y^2 = 36$

$$\frac{x^2}{9} + \frac{y^2}{4} = 1$$

ellipse: center $(0, 0)$
$a = 3$, $b = 2$
x-intercepts at: $(3, 0)$ and $(-3, 0)$
y-intercepts: $(0, 2)$ and $(0, -2)$

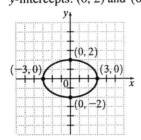

35. $y^2 = x^2 + 16$

$$\frac{y^2}{16} - \frac{x^2}{16} = 1$$

hyperbola: center $(0, 0)$
$a = 4$, $b = 4$
y-intercepts: $(4, 0)$ and $(-4, 0)$

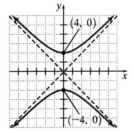

37. $y = -2x^2 + 4x - 3$

parabola: $\dfrac{-b}{2a} = \dfrac{-4}{-4} = 1$

$y = -2 + 4 - 3 = -1$
The vertex is $(1, -1)$.
opens downward

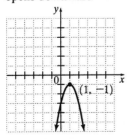

39. $20x^2 + 5y^2 = 100$

Graph $y_1 = \sqrt{\dfrac{100 - 20x^2}{5}} = 2\sqrt{5 - x^2}$

and $y_2 = -y_1$

41. $7y^2 - 3x^2 = 21$

Graph $y_1 = \sqrt{\dfrac{21 + 3x^2}{7}}$

and $y_2 = -y_1$

43. $18.8x^2 + 36.1y^2 = 205.8$

Graph $y_1 = \sqrt{\dfrac{205.8 - 18.8x^2}{36.1}}$

and $y_2 = -y_1$

45. $4.5x^2 - 6.7y^2 = 50.7$

Graph $y_1 = \sqrt{\dfrac{4.5x^2 - 50.7}{6.7}}$ and $y_2 = -y_1$

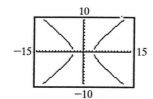

47. A: $a^2 = 36$, $b^2 = 13$
B: $a^2 = 4$, $b^2 = 4$
C: $a^2 = 25$, $b^2 = 16$
D: $a^2 = 39$, $b^2 = 25$
E: $a^2 = 17$, $b^2 = 81$
F: $a^2 = 36$, $b^2 = 36$
G: $a^2 = 16$, $b^2 = 65$
H: $a^2 = 144$, $b^2 = 140$

49. A: $d = \sqrt{36} = 6$
B: $d = \sqrt{4} = 2$
C: $d = \sqrt{25} = 5$
D: $d = \sqrt{25} = 5$
E: $d = \sqrt{81} = 9$
F: $d = \sqrt{36} = 6$
G: $d = \sqrt{16} = 4$
H: $d = \sqrt{144} = 12$

51. They are greater than 0 and less than 1.

53. They are greater than 1.

55. $\dfrac{(x - 1,782,000,000)^2}{3.42 \times 10^{23}} + \dfrac{(y - 356,400,000)^2}{1.368 \times 10^{22}} = 1$
center $(1,782,000,000,\ 356,400,000)$

57. $x < 5$ or $x < 1$
The solution is $(-\infty, \ 5)$.

59. $2x - 1 \geq 7$ and $-3x \leq -6$
$\qquad x \geq 4$ and $\qquad x \geq 2$
The solution is $[4, \ \infty)$.

61. $2x^3 - 4x^3 = (2 - 4)x^3 = -2x^3$

63. $(-5x^2)(x^2) = (-5)(1)x^{2+2} = -5x^4$

65. $\dfrac{(x+2)^2}{9} - \dfrac{(y-1)^2}{4} = 1$
center: $(-2, 1)$
$a = 3, b = 2$

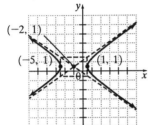

67. $\dfrac{(y+4)^2}{4} - \dfrac{x^2}{25} = 1$
center $(0, -4)$
$a = 5, b = 2$

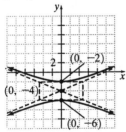

69. $\dfrac{(x-3)^2}{9} - \dfrac{(y-2)^2}{4} = 1$
center: $(3, 2)$
$a = 3, b = 2$

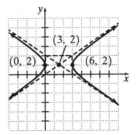

Exercise Set 10.3

1. $\begin{cases} x^2 + y^2 = 25 \\ 4x + 3y = 0 \end{cases}$
Solve equation 2 for y.
$3y = -4x$
$\quad y = \dfrac{-4x}{3}$
Substitute.
$$x^2 + \left(-\dfrac{4x}{3}\right)^2 = 25$$
$$x^2 + \dfrac{16x^2}{9} = 25$$
$$\dfrac{25}{9}x^2 = 25$$
$$\dfrac{x^2}{9} = 1$$
$$x^2 = 9$$
$$x = \pm\sqrt{9} = \pm 3$$
$x = 3: \ y = -\dfrac{4}{3}(3) = -4$
$x = -3: \ y = -\dfrac{4}{3}(-3) = 4$
The solutions are $(3, -4)$ and $(-3, 4)$.

3. $\begin{cases} x^2 + 4y^2 = 10 \\ y = x \end{cases}$

Substitute.

$x^2 + 4x^2 = 10$

$\qquad 5x^2 = 10$

$\qquad x^2 = 2$

$\qquad x = \pm\sqrt{2}$

$x = \sqrt{2};\ y = \sqrt{2}$

$x = -\sqrt{2};\ y = -\sqrt{2}$

The solutions are

$\left(\sqrt{2},\ \sqrt{2}\right)$ and $\left(-\sqrt{2},\ -\sqrt{2}\right)$.

5. $\begin{cases} y^2 = 4 - x \\ x - 2y = 4 \end{cases}$

$-2y = 4 - x$

Substitute.

$\qquad y^2 = -2y$

$y^2 + 2y = 0$

$y(y + 2) = 0$

$y = 0 \quad$ or $\quad y + 2 = 0$

$\qquad\qquad\qquad y = -2$

$y = 0:\ x - 2(0) = 4$

$\qquad\qquad x = 4$

$y = -2:\ x - 2(-2) = 4$

$\qquad\qquad x + 4 = 4$

$\qquad\qquad x = 0$

The solutions are (4, 0) and (0, −2).

7. $\begin{cases} x^2 + y^2 = 9 \\ 16x^2 - 4y^2 = 64 \end{cases}$

Divide equation two by 4.

$\begin{cases} x^2 + y^2 = 9 \\ 4x^2 - y^2 = 16 \end{cases}$

Add.

$5x^2 = 25$

$\quad x^2 = 5$

$\quad x = \pm\sqrt{5}$

Substitute back.

$5 + y^2 = 9$

$\qquad y^2 = 4$

$\qquad y = \pm 2$

The solutions are

$\left(\sqrt{5},\ 2\right),\left(\sqrt{5},\ -2\right),\left(-\sqrt{5},\ 2\right)$, and

$\left(-\sqrt{5},\ -2\right)$.

9. $\begin{cases} x^2 + 2y^2 = 2 \\ x - y = 2 \end{cases}$

$x = y + 2$

Substitute.

$\qquad (y + 2)^2 + 2y^2 = 2$

$y^2 + 4y + 4 + 2y^2 = 2$

$\qquad\quad 3y^2 + 4y + 4 = 2$

$\qquad\quad 3y^2 + 4y + 2 = 0$

$b^2 - 4ac = 4^2 - 4(3)(2) = 16 - 24 = -8 < 0$

Therefore, no real solutions exist.

The solution is $\varnothing$.

11. $\begin{cases} y = x^2 - 3 \\ 4x - y = 6 \end{cases}$

Substitute.

$4x - (x^2 - 3) = 6$

$4x - x^2 + 3 = 6$

$0 = x^2 - 4x + 3$

$0 = (x - 3)(x - 1)$

$x - 3 = 0 \quad \text{or} \quad x - 1 = 0$

$x = 3 \quad \text{or} \qquad x = 1$

$x = 3: \ y = 3^2 - 3 = 9 - 3 = 6$

$x = 1: \ y = 1^2 - 3 = 1 - 3 = -2$

The solutions are $(3, 6)$ and $(1, -2)$.

13. $\begin{cases} y = x^2 \\ 3x + y = 10 \end{cases}$

Substitute.

$3x + x^2 = 10$

$x^2 + 3x - 10 = 0$

$(x + 5)(x - 2) = 0$

$x + 5 = 0 \quad \text{or} \quad x - 2 = 0$

$x = -5 \quad \text{or} \qquad x = 2$

$x = -5: \ y = (-5)^2 = 25$

$x = 2: \ y = 2^2 = 4$

The solutions are $(-5, 25)$ and $(2, 4)$.

15. $\begin{cases} y = 2x^2 + 1 \\ x + y = -1 \end{cases}$

Substitute.

$x + 2x^2 + 1 = -1$

$2x^2 + x + 1 = -1$

$2x^2 + x + 2 = 0$

$b^2 - 4ac = 1^2 - 4(2)(2) = -15 < 0$

Therefore, no real solutions exist.

The solution is $\varnothing$.

17. $\begin{cases} y = x^2 - 4 \\ y = x^2 - 4x \end{cases}$

Substitute.

$x^2 - 4 = x^2 - 4x$

$-4 = -4x$

$x = 1$

$y = 1^2 - 4 = -3$

The solution is $(1, -3)$.

19. $\begin{cases} 2x^2 + 3y^2 = 14 \\ -x^2 + y^2 = 3 \end{cases}$

$y^2 = x^2 + 3$

Substitute.

$2x^2 + 3(x^2 + 3) = 14$

$2x^2 + 3x^2 + 9 = 14$

$5x^2 + 9 = 14$

$5x^2 = 5$

$x^2 = 1$

$x = \pm 1$

Substitute back.

$y^2 = 1 + 3$

$y^2 = 4$

$y = \pm 2$

The solutions are $(1, -2)$, $(1, 2)$, $(-1, -2)$, and $(-1, 2)$.

21. $\begin{cases} x^2 + y^2 = 1 \\ x^2 + (y + 3)^2 = 4 \end{cases}$

Subtract equation 1 from equation 2.

$(y + 3)^2 - y^2 = 3$

$y^2 + 6y + 9 - y^2 = 3$

$6y + 9 = 3$

$6y = -6$

$y = -1$

Substitute back.

$x^2 + (-1)^2 = 1$

$x^2 + 1 = 1$

$x^2 = 0$

$x = 0$

The solution is $(0, -1)$.

23. $\begin{cases} y = x^2 + 2 \\ y = -x^2 + 4 \end{cases}$

Substitute.

$x^2 + 2 = -x^2 + 4$

$2x^2 = 2$

$x^2 = 1$

$x = \pm 1$

Substitute back.

$y = 1 + 2 = 3$

The solutions are $(1, 3)$ and $(-1, 3)$.

25. $\begin{cases} 3x^2 + y^2 = 9 \\ 3x^2 - y^2 = 9 \end{cases}$

Subtract.

$2y^2 = 0$

$y^2 = 0$

$y = 0$

Substitute back.

$3x^2 + 0 = 9$

$3x^2 = 9$

$x^2 = 3$

$x = \pm\sqrt{3}$

The solutions are $\left(\sqrt{3}, 0\right)$ and $\left(-\sqrt{3}, 0\right)$.

27. $\begin{cases} x^2 + 3y^2 = 6 \\ x^2 - 3y^2 = 10 \end{cases}$

Add.

$2x^2 = 16$

$x^2 = 8$

$x = \pm\sqrt{8}$

$x = \pm 2\sqrt{2}$

Substitute back.

$8 + 3y^2 = 6$

$3y^2 = -2$

$y^2 = -\dfrac{2}{3}$

Therefore, no real solutions exist.
The solution is $\varnothing$.

29. $\begin{cases} x^2 + y^2 = 36 \\ y = \dfrac{1}{6}x^2 - 6 \end{cases}$

$y + 6 = \dfrac{1}{6}x^2$

$x^2 = 6(y + 6)$

Substitute.

$6(y + 6) + y^2 = 36$

$6y + 36 + y^2 = 36$

$6y + y^2 = 0$

$y(6 + y) = 0$

$y = 0$　or　$6 + y = 0$

$\qquad\qquad\qquad y = -6$

$y = 0:\ x^2 + 0^2 = 36$

$\qquad\qquad x^2 = 36$

$\qquad\qquad x = \pm 6$

$y = -6:\ x^2 + (-6)^2 = 36$

$\qquad\qquad x^2 + 36 = 36$

$\qquad\qquad x^2 = 0$

$\qquad\qquad x = 0$

The solutions are $(6, 0)$, $(-6, 0)$, and $(0, -6)$.

31. There can be 0, 1, 2, 3, or 4 real solutions. For the circle $x^2 + y^2 = 9$:

the parabola $y = -x^2 - 4$ does not intersect it.

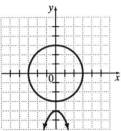

the parabola $y = x^2 + 3$ intersects it once.

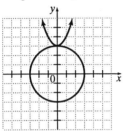

the parabola $y = x^2$ intersects it 2 times.

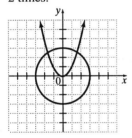

the parabola $y = x^2 - 3$ intersects it 3 times.

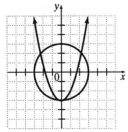

the parabola $y = x^2 - 4$ intersects it 4 times.

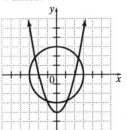

33. Let x and y be the numbers.
$$\begin{cases} x^2 + y^2 = 130 \\ x^2 - y^2 = 32 \end{cases}$$
Add.
$$2x^2 = 162$$
$$x^2 = 81$$
$$x = \pm 9$$
Substitute back.
$$\begin{array}{ll} 9^2 + y^2 = 130 & (-9)^2 + y^2 = 130 \\ \quad y^2 = 49 & \quad y^2 = 49 \\ \quad y = \pm 7 & \quad y = \pm 7 \end{array}$$
The numbers are 9 and 7, or 9 and –7, or –9 and 7, or –9 and –7.

35. Let x and y be the length and width of the keypad.
$$\begin{cases} 2x + 2y = 68 \\ \quad xy = 285 \end{cases}$$
Substitute $y = \dfrac{285}{x}$.
$$2x + 2\left(\frac{285}{x}\right) = 68$$
$$x + \frac{285}{x} = 34$$
$$x^2 - 34x + 285 = 0$$
$$(x - 19)(x - 15) = 0$$
$$x = 19 \text{ or } x = 15$$
Substitute back.
$$\begin{array}{ll} 2(19) + 2y = 68 & 2(15) + 2y = 68 \\ \quad 38 + 2y = 68 & \quad 30 + 2y = 68 \\ \quad\quad 2y = 30 & \quad\quad 2y = 38 \\ \quad\quad\quad y = 15 & \quad\quad\quad y = 19 \end{array}$$
The dimensions are 19 centimeters by 15 centimeters.

37. $p = -0.01x^2 - 0.2x + 9$

$p = 0.01x^2 - 0.1x + 3$

$-0.01x^2 - 0.2x + 9 = 0.01x^2 - 0.1x + 3$

$$0 = 0.02x^2 + 0.1x - 6$$

$$0 = x^2 + 5x - 300$$

$$0 = (x + 20)(x - 15)$$

$x + 20 = 0 \quad$ or $\quad x - 15 = 0$

$x = -20 \quad$ or $\quad x = 15$

Disregard the negative.

$p = -0.01(15)^2 - 0.2(15) + 9$

$p = 3.75$

The equilibrium quantity is 15 thousand compact discs and the price is \$3.75.

39. $\begin{cases} x^2 + 4y^2 = 10 \\ \quad\quad y = x \end{cases}$

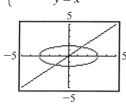

41. $\begin{cases} y = x^2 + 2 \\ y = -x^2 + 4 \end{cases}$

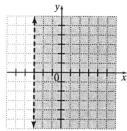

43. $x > -3$

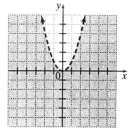

45. $y < 2x - 1$

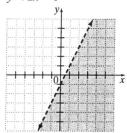

47. $P = x + (2x - 5) + (5x - 20) = 8x - 25$

The perimeter is $(8x - 25)$ inches.

49. $P = 2(x^2 + 3x + 1) + 2x^2$

$P = 2x^2 + 6x + 2 + 2x^2 = 4x^2 + 6x + 2$

The perimeter is $(4x^2 + 6x + 2)$ meters.

Exercise Set 10.4

1. $y < x^2$

First, graph the parabola with a dashed line.

Does $(0, 1)$ satisfy the inequality?

$1 < 0^2$

$1 < 0$ false

Shade portion of the graph which does not contain $(0, 1)$.

3. $x^2 + y^2 \geq 16$

First, graph the circle with a solid line.
Does (0, 0) satisfy the inequality?

$0^2 + 0^2 \geq 16$

$0 \geq 16$ false

Shade the portion of the graph which does not contain (0, 0).

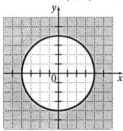

5. $\dfrac{x^2}{4} - y^2 < 1$

First, graph the hyperbola with a dashed line.
Does (–4, 0) satisfy the inequality?

$\dfrac{(-4)^2}{4} - 0^2 < 1$

$4 < 1$ false

Does (0, 0) satisfy the inequality?

$\dfrac{0^2}{4} - 0^2 < 1$

$0 < 1$ true

Does (4, 0) satisfy the inequality?

$\dfrac{4^2}{4} - 0^2 < 1$

$4 < 1$ false

Shade the portion of the graph containing (0, 0).

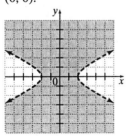

7. $y > (x-1)^2 - 3$

First, graph the parabola with a dashed line.
Does (0, 0) satisfy the inequality?

$0 > (0-1)^2 - 3$

$0 > -2$ true

Shade the portion of the graph containing (0, 0).

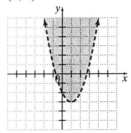

9. $x^2 + y^2 \leq 9$

First, graph the circle with a solid line.
Does (0, 0) satisfy the inequality?

$0^2 + 0^2 \leq 9$

$0 \leq 9$ true

Shade the portion of the graph containing (0, 0).

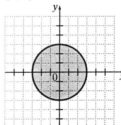

11. $y > -x^2 + 5$

First, graph the parabola with a dashed line. Does (0, 0) satisfy the inequality?

$0 > -0^2 + 5$

$0 > 5$ false

Shade the portion of the graph which does not contain (0, 0).

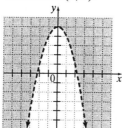

13. $\dfrac{x^2}{4} + \dfrac{y^2}{9} \le 1$

First, graph the ellipse with a solid line. Does (0, 0) satisfy the inequality?

$\dfrac{0^2}{4} + \dfrac{0^2}{9} \le 1$

$0 \le 1$ true

Shade the portion of the graph containing (0, 0).

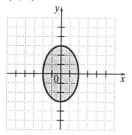

15. $\dfrac{y^2}{4} - x^2 \le 1$

First, graph the hyperbola with a solid line. Does (0, –4) satisfy the inequality?

$\dfrac{(-4)^2}{4} - 0 \le 1$

$4 \le 1$ false

Does (0, 0) satisfy the inequality?

$\dfrac{0^2}{4} - 0 \le 1$

$0 \le 1$ true

Does (0, 4) satisfy the inequality?

$\dfrac{4^2}{4} - 0 \le 1$

$4 \le 1$ false

Shade the portion of the graph containing (0, 0).

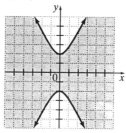

17. $y < (x - 2)^2 + 1$

Does (2, 0) satisfy the inequality?

$0 < (2 - 2)^2 + 1$

$0 < 1$ true

Shade the portion of the graph containing (2, 0).

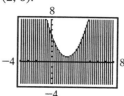

19. $y \le x^2 + x - 2$

Does (0, 0) satisfy the inequality?

$0 \le 0^2 + 0 - 2$

$0 \le -2$ false

Shade the portion of the graph which does not contain (0, 0).

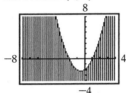

21. Answers may vary.

23. $\begin{cases} 2x - y < 2 \\ \quad y \le -x^2 \end{cases}$

First, graph $2x - y = 2$ with a dashed line.

Does (0, 0) satisfy $2x - y < 2$?

$2(0) - 0 < 2$

$0 < 2$ true

Shade the portion of the graph containing (0, 0).

Next, graph the parabola $y = -x^2$ with a solid line.

Does (0, 1) satisfy $y \le -x^2$?

$1 \le -0^2$

$1 \le 0$ false

Shade the portion of the graph which does not contain (0, 1).

The solution to the system is the overlapping region.

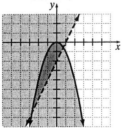

25. $\begin{cases} 4x + 3y \ge 12 \\ x^2 + y^2 < 16 \end{cases}$

First, graph the circle with a dashed line.

Does (0, 0) satisfy $x^2 + y^2 < 16$?

$0^2 + 0^2 < 16$

$0 < 16$ true

Shade the portion of the graph containing (0, 0).

Next, graph the line with a solid line.

Does (0, 0) satisfy $4x + 3y \ge 12$?

$4(0) + 3(0) \ge 12$

$0 \ge 12$ false

Shade the portion of the graph which does not contain (0, 0).

The solution to the system is the overlapping region.

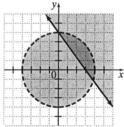

27. $\begin{cases} x^2 + y^2 \le 9 \\ x^2 + y^2 \ge 1 \end{cases}$

First, graph the circle with radius 1 with a solid line.

Does (0, 0) satisfy $x^2 + y^2 \ge 1$?

$0^2 + 0^2 \ge 1$

$\quad 0 \ge 1$ false

Shade the portion of the graph which does not contain (0, 0).

Next, graph the circle with radius 3 with a solid line.

Does (0, 0) satisfy $x^2 + y^2 \le 9$?

$0^2 + 0^2 \le 9$

$\quad 0 \le 9$ true

Shade the portion of the graph containing (0, 0).

The solution to the system is the overlapping region.

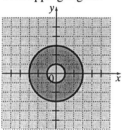

29. $\begin{cases} y > x^2 \\ y \ge 2x + 1 \end{cases}$

First, graph the parabola with a dashed line.

Does (0, 2) satisfy $y > x^2$?

$2 > 0^2$

$2 > 0$ true

Shade the portion of the graph containing (0, 2).

Next, graph the line with a solid line.

Does (0, 0) satisfy $y \ge 2x + 1$?

$0 \ge 2(0) + 1$

$0 \ge 1$ false

Shade the portion of the graph which does not contain (0, 0).

The solution to the system is the overlapping region.

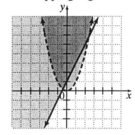

31. $\begin{cases} x > y^2 \\ y > 0 \end{cases}$

First graph the parabola with a dashed line.

Does (2, 0) satisfy $x > y^2$?

$2 > 0^2$

$2 > 0$ true

Shade the portion of the graph containing (2, 0).

Next, graph the line with a dashed line.

Does (0, 1) satisfy $y > 0$?

$1 > 0$ true

Shade the portion of the graph containing (0, 1).

The solution to the system is the overlapping region.

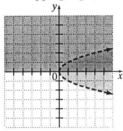

33. $\begin{cases} x^2 + y^2 > 9 \\ y > x^2 \end{cases}$

First, graph the circle with a dashed line.

Does (0, 0) satisfy $x^2 + y^2 > 9$?

$0^2 + 0^2 > 9$

$0 > 9$ false

Shade the portion of the graph which does not contain (0, 0).

Next, graph the parabola with a dashed line.

Does (0, 2) satisfy $y > x^2$?

$2 > 0^2$

$2 > 0$ true

Shade the portion of the graph containing (0, 2).

The solution to the system is the overlapping region.

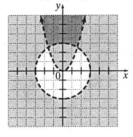

35. $\begin{cases} \dfrac{x^2}{4} + \dfrac{y^2}{9} \geq 1 \\ x^2 + y^2 \geq 4 \end{cases}$

First, graph the ellipse. Does (0, 0) satisfy

$\dfrac{x^2}{4} + \dfrac{y^2}{9} \geq 1$?

$\dfrac{0^2}{4} + \dfrac{0^2}{9} \geq 1$

$\quad 0 \geq 1$ false

Shade the portion of the graph which does not contain (0, 0).

Next, graph the circle. Does (0, 0) satisfy

$x^2 + y^2 \geq 4$?

$0^2 + 0^2 \geq 4$

$\quad 0 \geq 4$ false

Shade the portion of the graph which does not contain (0, 0).

The solution to the system is the overlapping region.

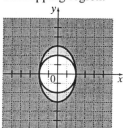

37. $\begin{cases} x^2 - y^2 \geq 1 \\ \quad\quad y \geq 0 \end{cases}$

First, graph the hyperbola. Does (−2, 0) satisfy $x^2 - y^2 \geq 1$?

$(-2)^2 - 0^2 \geq 1$

$\quad 4 \geq 1$ true

Does (0, 0) satisfy $x^2 + y^2 \geq 1$?

$0^2 + 0^2 \geq 1$

$\quad 0 \geq 1$ false

Does (2, 0) satisfy $x^2 + y^2 \geq 1$?

$2^2 + 0^2 \geq 1$

$\quad 4 \geq 1$ true

Shade the portion of the graph containing (−2, 0) and the portion of the graph containing (2, 0).

Next, graph the line. Does (0, 2) satisfy $y \geq 0$?

$2 \geq 0$ true

Shade the portion of the graph containing (0, 2).

The solution to the system is the overlapping region.

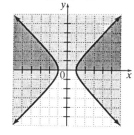

39. $\begin{cases} x+y \geq 1 \\ 2x+3y < 1 \\ x > -3 \end{cases}$

First, graph $x = -3$ with a dashed line.
Does $(0, 0)$ satisfy $x > -3$?
$0 > -3$ true
Shade the portion of the graph containing $(0, 0)$.
Next, graph $2x + 3y = 1$ with a dashed line.
Does $(0, 0)$ satisfy $2x + 3y < 1$?
$2(0) + 3(0) < 1$
$\qquad 0 < 1$ true
Shade the portion of the graph containing $(0, 0)$.
Next, graph $x + y = 1$ with a solid line.
Does $(0, 0)$ satisfy $x + y \geq 1$?
$0 + 0 \geq 1$
$\qquad 0 \geq 1$ false
Shade the portion of the graph which does not contain $(0, 0)$.
The solution to the system is the overlapping region.

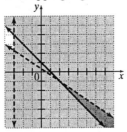

41. $\begin{cases} x^2 - y^2 < 1 \\ \dfrac{x^2}{16} + y^2 \leq 1 \\ x \geq -2 \end{cases}$

First, graph the hyperbola with a dashed line.
Does $(-2, 0)$ satisfy $x^2 - y^2 < 1$?
$(-2)^2 - 0^2 < 1$
$\qquad 4 < 1$ false
Does $(0, 0)$ satisfy $x^2 - y^2 < 1$?
$0^2 - 0^2 < 1$
$\qquad 0 < 1$ true
Does $(2, 0)$ satisfy $x^2 - y^2 < 1$?
$2^2 - 0^2 < 1$
$\qquad 4 < 1$ false
Shade the portion of the graph containing $(0, 0)$.
Next, graph the ellipse with a solid line.
Does $(0, 0)$ satisfy $\dfrac{x^2}{16} + y^2 \leq 1$?
$\dfrac{0^2}{16} + 0^2 \leq 1$
$\qquad 0 \leq 1$ true
Shade the portion of the graph containing $(0, 0)$.
Next, graph the line with a solid line.
Does $(0, 0)$ satisfy $x \geq -2$?
$0 \geq -2$ true
Shade the portion of the graph containing $(0, 0)$.
The solution to the system is the overlapping region.

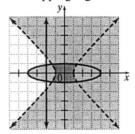

43. $\begin{cases} y \le x^2 \\ y \ge x + 2 \\ x \ge 0 \\ y \ge 0 \end{cases}$

First, graph the parabola with a solid line.

Does (0, 2) satisfy $y \le x^2$?

$2 \le 0^2$

$2 \le 0$ false

Shade the portion of the graph which does not contain (0, 2).

Next, graph $y = x + 2$ with a solid line.

Does (0, 0) satisfy $y \ge x + 2$?

$0 \ge 0 + 2$

$0 \ge 2$ false

Shade the portion of the graph which does not contain (0, 0).

Next, graph $x = 0$ with a solid line.

Does (1, 1) satisfy $x \ge 0$?

$1 \ge 0$ true

Shade the portion of the graph containing (1, 1).

Next, graph $y = 0$ with a solid line.

Does (1, 1) satisfy $y \ge 0$?

$1 \ge 0$ true

Shade the portion of the graph containing (1, 1).

The solution to the system is the overlapping region.

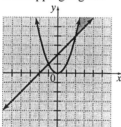

45. This is a function because no vertical line can cross the graph in more than one place.

47. This is not a function because a vertical line can cross the graph in two places.

49. $f(x) = 3x^2 - 2$

$\begin{aligned} f(-3) &= 3(-3)^2 - 2 \\ &= 3(9) - 2 \\ &= 25 \end{aligned}$

51. $f(x) = 3x^2 - 2$

$f(b) = 3b^2 - 2$

Chapter 10 Review

1. (–6, 3) and (8, 4)

$\begin{aligned} d &= \sqrt{[8 - (-6)]^2 + (4 - 3)^2} \\ &= \sqrt{14^2 + 1^2} \\ &= \sqrt{197} \end{aligned}$

The distance is $\sqrt{197}$ units.

2. (3, 5) and (8, 9)

$\begin{aligned} d &= \sqrt{(8 - 3)^2 + (9 - 5)^2} \\ &= \sqrt{5^2 + 4^2} \\ &= \sqrt{25 + 16} \\ &= \sqrt{41} \end{aligned}$

The distance is $\sqrt{41}$ units.

3. (–4, –6) and (–1, 5)

$\begin{aligned} d &= \sqrt{[-1 - (-4)]^2 + [5 - (-6)]^2} \\ &= \sqrt{3^2 + 11^2} \\ &= \sqrt{9 + 121} \\ &= \sqrt{130} \end{aligned}$

The distance is $\sqrt{130}$ units.

4. (–1, 5) and (2, –3)

$\begin{aligned} d &= \sqrt{[2 - (-1)]^2 + (-3 - 5)^2} \\ &= \sqrt{3^2 + (-8)^2} \\ &= \sqrt{9 + 64} \\ &= \sqrt{73} \end{aligned}$

The distance is $\sqrt{73}$ units.

5. $\left(-\sqrt{2},\ 0\right)$ and $\left(0,\ -4\sqrt{6}\right)$

$$d = \sqrt{\left[0-\left(-\sqrt{2}\right)\right]^2 + \left(-4\sqrt{6}-0\right)^2}$$
$$= \sqrt{\left(\sqrt{2}\right)^2 + \left(-4\sqrt{6}\right)^2}$$
$$= \sqrt{2+96}$$
$$= \sqrt{98}$$
$$= 7\sqrt{2}$$

The distance is $7\sqrt{2}$ units.

6. $\left(-\sqrt{5},\ -\sqrt{11}\right)$ and $\left(-\sqrt{5},\ -3\sqrt{11}\right)$

$$d = \sqrt{\left[-\sqrt{5}-\left(-\sqrt{5}\right)\right]^2 + \left[-3\sqrt{11}-\left(-\sqrt{11}\right)\right]^2}$$
$$= \sqrt{0 + \left(-2\sqrt{11}\right)^2}$$
$$= \sqrt{44}$$
$$= 2\sqrt{11}$$

The distance is $2\sqrt{11}$ units.

7. $(7.4, -8.6)$ and $(-1.2, 5.6)$

$$d = \sqrt{(-1.2-7.4)^2 + [5.6-(-8.6)]^2}$$
$$= \sqrt{(-8.6)^2 + (14.2)^2}$$
$$= \sqrt{275.6}$$
$$\approx 16.60$$

The distance is 16.60 units.

8. $(2.3, 1.8)$ and $(10.7, -9.2)$

$$d = \sqrt{(10.7-2.3)^2 + (-9.2-1.8)^2}$$
$$= \sqrt{(8.4)^2 + (-11)^2}$$
$$= \sqrt{191.56}$$
$$\approx 13.84$$

The distance is 13.84 units.

9. $(2, 6)$ and $(-12, 4)$

$$\left(\frac{2-12}{2},\ \frac{6+4}{2}\right)$$

The midpoint is $(-5, 5)$.

10. $(-3, 8)$ and $(11, 24)$

$$\left(\frac{-3+11}{2},\ \frac{8+24}{2}\right)$$

The midpoint is $(4, 16)$.

11. $(-6, -5)$ and $(-9, 7)$

$$\left(\frac{-6-9}{2},\ \frac{-5+7}{2}\right)$$

The midpoint is $\left(-\frac{15}{2},\ 1\right)$.

12. $(4, -6)$ and $(-15, 2)$

$$\left(\frac{4-15}{2},\ \frac{-6+2}{2}\right)$$

The midpoint is $\left(-\frac{11}{2},\ -2\right)$.

13. $\left(0,\ -\frac{3}{8}\right)$ and $\left(\frac{1}{10},\ 0\right)$

$$\left(\frac{0+\frac{1}{10}}{2},\ \frac{-\frac{3}{8}+0}{2}\right)$$

The midpoint is $\left(\frac{1}{20},\ -\frac{3}{16}\right)$.

14. $\left(\frac{3}{4},\ -\frac{1}{7}\right)$ and $\left(-\frac{1}{4},\ -\frac{3}{7}\right)$

$$\left(\frac{\frac{3}{4}-\frac{1}{4}}{2},\ \frac{-\frac{1}{7}-\frac{3}{7}}{2}\right)$$

The midpoint is $\left(\frac{1}{4},\ -\frac{2}{7}\right)$.

15. $\left(\sqrt{3},\ -2\sqrt{6}\right)$ and $\left(\sqrt{3},\ -4\sqrt{6}\right)$

$$\left(\frac{\sqrt{3}+\sqrt{3}}{2},\ \frac{-2\sqrt{6}-4\sqrt{6}}{2}\right)$$

The midpoint is $\left(\sqrt{3},\ -3\sqrt{6}\right)$.

16. $\left(-5\sqrt{3},\ 2\sqrt{7}\right)$ and $\left(-3\sqrt{3},\ 10\sqrt{7}\right)$

$$\left(\frac{-5\sqrt{3}-3\sqrt{3}}{2},\ \frac{2\sqrt{7}+10\sqrt{7}}{2}\right)$$

The midpoint is $\left(-4\sqrt{3},\ 6\sqrt{7}\right)$.

17. center $(-4, 4)$, radius 3

$$[x-(-4)]^2 + (y-4)^2 = 3^2$$
$$(x+4)^2 + (y-4)^2 = 9$$

18. center (5, 0), radius 5
$$(x-5)^2 + (y-0)^2 = 5^2$$
$$(x-5)^2 + y^2 = 25$$

19. center (−7, −9), radius $\sqrt{11}$
$$[x-(-7)]^2 + (y-(-9))^2 = \left(\sqrt{11}\right)^2$$
$$(x+7)^2 + (y+9)^2 = 11$$

20. center (0, 0), radius $\dfrac{7}{2}$
$$(x-0)^2 + (y-0)^2 = \left(\frac{7}{2}\right)^2$$
$$x^2 + y^2 = \frac{49}{4}$$

21. $x^2 + y^2 = 7$
or $(x-0)^2 + (y-0)^2 = \left(\sqrt{7}\right)^2$
The center is (0, 0).

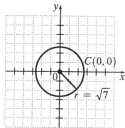

22. $x = 2(y-5)^2 + 4$
The vertex is (4, 5).

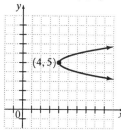

23. $x = -(y+2)^2 + 3$
The vertex is (3, −2).

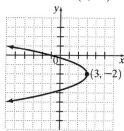

24. $(x-1)^2 + (y-2)^2 = 4$
The center is (1, 2). The radius is 2.

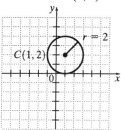

25. $y = -x^2 + 4x + 10$
$$y = -(x^2 - 4x) + 10$$
$$y = -(x^2 - 4x + 4) + 10 + 4$$
$$y = -(x-2)^2 + 14$$
The vertex is (2, 14).

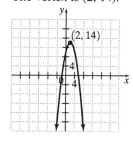

26. $x = -y^2 - 4y + 6$

$x = -(y^2 + 4y) + 6$

$x = -(y^2 + 4y + 4) + 6 + 4$

$x = -(y + 2)^2 + 10$

The vertex is $(10, -2)$.

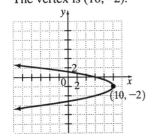

27. $x = \dfrac{1}{2}y^2 + 2y + 1$

$x = \dfrac{1}{2}(y^2 + 4y) + 1$

$x = \dfrac{1}{2}(y^2 + 4y + 4) + 1 - 2$

$x = \dfrac{1}{2}(y + 2)^2 - 1$

The vertex is $(-1, -2)$.

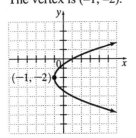

28. $y = -3x^2 + \dfrac{1}{2}x + 4$

$y = -3\left(x^2 - \dfrac{1}{6}x\right) + 4$

$y = -3\left(x^2 - \dfrac{1}{6}x + \dfrac{1}{144}\right) + 4 + \dfrac{3}{144}$

$y = -3\left(x - \dfrac{1}{12}\right)^2 + \dfrac{193}{48}$

The vertex is $\left(\dfrac{1}{12}, \dfrac{193}{48}\right)$.

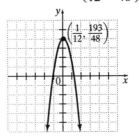

29. $x^2 + y^2 + 2x + y = \dfrac{3}{4}$

$(x^2 + 2x + 1) + \left(y^2 + y + \dfrac{1}{4}\right) = \dfrac{3}{4} + 1 + \dfrac{1}{4}$

$(x + 1)^2 + \left(y + \dfrac{1}{2}\right)^2 = 2$

$(x + 1)^2 + \left(y + \dfrac{1}{2}\right)^2 = \left(\sqrt{2}\right)^2$

The center is $\left(-1, -\dfrac{1}{2}\right)$.

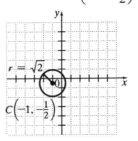

30.
$$x^2 + y^2 + 3y = \frac{7}{4}$$
$$x^2 + \left(y^2 + 3y + \frac{9}{4}\right) = \frac{7}{4} + \frac{9}{4}$$
$$x^2 + \left(y + \frac{3}{2}\right)^2 = 4$$

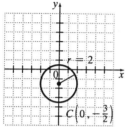

31.
$$4x^2 + 4y^2 + 16x + 8y = 1$$
$$x^2 + y^2 + 4x + 2y = \frac{1}{4}$$
$$x^2 + 4x + y^2 + 2y = \frac{1}{4}$$
$$(x^2 + 4x + 4) + (y^2 + 2y + 1) = \frac{1}{4} + 4 + 1$$
$$(x+2)^2 + (y+1)^2 = \frac{21}{4}$$
$$(x+2)^2 + (y+1)^2 = \left(\frac{\sqrt{21}}{2}\right)^2$$

The center is (–2, –1).

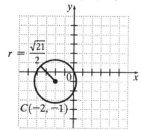

32.
$$3x^2 + 6x + 3y^2 = 9$$
$$x^2 + 2x + y^2 = 3$$
$$(x^2 + 2x + 1) + y^2 = 3 + 1$$
$$(x+1)^2 + y^2 = 4$$
The center is (–1, 0).

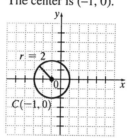

33. $y = x^2 + 6x + 9$
$$y = (x+3)^2$$
The vertex is (–3, 0).

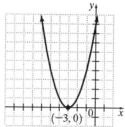

34. $x = y^2 + 6y + 9$
$$x = (y+3)^2$$
The vertex is (0, –3).

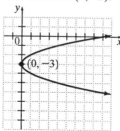

35. The center is (5.6, –2.4).

$$\text{radius} = \frac{6.2}{2} = 3.1$$

$$(x - 5.6)^2 + [y - (-2.4)]^2 = 3.1^2$$

$$(x - 5.6)^2 + (y + 2.4)^2 = 9.61$$

36. $x^2 + \dfrac{y^2}{4} = 1$

$a = 1,\ b = 2$, The center is (0, 0).

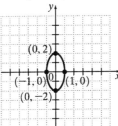

37. $x^2 - \dfrac{y^2}{4} = 1$

$a = 1,\ b = 2$

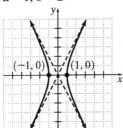

38. $\dfrac{y^2}{4} - \dfrac{x^2}{16} = 1$

$a = 4,\ b = 2$

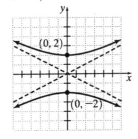

39. $\dfrac{y^2}{4} + \dfrac{x^2}{16} = 1$

$a = 4,\ b = 2$

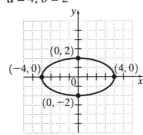

40. $\dfrac{x^2}{5} + \dfrac{y^2}{5} = 1$

Center (0, 0), radius $= \sqrt{5}$

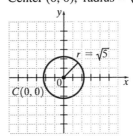

41. $\dfrac{x^2}{5} - \dfrac{y^2}{5} = 1$

$a = \sqrt{5},\ b = \sqrt{5}$

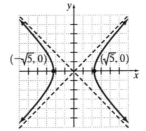

42. $-5x^2 + 25y^2 = 125$

$$\frac{y^2}{5} - \frac{x^2}{25} = 1$$

$a = 5, \; b = \sqrt{5}$

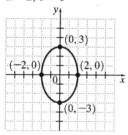

43. $4y^2 + 9x^2 = 36$

$$\frac{y^2}{9} + \frac{x^2}{4} = 1$$

$a = 2, \; b = 3$

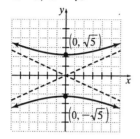

44. $\dfrac{(x-2)^2}{4} + (y-1)^2 = 1$

$a = 2, \; b = 1$, center $(2, 1)$

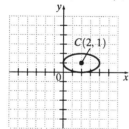

45. $\dfrac{(x+3)^2}{9} + \dfrac{(y-4)^2}{25} = 1$

$a = 3, \; b = 5$, center $(-3, 4)$

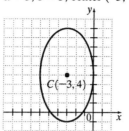

46. $x^2 - y^2 = 1$

Graph $y_1 = \sqrt{x^2 - 1}$ and $y_2 = -y_1$.

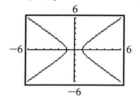

47. $36y^2 - 49x^2 = 1764$

Graph $y_1 = \dfrac{7\sqrt{x^2 + 36}}{6}$ and $y_2 = -y_1$.

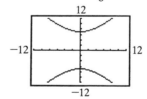

48. $y^2 = x^2 + 9$

Graph $y_1 = \sqrt{x^2 + 9}$ and $y_2 = -y_1$.

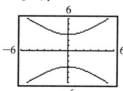

49. $x^2 = 4y^2 - 16$

Graph $y_1 = \dfrac{\sqrt{x^2 + 16}}{2}$ and $y_2 = -y_1$.

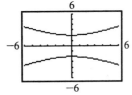

50. $100 - 25x^2 = 4y^2$

Graph $y_1 = \dfrac{5\sqrt{4 - x^2}}{2}$ and $y_2 = -y_1$.

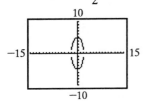

51. parabola

$y = x^2 + 4x + 6$

$y = x^2 + 4x + 4 + 6 - 4$

$y = (x + 2)^2 + 2$

vertex: $(-2, 2)$

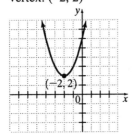

52. hyperbola

$y^2 = x^2 + 6$

$\dfrac{y^2}{6} - \dfrac{x^2}{6} = 1$

$\dfrac{(y - 0)^2}{\left(\sqrt{6}\right)^2} - \dfrac{(x - 0)^2}{\left(\sqrt{6}\right)^2} = 1$

$a = \sqrt{6},\ \ b = \sqrt{6}$, center $(0, 0)$

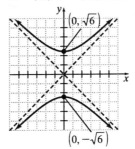

53. circle

$y^2 + x^2 = 4x + 6$

$x^2 - 4x + y^2 = 6$

$x^2 - 4x + 4 + y^2 = 6 + 4$

$(x - 2)^2 + y^2 = 10$

$(x - 2)^2 + (y - 0)^2 = \left(\sqrt{10}\right)^2$

center $(2, 0)$, radius $= \sqrt{10}$

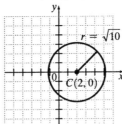

54. ellipse

$$y^2 + 2x^2 = 4x + 6$$
$$y^2 + 2(x^2 - 2x + 1) = 6 + 2$$
$$y^2 + 2(x - 1)^2 = 8$$
$$\frac{y^2}{8} + \frac{(x-1)^2}{4} = 1$$
$$\frac{(x-1)^2}{2^2} + \frac{(y-0)^2}{\left(2\sqrt{2}\right)^2} = 1$$

$a = 2$, $b = 2\sqrt{2}$, center $(1, 0)$

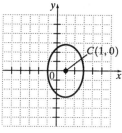

55. circle

$$x^2 + y^2 - 8y = 0$$
$$x^2 + y^2 - 8y + 16 = 0 + 16$$
$$x^2 + (y - 4)^2 = 16$$
$$(x - 0)^2 + (y - 4)^2 = 4^2$$

center $(0, 4)$, radius $= 4$

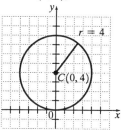

56. parabola

$$x - 4y = y^2$$
$$x = y^2 + 4y + 4 - 4$$
$$x = (y + 2)^2 - 4$$

vertex: $(-4, -2)$

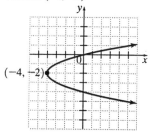

57. hyperbola

$$x^2 - 4 = y^2$$
$$x^2 - y^2 = 4$$
$$\frac{x^2}{4} - \frac{y^2}{4} = 1$$
$$\frac{(x - 0)^2}{2^2} - \frac{(y - 0)^2}{2^2} = 1$$

$a = 2$, $b = 2$, center $(0, 0)$

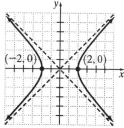

58. circle

$$x^2 = 4 - y^2$$
$$x^2 + y^2 = 4$$

center $(0, 0)$, radius $= 2$

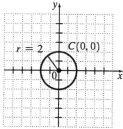

59. ellipse

$$6(x-2)^2 + 9(y+5)^2 = 36$$

$$\frac{(x-2)^2}{6} + \frac{(y+5)^2}{4} = 1$$

$$\frac{(x-2)^2}{\left(\sqrt{6}\right)^2} + \frac{(y+5)^2}{2^2} = 1$$

$a = \sqrt{6}$, $b = 2$, center $(2, -5)$

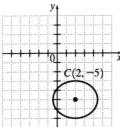

60. hyperbola

$$36y^2 = 576 + 16x^2$$

$$\frac{y^2}{16} - \frac{x^2}{36} = 1$$

$$\frac{(y-0)^2}{4^2} - \frac{(x-0)^2}{6^2} = 1$$

$a = 6$, $b = 4$, center $(0, 0)$

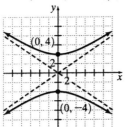

61. hyperbola

$$\frac{x^2}{16} - \frac{y^2}{25} = 1$$

$$\frac{(x-0)^2}{4^2} - \frac{(y-0)^2}{5^2} = 1$$

$a = 4$, $b = 5$, center $(0, 0)$

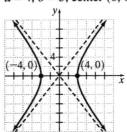

62. circle

$$3(x-7)^2 + 3(y+4)^2 = 1$$

$$(x-7)^2 + (y+4)^2 = \frac{1}{3}$$

center $(7, -4)$, radius $= \frac{\sqrt{3}}{3}$

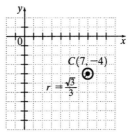

63. $\dfrac{y^2}{4} + \dfrac{x^2}{16} = 1$

$4y^2 + x^2 = 16$

$4y^2 = 16 - x^2$

$y^2 = \dfrac{16 - x^2}{4}$

$y = \pm\sqrt{\dfrac{16 - x^2}{4}}$

$y = \pm\dfrac{\sqrt{16 - x^2}}{2}$

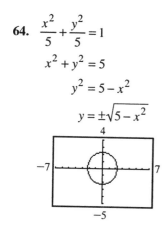

64. $\dfrac{x^2}{5} + \dfrac{y^2}{5} = 1$

$x^2 + y^2 = 5$

$y^2 = 5 - x^2$

$y = \pm\sqrt{5 - x^2}$

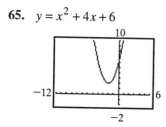

65. $y = x^2 + 4x + 6$

66. $x^2 = 4 - y^2$

$y^2 = 4 - x^2$

$y = \pm\sqrt{4 - x^2}$

67. $\begin{cases} y = 2x - 4 \\ y^2 = 4x \end{cases}$

Substituting (1) in (2) gives

$(2x - 4)^2 = 4x$

$4x^2 - 16x + 16 = 4x$

$4x^2 - 20x + 16 = 0$

$x^2 - 5x + 4 = 0$

$(x - 1)(x - 4) = 0$

$x - 1 = 0$ or $x - 4 = 0$

$x = 1$ or $x = 4$

$y = -2$ or $y = 4$

The solutions are $(1, -2)$ and $(4, 4)$.

68. $\begin{cases} x^2 + y^2 = 4 \\ x - y = 4 \text{ or } x = y + 4 \end{cases}$

Substitute.

$(y + 4)^2 + y^2 = 4$

$y^2 + 8y + 16 + y^2 = 4$

$2y^2 + 8y + 16 = 4$

$2y^2 + 8y + 12 = 0$

$y^2 + 4y + 6 = 0$

$b^2 - 4ac = 4^2 - 4 \cdot 1 \cdot 6 = 16 - 24 = -8$

Therefore, no real solutions exist.

The solution is $\varnothing$.

69. $\begin{cases} y = x + 2 \\ y = x^2 \end{cases}$

Substituting (1) in (2) gives

$x + 2 = x^2$

$0 = x^2 - x - 2$

$0 = (x + 1)(x - 2)$

$x + 1 = 0$ or $x - 2 = 0$

$x = -1$ or $x = 2$

$y = 1$ or $y = 4$

The solutions are $(-1, 1)$ and $(2, 4)$.

70. $\begin{cases} y = x^2 - 5x + 1 \\ y = -x + 6 \end{cases}$

Substitute.

$-x + 6 = x^2 - 5x + 1$

$0 = x^2 - 4x - 5$

$0 = (x - 5)(x + 1)$

$x - 5 = 0$ or $x + 1 = 0$

$x = 5$ or $x = -1$

$x = 5: y = -5 + 6 = 1$

$x = -1: y = -(-1) + 6 = 1 + 6 = 7$

The solutions are $(5, 1)$ and $(-1, 7)$.

71. $\begin{cases} 4x - y^2 = 0 \\ 2x^2 + y^2 = 16 \end{cases}$

From (1) we have $y^2 = 4x$. Substituting in (2) gives

$2x^2 + 4x = 16$

$2x^2 + 4x - 16 = 0$

$x^2 + 2x - 8 = 0$

$(x - 2)(x + 4) = 0$

$x - 2 = 0$ or $x + 4 = 0$

$x = 2$ or $x = -4$

$x = 2:\ 4(2) - y^2 = 0$

$y^2 = 8$

$y = \pm 2\sqrt{2}$

$x = -4:\ 4(-4) - y^2 = 0$

$y^2 = -16$ reject

The solutions are $\left(2, 2\sqrt{2}\right)$ and $\left(2, -2\sqrt{2}\right)$.

72. $\begin{cases} x^2 + 4y^2 = 16 \\ x^2 + y^2 = 4 \end{cases}$

Subtract.

$3y^2 = 12$

$y^2 = 4$

$y = \pm 2$

Substitute back.

$x^2 + 4 = 4$

$x^2 = 0$

$x = 0$

The solutions are $(0, 2)$ and $(0, -2)$.

73. $\begin{cases} x^2 + y^2 = 10 \\ 9x^2 + y^2 = 18 \end{cases}$

Subtracting (1) from (2) we have

$8x^2 = 8$

$x^2 = 1$

$x = \pm 1$

Substitution gives

$1 + y^2 = 10$

$y^2 = 9$

$y = \pm 3$

The solutions are $(-1, 3)$, $(-1, -3)$, $(1, 3)$, and $(1, -3)$.

74. $\begin{cases} x^2 + 2y = 9 \\ 5x - 2y = 5 \end{cases}$

Add.

$$x^2 + 5x = 14$$
$$x^2 + 5x - 14 = 0$$
$$(x + 7)(x - 2) = 0$$
$$x + 7 = 0 \quad \text{or} \quad x - 2 = 0$$
$$x = -7 \quad \text{or} \quad x = 2$$
$$x = -7: \ (-7)^2 + 2y = 9$$
$$49 + 2y = 9$$
$$2y = -40$$
$$y = -20$$
$$x = 2: \ 2^2 + 2y = 9$$
$$4 + 2y = 9$$
$$2y = 5$$
$$y = \frac{5}{2}$$

The solutions are $(-7, -20)$ and $\left(2, \frac{5}{2}\right)$.

75. $\begin{cases} y = 3x^2 + 5x - 4 \\ y = 3x^2 - x + 2 \end{cases}$

Subtracting (2) from (1) gives
$$0 = 6x - 6$$
$$6x = 6$$
$$x = 1$$
Substitution gives
$$y = 3(1)^2 - 1 + 2 = 4.$$
The solution is $(1, 4)$.

76. $\begin{cases} x^2 - 3y^2 = 1 \text{ or } x^2 = 3y^2 + 1 \\ 4x^2 + 5y^2 = 21 \end{cases}$

Substitute.

$$4(3y^2 + 1) + 5y^2 = 21$$
$$12y^2 + 4 + 5y^2 = 21$$
$$17y^2 + 4 = 21$$
$$17y^2 = 17$$
$$y^2 = 1$$
$$y = \pm 1$$

Substitute back.

$$x^2 - 3(1) = 1$$
$$x^2 - 3 = 1$$
$$x^2 = 4$$
$$x = \pm 2$$

The solutions are $(2, 1)$, $(2, -1)$, $(-2, 1)$, and $(-2, -1)$.

77. $\begin{cases} xy = 150 \\ 2x + 2y = 50 \end{cases}$

$$x = \frac{150}{y}$$
$$2\left(\frac{150}{y}\right) + 2y = 50$$
$$\frac{150}{y} + y = 25$$
$$150 + y^2 = 25y$$
$$y^2 - 25y + 150 = 0$$
$$(y - 15)(y - 10) = 0$$
$$y = 15 \text{ or } y = 10$$

$$2x + 2(15) = 50 \qquad 2x + 2(10) = 50$$
$$2x = 20 \qquad\qquad 2x = 30$$
$$x = 10 \qquad\qquad x = 15$$

The room is 15 feet by 10 feet.

78. Four real solutions

79. $y \le -x^2 + 3$

Graph $y = -x^2 + 3$ with a solid line.

Does (0, 0) satisfy $y \le x^2 + 3$?

$0 \le -0^2 + 3$

$0 \le 3$ true

Shade the portion of the graph containing (0, 0).

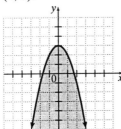

80. $x^2 + y^2 < 9$

Graph $x^2 + y^2 = 9$ with a dashed line.

Does (0, 0) satisfy $x^2 + y^2 < 9$?

$0 + 0 < 9$

$0 < 9$ true

Shade the portion of the graph containing (0, 0).

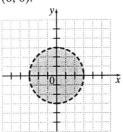

81. $x^2 - y^2 < 1$

Graph $x^2 - y^2 = 1$ with a dashed line.

Does (0, 0) satisfy the inequality?

$0 - 0 < 1$

$0 < 1$ true

Does (2, 0) satisfy the inequality?

$2^2 - 0^2 < 1$

$4 < 1$ false

Does $(-2, 0)$ satisfy the inequality?

$(-2)^2 - 0^2 < 1$

$4 < 1$ false

Shade the portion of the graph containing (0, 0).

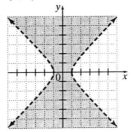

82. $\dfrac{x^2}{4} + \dfrac{y^2}{9} \ge 1$

Graph $\dfrac{x^2}{4} + \dfrac{y^2}{9} = 1$ with a solid line.

Does (0, 0) satisfy the inequality?

$\dfrac{0^2}{4} + \dfrac{0^2}{9} \ge 1$

$0 \ge 1$ false

Shade the portion of the graph that does not contain (0, 0).

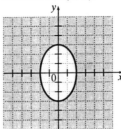

83. $\begin{cases} 2x \leq 4 \\ x + y \geq 1 \end{cases}$

First, graph $2x = 4$ with a solid line.
Does $(0, 0)$ satisfy the inequality?
$2 \cdot 0 \leq 4$

$0 \leq 4$ true

Shade the portion of the graph containing $(0, 0)$. Next, graph $x + y = 1$ with a solid line. Does $(0, 0)$ satisfy the inequality?
$0 + 0 \geq 1$

$0 \geq 1$ false

Shade the portion of the graph that does not contain $(0, 0)$. The solution to the system is the overlapping region.

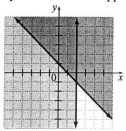

84. $\begin{cases} 3x + 4y \leq 12 \\ x - 2y > 6 \end{cases}$

First, graph $3x + 4y = 12$ with a solid line.
Does $(0, 0)$ satisfy the inequality?
$3(0) + 4(0) \leq 12$

$0 \leq 12$ true

Shade the portion of the graph containing $(0, 0)$. Next, graph $x - 2y = 6$ with a dashed line. Does $(0, 0)$ satisfy the inequality?
$0 - 2(0) > 6$

$0 > 6$ false

Shade the portion of the graph that does not contain $(0, 0)$. The solution to the system is the overlapping region.

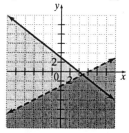

85. $\begin{cases} y > x^2 \\ x + y \geq 3 \end{cases}$

First, graph $y = x^2$ with a dashed line.
Does $(1, 0)$ satisfy the inequality?
$0 > 1^2$

$0 > 1$ false

Shade the portion of the graph that does not contain $(1\ 0)$. Next, graph $x + y = 3$ with a solid line. Does $(0, 0)$ satisfy the inequality?
$0 + 0 \geq 3$

$0 \geq 3$ false

Shade the portion of the graph that does not contain $(0, 0)$. The solution to the system is the overlapping region.

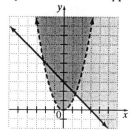

86. $\begin{cases} x^2 + y^2 \leq 16 \\ x^2 + y^2 \geq 4 \end{cases}$

First, graph $x^2 + y^2 = 16$ with a solid line.
Does (0, 0) satisfy the inequality?
$0^2 + 0^2 \leq 16$
$\quad 0 \leq 16$ true
Shade the portion of the graph containing

(0, 0). Next, graph $x^2 + y^2 = 4$ with a
solid line.
Does (0, 0) satisfy the inequality?
$0^2 + 0^2 \geq 4$
$\quad 0 \geq 4$ false
Shade the portion of the graph that does
not contain (0, 0).The solution to the
system is the overlapping region.

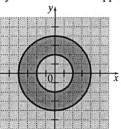

87. $\begin{cases} x^2 + y^2 < 4 \\ x^2 - y^2 \leq 1 \end{cases}$

First, graph $x^2 + y^2 = 4$ with a dashed
line.
Does (0, 0) satisfy the inequality?
$0^2 + 0^2 < 4$
$\quad 0 < 4$ true
Shade the portion of the graph containing
(0, 0).

Next, graph $x^2 - y^2 = 1$ with a solid line.
Does (0, 0) satisfy the inequality?
$0^2 - 0^2 \leq 1$
$\quad 0 \leq 1$ true
Does (2, 0) satisfy the inequality?
$2^2 - 0^2 \leq 1$
$\quad 4 - 0 \leq 1$
$\quad 4 \leq 1$ false
Does $(-2, 0)$ satisfy the inequality?
$(-2)^2 - 0^2 \leq 1$
$\quad 4 - 0 \leq 1$
$\quad 4 \leq 1$ false
Shade the portion of the graph that
contains (0, 0).
The solution to the system is the
overlapping region.

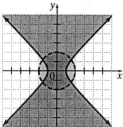

88. $\begin{cases} x^2 + y^2 < 4 \\ \quad y \geq x^2 - 1 \\ \quad x \geq 0 \end{cases}$

First, graph $x^2 + y^2 = 4$ with a dashed line.
Does (0, 0) satisfy the inequality?
$0^2 + 0^2 < 4$
$\quad 0 < 4$ true
Shade the portion of the graph containing (0, 0).

Next, graph $y = x^2 - 1$ with a solid line.
Does (0, 0) satisfy the inequality?
$0 \geq 0^2 - 1$
$0 \geq -1$ true
Shade the portion of the graph containing (0, 0).
Next, graph $x = 0$ with a solid line.
Does (1, 0) satisfy the inequality?
$1 \geq 0$ true
Shade the portion of the graph containing (1, 0). The solution to the system is the overlapping region.

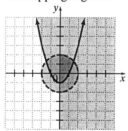

Chapter 10 Test

1. $(-6, 3)$ and $(-8, -7)$
$d = \sqrt{(-8+6)^2 + (-7-3)^2}$
$\quad = \sqrt{(-2)^2 + (-10)^2}$
$\quad = \sqrt{4 + 100}$
$\quad = \sqrt{104}$
$\quad = 2\sqrt{26}$
The distance is $2\sqrt{26}$ units.

2. $\left(-2\sqrt{5}, \sqrt{10}\right)$ and $\left(-\sqrt{5}, 4\sqrt{10}\right)$
$d = \sqrt{\left(-\sqrt{5} + 2\sqrt{5}\right)^2 + \left(4\sqrt{10} - \sqrt{10}\right)^2}$
$\quad = \sqrt{\left(\sqrt{5}\right)^2 + \left(3\sqrt{10}\right)^2}$
$\quad = \sqrt{5 + 90}$
$\quad = \sqrt{95}$
The distance is $\sqrt{95}$ units.

3. $(-2, -5)$ and $(-6, 12)$
$\left(\dfrac{-2-6}{2}, \dfrac{-5+12}{2}\right)$
The midpoint of the line segment is $\left(-4, \dfrac{7}{2}\right)$.

4. $\left(-\dfrac{2}{3}, -\dfrac{1}{5}\right)$ and $\left(-\dfrac{1}{3}, \dfrac{4}{5}\right)$
$\left(\dfrac{-\frac{2}{3} - \frac{1}{3}}{2}, \dfrac{-\frac{1}{5} + \frac{4}{5}}{2}\right)$
The midpoint of the line segment is $\left(-\dfrac{1}{2}, \dfrac{3}{10}\right)$.

5. $\qquad x^2 + y^2 = 36$
$(x - 0)^2 + (y - 0)^2 = 6^2$
Circle: center (0, 0), radius = 6

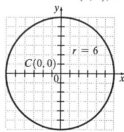

6.
$$x^2 - y^2 = 36$$
$$\frac{x^2}{36} - \frac{y^2}{36} = 1$$
$$\frac{(x-0)^2}{6^2} - \frac{(y-0)^2}{6^2} = 1$$
Hyperbola:
$a = 6$, $b = 6$, center$(0, 0)$

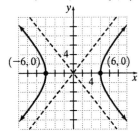

7.
$$16x^2 + 9y^2 = 144$$
$$\frac{x^2}{9} + \frac{y^2}{16} = 1$$
$$\frac{(x-0)^2}{3^2} + \frac{(y-0)^2}{4^2} = 1$$
Ellipse: $a = 3$, $b = 4$, center $(0, 0)$

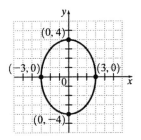

8. $y = x^2 - 8x + 16$
$$y = (x-4)^2 + 0$$
Parabola: The vertex is $(4, 0)$.

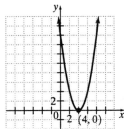

9.
$$x^2 + y^2 + 6x = 16$$
$$x^2 + 6x + y^2 = 16$$
$$(x^2 + 6x + 9) + y^2 = 16 + 9$$
$$(x+3)^2 + y^2 = 5^2$$
Circle: center$(-3, 0)$, radius 5

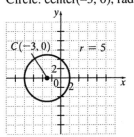

10. $x = y^2 + 8y - 3$
$$x = (y^2 + 8y + 16) - 3 - 16$$
$$x = (y+4)^2 - 19$$
Parabola: The vertex is $(-19, -4)$.

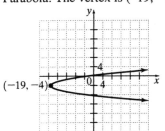

11. $\dfrac{(x-4)^2}{16} + \dfrac{(y-3)^2}{9} = 1$
$$\frac{(x-4)^2}{4^2} + \frac{(y-3)^2}{3^2} = 1$$
Ellipse: $a = 4$, $b = 3$, center $(4, 3)$

12. $y^2 - x^2 = 1$

$$\frac{(y-0)^2}{1^2} - \frac{(x-0)^2}{1^2} = 1$$

hyperbola: $a = 1$, $b = 1$,
center $(0, 0)$

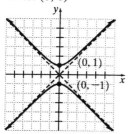

13. $\begin{cases} x^2 + y^2 = 169 \\ 5x + 12y = 0 \end{cases}$

$12y = -5x$

$$y = -\frac{5x}{12}$$

Substitute.

$$x^2 + \left(-\frac{5x}{12}\right)^2 = 169$$

$$x^2 + \frac{25x^2}{144} = 169$$

$$\frac{169x^2}{144} = 169$$

$$x^2 = 144$$

$$x = \pm 12$$

Substitute back.

$x = 12$: $y = -\dfrac{5}{12}(12) = -5$

$x = -12$: $y = -\dfrac{5}{12}(-12) = 5$

The solutions are $(12, -5)$ and $(-12, 5)$.

14. $\begin{cases} x^2 + y^2 = 26 \\ x^2 - y^2 = 24 \end{cases}$

Add.

$$2x^2 = 50$$

$$x^2 = 25$$

$$x = \pm 5$$

Substitute back.

$$25 + y^2 = 26$$

$$y^2 = 1$$

$$y = \pm 1$$

The solutions are $(5, 1)$, $(5, -1)$, $(-5, 1)$,
and $(-5, -1)$.

15. $\begin{cases} y = x^2 - 5x + 6 \\ y = 2x \end{cases}$

Substitute.

$$2x = x^2 - 5x + 6$$

$$0 = x^2 - 7x + 6$$

$$0 = (x-1)(x-6)$$

$$x - 1 = 0 \quad \text{or} \quad x - 6 = 0$$

$$x = 1 \quad \text{or} \quad x = 6$$

$x = 1$: $y = 2(1) = 2$
$x = 6$: $y = 2(6) = 12$

The solutions are $(1, 2)$ and $(6, 12)$.

16. $\begin{cases} x^2 + 4y^2 = 5 \\ y = x \end{cases}$

Substitute.

$$x^2 + 4x^2 = 5$$

$$5x^2 = 5$$

$$x^2 = 1$$

$$x = \pm 1$$

$x = 1$: $y = 1$
$x = -1$: $y = -1$

The solutions are $(1, 1)$ and $(-1, -1)$.

17. $\begin{cases} 2x + 5y \geq 10 \\ \quad\ y \geq x^2 + 1 \end{cases}$

First, graph $2x + 5y = 10$ with a solid line.
Does (0, 0) satisfy the inequality?
$2(0) + 5(0) \geq 10$
$\qquad 0 \geq 10$ false
Shade the portion of the graph that does
not contain (0, 0).

Next, graph $y = x^2 - 1$ with a solid line.
Does (0, 0) satisfy the inequality?
$0 \geq 0^2 + 1$

$0 \geq 1$ false
Shade the portion of the graph that does
not contain (0, 0).
The solution to the system is the
overlapping region.

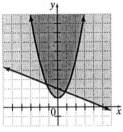

18. $\begin{cases} \dfrac{x^2}{4} + y^2 \leq 1 \\ \quad\ x + y > 1 \end{cases}$

First, graph $\dfrac{x^2}{4} + y^2 = 1$ with a solid line.

Does (0, 0) satisfy the inequality?

$\dfrac{0^2}{4} + 0^2 \leq 1$

$\qquad 0 \leq 1$ true
Shade the portion of the graph containing
(0, 0).
Next, graph $x + y = 1$ with a dashed line.
Does (0, 0) satisfy the inequality?
$0 + 0 > 1$
$\qquad 0 > 1$ false
Shade the portion of the graph that does
not contain (0, 0).
The solution to the system is the
overlapping region.

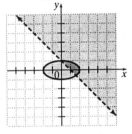

19. $\begin{cases} x^2 + y^2 > 1 \\ \dfrac{x^2}{4} - y^2 \ge 1 \end{cases}$

First, graph $x^2 + y^2 = 1$ with a dashed line.
Does (0, 0) satisfy the inequality?
$0^2 + 0^2 > 0$
$\qquad 0 > 1$ false
Shade the portion of the graph that does not contain (0, 0).

Next, graph $\dfrac{x^2}{4} - y^2 = 1$ with a solid line.
Does (0, 0) satisfy the inequality?
$\dfrac{0^2}{4} - 0^2 \ge 1$
$\qquad 0 \ge 1$ false
Does (3, 0) satisfy the inequality?
$\dfrac{3^2}{4} - 0^2 \ge 1$
$\qquad \dfrac{9}{4} \ge 1$ true
Does $(-3, 0)$ satisfy the inequality?
$\dfrac{(-3)^2}{4} - 0^2 \ge 1$
$\qquad \dfrac{9}{4} \ge 1$ true
Shade the portions of the graph containing
$(3, 0)$ and $(-3, 0)$.
The solution to the system is the overlapping region.

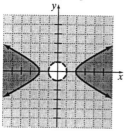

20. $\begin{cases} x^2 + y^2 \ge 4 \\ x^2 + y^2 < 16 \\ \qquad y \ge 0 \end{cases}$

First, graph $x^2 + y^2 = 4$ with a solid line.
Does (0, 0) satisfy the inequality?
$0^2 + 0^2 \ge 4$
$\qquad 0 \ge 4$ false
Shade the portion of the graph that does not contain (0, 0).
Next, graph $x^2 + y^2 = 16$ with a dashed line.
Does (0, 0) satisfy the inequality?
$0^2 + 0^2 < 16$
$\qquad 0 < 16$ true
Shade the portion of the graph containing (0, 0).
Next, graph $y = 0$ with a solid line.
Does (0, 2) satisfy the inequality?
$2 \ge 0$ true
Shade the portion of the graph containing (0, 2).
The solution to the system is the overlapping region.

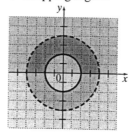

21. Graph B; vertex in second quadrant, opens to the right.

22. $100x^2 + 225y^2 = 22,500$

$$\frac{x^2}{225} + \frac{y^2}{100} = 0$$

$a = \sqrt{225} = 15$

$b = \sqrt{100} = 10$

Width = 15 + 15 = 30

Height = 10

The width is 30 feet and the height is 10 feet.

Chapter 11

Exercise Set 11.1

1. $a_n = n + 4$

$a_1 = 1 + 4 = 5$

$a_2 = 2 + 4 = 6$

$a_3 = 3 + 4 = 7$

$a_4 = 4 + 4 = 8$

$a_5 = 5 + 4 = 9$

Thus, the first five terms of the sequence $a_n = n + 4$ are 5, 6, 7, 8, 9.

3. $a_n = (-1)^n$

$a_1 = (-1)^1 = -1$

$a_2 = (-1)^2 = 1$

$a_3 = (-1)^3 = -1$

$a_4 = (-1)^4 = 1$

$a_5 = (-1)^5 = -1$

Thus, the first five terms of the sequence $a_n = (-1)^n$ are $-1, 1, -1, 1, -1$.

5. $a_n = \dfrac{1}{n+3}$

$a_1 = \dfrac{1}{1+3} = \dfrac{1}{4}$

$a_2 = \dfrac{1}{2+3} = \dfrac{1}{5}$

$a_3 = \dfrac{1}{3+3} = \dfrac{1}{6}$

$a_4 = \dfrac{1}{4+3} = \dfrac{1}{7}$

$a_5 = \dfrac{1}{5+3} = \dfrac{1}{8}$

Thus, the first five terms of the sequence $a_n = \dfrac{1}{n+3}$ are $\dfrac{1}{4}, \dfrac{1}{5}, \dfrac{1}{6}, \dfrac{1}{7}, \dfrac{1}{8}$.

7. $a_n = 2n$

$a_1 = 2(1) = 2$

$a_2 = 2(2) = 4$

$a_3 = 2(3) = 6$

$a_4 = 2(4) = 8$

$a_5 = 2(5) = 10$

Thus, the first five terms of the sequence $a_n = 2n$ are 2, 4, 6, 8, 10.

9. $a_n = -n^2$

$a_1 = -1^2 = -1$

$a_2 = -2^2 = -4$

$a_3 = -3^2 = -9$

$a_4 = -4^2 = -16$

$a_5 = -5^2 = -25$

Thus, the first five terms of the sequence $a_n = -n^2$ are $-1, -4, -9, -16, -25$.

11. $a_n = 2^n$

$a_1 = 2^1 = 2$

$a_2 = 2^2 = 4$

$a_3 = 2^3 = 8$

$a_4 = 2^4 = 16$

$a_5 = 2^5 = 32$

Thus, the first five terms of the sequence $a_n = 2^n$ are 2, 4, 8, 16, 32.

13. $a_n = 2n + 5$

$a_1 = 2(1) + 5 = 2 + 5 = 7$

$a_2 = 2(2) + 5 = 4 + 5 = 9$

$a_3 = 2(3) + 5 = 6 + 5 = 11$

$a_4 = 2(4) + 5 = 8 + 5 = 13$

$a_5 = 2(5) + 5 = 10 + 5 = 15$

Thus, the first five terms of the sequence $a_n = 2n + 5$ are 7, 9, 11, 13, 15.

15. $a_n = (-1)^n n^2$

$a_1 = (-1)^1 (1)^2 = -1(1) = -1$

$a_2 = (-1)^2 (2)^2 = 1(4) = 4$

$a_3 = (-1)^3 (3)^2 = -1(9) = -9$

$a_4 = (-1)^4 (4)^2 = 1(16) = 16$

$a_5 = (-1)^5 (5)^2 = -1(25) = -25$

Thus, the first five terms of the sequence
$a_n = (-1)^n n^2$ are $-1, 4, -9, 16, -25$.

17. $a_n = 3n^2$

$a_5 = 3(5)^2 = 3(25) = 75$

19. $a_n = 6n - 2$

$a_{20} = 6(20) - 2 = 120 - 2 = 118$

21. $a_n = \dfrac{n+3}{n}$

$a_{15} = \dfrac{15+3}{15} = \dfrac{18}{15} = \dfrac{6}{5}$

23. $a_n = (-3)^n$

$a_6 = (-3)^6 = 729$

25. $a_n = \dfrac{n-2}{n+1}$

$a_6 = \dfrac{6-2}{6+1} = \dfrac{4}{7}$

27. $a_n = \dfrac{(-1)^n}{n}$

$a_8 = \dfrac{(-1)^8}{8} = \dfrac{1}{8}$

29. $a_n = -n^2 + 5$

$a_{10} = -10^2 + 5 = -100 + 5 = -95$

31. $a_n = \dfrac{(-1)^n}{n+6}$

$a_{19} = \dfrac{(-1)^{19}}{19+6} = -\dfrac{1}{25}$

33. 3, 7, 11, 15, or
$4(1) - 1, 4(2) - 1, 4(3) - 1, 4(4) - 1$
In general, $a_n = 4n - 1$.

35. $-2, -4, -8, -16$ or $-2^1, -2^2, -2^3, -2^4$
In general, $a_n = -2^n$.

37. $\dfrac{1}{3}, \dfrac{1}{9}, \dfrac{1}{27}, \dfrac{1}{81}$ or $\dfrac{1}{3^1}, \dfrac{1}{3^2}, \dfrac{1}{3^3}, \dfrac{1}{3^4}$

In general, $a_n = \dfrac{1}{3^n}$.

39. $a_n = 32n - 16$

$a_2 = 32(2) - 16 = 64 - 16 = 48$

$a_3 = 32(3) - 16 = 96 - 16 = 80$

$a_4 = 32(4) - 16 = 128 - 16 = 112$

The Thermos will fall 48 feet, 80 feet, and
112 feet in the second, third, and fourth
seconds, respectively.

41. $0.10, 0.20, 0.40$ or $0.10, 0.10(2), 0.10(2)^2$
In general, $a_n = 0.10(2)^{n-1}$.

$a_{14} = 0.10(2)^{13} = 819.20$
Mark will receive $819.20.

43. $a_n = 75(2)^{n-1}$

$a_6 = 75(2)^5 = 75(32) = 2400$
There were 2400 cases at the beginning of
the sixth year.
$a_1 = 75(2)^0 = 75(1) = 75$
There were 75 cases at the beginning of
the first year.

45. $a_n = \frac{1}{2}a_{n-1}$ for $n > 1$, $a_1 = 800$

In 2000, $n = 1$ and $a_1 = 800$.

In 2001, $n = 2$ and $a_2 = \frac{1}{2}(800) = 400$.

In 2002, $n = 3$ and $a_3 = \frac{1}{2}(400) = 200$.

In 2003, $n = 4$ and $a_4 = \frac{1}{2}(200) = 100$.

In 2004, $n = 5$ and $a_5 = \frac{1}{2}(100) = 50$.

The population estimate for 2004 is 50 sparrows.

Continuing the sequence:

in 2005, $n = 6$ and $a_6 = \frac{1}{2}(50) = 25$;

in 2006, $n = 7$ and $a_7 = \frac{1}{2}(25) = 12$;

in 2007, $n = 8$ and $a_8 = \frac{1}{2}(12) = 6$;

in 2008, $n = 9$ and $a_9 = \frac{1}{2}(6) = 3$;

in 2009, $n = 10$ and $a_{10} = \frac{1}{2}(3) \approx 1$;

in 2010, $n = 11$ and $a_{11} = \frac{1}{2}(1) \approx 0$.

The population is estimated to become extinct in 2010.

47. $a_n = \frac{1}{\sqrt{n}}$

$a_1 = \frac{1}{\sqrt{1}} = \frac{1}{1} = 1$

$a_2 = \frac{1}{\sqrt{2}} \approx 0.7071$

$a_3 = \frac{1}{\sqrt{3}} \approx 0.5774$

$a_4 = \frac{1}{\sqrt{4}} = \frac{1}{2} = 0.5$

$a_5 = \frac{1}{\sqrt{5}} \approx 0.4472$

Thus, the first five terms of the sequence $a_n = \frac{1}{\sqrt{n}}$ are approximately

$1, 0.7071, 0.5774, 0.5, 0.4472$.

49. $a_n = \left(1 + \frac{1}{n}\right)^n$

$a_1 = \left(1 + \frac{1}{1}\right)^1 = (2)^1 = 2$

$a_2 = \left(1 + \frac{1}{2}\right)^2 = \left(\frac{3}{2}\right)^2 = 2.25$

$a_3 = \left(1 + \frac{1}{3}\right)^3 = \left(\frac{4}{3}\right)^3 \approx 2.3704$

$a_4 = \left(1 + \frac{1}{4}\right)^4 = \left(\frac{5}{4}\right)^4 \approx 2.4414$

$a_5 = \left(1 + \frac{1}{5}\right)^5 = \left(\frac{6}{5}\right)^5 \approx 2.4883$

Thus, the first five terms of the sequence $a_n = \left(1 + \frac{1}{n}\right)^n$ are approximately 2, 2.25, 2.3704, 2.4414, 2.4883.

51. $f(x) = (x-1)^2 + 3$

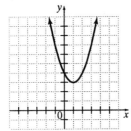

53. $f(x) = 2(x+4)^2 + 2$

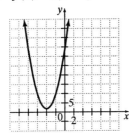

55. $(-4, -1)$ and $(-7, -3)$

$$d = \sqrt{[-7-(-4)]^2 + [-3-(-1)]^2}$$
$$= \sqrt{(-7+4)^2 + (-3+1)^2}$$
$$= \sqrt{(-3)^2 + (-2)^2}$$
$$= \sqrt{9+4} = \sqrt{13}$$

The distance is $\sqrt{13}$ units.

57. $(2, -7)$ and $(-3, -3)$

$$d = \sqrt{(-3-2)^2 + [-3-(-7)]^2}$$
$$= \sqrt{(-5)^2 + (-3+7)^2}$$
$$= \sqrt{(-5)^2 + (4)^2}$$
$$= \sqrt{25+16} = \sqrt{41}$$

The distance is $\sqrt{41}$ units.

Exercise Set 11.2

1. $a_1 = 4$
$a_2 = 4 + 2 = 6$
$a_3 = 6 + 2 = 8$
$a_4 = 8 + 2 = 10$
$a_5 = 10 + 2 = 12$
The first five terms are 4, 6, 8, 10, 12.

3. $a_1 = 6$
$a_2 = 6 + (-2) = 4$
$a_3 = 4 + (-2) = 2$
$a_4 = 2 + (-2) = 0$
$a_5 = 0 + (-2) = -2$
The first five terms are 6, 4, 2, 0, –2.

5. $a_1 = 1$
$a_2 = 1(3) = 3$
$a_3 = 3(3) = 9$
$a_4 = 9(3) = 27$
$a_5 = 27(3) = 81$
The first five terms are 1, 3, 9, 27, 81.

7. $a_1 = 48$
$$a_2 = 48\left(\frac{1}{2}\right) = 24$$
$$a_3 = 24\left(\frac{1}{2}\right) = 12$$
$$a_4 = 12\left(\frac{1}{2}\right) = 6$$
$$a_5 = 6\left(\frac{1}{2}\right) = 3$$
The first five terms are 48, 24, 12, 6, 3.

9. $a_n = a_1 + (n-1)d$
$a_1 = 12, \ d = 3$
$a_n = 12 + (n-1)3 = 12 + 3n - 3 = 3n + 9$
$a_8 = 3(8) + 9 = 24 + 9 = 33$

11. $a_n = a_1 r^{n-1}$
$a_1 = 7, \ r = -5$
$a_n = 7(-5)^{n-1}$
$a_4 = 7(-5)^{4-1} = 7(-5)^3 = 7(-125) = -875$

13. $a_n = a_1 + (n-1)d$
$a_1 = -4, \ d = -4$
$a_n = -4 + (n-1)(-4) = -4 - 4n + 4 = -4n$
$a_{15} = -4(15) = -60$

15. $0, 12, 24, \ldots$

$a_n = a_1 + (n-1)d$

$a_9 = a_1 + (9-1)d = a_1 + 8d$

$a_1 = 0, \; d = a_2 - a_1 = 12 - 0 = 12$

$a_9 = a_1 + 8d = 0 + 8(12) = 96$

17. $20, 18, 16, \ldots$

$a_n = a_1 + (n-1)d$

$a_{25} = a_1 + (25-1)d = a_1 + 24d$

$a_1 = 20, \; d = a_2 - a_1 = 18 - 20 = -2$

$a_{25} = a_1 + 24d = 20 + 24(-2) = 20 - 48 = -28$

19. $2, -10, 50, \ldots$

$a_n = a_1 r^{n-1}$

$a_1 = 2, \; r = a_2 \div a_1 = -10 \div 2 = -5$

$a_n = 2(-5)^{n-1}$

$a_5 = 2(-5)^{5-1} = 2(-5)^4 = 2(625) = 1250$

21. $a_4 = 19, \; a_{15} = 52$

$\begin{cases} a_4 = a_1 + (4-1)d \\ a_{15} = a_1 + (15-1)d \end{cases}$ or $\begin{cases} 19 = a_1 + 3d \\ 52 = a_1 + 14d \end{cases}$

Multiply both sides of the second equation by -1 in order to solve for d.

$\begin{aligned} 19 &= a_1 + 3d \\ -52 &= -a_1 - 14d \\ \hline -33 &= -11d \end{aligned}$

$3 = d$

To find a_1, substitute $d = 3$ into the first equation.

$19 = a_1 + 3(3)$

$19 = a_1 + 9$

$10 = a_1$

Thus, $a_1 = 10$ and $d = 3$, so

$a_n = 10 + (n-1)3$

$ = 10 + 3n - 3$

$ = 7 + 3n$

and $a_8 = 7 + 3(8) = 7 + 24 = 31$.

23. $a_2 = -1, \; a_4 = 5$

$\begin{cases} a_2 = a_1 + (2-1)d \\ a_4 = a_1 + (4-1)d \end{cases}$ or $\begin{cases} -1 = a_1 + d \\ 5 = a_1 + 3d \end{cases}$

Multiply both sides of the second equation by -1 in order to solve for d.

$\begin{aligned} -1 &= a_1 + d \\ -5 &= -a_1 - 3d \\ \hline -6 &= -2d \end{aligned}$

$3 = d$

To find a_1, substitute $d = 3$ into the first equation.

$-1 = a_1 + 3$

$-4 = a_1$

Thus, $a_1 = -4$ and $d = 3$, so

$a_n = -4(n-1)3$

$ = -4 + 3n - 3$

$ = -7 + 3n$

and $a_9 = -7 + 3(9) = -7 + 27 = 20$.

25. $a_2 = -\dfrac{4}{3}, \; a_3 = \dfrac{8}{3}$

Notice that $\dfrac{8}{3} \div -\dfrac{4}{3} = \dfrac{8}{3} \cdot -\dfrac{3}{4} = -2$,

so $r = -2$.

$a_2 = a_1(-2)^{2-1}$

$-\dfrac{4}{3} = a_1(-2)$

$\dfrac{2}{3} = a_1$

The first term is $\dfrac{2}{3}$ and the common ratio is -2.

27. Answers may vary.

29. $2, 4, 6$ is an arithmetic sequence.

$a_1 = 2$ and $d = a_2 - a_1 = 4 - 2 = 2$

31. $5, 10, 20$ is a geometric sequence.

$a_1 = 5$ and $r = a_2 \div a_1 = 10 \div 5 = 2$

33. $\dfrac{1}{2}, \dfrac{1}{10}, \dfrac{1}{50}$ is a geometric sequence.

$a_1 = \dfrac{1}{2}$ and

$r = a_2 \div a_1 = \dfrac{1}{10} \div \dfrac{1}{2} = \dfrac{1}{10} \cdot \dfrac{2}{1} = \dfrac{1}{5}$

35. $x, 5x, 25x$ is a geometric sequence.
$a_1 = x$ and $r = a_2 \div a_1 = 5x \div x = 5$

37. $p, \; p+4, \; p+8$ is an arithmetic sequence.
$a_1 = p$ and $d = a_2 - a_1 = (p+4) - p = 4$

39. $a_n = a_1 + (n-1)d$

$a_1 = 14, \; d = \dfrac{1}{4}$

$a_n = 14 + (n-1)\dfrac{1}{4} = 14 + \dfrac{1}{4}n - \dfrac{1}{4} = \dfrac{1}{4}n + \dfrac{55}{4}$

$a_{21} = \dfrac{1}{4}(21) + \dfrac{55}{4} = \dfrac{21}{4} + \dfrac{55}{4} = \dfrac{76}{4}$

41. $a_n = a_1 r^{n-1}$

$a_1 = 3, \; r = -\dfrac{2}{3}$

$a_n = 3\left(-\dfrac{2}{3}\right)^{n-1}$

$a_4 = 3\left(-\dfrac{2}{3}\right)^{4-1} = 3\left(-\dfrac{2}{3}\right)^3 = 3\left(-\dfrac{8}{27}\right) = -\dfrac{8}{9}$

43. $\dfrac{3}{2}, \; 2, \; \dfrac{5}{2}, \ldots$

$a_n = a_1 + (n-1)d$

$a_{15} = a_1 + (15-1)d = a_1 + 14d$

$a_1 = \dfrac{3}{2}, \; d = a_2 - a_1 = 2 - \dfrac{3}{2} = \dfrac{1}{2}$

$a_{15} = a_1 + 14d = \dfrac{3}{2} + 14\left(\dfrac{1}{2}\right) = \dfrac{3}{2} + \dfrac{14}{2} = \dfrac{17}{2}$

45. $24, \; 8, \; \dfrac{8}{3}, \ldots$

$a_n = a_1 r^{n-1}$

$a_1 = 24, \; r = a_2 \div a_1 = 8 \div 24 = \dfrac{8}{24} = \dfrac{1}{3}$

$a_n = 24\left(\dfrac{1}{3}\right)^{n-1}$

$a_6 = 24\left(\dfrac{1}{3}\right)^{6-1} = 24\left(\dfrac{1}{3}\right)^5 = 24\left(\dfrac{1}{243}\right) = \dfrac{8}{81}$

47. $a_3 = 2, \; a_{17} = -40$

$\begin{cases} a_3 = a_1 + (3-1)d \\ a_{17} = a_1 + (17-1)d \end{cases}$ or $\begin{cases} 2 = a_1 + 2d \\ -40 = a_1 + 16d \end{cases}$

Multiply both sides of the second equation by -1 in order to solve for d.

$\begin{array}{r} 2 = \quad a_1 + 2d \\ 40 = -a_1 - 16d \\ \hline 42 = \quad\quad -14d \end{array}$

$-3 = d$

To find a_1, substitute $d = -3$ into the first equation.

$2 = a_1 + 2(-3)$

$2 = a_1 - 6$

$8 = a_1$

Thus, $a_1 = 8$ and $d = -3$, so

$a_n = 8 + (n-1)(-3)$

$\quad = 8 - 3n + 3$

$\quad = 11 - 3n$

and $a_{10} = 11 - 3(10) = 11 - 30 = -19$.

49. $54, 58, 62, \ldots$

$a_1 = 54, \; d = a_2 - a_1 = 58 - 54 = 4$

The general term of the sequence is

$a_n = 54 + (n-1)4$ or $a_n = 4n + 50$.

$a_{20} = 4(20) + 50 = 80 + 50 = 130$

There are 130 seats in the twentieth row.

51. $a_1 = 6$, $r = 3$

$a_n = a_1 r^{n-1} = 6(3)^{n-1} = 2 \cdot 3 \cdot 3^{n-1} = 2(3)^n$

The general term of the sequence is

$a_n = 6(3)^{n-1}$ or $a_n = 2(3)^n$.

53. $a_1 = 486$, $r = \frac{1}{3}$

$a_n = a_1 r^{n-1}$

Initial height $= a_1 = 486\left(\frac{1}{3}\right)^{1-1} = 486$

Rebound 1 $= a_2 = 486\left(\frac{1}{3}\right)^{2-1} = 162$

Rebound 2 $= a_3 = 486\left(\frac{1}{3}\right)^{3-1} = 54$

Rebound 3 $= a_4 = 486\left(\frac{1}{3}\right)^{4-1} = 18$

Rebound 4 $= a_5 = 486\left(\frac{1}{3}\right)^{5-1} = 6$

The first five terms of the sequence are 486, 162, 54, 18, 6.

$a_n = 486\left(\frac{1}{3}\right)^{n-1}$

Solve: $486\left(\frac{1}{3}\right)^{n-1} = 1$

$\left(\frac{1}{3}\right)^{n-1} = \frac{1}{486}$

$\frac{1}{3^{n-1}} = \frac{1}{486}$

$3^{n-1} = 486$

$3^n = 1458$

$n = \log_3 1458$

$n = \frac{\ln 1458}{\ln 3} \approx 6.6$

Since a_7 is less than one foot and a_7 corresponds to the 6th rebound, the ball will rebound less than one foot on the 6th bounce.

55. $a_1 = 4000$, $d = 125$

$a_n = a_1 + (n-1)d$

$a_n = 4000 + (n-1)125$ or

$a_n = 3875 + 125n$

$a_{12} = 3875 + 125(12) = 3875 + 1500 = 5375$

His salary for his last month of training is \$5375.

57. $a_1 = 400$, $r = \frac{1}{2}$

12 hrs = 4(3 hrs), so we seek the fourth term after a_1, namely a_5.

$a_n = a_1 r^{n-1}$

$a_n = 400\left(\frac{1}{2}\right)^{n-1}$

$a_5 = 400\left(\frac{1}{2}\right)^4 = \frac{400}{16} = 25$

25 grams of the radioactive material remain after 12 hours.

59. $a_1 = \$11,782.40$, $r = 0.5$

$a_n = 11,782.40(0.5)^{n-1}$

$a_1 = \$11,782.40$

$a_2 = 11,782.40(0.5)^{2-1} = 11,782.40(0.5)^1 = 5891.20$

$a_3 = 11,782.40(0.5)^{3-1} = 11,782.40(0.5)^2 = 2945.60$

$a_4 = 11,782.40(0.5)^{4-1} = 11,782.40(0.5)^3 = 1472.80$

The first four terms of the sequence are $\$11,782.40$, $\$5891.20$, $\$2945.60$, $\$1472.80$.

61. $a_1 = 19.652$, $d = -0.034$

$a_n = 19.652 + (n-1)(-0.034)$

$a_1 = 19.652$

$a_2 = 19.652 + (2-1)(-0.034) = 19.652 + 1(-0.034) = 19.618$

$a_3 = 19.652 + (3-1)(-0.034) = 19.652 + 2(-0.034) = 19.584$

$a_4 = 19.652 + (4-1)(-0.034) = 19.652 + 3(-0.034) = 19.550$

The first four terms of the sequence are 19.652, 19.618, 19.584, 19.550.

63. Answers may vary.

65. $\dfrac{1}{3(1)} + \dfrac{1}{3(2)} + \dfrac{1}{3(3)} = \dfrac{1}{3}\left(\dfrac{1}{1} + \dfrac{1}{2} + \dfrac{1}{3}\right) = \dfrac{1}{3}\left(\dfrac{6}{6} + \dfrac{3}{6} + \dfrac{2}{6}\right) = \dfrac{1}{3}\left(\dfrac{11}{6}\right) = \dfrac{11}{18}$

67. $3^0 + 3^1 + 3^2 + 3^3 = 1 + 3 + 9 + 27 = 40$

69. $\dfrac{8-1}{8+1} + \dfrac{8-2}{8+2} + \dfrac{8-3}{8+3} = \dfrac{7}{9} + \dfrac{6}{10} + \dfrac{5}{11} = \dfrac{770}{990} + \dfrac{594}{990} + \dfrac{450}{990} = \dfrac{1814}{990} = \dfrac{907}{495}$

Exercise Set 11.3

1. $\displaystyle\sum_{i=1}^{4}(i-3) = (1-3)+(2-3)+(3-3)+(4-3) = -2 + (-1) + 0 + 1 = -2$

3. $\displaystyle\sum_{i=4}^{7}(2i+4) = [2(4)+4] + [2(5)+4] + [2(6)+4] + [2(7)+4] = 12 + 14 + 16 + 18 = 60$

5. $\displaystyle\sum_{i=2}^{4}(i^2-3) = (2^2-3)+(3^2-3)+(4^2-3) = 1 + 6 + 13 = 20$

7. $\displaystyle\sum_{i=1}^{3}\left(\dfrac{1}{i+5}\right) = \dfrac{1}{1+5} + \dfrac{1}{2+5} + \dfrac{1}{3+5} = \dfrac{1}{6} + \dfrac{1}{7} + \dfrac{1}{8} = \dfrac{28}{168} + \dfrac{24}{168} + \dfrac{21}{168} = \dfrac{73}{168}$

9. $\displaystyle\sum_{i=1}^{3}\dfrac{1}{6i} = \dfrac{1}{6(1)} + \dfrac{1}{6(2)} + \dfrac{1}{6(3)} = \dfrac{1}{6} + \dfrac{1}{12} + \dfrac{1}{18} = \dfrac{6+3+2}{36} = \dfrac{11}{36}$

11. $\displaystyle\sum_{i=2}^{6} 3i = 3(2) + 3(3) + 3(4) + 3(5) + 3(6) = 6 + 9 + 12 + 15 + 18 = 60$

13. $\displaystyle\sum_{i=3}^{5} i(i+2) = 3(3+2) + 4(4+2) + 5(5+2) = 15 + 24 + 35 = 74$

15. $\displaystyle\sum_{i=1}^{5} 2^i = 2^1 + 2^2 + 2^3 + 2^4 + 2^5 = 2 + 4 + 8 + 16 + 32 = 62$

17. $\displaystyle\sum_{i=1}^{4} \frac{4i}{i+3} = \frac{4(1)}{1+3} + \frac{4(2)}{2+3} + \frac{4(3)}{3+3} + \frac{4(4)}{4+3} = 1 + \frac{8}{5} + 2 + \frac{16}{7} = \frac{105}{35} + \frac{56}{35} + \frac{80}{35} = \frac{241}{35}$

19. $1 + 3 + 5 + 7 + 9 = [(2) - 1] + [2(2) - 1] + [2(3) - 1] + [2(4) - 1] + [2(5) - 1] = \displaystyle\sum_{i=1}^{5}(2i - 1)$

21. $4 + 12 + 36 + 108 = 4 + 4(3) + 4(3^2) + 4(3^3) = \displaystyle\sum_{i=1}^{4} 4(3)^{i-1}$

23. $12 + 9 + 6 + 3 + 0 + (-3)$
$= [-3(1) + 15] + [-3(2) + 15] + [-3(3) + 15] + [-3(4) + 15] + [-3(5) + 15] + [-3(6) + 15]$
$= \displaystyle\sum_{i=1}^{6}(-3i + 15)$

25. $12 + 4 + \dfrac{4}{3} + \dfrac{4}{9} = \dfrac{4}{3^{-1}} + \dfrac{4}{3^0} + \dfrac{4}{3^1} + \dfrac{4}{3^2} = \displaystyle\sum_{i=1}^{4} \frac{4}{3^{i-2}}$

27. $1 + 4 + 9 + 16 + 25 + 36 + 49 = 1^2 + 2^2 + 3^2 + 4^2 + 5^2 + 6^2 + 7^2 = \displaystyle\sum_{i=1}^{7} i^2$

29. $S_2 = \displaystyle\sum_{i=1}^{2}(i+2)(i-5) = (1+2)(1-5) + (2+2)(2-5) = 3(-4) + 4(-3) = -12 + (-12) = -24$

31. $S_2 = \displaystyle\sum_{i=1}^{2} i(i-6) = 1(1-6) + 2(2-6) = 1(-5) + 2(-4) = -5 + (-8) = -13$

33. $S_4 = \sum_{i=1}^{4} (i+3)(i+1)$

$= (1+3)(1+1) + (2+3)(2+1) + (3+3)(3+1) + (4+3)(4+1)$
$= 4(2) + 5(3) + 6(4) + 7(5)$
$= 8 + 15 + 24 + 35$
$= 82$

35. $S_4 = \sum_{i=1}^{4} (-2i) = -2(1) + (-2)(2) + (-2)(3) + (-2)(4) = -2 - 4 - 6 - 8 = -20$

37. $S_3 = \sum_{i=1}^{3} \left(-\frac{i}{3}\right) = \left(-\frac{1}{3}\right) + \left(-\frac{2}{3}\right) + \left(-\frac{3}{3}\right) = -\frac{6}{3} = -2$

39. 1, 2, 3, ..., 10

$a_n = n$

$S_{10} = \sum_{i=1}^{10} i = 1 + 2 + 3 + \cdots + 10 = 55$

A total of 55 trees were planted.

41. $a_1 = 6, \ r = 2$

$a_n = a_1 r^{n-1}$

The general term of the sequence is $a_n = 6 \cdot 2^{n-1}$.

$a_5 = 6 \cdot 2^{5-1} = 6 \cdot 2^4 = 6 \cdot 16 = 96$

There will be 96 fungus units at the beginning of the fifth day.

43. $a_1 = 50, r = 2$

$a_n = a_1 r^n$

The general term of the sequence is $a_n = 50(2)^n$, where n represents the number of 12-hour periods.

$a_4 = 50(2)^4 = 50(16) = 800$

There are 800 bacteria after 48 hours.

45. $a_n = (n+1)(n+2) = n^2 + 3n + 2$

$a_4 = 4^2 + 3(4) + 2 = 30$

30 opossums were killed in the fourth month.

$S_4 = \sum_{i=1}^{4} (i+1)(i+2)$

$= (1+1)(1+2) + (2+1)(2+2) + (3+1)(3+2) + (4+1)(4+2)$
$= 2(3) + 3(4) + 4(5) + 5(6)$
$= 6 + 12 + 20 + 30$
$= 68$

68 opossums were killed in the first four months.

47. $a_n = 100(0.5)^n$

$a_4 = 100(0.5)^4 = 100(0.0625) = 6.25$

There were 6.25 pounds of decay in the fourth year.

$$S_4 = \sum_{i=1}^{4} 100(0.5)^i$$

$$= 100(0.5)^1 + 100(0.5)^2 + 100(0.5)^3 + 100(0.5)^4$$
$$= 100(0.5) + 100(0.25) + 100(0.125) + 100(0.0625)$$
$$= 50 + 25 + 12.5 + 6.25$$
$$= 93.75$$

There were 93.75 pounds of decay in the first four years.

49. $a_1 = 40$, $r = \dfrac{4}{5}$

$$a_n = a_1 r^{n-1}$$

$$a_n = 40\left(\frac{4}{5}\right)^{n-1}$$

$$a_5 = 40\left(\frac{4}{5}\right)^{5-1} = 40\left(\frac{4}{5}\right)^4 = 16.384 \text{ or } 16.4$$

The length of the fifth swing is approximately 16.4 inches.

$$S_5 = \sum_{i=1}^{5} 40\left(\frac{4}{5}\right)^{i-1}$$

$$= 40\left(\frac{4}{5}\right)^{1-1} + 40\left(\frac{4}{5}\right)^{2-1} + 40\left(\frac{4}{5}\right)^{3-1} + 40\left(\frac{4}{5}\right)^{4-1} + 40\left(\frac{4}{5}\right)^{5-1}$$

$$= 40\left(\frac{4}{5}\right)^0 + 40\left(\frac{4}{5}\right)^1 + 40\left(\frac{4}{5}\right)^2 + 40\left(\frac{4}{5}\right)^3 + 40\left(\frac{4}{5}\right)^4$$

$$= 40 + \frac{160}{5} + \frac{640}{25} + \frac{2560}{125} + \frac{10,240}{625}$$

$$= \frac{16,808}{125}$$

$$= 134.464$$

$$\approx 134.5$$

The total length swung during the first five swings is 134.5 inches.

51. a. $\displaystyle\sum_{i=1}^{7} i + i^2 = (1+1^2)+(2+2^2)+(3+3^2)+(4+4^2)+(5+5^2)+(6+6^2)+(7+7^2)$

$$= 2+6+12+20+30+42+56$$

b. $\displaystyle\sum_{i=1}^{7} i + \sum_{i=1}^{7} i^2 = (1+2+3+4+5+6+7)+(1+4+9+16+25+36+49)$

c. They are equal; 168

d. True; consider the commutative property of addition.

53. $\dfrac{5}{1-\frac{1}{2}} = \dfrac{5}{\frac{1}{2}} = 5 \cdot \dfrac{2}{1} = 10$

55. $\dfrac{\frac{1}{3}}{1-\frac{1}{10}} = \dfrac{\frac{1}{3}}{\frac{9}{10}} = \dfrac{1}{3} \cdot \dfrac{10}{9} = \dfrac{10}{27}$

57. $\dfrac{3(1-2^4)}{1-2} = \dfrac{3(1-16)}{-1}$

$$= \dfrac{3(-15)}{-1}$$

$$= \dfrac{-45}{-1}$$

$$= 45$$

59. $\dfrac{10}{2}(3+15) = \dfrac{10}{2}(18) = \dfrac{180}{2} = 90$

Exercise Set 11.4

1. 1, 3, 5, 7, . . .
The first term of this arithmetic sequence is $a_1 = 1$ and the sixth term is $a_6 = 11$.
$$S_6 = \dfrac{6}{2}(a_1+a_6) = 3(1+11) = 3(12) = 36$$

3. 4, 12, 36, . . .
This is a geometric sequence with $a_1 = 4$ and $r = 3$.
$$S_5 = \dfrac{a_1(1-r^5)}{1-r}$$
$$= \dfrac{4(1-3^5)}{1-3}$$
$$= -2(1-243)$$
$$= -2(-242)$$
$$= 484$$

5. 3, 6, 9, . . .
The first term of this arithmetic sequence is $a_1 = 3$ and the sixth term is $a_6 = 18$.
$$S_6 = \dfrac{6}{2}(a_1+a_6) = 3(3+18) = 3(21) = 63$$

7. $2, \dfrac{2}{5}, \dfrac{2}{25}, \ldots$

This is a geometric sequence with $a_1 = 2$

and $r = \dfrac{1}{5}$.

$$S_4 = \dfrac{a_1(1 - r^4)}{1 - r}$$

$$= \dfrac{2\left[1 - \left(\frac{1}{5}\right)^4\right]}{1 - \frac{1}{5}}$$

$$= \dfrac{5}{2}\left(1 - \dfrac{1}{625}\right)$$

$$= \dfrac{5}{2}\left(\dfrac{624}{625}\right)$$

$$= \dfrac{312}{125}$$

$$= 2.496$$

9. $1, 2, 3, \ldots, 10$

The first term of this arithmetic sequence is $a_1 = 1$ and the tenth term is $a_{10} = 10$.

$$S_{10} = \dfrac{10}{2}(a_1 + a_{10}) = 5(1 + 10) = 5(11) = 55$$

11. $1, 3, 5, 7$

The first term of this arithmetic sequence is $a_1 = 1$ and the fourth term is $a_4 = 7$.

$$S_4 = \dfrac{4}{2}(a_1 + a_4) = 2(1 + 7) = 2(8) = 16$$

13. $12, 6, 3, \ldots$

$a_1 = 12$, $r = a_2 \div a_1 = 6 \div 12 = \dfrac{6}{12} = \dfrac{1}{2}$

$$S_\infty = \dfrac{a_1}{1 - r} = \dfrac{12}{1 - \frac{1}{2}} = \dfrac{12}{\frac{1}{2}} = 24$$

15. $\dfrac{1}{10}, \dfrac{1}{100}, \dfrac{1}{1000}, \ldots$

$a_1 = \dfrac{1}{10}$,

$$r = a_2 \div a_1 = \dfrac{1}{100} \div \dfrac{1}{10} = \dfrac{1}{100} \cdot \dfrac{10}{1} = \dfrac{1}{10}$$

$$S_\infty = \dfrac{a_1}{1 - r} = \dfrac{\frac{1}{10}}{1 - \frac{1}{10}} = \dfrac{\frac{1}{10}}{\frac{9}{10}} = \dfrac{1}{9}$$

17. $-10, -5, -\dfrac{5}{2}, \ldots$

$a_1 = -10$,

$$r = a_2 \div a_1 = -5 \div (-10) = \dfrac{-5}{-10} = \dfrac{1}{2}$$

$$S_\infty = \dfrac{a_1}{1 - r} = \dfrac{-10}{1 - \frac{1}{2}} = \dfrac{-10}{\frac{1}{2}} = -20$$

19. $2, -\dfrac{1}{4}, \dfrac{1}{32}, \ldots$

$a_1 = 2$,

$$r = a_2 \div a_1 = -\dfrac{1}{4} \div 2 = -\dfrac{1}{4} \cdot \dfrac{1}{2} = -\dfrac{1}{8}$$

$$S_\infty = \dfrac{a_1}{1 - r} = \dfrac{2}{1 - \left(-\frac{1}{8}\right)} = \dfrac{2}{\frac{9}{8}} = \dfrac{16}{9}$$

21. $\dfrac{2}{3}, -\dfrac{1}{3}, \dfrac{1}{6}, \ldots$

$a_1 = \dfrac{2}{3}$,

$$r = a_2 \div a_1 = -\dfrac{1}{3} \div \dfrac{2}{3} = -\dfrac{1}{3} \cdot \dfrac{3}{2} = -\dfrac{1}{2}$$

$$S_\infty = \dfrac{a_1}{1 - r} = \dfrac{\frac{2}{3}}{1 - \left(-\frac{1}{2}\right)} = \dfrac{\frac{2}{3}}{\frac{3}{2}} = \dfrac{4}{9}$$

23. $-4, 1, 6, \ldots, 41$

The first term of this arithmetic sequence is $a_1 = -4$ and the tenth term is $a_{10} = 41$.

$$S_{10} = \dfrac{10}{2}(a_1 + a_{10})$$

$$= 5(-4 + 41)$$

$$= 5(37)$$

$$= 185$$

25. $3, \dfrac{3}{2}, \dfrac{3}{4}, \ldots$

This is a geometric sequence with $a_1 = 3$

and $r = \dfrac{1}{2}$.

$$S_7 = \frac{a_1(1 - r^7)}{1 - r}$$

$$= \frac{3\left[1 - \left(\frac{1}{2}\right)^7\right]}{1 - \frac{1}{2}}$$

$$= 6\left(1 - \frac{1}{128}\right)$$

$$= 6\left(\frac{127}{128}\right)$$

$$= \frac{381}{64}$$

27. $-12, \; 6, \; -3, \; \ldots$

This ia a geometric sequence with

$a_1 = -12$ and $r = -\dfrac{1}{2}$.

$$S_5 = \frac{a_1(1 - r^5)}{1 - r}$$

$$= \frac{-12\left[1 - \left(-\frac{1}{2}\right)^5\right]}{1 - \left(-\frac{1}{2}\right)}$$

$$= -8\left(1 + \frac{1}{32}\right)$$

$$= -8\left(\frac{33}{32}\right)$$

$$= -\frac{33}{4} \quad \text{or} \quad -8.25$$

29. $\dfrac{1}{2}, \dfrac{1}{4}, \; 0, \; \ldots, \; -\dfrac{17}{4}$

The first term of this arithmetic sequence

is $a_1 = \dfrac{1}{2}$ and the twentieth term is

$a_{20} = -\dfrac{17}{4}$.

$$S_{20} = \frac{20}{2}(a_1 + a_{20})$$

$$= 10\left(\frac{1}{2} - \frac{17}{4}\right)$$

$$= 10\left(-\frac{15}{4}\right)$$

$$= -\frac{75}{2}$$

31. This is a geometric sequence with

$a_1 = 8$ and $r = -\dfrac{2}{3}$.

$$S_3 = \frac{a_1(1 - r^3)}{1 - r}$$

$$= \frac{8\left[1 - \left(-\frac{2}{3}\right)^3\right]}{1 - \left(-\frac{2}{3}\right)}$$

$$= \frac{24}{5}\left(1 + \frac{8}{27}\right)$$

$$= \frac{24}{5}\left(\frac{35}{27}\right)$$

$$= \frac{8}{1}\left(\frac{7}{9}\right)$$

$$= \frac{56}{9}$$

33. The first five terms of the sequence are
4000, 3950, 3900, 3850, 3800.

$a_1 = 4000, \ d = -50, \ n = 12$

$a_n = a_1 + (n-1)d$

$a_n = 4000 + (12-1)(-50)$

$\quad = 4000 + 11(-50)$

$\quad = 4000 - 550$

$\quad = 3450$

There were 3450 car sales predicted for the twelfth month.

$S_{12} = \dfrac{12}{2}(a_1 + a_{12})$

$\quad = \dfrac{12}{2}(4000 + 3450)$

$\quad = 6(7450)$

$\quad = 44,700$

$44,700$ cars were predicted to be sold in the first 12 months.

35. Firm A:

The first term of this arithmetic sequence is $a_1 = 22,000$ and the tenth term is $a_{10} = 31,000$.

$S_{10} = \dfrac{10}{2}(a_1 + a_{10})$

$\quad = 5(22,000 + 31,000)$

$\quad = 5(53,000)$

$\quad = \$265,000$

Firm B:

The first term of this arithmetic sequence is $a_1 = 20,000$ and the tenth term is $a_{10} = 30,800$.

$S_{10} = \dfrac{10}{2}(a_1 + a_{10})$

$\quad = \dfrac{10}{2}(20,000 + 30,800)$

$\quad = 5(50,800)$

$\quad = \$254,000$

Thus, Firm A is making the more profitable offer.

37. $a_1 = 30,000, \ r = 1.10$

$a_n = 30,000(1.10)^{n-1}$

$a_4 = 30,000(1.10)^{4-1}$

$\quad = 30,000(1.10)^3$

$\quad = 39,930$

She made \$39,930 during her fourth year of business.

$S_4 = \dfrac{a_1(1 - r^4)}{1 - r}$

$\quad = \dfrac{30,000(1 - 1.10^4)}{1 - 1.10}$

$\quad = 139,230$

Her total earnings were \$139,230 during the first four years.

39. $a_1 = 30, \ r = 0.9$

$a_n = 30(0.9)^{n-1}$

$a_5 = 30(0.9)^{5-1}$

$\quad = 30(0.9)^4$

$\quad = 19.683$

He takes approximately 20 minutes to assemble the fifth computer.

$S_5 = \dfrac{a_1(1 - r^5)}{1 - r} = \dfrac{30(1 - 0.9^5)}{1 - 0.9} = 122.853$

It takes approximately 123 minutes to assemble the first 5 computers.

41. $a_1 = 20, \ r = \dfrac{4}{5}$

$S_\infty = \dfrac{20}{1 - \frac{4}{5}} = \dfrac{20}{\frac{1}{5}} = 100$

We double the number (to account for the flight up as well as down) and subtract 20 (since the first bounce was preceded by only a downward flight). Thus, the ball traveled a total distance of $2(100) - 20 = 180$ feet.

43. Player A:
The first term of this arithmetic sequence is $a_1 = 1$ and the ninth term is $a_9 = 9$.
$$S_9 = \frac{9}{2}(a_1 + a_9) = \frac{9}{2}(1 + 9) = \frac{9}{2}(10) = 45$$
Player A's score is 45.

Player B:
The first term of this arithmetic sequence is $a_1 = 10$ and the sixth term is $a_6 = 15$.
$$S_6 = \frac{6}{2}(a_1 + a_6) = 3(10 + 15) = 3(25) = 75$$
Player B's score is 75 points.

45. The first term of this arithmetic sequence is $a_1 = 200$ and the common difference is $d = -5$.
$$\begin{aligned}
a_n &= a_1 + (n-1)d \\
&= 200 + (n-1)(-5) \\
&= 200 - 5n + 5 \\
&= 205 - 5n
\end{aligned}$$
$$a_{20} = 205 - 5(20) = 205 - 100 = 105$$
The rental fee for the twentieth day was $105.
$$\begin{aligned}
S_{20} &= \frac{20}{2}(a_1 + a_{20}) \\
&= 10(200 + 105) \\
&= 10(305) \\
&= 3050
\end{aligned}$$
Thus, $3050 rent is paid for 20 days during the holiday rush.

47. $a_1 = 0.01, \; r = 2$
$$\begin{aligned}
S_{30} &= \frac{a_1(1 - r^{30})}{1 - r} \\
&= \frac{0.01(1 - 2^{30})}{1 - 2} \\
&= 10,737,418.23
\end{aligned}$$
He would pay $10,737,418.23 in room and board for the 30 days.

49. $0.88\overline{8} = 0.8 + 0.08 + 0.008 + \cdots$
$$= \frac{8}{10} + \frac{8}{100} + \frac{8}{1000} + \cdots$$
This is a geometric series with $a_1 = \dfrac{8}{10}$
and $r = \dfrac{1}{10}$.
$$S_\infty = \frac{\frac{8}{10}}{1 - \frac{1}{10}} = \frac{\frac{8}{10}}{\frac{9}{10}} = \frac{8}{10} \cdot \frac{10}{9} = \frac{8}{9}$$

51. Answers may vary.

53. $6 \cdot 5 \cdot 4 \cdot 3 \cdot 2 \cdot 1 = (30)(12)(2) = 720$

55. $\dfrac{3 \cdot 2 \cdot 1}{2 \cdot 1} = 3$

57. $\begin{aligned}[t] (x+5)^2 &= x^2 + 2(5x) + 25 \\ &= x^2 + 10x + 25 \end{aligned}$

59. $(2x - 1)^3$
$$\begin{aligned}
&= (2x - 1)(2x - 1)^2 \\
&= (2x - 1)(4x^2 - 4x + 1) \\
&= 8x^3 - 8x^2 + 2x - 4x^2 + 4x - 1 \\
&= 8x^3 - 12x^2 + 6x - 1
\end{aligned}$$

Exercise Set 11.5

1. $(m+n)^3 = 1 \cdot m^3 + 3 \cdot m^2n + 3 \cdot mn^2 + 1 \cdot n^3 = m^3 + 3m^2n + 3mn^2 + n^3$

3. $(c+d)^5 = 1 \cdot c^5 + 5 \cdot c^4d + 10 \cdot c^3d^2 + 10 \cdot c^2d^3 + 5 \cdot cd^4 + 1 \cdot d^5$
    $$= c^5 + 5c^4d + 10c^3d^2 + 10c^2d^3 + 5cd^4 + d^5$$

5. $(y-x)^5 = [y + (-x)]^5$
    $$= 1 \cdot y^5 + 5 \cdot y^4(-x) + 10 \cdot y^3(-x)^2 + 10 \cdot y^2(-x)^3 + 5 \cdot y(-x)^4 + 1 \cdot (-x)^5$$
    $$= y^5 - 5y^4x + 10y^3x^2 - 10y^2x^3 + 5yx^4 - x^5$$

7. Answers may vary.

9. $\dfrac{8!}{7!} = \dfrac{8 \cdot 7!}{7!} = 8$

11. $\dfrac{7!}{5!} = \dfrac{7 \cdot 6 \cdot 5!}{5!} = 7 \cdot 6 = 42$

13. $\dfrac{10!}{7!2!} = \dfrac{10 \cdot 9 \cdot 8 \cdot 7!}{7!2!} = \dfrac{10 \cdot 9 \cdot 8}{2 \cdot 1} = \dfrac{720}{2} = 360$

15. $\dfrac{8!}{6!0!} = \dfrac{8 \cdot 7 \cdot 6!}{6! \cdot 1} = 8 \cdot 7 = 56$

17. Let $a = a$, $b = b$, and $n = 7$ in the binomial theorem.
    $(a+b)^7$
    $$= a^7 + 7a^6b + \frac{7 \cdot 6}{2!}a^5b^2 + \frac{7 \cdot 6 \cdot 5}{3!}a^4b^3 + \frac{7 \cdot 6 \cdot 5 \cdot 4}{4!}a^3b^4 + \frac{7 \cdot 6 \cdot 5 \cdot 4 \cdot 3}{5!}a^2b^5 + \frac{7 \cdot 6 \cdot 5 \cdot 4 \cdot 3 \cdot 2}{6!}ab^6 + b^7$$
    $$= a^7 + 7a^6b + 21a^5b^2 + 35a^4b^3 + 35a^3b^4 + 21a^2b^5 + 7ab^6 + b^7$$

19. Let $a = a$, $b = 2b$, and $n = 5$ in the binomial theorem.
    $$(a+2b)^5 = a^5 + 5a^4(2b) + \frac{5 \cdot 4}{2!}a^3(2b)^2 + \frac{5 \cdot 4 \cdot 3}{3!}a^2(2b)^3 + \frac{5 \cdot 4 \cdot 3 \cdot 2}{4!}a(2b)^4 + (2b)^5$$
    $$= a^5 + 10a^4b + 40a^3b^2 + 80a^2b^3 + 80ab^4 + 32b^5$$

21. Let $a = q$, $b = r$, and $n = 9$ in the binomial theorem.
    $$(q+r)^9 = q^9 + 9q^8r + \frac{9 \cdot 8}{2!}q^7r^2 + \frac{9 \cdot 8 \cdot 7}{3!}q^6r^3 + \frac{9 \cdot 8 \cdot 7 \cdot 6}{4!}q^5r^4 + \frac{9 \cdot 8 \cdot 7 \cdot 6 \cdot 5}{5!}q^4r^5$$
    $$+ \frac{9 \cdot 8 \cdot 7 \cdot 6 \cdot 5 \cdot 4}{6!}q^3r^6 + \frac{9 \cdot 8 \cdot 7 \cdot 6 \cdot 5 \cdot 4 \cdot 3}{7!}q^2r^7 + \frac{9 \cdot 8 \cdot 7 \cdot 6 \cdot 5 \cdot 4 \cdot 3 \cdot 2}{8!}qr^8 + r^9$$
    $$= q^9 + 9q^8r + 36q^7r^2 + 84q^6r^3 + 126q^5r^4 + 126q^4r^5 + 84q^3r^6 + 36q^2r^7 + 9qr^8 + r^9$$

23. Let $a = 4a$, $b = b$, and $n = 5$ in the binomial theorem.
    $$(4a+b)^5 = (4a)^5 + 5(4a)^4b + \frac{5 \cdot 4}{2!}(4a)^3b^2 + \frac{5 \cdot 4 \cdot 3}{3!}(4a)^2b^3 + \frac{5 \cdot 4 \cdot 3 \cdot 2}{4!}(4a)b^4 + b^5$$
    $$= 1024a^5 + 1280a^4b + 640a^3b^2 + 160a^2b^3 + 20ab^4 + b^5$$

25. Let $a = 5a$, $b = -2b$, and $n = 4$ in the binomial theorem.

$$(5a - 2b)^4 = (5a)^4 + 4(5a)^3(-2b) + \frac{4 \cdot 3}{2!}(5a)^2(-2b)^2 + \frac{4 \cdot 3 \cdot 2}{3!}(5a)(-2b)^3 + (-2b)^4$$

$$= 625a^4 - 1000a^3b + 600a^2b^2 - 160ab^3 + 16b^4$$

27. Let $a = 2a$, $b = 3b$, and $n = 3$ in the binomial theorem.

$$(2a + 3b)^3 = (2a)^3 + 3(2a)^2(3b) + \frac{3 \cdot 2}{2!}(2a)(3b)^2 + (3b)^3 = 8a^3 + 36a^2b + 54ab^2 + 27b^3$$

29. Let $a = x$, $b = 2$, and $n = 5$ in the binomial theorem.

$$(x + 2)^5 = x^5 + 5x^4(2) + \frac{5 \cdot 4}{2!}x^3(2^2) + \frac{5 \cdot 4 \cdot 3}{3!}x^2(2^3) + \frac{5 \cdot 4 \cdot 3 \cdot 2}{4!}x(2^4) + 2^5$$

$$= x^5 + 10x^4 + 40x^3 + 80x^2 + 80x + 32$$

31. Let $a = c$, $b = -d$, $n = 5$ and $r = 4$ in the formula given.

$$\frac{5!}{4!(5 - 4)!}c^{5-4}(-d)^4 = 5cd^4$$

33. Let $a = 2c$, $b = d$, $n = 7$ and $r = 7$ in the formula given.

$$\frac{7!}{7!(7 - 7)!}(2c)^{7-7}d^7 = d^7$$

35. Let $a = 2r$, $b = -s$, $n = 5$ and $r = 3$ in the formula given.

$$\frac{5!}{3!(5 - 3)!}(2r)^{5-3}(-s)^3 = -40r^2s^3$$

37. Let $a = x$, $b = y$, $n = 4$ and $r = 2$ in the formula given.

$$\frac{4!}{2!(4 - 2)!}x^{4-2}y^2 = 6x^2y^2$$

39. Let $a = a$, $b = 3b$, $n = 10$ and $r = 1$ in the formula given.

$$\frac{10!}{1!(10 - 1)!}a^{10-1}(3b)^1 = 30a^9b$$

41. $f(x) = |x|$

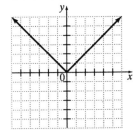

This function is not one-to-one since it does not pass the horizontal line test.

43. $H(x) = 2x + 3$

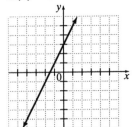

This function is one-to-one since it passes the vertical and horizontal line tests.

45. $f(x) = x^2 + 3$

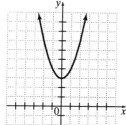

This function is not one-to-one since it does not pass the horizontal line test.

Chapter 11 Review

1. $a_n = -3n^2$

$a_1 = -3(1)^2 = -3$

$a_2 = -3(2)^2 = -12$

$a_3 = -3(3)^2 = -27$

$a_4 = -3(4)^2 = -48$

$a_5 = -3(5)^2 = -75$

The first five terms of the sequence are $-3, -12, -27, -48, -75$.

2. $a_n = n^2 + 2n$

$a_1 = 1^2 + 2(1) = 1 + 2 = 3$

$a_2 = 2^2 + 2(2) = 4 + 4 = 8$

$a_3 = 3^2 + 2(3) = 9 + 6 = 15$

$a_4 = 4^2 + 2(4) = 16 + 8 = 24$

$a_5 = 5^2 + 2(5) = 25 + 10 = 35$

The first five terms of the sequence are $3, 8, 15, 24, 35$.

3. $a_n = \dfrac{(-1)^n}{100}$

$a_{100} = \dfrac{(-1)^{100}}{100} = \dfrac{1}{100}$

4. $a_n = \dfrac{2n}{(-1)^2}$

$a_{50} = \dfrac{2(50)}{(-1)^2} = 100$

5. $\dfrac{1}{6}, \dfrac{1}{12}, \dfrac{1}{18}, \ldots$

or $\dfrac{1}{6\cdot1}, \dfrac{1}{6\cdot2}, \dfrac{1}{6\cdot3}, \ldots$

In general, $a_n = \dfrac{1}{6n}$.

6. $-1, 4, -9, 16, \ldots$

or $(-1)^1 \cdot 1^2, (-1)^2 \cdot 2^2, (-1)^3 \cdot 3^2,$

$(-1)^4 \cdot 4^2, \ldots$

In general, $a_n = (-1)^n n^2$.

7. $a_n = 32n - 16$

$a_5 = 32(5) - 16 = 160 - 16 = 144$

$a_6 = 32(6) - 16 = 192 - 16 = 176$

$a_7 = 32(7) - 16 = 224 - 16 = 208$

The olive fell 144 feet, 176 feet, and 208 feet during the fifth, sixth, and seventh seconds, respectively.

8. $a_n = 100(2)^{n-1}$

$10,000 = 100(2)^{n-1}$

$100 = 2^{n-1}$

$\log 100 = (n-1)\log 2$

$n = \dfrac{\log 100}{\log 2} + 1 \approx 7.6$

The culture will measure at least 10,000 on the eighth day.

$a_0 = 100(2)^{0-1} = 50$

Originally, the measure of the culture was 50.

9. $a_1 = 450$

$a_2 = 3(450) = 1350$

$a_3 = 3(1350) = 4050$

$a_4 = 3(4050) = 12,150$

$a_5 = 3(12,150) = 36,450$

This predicts that in 2003 (when $n = 5$) the number of infected people will be 36,450.

10. $a_1 = 50$, $d = 8$

$$a_n = a_1 + (n-1)d$$
$$= 50 + (n-1)8$$
$$= 50 + 8n - 8$$
$$= 8n + 42$$

$a_1 = 50$
$a_2 = 8(2) + 42 = 58$
$a_3 = 8(3) + 42 = 66$
$a_4 = 8(4) + 42 = 74$
$a_5 = 8(5) + 42 = 82$
$a_6 = 8(6) + 42 = 90$
$a_7 = 8(7) + 42 = 98$
$a_8 = 8(8) + 42 = 106$
$a_9 = 8(9) + 42 = 114$
$a_{10} = 8(10) + 42 = 122$

There are 122 seats in the tenth row.

11. $a_1 = -2$, $r = \frac{2}{3}$

$$a_n = a_1 r^{n-1}$$

$a_1 = -2$

$$a_2 = (-2)\left(\frac{2}{3}\right)^{2-1} = -\frac{4}{3}$$

$$a_3 = (-2)\left(\frac{2}{3}\right)^{3-1} = -\frac{8}{9}$$

$$a_4 = (-2)\left(\frac{2}{3}\right)^{4-1} = -\frac{16}{27}$$

$$a_5 = (-2)\left(\frac{2}{3}\right)^{5-1} = -\frac{32}{81}$$

The first five terms of the sequence are
$$-2, \ -\frac{4}{3}, \ -\frac{8}{9}, \ -\frac{16}{27}, \ -\frac{32}{81}.$$

12. $a_1 = 12$, $d = -1.5$

$$a_n = a_1 + (n-1)d$$
$$a_n = 12 + (n-1)(-1.5)$$
$$= 12 - 1.5n + 1.5$$
$$= 13.5 - 1.5n$$

$a_1 = 12$
$a_2 = 13.5 - 1.5(2) = 10.5$
$a_3 = 13.5 - 1.5(3) = 9$
$a_4 = 13.5 - 1.5(4) = 7.5$
$a_5 = 13.5 - 1.5(5) = 6$

The first five terms of the sequence are
12, 10.5, 9, 7.5, 6.

13. $a_1 = -5$, $d = 4$

$$a_n = a_1 + (n-1)d$$
$$a_n = -5 + (n-1)4$$
$$= -5 + 4n - 4$$
$$= 4n - 9$$

$$a_{30} = 4(30) - 9$$
$$= 120 - 9$$
$$= 111$$

14. $a_1 = 2$, $d = \frac{3}{4}$

$$a_n = a_1 + (n-1)d$$
$$a_n = 2 + (n-1)\frac{3}{4} = 2 + \frac{3}{4}n - \frac{3}{4} = \frac{3}{4}n - \frac{5}{4}$$
$$a_{11} = \frac{3}{4}(11) - \frac{5}{4} = \frac{33}{4} - \frac{5}{4} = \frac{28}{4} = \frac{19}{2}$$

15. 12, 7, 2, . . .

$a_1 = 12$, $d = -5$
$$a_n = a_1 + (n-1)d$$
$$a_n = 12 + (n-1)(-5) = 12 - 5n + 5 = 17 - 5n$$
$$a_{20} = 17 - 5(20) = 17 - 100 = -83$$

16. 4, 6, 9, . . .

$a_1 = 4$, $r = \frac{3}{2}$

$$a_n = a_1 r^{n-1}$$

$$a_n = 4\left(\frac{3}{2}\right)^{n-1}$$

$$a_6 = 4\left(\frac{3}{2}\right)^{6-1} = 4\left(\frac{3}{2}\right)^5 = \frac{4(243)}{32} = \frac{243}{8}$$

17. $a_4 = 18$, $a_{20} = 98$

$$\begin{cases} a_4 = a_1 + (4-1)d \\ a_{20} = a_1 + (20-1)d \end{cases} \text{ or } \begin{cases} 18 = a_1 + 3d \\ 98 = a_1 + 19d \end{cases}$$

Multiply both sides of the second equation by -1 in order to solve for d.

$$18 = \quad a_1 + \ 3d$$
$$\underline{-98 = \quad -a_1 - 19d}$$
$$-80 = \quad\quad\ -16d$$
$$5 = d$$

To find a_1, substitute $d = 5$ into the first equation.

$18 = a_1 + 3(5)$

$18 = a_1 + 15$

$3 = a_1$

The first term is $a_1 = 3$ and the common difference is $d = 5$.

18. $a_3 = -48$, $a_4 = 192$

Notice that $192 \div (-48) = -4$, so $r = -4$.

$a_3 = a_1(-4)^{3-1}$

$-48 = a_1(-4)^2$

$-3 = a_1$

The first term is $a_1 = -3$ and the common ratio is $r = -4$.

19. $\dfrac{3}{10}, \dfrac{3}{100}, \dfrac{3}{1000}, \ldots$ or

$\dfrac{3}{10}, \dfrac{3}{10^2}, \dfrac{3}{10^3}, \ldots$

In general, $a_n = \dfrac{3}{10^n}$.

20. $50, 58, 66, \ldots$

$a_1 = 50$, $d = a_2 - a_1 = 58 - 50 = 8$

$a_n = a_1 + (n-1)d$

In general, $a_n = 50 + (n-1)8$ or

$a_n = 42 + 8n$.

21. $\dfrac{8}{3}, 4, 6, \ldots$ is a geometric sequence.

$a_1 = \dfrac{8}{3}$, $r = a_2 \div a_1 = 4 \div \dfrac{8}{3} = 4 \cdot \dfrac{3}{8} = \dfrac{3}{2}$

22. $-10.5, -6.1, -1.7$ is an arithmetic sequence.

$a_1 = -10.5$,

$d = a_2 - a_1 = -6.1 - (-10.5) = 4.4$

23. $7x, -14x, 28x$ is a geometric sequence.

$a_1 = 7x$, $r = a_2 \div a_1 = -14x \div 7x = -2$

24. $3x^2, 9x^4, 81x^8, \ldots$ is neither an arithmetic nor a geometric sequence.

25. $a_1 = 8$, $r = 0.75$

$a_n = a_1 r^{n-1}$

$a_1 = 8$

$a_2 = 8(0.75)^1 = 6$

$a_3 = 8(0.75)^2 = 4.5$

$a_4 = 8(0.75)^3 \approx 3.4$

$a_5 = 8(0.75)^4 \approx 2.5$

$a_6 = 8(0.75)^5 \approx 1.9$

$8, 6, 4.5, 3.4, 2.5, 1.9$

Yes, a ball that rebounds to a height of 2.5 feet after the fifth bounce is good since $2.5 \geq 1.9$.

26. $25, 21, \ldots$

$a_1 = 25$, $d = -4$

In general, $a_n = 25 + (n-1)(-4)$ or

$a_n = 29 - 4n$.

$a_n = 29 - 4n$ is positive only for $n = 1$ to $n = 7$. Thus, the top row contains

$a_7 = 29 - 4(7) = 29 - 28 = 1$ can.

27. $1, 2, 4, \ldots$

$a_1 = 1$, $r = 2$

In general, $a_n = 1 \cdot 2^{n-1}$ or $a_n = 2^{n-1}$.

$a_{10} = 2^{10-1} = 2^9 = 512$

You will save $512 on the tenth day.

$a_{30} = 2^{30-1} = 2^{29} = 536,870,912$

You will save $536,870,912 on the thirtieth day.

28. $a_n = a_1 r^{n-1}$

$a_1 = 30, \ r = 0.7$

$a_n = 30(0.7)^{n-1}$

$a_5 = 30(0.7)^{5-1} = 30(0.7)^4 = 7.203$

The length of the arc on the fifth swing is 7.203 inches.

29. $a_1 = 900, \ d = 150$

In general, $a_n = 900 + (n-1)150$ or

$a_n = 150n + 750.$

$a_6 = 900 + (6-1)150 = 900 + 750 = 1650$

Her salary at the end of her training is $1650 per month.

30. $\dfrac{1}{512}, \ \dfrac{1}{256}, \ \dfrac{1}{128}, \ \ldots$

First fold: $a_1 = \dfrac{1}{256}, \ r = a_2 \div a_1 = \dfrac{1}{128} \div \dfrac{1}{256} = \dfrac{1}{128} \cdot \dfrac{256}{1} = 2$

$a_n = a_1 r^{n-1}$

$a_n = \dfrac{1}{256}(2)^{n-1}$

$a_{15} = \dfrac{1}{256}(2)^{15-1} = 64$

The thickness of the stack after 15 folds is 64 inches.

31. $\displaystyle\sum_{i=1}^{5}(2i-1) = [2(1)-1] + [2(2)-1] + [2(3)-1] + [2(4)-1] + [2(5)-1] = 1 + 3 + 5 + 7 + 9 = 25$

32. $\displaystyle\sum_{i=1}^{5}i(i+2) = 1(1+2) + 2(2+2) + 3(3+2) + 4(4+2) + 5(5+2) = 3 + 8 + 15 + 24 + 35 = 85$

33. $\displaystyle\sum_{i=2}^{4}\dfrac{(-1)^i}{2i} = \dfrac{(-1)^2}{2(2)} + \dfrac{(-1)^3}{2(3)} + \dfrac{(-1)^4}{2(4)} = \dfrac{1}{4} - \dfrac{1}{6} + \dfrac{1}{8} = \dfrac{6-4+3}{24} = \dfrac{5}{24}$

34. $\displaystyle\sum_{i=3}^{5}5(-1)^{i-1} = 5(-1)^{3-1} + 5(-1)^{4-1} + 5(-1)^{5-1} = 5(1) + 5(-1) + 5(1) = 5$

35. $a_n = (n-3)(n+2)$

$S_4 = \displaystyle\sum_{i=1}^{4}(i-3)(i+2) = (1-3)(1+2) + (2-3)(2+2) + (3-3)(3+2) + (4-3)(4+2) = -6 - 4 + 0 + 6 = -4$

36. $a_n = n^2$

$$S_6 = \sum_{i=1}^{6} i^2 = (1)^2 + (2)^2 + (3)^2 + (4)^2 + (5)^2 + (6)^2 = 1 + 4 + 9 + 16 + 25 + 36 = 91$$

37. $a_n = -8 + (n-1)3 = -8 + 3n - 3 = 3n - 11$

$$S_5 = \sum_{i=1}^{5} (3i - 11)$$
$$= [3(1) - 11] + [3(2) - 11] + [3(3) - 11] + [3(4) - 11] + [3(5) - 11]$$
$$= (3 - 11) + (6 - 11) + (9 - 11) + (12 - 11) + (15 - 11)$$
$$= -8 + (-5) + (-2) + 1 + 4$$
$$= -10$$

38. $a_n = 5(4)^{n-1}$

$$S_3 = \sum_{i=1}^{3} 5(4)^{i-1} = 5(4)^0 + 5(4)^1 + 5(4)^2 = 5 + 20 + 80 = 105$$

39. $1 + 3 + 9 + 27 + 81 + 243 = 3^0 + 3^1 + 3^2 + 3^3 + 3^4 + 3^5 = \sum_{i=1}^{6} 3^{i-1}$

40. $6 + 2 + (-2) + (-6) + (-10) + (-14) + (-18)$
$$= [10 - 4(1)] + [10 - 4(2)] + [10 - 4(3)] + [10 - 4(4)] + [10 - 4(5)] + [10 - 4(6)] + [10 - 4(7)]$$
$$= \sum_{i=1}^{7} (10 - 4i)$$

41. $\dfrac{1}{4} + \dfrac{1}{16} + \dfrac{1}{64} + \dfrac{1}{256} = \dfrac{1}{4^1} + \dfrac{1}{4^2} + \dfrac{1}{4^3} + \dfrac{1}{4^4} = \sum_{i=1}^{4} \dfrac{1}{4^i}$

42. $1 + \left(-\dfrac{3}{2}\right) + \dfrac{9}{4} = \left(-\dfrac{3}{2}\right)^0 + \left(-\dfrac{3}{2}\right)^1 + \left(-\dfrac{3}{2}\right)^2 = \sum_{i=1}^{3} \left(-\dfrac{3}{2}\right)^{i-1}$

43. $a_1 = 20,\ r = 2$

$a_n = 20(2)^n$ represents the number of yeast, where n represents the number of 8-hour periods.
Since $48 = 6(8)$ here, $n = 6$.
$a_6 = 20(2)^6 = 1280$
The total yeast after 48 hours is 1280.

44. $a_n = n^2 + 2n - 1$

$a_4 = (4)^2 + 2(4) - 1 = 23$

23 cranes were born in the fourth year.

$S_4 = \sum_{i=1}^{4} (i^2 + 2i - 1) = (1 + 2 - 1) + (4 + 4 - 1) + (9 + 6 - 1) + (16 + 8 - 1) = 2 + 7 + 14 + 23 = 46$

46 cranes were born in the first four years.

45. For job A:

$a_1 = 39,500, \ d = 2200$

$a_n = a_1 + (n-1)d$

$\quad = 39,500 + (n-1)2200$

$\quad = 39,500 + 2200n - 2200$

$\quad = 2200n + 37,300$

$a_5 = 2200(5) + 37,300$

$\quad = 11,000 + 37,300$

$\quad = 48,300$

The salary for the fifth year under job A is
$48,300.

For job B:

$a_1 = 41,000, \ d = 1400$

$a_n = a_1 + (n-1)d$

$\quad = 41,000 + (n-1)1400$

$\quad = 41,000 + 1400n - 1400$

$\quad = 1400n + 39,600$

$a_5 = 1400(5) + 39,600$

$\quad = 7000 + 39,600$

$\quad = 46,600$

The salary for the fifth year under job B
is $46,600. For the fifth year, job *A* has a
higher salary.

46. $a_n = 200(0.5)^n$

$a_3 = 200(0.5)^3 = 25$

The amount of decay in the third year
is 25 kilograms.

$S_3 = \sum_{i=1}^{3} 200(0.5)^i$

$\quad = 200(0.5) + 200(0.5)^2 + 200(0.5)^3$

$\quad = 100 + 50 + 25$

$\quad = 175$

The total amount of decay in the first three
years is 175 kilograms.

47. 15, 19, 23, . . .

$a_1 = 15, \ d = a_2 - a_1 = 19 - 15 = 4$

$a_n = a_1 + (n-1)d$

$\quad = 15 + (n-1)4$

$\quad = 15 + 4n - 4$

$\quad = 4n + 11$

$a_6 = 4(6) + 11 = 35$

$S_n = \frac{n}{2}(a_1 + a_n)$

$S_6 = \frac{6}{2}(15 + 35) = 3(50) = 150$

48. 5, −10, 20, . . .

$a_1 = 5, \ r = a_2 \div a_1 = -10 \div 5 = -2$

$S_n = \frac{a_1(1 - r^n)}{1 - r}$

$S_9 = \frac{5[1 - (-2)^9]}{1 - (-2)} = 855$

49. 1, 3, 5, . . .

$a_1 = 1, \ d = a_2 - a_1 = 3 - 1 = 2$

$a_n = a_1 + (n-1)d$

$\quad = 1 + (n-1)2$

$\quad = 1 + 2n - 2$

$\quad = 2n - 1$

$a_{30} = 2(30) - 1 = 60 - 1 = 59$

$S_n = \frac{n}{2}(a_1 + a_n)$

$S_{30} = \frac{30}{2}(1 + 59) = 15(60) = 900$

50. $7, \ 14, \ 21, \ 28, \ \ldots$

$a_1 = 7, \ d = a_2 - a_1 = 14 - 7 = 7$

$\begin{aligned} a_n &= a_1 + (n-1)d \\ &= 7 + (n-1)7 \\ &= 7 + 7n - 7 \\ &= 7n \end{aligned}$

$a_{20} = 7(20) = 140$

$S_n = \dfrac{n}{2}(a_1 + a_n)$

$S_{20} = \dfrac{20}{2}(7 + 140) = 1470$

51. $8, \ 5, \ 2, \ \ldots$

$a_1 = 8, \ d = a_2 - a_1 = 5 - 8 = -3$

$\begin{aligned} a_n &= a_1 + (n-1)d \\ &= 8 + (n-1)(-3) \\ &= 8 - 3n + 3 \\ &= 11 - 3n \end{aligned}$

$a_{20} = 11 - 3(20) = -49$

$S_n = \dfrac{n}{2}(a_1 + a_n)$

$S_{20} = \dfrac{20}{2}(8 - 49) = 10(-41) = -410$

52. $\dfrac{3}{4}, \ \dfrac{9}{4}, \ \dfrac{27}{4}, \ \ldots$

$a_1 = \dfrac{3}{4}, \ r = a_2 \div a_1 = \dfrac{9}{4} \div \dfrac{3}{4} = \dfrac{9}{4} \cdot \dfrac{4}{3} = 3$

$S_n = \dfrac{a_1(1 - r^n)}{1 - r}$

$S_8 = \dfrac{\frac{3}{4}(1 - 3^8)}{1 - 3} = 2460$

53. $a_1 = 6, \ r = 5$

$S_n = \dfrac{a_1(1 - r^n)}{1 - r}$

$S_4 = \dfrac{6(1 - 5^4)}{1 - 5}$

$= \dfrac{-3}{2}(1 - 625)$

$= \dfrac{3}{2}(624)$

$= 936$

54. $a_1 = -3, \ d = -6$

$\begin{aligned} a_n &= a_1 + (n-1)d \\ &= -3 + (n-1)(-6) \\ &= -3 - 6n + 6 \\ &= 3 - 6n \end{aligned}$

$a_{100} = 3 - 6(100) = -597$

$S_n = \dfrac{n}{2}(a_1 + a_n)$

$\begin{aligned} S_{100} &= \dfrac{100}{2}[-3 + (-597)] \\ &= 50(-600) \\ &= -30,000 \end{aligned}$

55. $5, \ \dfrac{5}{2}, \ \dfrac{5}{4}, \ \ldots$

$a_1 = 5, \ r = a_2 \div a_1 = \dfrac{5}{2} \div 5 = \dfrac{5}{2} \cdot \dfrac{1}{5} = \dfrac{1}{2}$

$S_\infty = \dfrac{a_1}{1 - r} = \dfrac{5}{1 - \frac{1}{2}} = \dfrac{5}{\frac{1}{2}} = 10$

56. $18, \ -2, \ \dfrac{2}{9}, \ \ldots$

$a_1 = 18, \ r = a_2 \div a_1 = -2 \div 18 = -\dfrac{2}{18} = -\dfrac{1}{9}$

$S_\infty = \dfrac{a_1}{1 - r} = \dfrac{18}{1 + \frac{1}{9}} = \dfrac{81}{5}$

57. $-20, \ -4, \ -\dfrac{4}{5}, \ \ldots$

$a_1 = -20,$

$r = a_2 \div a_1 = -4 \div (-20) = \dfrac{-4}{-20} = \dfrac{1}{5}$

$S_\infty = \dfrac{a_1}{1 - r} = \dfrac{-20}{1 - \frac{1}{5}} = \dfrac{-20}{\frac{4}{5}} = -25$

58. $0.2, \ 0.02, \ 0.002, \ \ldots$

$a_1 = 0.2, \ r = a_2 \div a_1 = 0.02 \div 0.2 = 0.1$

$S_\infty = \dfrac{a_1}{1 - r} = \dfrac{0.2}{1 - 0.1} = \dfrac{2}{9}$

59. $a_1 = 20,000, \ r = 1.15$

$a_n = 20,000(1.15)^{n-1}$

$a_4 = 20,000(1.15)^{4-1} = 30,418$

$30,418 was earned in his fourth year.

$S_4 = \dfrac{20,000(1 - 1.15^4)}{1 - 1.15} = 99,868$

$99,868 was earned in his first four years.

60. $a_1 = 40, \ r = 0.8$

$a_n = 40(0.8)^{n-1}$

$a_4 = 40(0.8)^{4-1} = 20.48$

It takes him 20.48 minutes to assemble the fourth television.

$S_n = \dfrac{a_1(1 - r^n)}{1 - r}$

$S_4 = \dfrac{40(1 - 0.8^4)}{1 - 0.8} = 118$

It takes him 118 minutes to assemble the first four televisions.

61. $a_1 = 100, \ d = -7$

$a_n = a_1 + (n-1)d$

$\quad = 100 + (n-1)(-7)$

$\quad = 100 - 7n + 7$

$\quad = 107 - 7n$

$a_7 = 107 - 7(7) = 107 - 49 = 58$

The farmer paid $58 rent for the seventh day.

$S_n = \dfrac{n}{2}(a_1 + a_n)$

$S_7 = \dfrac{7}{2}(100 + 58) = 553$

The farmer paid $553 rent for the first seven days.

62. $a_1 = 15, \ r = 0.8$

$S_\infty = \dfrac{a_1}{1 - r} = \dfrac{15}{1 - 0.8} = 75$

We double the number (to account for the flight up as well as down) and subtract 15 (since the first bounce was preceded by only a downward flight). Thus, the ball traveled a total distance of $2(75) - 15 = 135$ feet.

63. $1800, \ 600, \ 200, \ \ldots$

$a_1 = 1800,$

$r = a_2 \div a_1 = 600 \div 1800 = \dfrac{600}{1800} = \dfrac{1}{3}$

$S_6 = 1800 \dfrac{\left[1 - \left(\frac{1}{3}\right)^6 \right]}{1 - \frac{1}{3}}$

$\quad = 2700\left(1 - \dfrac{1}{729} \right)$

$\quad = 2700\left(\dfrac{728}{729} \right)$

$\quad = \dfrac{72,800}{27}$

$\quad \approx 2696$

Approximately 2696 mosquitoes were killed during the first six days after the spraying.

64. $1800, \ 600, \ 200, \ \ldots$

For which n is $a_n > 1$?

$$a_n = a_1 r^{n-1} = 1800\left(\frac{1}{3}\right)^{n-1}$$

Solve: $1800\left(\frac{1}{3}\right)^{n-1} > 1$

$$(n-1)\log\left(\frac{1}{3}\right) > \log\frac{1}{1800}$$
$$(n-1)(-0.4771213) > (-3.2552725)$$
$$n - 1 < 6.8$$
$$n < 7.8$$

It is no longer effective on the eighth day. About 2700 mosquitoes were killed.

65. $0.55\overline{5} = 0.5 + 0.05 + 0.005 + \ldots$

$$= \frac{5}{10} + \frac{5}{100} + \frac{5}{1000} + \ldots$$

This is a geometric series with $a_1 = \dfrac{5}{10}$

and $r = \dfrac{1}{10}$.

$$S_\infty = \frac{\frac{5}{10}}{1 - \frac{1}{10}} = \frac{\frac{5}{10}}{\frac{9}{10}} = \frac{5}{9}$$

Thus, $0.55\overline{5} = \dfrac{5}{9}$.

66. $27, \ 30, \ 33, \ \ldots$

$a_1 = 27, \ d = a_2 - a_1 = 30 - 27 = 3$

$$\begin{aligned}
a_n &= a_1 + (n-1)d \\
&= 27 + (n-1)3 \\
&= 27 + 3n - 3 \\
&= 3n + 24
\end{aligned}$$

$a_{20} = 3(20) + 24 = 84$

$S_n = \dfrac{n}{2}(a_1 + a_n)$

$S_{20} = \dfrac{20}{2}(27 + 84) = 1110$

There are 1110 seats in the theater.

67. $(x+z)^5 = 1 \cdot x^5 + 5 \cdot x^4 z + 10 \cdot x^3 z^2 + 10 \cdot x^2 z^3 + 5 \cdot xz^4 + 1 \cdot z^5$

$\qquad = x^5 + 5x^4 z + 10x^3 z^2 + 10x^2 z^3 + 5xz^4 + z^5$

68. $(y-r)^6 = 1 \cdot y^6 + 6 \cdot y^5(-r) + 15 \cdot y^4(-r)^2 + 20 \cdot y^3(-r)^3 + 15 \cdot y^2(-r)^4 + 6 \cdot y(-r)^5 + 1 \cdot (-r)^6$

$\qquad = y^6 - 6y^5 r + 15y^4 r^2 - 20y^3 r^3 + 15y^2 r^4 - 6yr^5 + r^6$

69. $(2x+y)^4 = 1 \cdot (2x)^4 + 4 \cdot (2x)^3 y + 6 \cdot (2x)^2 y^2 + 4 \cdot (2x)y^3 + 1 \cdot y^4$

$\qquad = 16x^4 + 32x^3 y + 24x^2 y^2 + 8xy^3 + y^4$

70. $(3y-z)^4 = 1 \cdot (3y)^4 + 4 \cdot (3y)^3(-z) + 6 \cdot (3y)^2(-z)^2 + 4 \cdot (3y)(-z)^3 + 1 \cdot (-z)^4$

$\qquad = 81y^4 - 108y^3 z + 54y^2 z^2 - 12yz^3 + z^4$

71. $(b+c)^8 = b^8 + 8b^7 c + \dfrac{8 \cdot 7}{2!} b^6 c^2 + \dfrac{8 \cdot 7 \cdot 6}{3!} b^5 c^3 + \dfrac{8 \cdot 7 \cdot 6 \cdot 5}{4!} b^4 c^4 + \dfrac{8 \cdot 7 \cdot 6 \cdot 5 \cdot 4}{5!} b^3 c^5$

$\qquad + \dfrac{8 \cdot 7 \cdot 6 \cdot 5 \cdot 4 \cdot 3}{6!} b^2 c^6 + \dfrac{8 \cdot 7 \cdot 6 \cdot 5 \cdot 4 \cdot 3 \cdot 2}{7!} bc^7 + c^8$

$\qquad = b^8 + 8b^7 c + 28b^6 c^2 + 56b^5 c^3 + 70b^4 c^4 + 56b^3 c^5 + 28b^2 c^6 + 8bc^7 + c^8$

72. $(x-w)^7 = \left[x + (-w)\right]^7$

$\qquad = x^7 + 7x^6(-w)^1 + \dfrac{7 \cdot 6}{2!} x^5(-w)^2 + \dfrac{7 \cdot 6 \cdot 5}{3!} x^4(-w)^3 + \dfrac{7 \cdot 6 \cdot 5 \cdot 4}{4!} x^3(-w)^4$

$\qquad + \dfrac{7 \cdot 6 \cdot 5 \cdot 4 \cdot 3}{5!} x^2(-w)^5 + \dfrac{7 \cdot 6 \cdot 5 \cdot 4 \cdot 3 \cdot 2}{6!} x^1(-w)^6 + (-w)^7$

$\qquad = x^7 - 7x^6 w + 21x^5 w^2 - 35x^4 w^3 + 35x^3 w^4 - 21x^2 w^5 + 7xw^6 - w^7$

73. $(4m-n)^4 = \left[4m + (-n)\right]^4$

$\qquad = (4m)^4 + 4(4m)^3(-n) + \dfrac{4 \cdot 3}{2!}(4m)^2(-n)^2 + \dfrac{4 \cdot 3 \cdot 2}{3!}(4m)(-n)^3 + (-n)^4$

$\qquad = 256m^4 - 256m^3 n + 96m^2 n^2 - 16mn^3 + n^4$

74. $(p-2r)^5 = p^5 + 5p^4(-2r) + \dfrac{5 \cdot 4}{2!} p^3(-2r)^2 + \dfrac{5 \cdot 4 \cdot 3}{3!} p^2(-2r)^3 + \dfrac{5 \cdot 4 \cdot 3 \cdot 2}{4!} p(-2r)^4 + (-2r)^5$

$\qquad = p^5 - 10p^4 r + 40p^3 r^2 - 80p^2 r^3 + 80pr^4 - 32r^5$

75. Let $n = 7$, $a = a$, $b = b$ and $r = 3$ in the formula given.

$\qquad \dfrac{7!}{3!(7-3)!} a^{7-3} b^3 = 35a^4 b^3$

76. Let $n = 10$, $a = y$, $b = 2z$ and $r = 10$ in the formula given.

$\qquad \dfrac{10!}{10!0!} y^{10-10}(2z)^{10} = 1024z^{10}$

Chapter 11 Test

1. $a_n = \dfrac{(-1)^n}{n+4}$

$a_1 = \dfrac{(-1)^1}{1+4} = -\dfrac{1}{5}$

$a_2 = \dfrac{(-1)^2}{2+4} = \dfrac{1}{6}$

$a_3 = \dfrac{(-1)^3}{3+4} = -\dfrac{1}{7}$

$a_4 = \dfrac{(-1)^4}{4+4} = \dfrac{1}{8}$

$a_5 = \dfrac{(-1)^5}{5+4} = -\dfrac{1}{9}$

The first five terms of the sequence are

$-\dfrac{1}{5}, \dfrac{1}{6}, -\dfrac{1}{7}, \dfrac{1}{8}, -\dfrac{1}{9}.$

2. $a_n = \dfrac{3}{(-1)^n}$

$a_1 = \dfrac{3}{(-1)^1} = -3$

$a_2 = \dfrac{3}{(-1)^2} = 3$

$a_3 = \dfrac{3}{(-1)^3} = -3$

$a_4 = \dfrac{3}{(-1)^4} = 3$

$a_5 = \dfrac{3}{(-1)^5} = -3$

The first five terms of the sequence are
$-3, 3, -3, 3, -3.$

3. $a_n = 10 + 3(n-1)$
$a_{80} = 10 + 3(80-1) = 10 + 237 = 247$

4. $a_n = (n+1)(n-1)(-1)^n$

$a_{200} = (200+1)(200-1)(-1)^{200}$

$= 200^2 - 1^2$

$= 40000 - 1$

$= 39,999$

5. $\dfrac{2}{5}, \dfrac{2}{25}, \dfrac{2}{125}, \ldots$ or $\dfrac{2}{5^1}, \dfrac{2}{5^2}, \dfrac{2}{5^3}, \ldots$

In general, $a_n = \dfrac{2}{5_n}$ or $a_n = \dfrac{2}{5}\left(\dfrac{1}{5}\right)^{n-1}.$

6. $-9, 18, -27, 36, \ldots$

or $(-1)^1 9 \cdot 1, \ (-1)^2 9 \cdot 2, \ (-1)^3 9 \cdot 3,$

$(-1)^4 9 \cdot 4, \ \ldots$

In general, $a_n = (-1)^n 9n.$

7. $a_n = 5(2)^{n-1}$

$a_1 = 5, \ r = 2$

$S_n = \dfrac{a_1(1-r^n)}{1-r}$

$S_5 = \dfrac{5(1-2^5)}{1-2} = -5(1-32) = 155$

8. $a_n = 18 + (n-1)(-2)$

$= 18 - 2n + 2$

$= 20 - 2n$

$a_{30} = 20 - 2(30) = -40$

$S_n = \dfrac{n}{2}(a_1 + a_n)$

$S_{30} = \dfrac{30}{2}[18 + (-40)] = 15(-22) = -330$

9. $a_1 = 24, \ r = \dfrac{1}{6}$

$S_\infty = \dfrac{a_1}{1-r} = \dfrac{24}{1-\frac{1}{6}} = \dfrac{24}{\frac{5}{6}} = \dfrac{144}{5}$

10. $\dfrac{3}{2}, -\dfrac{3}{4}, \dfrac{3}{8}, \ldots$

$a_1 = \dfrac{3}{2}, \; r = a_2 \div a_1 = -\dfrac{3}{4} \div \dfrac{3}{2} = -\dfrac{3}{4} \cdot \dfrac{2}{3} = -\dfrac{1}{2}$

$S_\infty = \dfrac{a_1}{1-r} = \dfrac{\frac{3}{2}}{1-\left(-\frac{1}{2}\right)} = \dfrac{\frac{3}{2}}{\frac{3}{2}} = 1$

11. $\displaystyle\sum_{i=1}^{4} i(i-2) = 1(1-2) + 2(2-2) + 3(3-2) + 4(4-2) = -1 + 0 + 3 + 8 = 10$

12. $\displaystyle\sum_{i=2}^{4} 5(2)^i(-1)^{i-1} = 5(2)^2(-1)^{2-1} + 5(2)^3(-1)^{3-1} + 5(2)^4(-1)^{4-1} = -20 + 40 - 80 = -60$

13. $(a-b)^6 = [a+(-b)]^6$

$\quad = 1 \cdot a^6 + 6 \cdot a^5(-b) + 15 \cdot a^4(-b)^2 + 20 \cdot a^3(-b)^3 + 15 \cdot a^2(-b)^4 + 6 \cdot a(-b)^5 + 1 \cdot (-b)^6$

$\quad = a^6 - 6a^5b + 15a^4b^2 - 20a^3b^3 + 15a^2b^4 - 6ab^5 + b^6$

14. $(2x+y)^5 = 1 \cdot (2x)^5 + 5 \cdot (2x)^4 y + 10 \cdot (2x)^3 y^2 + 10 \cdot (2x)^2 y^3 + 5 \cdot (2x)y^4 + 1 \cdot y^5$

$\quad = 32x^5 + 80x^4 y + 80x^3 y^2 + 40x^2 y^3 + 10xy^4 + 1 \cdot y^5$

15. $(y+z)^8 = y^8 + 8y^7 z + \dfrac{8 \cdot 7}{2!} y^6 x^2 + \dfrac{8 \cdot 7 \cdot 6}{3!} y^5 z^3 + \dfrac{8 \cdot 7 \cdot 6 \cdot 5}{4!} y^4 z^4 + \dfrac{8 \cdot 7 \cdot 6 \cdot 5 \cdot 4}{5!} y^3 z^5$

$\quad + \dfrac{8 \cdot 7 \cdot 6 \cdot 5 \cdot 4 \cdot 3}{6!} y^2 z^6 + \dfrac{8 \cdot 7 \cdot 6 \cdot 5 \cdot 4 \cdot 3 \cdot 2}{7!} yz^7 + z^8$

$\quad = y^8 + 8y^7 z + 28y^6 z^2 + 56y^5 z^3 + 70y^4 z^4 + 56y^3 z^5 + 28y^2 z^6 + 8yz^7 + z^8$

16. $(2p+r)^7 = (2p)^7 + 7(2p)^6 r + \dfrac{7 \cdot 6}{2!}(2p)^5 r^2 + \dfrac{7 \cdot 6 \cdot 5}{3!}(2p)^4 r^3 + \dfrac{7 \cdot 6 \cdot 5 \cdot 4}{4!}(2p)^3 r^4$

$\quad + \dfrac{7 \cdot 6 \cdot 5 \cdot 4 \cdot 3}{5!}(2p)^2 r^5 + \dfrac{7 \cdot 6 \cdot 5 \cdot 4 \cdot 3 \cdot 2}{6!}(2p)r^6 + r^7$

$\quad = 128p^7 + 448p^6 + 672p^5 r^2 + 560p^4 r^3 + 280p^3 r^4 + 84p^2 r^5 + 14pr^6 + r^7$

17. $a_n = 250 + 75(n-1) = 250 + 75n - 75 = 75n + 175$

$a_{10} = 75(10) + 175 = 925$

There were 925 people in the town at the beginning of the tenth year.

$a_1 = 75(1) + 175 = 250$

There were 250 people in the town at the beginning of the first year.

18. $1, 3, 5, \ldots$

$a_1 = 1, \, d = 2$

$a_n = a_1 + (n-1)d$

$\quad = 1 + (n-1)2$

$\quad = 1 + 2n - 2$

$\quad = 2n - 1$

We want $1 + 3 + 5 + \ldots + 15$.

$S_n = \dfrac{n}{2}(a_1 + a_n)$

$S_8 = \dfrac{8}{2}(1 + 15) = 4(16) = 64$

There were 64 shrubs planted in the 8 rows.

19. $a_1 = 80, \, r = \dfrac{3}{4}$

$a_n = a_1 r^{n-1}$

$a_n = 80\left(\dfrac{3}{4}\right)^{n-1}$

$a_4 = 80\left(\dfrac{3}{4}\right)^{4-1} = 80\left(\dfrac{27}{64}\right) = \dfrac{135}{4}$ or 33.75

The arc length is 33.75 centimeters on the fourth swing.

$S_n = \dfrac{a_1(1 - r^n)}{1 - r}$

$S_8 = \dfrac{80\left(1 - \left(\frac{3}{4}\right)^4\right)}{1 - \frac{3}{4}}$

$\quad = 320\left(1 - \dfrac{81}{256}\right)$

$\quad = 320\left(\dfrac{175}{256}\right)$

$\quad = \dfrac{875}{4}$

$\quad = 218.75$

The total arc length is 218.75 centimeters for the first 4 swings.

20. $a_1 = 80, \, r = \dfrac{3}{4}$

$S_\infty = \dfrac{a_1}{1 - r} = \dfrac{80}{1 - \frac{3}{4}} = \dfrac{80}{\frac{1}{4}} = 320$

The total of the arc lengths is 320 centimeters before the pendulum comes to rest.

21. $16, 48, 80, \ldots$

$a_1 = 16, \, d = a_2 - a_1 = 48 - 16 = 32$

$a_n = a_1 + (n-1)d$

$\quad = 16 + (n-1)32$

$\quad = 16 + 32n - 32$

$\quad = 32n - 16$

$a_{10} = 32(10) - 16 = 320 - 16 = 304$

He falls 304 feet during the tenth second.

$S_n = \dfrac{n}{2}(a_1 + a_n)$

$S_{10} = \dfrac{10}{2}(16 + 304) = 5(320) = 1600$

He falls 1600 feet during the first 10 seconds.

22. $0.42\overline{42} = 0.42 + 0.0042 + 0.000042 + \ldots$

$a_1 = 0.42, \, r = 0.01$

$S_\infty = \dfrac{a_1}{1 - r} = \dfrac{0.42}{1 - 0.01} = \dfrac{0.42}{0.99} = \dfrac{42}{99} = \dfrac{14}{33}$

Thus, $0.42\overline{42} = \dfrac{14}{33}$.

Appendices

Appendix A Exercise Set

1. $90° - 19° = 71°$

3. $90° - 70.8° = 19.2°$

5. $90° - 11\frac{1}{4}° = 78\frac{3}{4}°$

7. $180° - 150° = 30°$

9. $180° - 30.2° = 149.8°$

11. $180° - 79\frac{1}{2}° = 100\frac{1}{2}°$

13. $m\angle 1 = 110°$ since $\angle 1$ and the given angle are vertical angles.
$m\angle 2 = m\angle 3 = 70°$ since they are supplementary to $\angle 1$ and the given angle.
$m\angle 4 = 70°$ and $m\angle 5 = 110°$ since they are alternate interior angles with $\angle 3$ and the given angle, respectively.
$m\angle 6 = 70°$ since $\angle 4$ and $\angle 6$ are vertical angles.
$m\angle 7 = 110°$ since $\angle 5$ and $\angle 7$ are vertical angles.

15. $180° - 11° - 79° = 90°$

17. $180° - 25° - 65° = 90°$

19. $180° - 30° - 60° = 90°$

21. $90° - 45° = 45°$, so the other two angles are $45°$ and $90°$.

23. $90° - 17° = 73°$, so the other two angles are $73°$ and $90°$.

25. $90° - 39\frac{3}{4}° = 50\frac{1}{4}°$, so the other two angles are $50\frac{1}{4}°$ and $90°$.

27.
$$\frac{12}{4} = \frac{18}{x}$$
$$4x\left(\frac{12}{4}\right) = 4x\left(\frac{18}{x}\right)$$
$$12x = 72$$
$$x = 6$$

29.
$$\frac{6}{9} = \frac{3}{x}$$
$$9x\left(\frac{6}{9}\right) = 9x\left(\frac{3}{x}\right)$$
$$6x = 27$$
$$x = 4.5$$

31. $6^2 + 8^2 = c^2$
$$36 + 64 = c^2$$
$$100 = c^2$$
$$c = 10$$

33. $5^2 + b^2 = 13^2$
$$25 + b^2 = 169$$
$$b^2 = 144$$
$$b = 12$$

Appendix C Exercise Set

1. $V = lwh = (6 \text{ in.})(4 \text{ in.})(3 \text{ in.}) = 72$ cu. in.
$SA = 2lh + 2wh + 2lw$
$\quad = 2(6 \text{ in.})(3 \text{ in.}) + 2(4 \text{ in.})(3 \text{ in.}) + 2(6 \text{ in.})(4 \text{ in.})$
$\quad = 36$ sq. in. $+ 24$ sq. in. $+ 48$ sq. in.
$\quad = 108$ sq. in.

3. $V = s^3 = (8 \text{ cm})^3 = 512$ cu. cm
$SA = 6s^2 = 6(8 \text{ cm})^2 = 384$ sq. cm

5. $V = \dfrac{1}{3}\pi r^2 h = \dfrac{1}{3}\pi(2 \text{ yd})^2(3 \text{ yd}) = 4\pi$ cu. yd $\approx 4\left(\dfrac{22}{7}\right)$ cu. yd $= 12\dfrac{4}{7}$ cu. yd

$SA = \pi r\sqrt{r^2 + h^2} + \pi r^2$

$\quad = \pi(2 \text{ yd})\left(\sqrt{(2 \text{ yd})^2 + (3 \text{ yd})^2}\right) + \pi(2 \text{ yd})^2$

$\quad = \pi(2 \text{ yd})(\sqrt{13} \text{ yd}) + \pi(4 \text{ sq. yd})$

$\quad = \pi(2\sqrt{13} + 4)$ sq. yd

$\quad \approx 35.20$ sq. yd

7. $V = \dfrac{4}{3}\pi r^3 = \dfrac{4}{3}\pi(5 \text{ in.})^3 = \dfrac{500}{3}\pi$ cu. in. $\approx \dfrac{500}{3}\left(\dfrac{22}{7}\right)$ cu. in. $= 523\dfrac{17}{21}$ cu. in.

$SA = 4\pi r^2 = 4\pi(5 \text{ in.})^2 = 100\pi$ sq. in. $\approx 100\left(\dfrac{22}{7}\right)$ sq. in. $= 314\dfrac{2}{7}$ sq. in.

9. $V = \dfrac{1}{3}s^2 h = \dfrac{1}{3}(6 \text{ cm})^2(4 \text{ cm}) = 48$ cu. cm

$SA = B + \dfrac{1}{2}pl = (6 \text{ cm})^2 + \dfrac{1}{2}(24 \text{ cm})(5 \text{ cm}) = 36$ sq. cm $+ 60$ sq. cm $= 96$ sq. cm

11. $V = s^3 = \left(1\dfrac{1}{3} \text{ in.}\right)^3 = \dfrac{64}{27}$ cu. in. $= 2\dfrac{10}{27}$ cu. in.

13. $SA = 2lh + 2wh + 2lw$
$\quad = 2(2 \text{ ft})(1.4 \text{ ft}) + 2(2 \text{ ft})(3 \text{ ft}) + 2(1.4 \text{ ft})(3 \text{ ft})$
$\quad = 5.6$ sq. ft $+ 12$ sq. ft $+ 8.4$ sq. ft
$\quad = 26$ sq. ft

15. $V = \dfrac{1}{3}s^2 h = \dfrac{1}{3}(5 \text{ in.})^2(1.3 \text{ in.}) = 10\dfrac{5}{6}$ or $10.8\overline{3}$ cu. in.

17. $V = \dfrac{1}{3}s^2 h = \dfrac{1}{3}(12 \text{ cm})^2(20 \text{ cm}) = 960$ cu. cm

19. $SA = 4\pi r^2 = 4\pi(7 \text{ in.})^2 = 196\pi$ sq. in.

21. $V = lwh$

$$= (2 \text{ ft})\left(2\frac{1}{2} \text{ ft}\right)\left(1\frac{1}{2} \text{ ft}\right)$$

$$= 7\frac{1}{2} \text{ cu. ft}$$

23. $V = \frac{1}{3}\pi r^2 h$

$$\approx \frac{1}{3}\left(\frac{22}{7}\right)(2 \text{ cm})^2(3 \text{ cm})$$

$$= 12\frac{4}{7} \text{ cu. cm}$$

Appendix D

Practice Problem 1

a.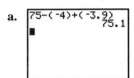

b.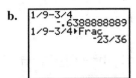

c.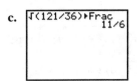

d.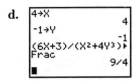

Practice Problem 3

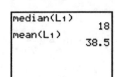

The median of this data is 18, and the mean is 38.5.

Practice Problem 5

a. $4x + y = 5$
$$y = 5 - 4x$$

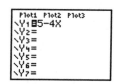

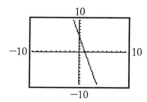

b. $6x - 3y = 36$
$$-3y = 36 - 6x$$
$$y = \frac{36 - 6x}{-3}$$
$$y = -12 + 2x$$

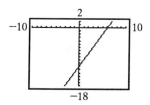

Practice Problem 7

a.

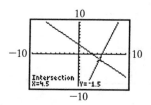

b.

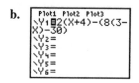

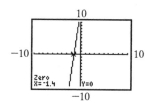

Practice Problem 9

Practice Problem 11

a. $5x^2 + y^2 = 36$
$$y^2 = 36 - 5x^2$$
$$y = \pm\sqrt{36 - 5x^2}$$

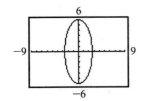

b. $\dfrac{y^2}{9} - \dfrac{x^2}{25} = 1$
$$25y^2 - 9x^2 = 225$$
$$25y^2 = 225 + 9x^2$$
$$5y = \pm\sqrt{225 + 9x^2}$$
$$y = \pm 0.2\sqrt{225 + 9x^2}$$

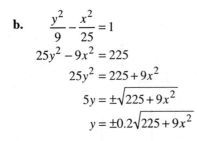